Atomic Age America

Martin V. Melosi
University of Houston

Boston Columbus Indianapolis New York San Francisco Upper Saddle River
Amsterdam Cape Town Dubai London Madrid Milan Munich Paris Montréal Toronto
Delhi Mexico City São Paulo Sydney Hong Kong Seoul Singapore Taipei Tokyo

Editorial Director: Craig Campanella
Editor in Chief: Dickson Musslewhite
Publisher: Charlyce Jones Owen
Associate Editor: Emsal Hasan
Editorial Assistant: Maureen Diana
Director of Marketing: Brandy Dawson
Senior Marketing Manager: Maureen Prado Roberts
Senior Marketing Assistant: Christine Liva
Production Manager: Fran Russello
Image Permission Coordinator: Beth Brenzel
Cover Art: (Nagasaki, Japan under atomic bomb attack) Library of Congress, Prints & Photographs Division,
    [LC-USZ62-39852], (Nuclear power plant) Photographs in the Carol M. Highsmith Archive, Library of Congress,
    Prints and Photographs Division, [LC-DIG-highsm-13019]
Media Director: Brian Hyland
Media Editor: Andrea Messineo
Full-Service Project Management: Anju Joshi/PreMediaGlobal
Printer/Binder: Courier Companies, Inc.
Text Font: 10/12 Minion Pro

Credits and acknowledgments borrowed from other sources and reproduced, with permission, in this textbook appear **on the appropriate page within text**.

Library of Congress Cataloging-in-Publication Data
Melosi, Martin V.
    Atomic age America / Martin V. Melosi.
        p. cm.
    Includes index.
    ISBN-13: 978-0-205-74254-7
    ISBN-10: 0-205-74254-8
    1. Nuclear weapons—United States—History.   2. Nuclear energy—United States—History.   3. Nuclear power plants—United States.   4. Arms race—History—20th century.   5. Deterrence (strategy)   6. Nuclear nonproliferation.   7. Cold War.   I. Title.
    U264.3.M45 2013
    355.02'170973—dc23
                                                                                2012026596

10 9 8 7 6 5 4 3 2 1

ISBN 10: 0-205-74254-8
ISBN 13: 978-0-205-74254-7

For Gianna

CONTENTS

Chapter 4   The Cold War and Atomic Diplomacy: Deterrence, Espionage, and the Superbomb   85

PREFACE

This book was a real journey for me. I taught (and still teach) a course entitled "Atomic Age America" at the University of Houston, and also taught it once at the University of Southern Denmark. It was designed for both graduate and undergraduate students. I decided to develop the course to be a one-time special topics seminar, which grew out of a few chapters about atomic energy in my book *Coping with Abundance: Energy and Environment in Industrial America* (1985). I liked the fact that studying the subject of atomic energy opened up so many teaching possibilities in a variety of fields. From the onset I also believed that the subject matter begged for film to be integrated into what was primarily a small-class setting in which we read many pertinent books and articles and carried out spirited discussion. Commercial and documentary film offer an opportunity to graphically portray different historical eras, the drama and horror of nuclear weapons, and compelling stories related to a new and seemingly exotic energy source. Ultimately I showed one film per week, rotating them each time I taught the course. Students loved seeing the films, but at first were wary of a class that included so much science and technology. They were relieved to discover that we would examine atomic energy from a variety of vantage points, including political, diplomatic, economic, cultural, environmental, and technical. Obviously the films were the initial draw, but the students soon discovered that I was using them as documents to complement the written word, and did not base my choice of visual material on what was a "good movie" in any purely aesthetic or critical sense. The integration of traditional documentary evidence, books, and articles along with film (even *The Simpsons*) broadened the scope and depth of the course immeasurably.

Still I was frustrated by not having a narrative spine in the form of a basic text upon which to build the course. I first considered editing a volume that would incorporate the work of well-known authors in the field. I quickly realized that I would not be able or willing to dictate to them precisely what I was looking for, that is, the integration of atomic energy with core issues in American and world events over a span of many years. Therefore, I decided to venture out alone and to write the book myself.

ACKNOWLEDGMENTS

I began my career as a diplomatic historian, and I must credit two individuals who were central to my training: recently deceased Jules A. Karlin at the University of Montana and Robert A. Divine—my dissertation supervisor—at the University of Texas, Austin. Since about the mid-1970s, however, my chosen research has focused on urban environmental history, energy history, and the history of technology. Not considering myself an expert on nuclear weapons or nuclear power in general, I rather naively believed that I could merge all of my various interests and training into this study and pick up what else I needed (particularly the history of science) on the fly. I had to seriously update my knowledge of diplomatic history and world affairs, and blend it with my expertise in energy and environmental history in such a way as to be credible. Whether this was arrogance or stupidity, I leave to the readers to decide. I nonetheless am happy with the experience of working on this very complex and expansive study, simply because it is complex and expansive. As with most writers, I learned more than I ever will be able to impart to others in these pages. My sense of accomplishment comes from sticking with the project and enjoying the ride.

I very much would like to thank my editor, Charlyce Jones-Owen for time spent with me on the project, and to the group of reviewers she lined up to critique the proposal and the chapters themselves: Amy E. Foster, University of Central Florida; Donald Gawronski, Mesa Community College; Lori Clune, California State University, Fresno; Craig McConnell, California State University, Fullerton; David J. Snyder, Central Michigan University; Patrick McCray, University of California, Santa Barbara; Steven Conn, Ohio State University; John Turner, University of South Alabama; Daniel Murphy, Hanover College; and Todd Good, University of Wisconsin, Stevens Point.

I am particularly indebted to Sam Walker, friend and recently retired chief historian for the Nuclear Regulatory Commission, who read and commented on every chapter. His wise counsel saved me much embarrassment and forced me to rethink many elements of key sections. I know of no one with the breadth and depth of knowledge on the subject of this book, nor anyone as even-handed in his historical assessments. I did not take all of his advice, but I very carefully weighed everything he said. Sometimes I forged ahead to make my own way and my own mistakes. That Sam and I also shared stories about our new granddaughters during the process was an added bonus. Colleagues and students at the University of Houston also provided valuable insights, especially my friends and co-workers Joe Pratt and Kathy Brosnan, and my very able PhD students, Joe Stromberg and Julie Cohn. Thanks also to our program coordinator, Kristin Deville, for her many kindnesses. I also benefited from presentations of the book at the University of Oklahoma, the United States Air Force Academy, the University of Minnesota-Morris, and at the American Historical Association meeting in Chicago.

My wife Carolyn continues to possess incredible patience in dealing with my moods when I am in writing mode (and other times as well). I could not do this work without her. And as the Dedication proclaims, this book is for my precious granddaughter Gianna, who already at the age of one and a half years possesses many of the wonderful traits of my daughters Adria (her mother) and Gina (her aunt). However, she does not as yet have the wine expertise of her father Steven. My hope is that students of all ages will enjoy *Atomic Age America*.

Martin V. Melosi
University of Houston

THE MOMENT YOU KNOW

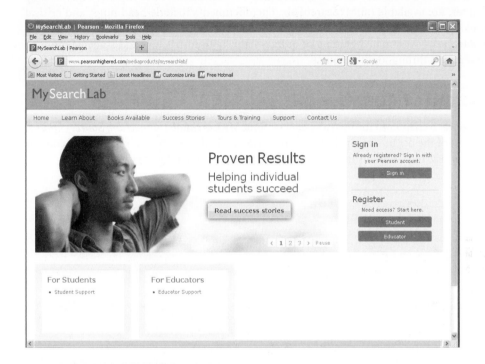

www.mysearchlab.com

MYSEARCHLAB WITH ETEXT delivers proven results in helping individual students succeed. Its automatically graded assessments and interactive eText provide engaging experiences that personalize, stimulate, and measure learning for each student. And, it comes from a trusted partner with educational expertise and a deep commitment to helping students, instructors, and departments achieve their goals.

PERSONALIZE LEARNING

Writing, Research and Citing Sources

- Step by step tutorials present complete overviews of the research and writing process.
- Instructors and students receive access to the EBSCO ContentSelect database, census data from Social Explorer, Associated Press news feeds, and the Pearson bookshelf. Pearson SourceCheck helps students and instructors monitor originality and avoid plagiarism.

ETEXT AND MORE

- **Pearson eText**—An e-book version of *Atomic Age America* is included in MySearchLab. As with the printed text, students can highlight and add their own notes as they read their interactive text online.

- **Chapter quizzes and flashcards**—Chapter and key term reviews are available for each chapter online and offer immediate feedback.
- **Primary and Secondary Source Documents**—A collection of documents, organized by chapter, are available on MySearchLab. The documents include head notes and critical thinking questions.
- **Gradebook**—Automated grading of quizzes helps both instructors and students monitor their results throughout the course.

MYSEARCHLAB CONNECTIONS

At the end of each chapter in the text, a special section, *MySearchLab Connection: Sources Online*, provides a list of the documents included on the MySearchLab website that relate to the content of the chapter. See the Contents for a full list of the sources listed in the text.

ABOUT THE AUTHOR

Martin V. Melosi is Hugh Roy and Lillie Cranz Cullen University Professor and Director of the Center for Public History at the University of Houston. His primary fields of study are environmental, urban, and energy history. He is the author or editor of nineteen books and more than 90 articles and book chapters, including the award-winning *The Sanitary City* (2000). In 2000–01 he held the Fulbright Chair in American Studies at the University of 19 Southern Denmark, and has been a visiting professor at the University of Paris, University of Helsinki, Tampere Technical University, Peking University, and Shanghai University. He is past-president of the American Society for Environmental History, the Public Works Historical Society, the Urban History Association, and the National Council on Public History.

A Most Controversial Technology

Not only will atomic power be released, but someday we will harness the rise and fall of the tides and imprison the rays of the sun.

THOMAS ALVA EDISON (1847–1931) INVENTOR

When you see something that is technically sweet, you go ahead and do it and you argue about what to do about it only after you have had your technical success. That is the way it was with the atomic bomb.

J. ROBERT OPPENHEIMER (1904–1967) FATHER OF THE A-BOMB

The world has achieved brilliance without wisdom, power without conscience. Ours is a world of nuclear giants and ethical infants. We know more about war than we know about peace, more about killing that we know about living.

GENERAL OMAR NELSON BRADLEY (1893–1981)
COMMANDER, US FORCES IN NORMANDY, WORLD WAR II

The future of the world, dependent as it is upon atomic energy, requires more understanding and knowledge about the atom.

WILLARD LIBBY (1908–1980)
AMERICAN PHYSICAL CHEMIST

The only use for an atomic bomb is to keep somebody else from using one.

GEORGE WALD (1906–1997)
AMERICAN SCIENTIST AND NOBEL LAUREATE

> *It's ridiculous that time and time again we need a radioactive cloud coming out of a nuclear power-station to remind us that atomic energy is extraordinarily dangerous.*
>
> PIERRE SCHAEFFER (1910–1995)
> FRENCH COMPOSER, ENGINEER, MUSICOLOGIST

> *The Atomic Age is here to stay—but are we?*
>
> BENNETT CERF (1898–1971)
> CO-FOUNDER OF RANDOM HOUSE

Atomic theory began more than 2,000 years ago as curiosity about what made up the world. It was at first a philosophical inquiry more than a scientific one. The development of atomic science eventually allowed humans to release extraordinary amounts of energy from the smallest particles in nature hoping to harness such power toward myriad practical ends. The immediate results were weapons and reactors, and also wonder and fear.

Atomic energy spawned several transforming technologies. This book primarily deals with two major technologies: nuclear weapons and nuclear power.[1] Atomic weapons first appeared in literary fiction like H.G. Wells's *The World Set Free* (1913), rather than in the real world. [2] Wells wrote not only about atomic bombs, but atomic energy in general as well as atomic airplanes, atomic smelting, atomic hay lorries, and atomic engineering. He began the book with these lines: "The history of mankind is the history of the attainment of external power. Man is the tool-using, fire-making animal. From the outset of his terrestrial career we find him supplementing the natural strength and bodily weapons of a beast by the heat of burning and the rough implement of stone. So he passed beyond the ape. From that he expands…" By looking at the development and impact of nuclear weapons and nuclear power, *Atomic Age America* builds upon Wells' observation about the human quest for external power and the role of the human as "the tool-using, fire-making animal" in search of what that implies in recent history.

Between nuclear weapons and nuclear power, the former is clearly the most revolutionary. The building of atomic bombs in the 1940s changed the face of total war. Thermonuclear weapons that followed the first atomic bombs made real the possibility of turning the earth into a cinder. These weapons were "the shock of the new" and tools for conducting war, but also gave pause for reason to avoid it.[3] This was less true when nuclear weapons first were invented. In the context of World War II, the race to build the bomb came with an implicit objective of gaining advantage over Germany, and possibly using the weapon first. That the United States dropped atomic bombs on Japan after Germany's surrender offers a strong indication that the Manhattan Project's success made it easier to shift targets than to reconsider policy. The race for the bomb, unfortunately, gave little time to reflect on the weapon's terrible potential. While those who acquired nuclear weapons (with some exceptions) were not so willing to give them up, in the long run they considered ways of trying to make sure that they never needed to use them.[4] Yet the threat was always there, and many came to believe that it was not prudent to assume that rational behavior guided the finger on the button in all cases.

Interest in production of nuclear power, of course, focused on generation rather than destruction, built on the hope that humankind would find an answer to its essential energy wants. If nuclear power was not "too cheap to meter," it seemed too tempting to ignore. Nuclear power offered the prospect of low-cost and plentiful energy, which was a dream expressed well before the atomic age and well before harnessing the atom. In most respects, abundant nuclear-generated power remains a future goal, more than a present reality. While about 14 percent of the world's electricity in recent years comes from 432 nuclear power plants, controversy rages as to whether nuclear power can ever (or should ever) dominate the energy market. The biggest issue is safety, and historical markers such as Three Mile Island, Chernobyl, and Fukushima rivet attention on the risks and benefits of a nuclear energy economy.

Radiation is the issue that inextricably links nuclear weapons and nuclear power historically. It makes them curious, mysterious, unique, and dangerous. Historian M. Joshua Silverman made a good case:

> For the historian, the study of nuclear technologies is a study of contrasts. On the one hand, nuclear technologies represent some of the most impressive scientific and technical achievements of the modern age. They have embodied humanity's greatest inspirations, including world peace, improved health care, and unlimited energy. On the other hand, nuclear technologies have aroused seemingly intractable conflict and social turmoil, giving tangible form to some of humanity's greatest fears, including nuclear annihilation, genetic mutations, and environmental disaster.[5]

Or as scientists and historian Spencer Weart concluded, "Atomic energy was coming to stand for things more important than atomic energy itself. First of all it was coming to stand for all the powers of science, powers for the better—or perhaps for the worse."[6] In modern times, especially, many regard "technological advances" as making life better and improving the lot of humans.[7] Nuclear technologies, however, were never greeted with unvarnished optimism, but were often regarded as a mixed blessing. For every Pollyanna there was a detractor; for every defender there was a naysayer.

The importance of nuclear technologies should not lead to the conclusion that their persistence and impacts were inevitable, rational, or preordained. Historian of technology David Nye rightly argued that "A technology is not merely a system of machines with certain functions, rather, it is an expression of a social world...No technology exists in isolation."[8] Saying this also is to admit that "there is no single, no logical, and no necessary" relationship between people and machines. It is important to avoid the extremes of technological determinism: On the "hard" end of the spectrum is the notion that technology has agency, that is, technology may be autonomous or have on its own ability to effect change.[9] The other end of the spectrum (or "soft" determinism) presents the history of technology as essentially a story of human actions. In this case agency is based on a combination of social, economic, political, and cultural variables which strip technology of the ability to initiate change.[10] To avoid the extremes of determinism is to view nuclear technologies not in isolation, but within the historical context in which they were created and employed. This is a goal of *Atomic Age America*, particularly to place nuclear weapons and nuclear power within the context of war and peace in the latter half of the twentieth century and into the twenty-first. It is important to evaluate the choices that government officials, scientists, and others made in promoting or rejecting these technologies and why they made them. Much too often, atomic and hydrogen bombs and nuclear power, in particular, have been set apart from basic historical narratives as if they existed in a special box, treated as anomalies more than integral features of the central thrust of history. They are too important to isolate from other factors that change the world.

Placing nuclear weapons and nuclear power in the stream of history is no easy task, however. Yet connecting them to a larger context may help answer the question of how they influenced current events and to what purpose. The "big issues" clearly revolve around the existence of weapons that had the capacity to destroy human life and completely render the planet uninhabitable. While people did not acknowledge the burden of an apocalypse every day of every year, warfare itself took on an entirely new meaning. Questions remained as to whether it was possible to contain conflict any longer in modern society. In the case of nuclear power, the end goal was not destruction, or keeping people from destroying themselves, but harnessing an unlimited source of energy. Not only was there a question about "Can it be done?" but also "Should it be done?" It is not that people eschewed the possibility of endless, cheap energy, but at what cost? This is where fear of bombs intersected with fear of reactors. Radioactive material changed the whole discourse over risk, not only in environmental terms but also in ethical terms. What nuclear weapons and nuclear reactors share is their radioactive residue, and the question of whether we can live with that reality. When the blast occurs or when the reactor core melts, the impact of those technologies remains almost perpetual in human years if not geological years. The unending debate over radiation and radioactivity has been to determine how much is too much. Are the rewards worth the risk, but also what is the nature of that risk? The answers seem substantially clearer when discussing bombs versus reactors. And thus the story of nuclear energy always returns to hope, promise, and risk.

While the book is not primarily global in scope, it does try to place the American experience in a world context for comparative purposes, and to avoid creating the impression that atomic energy disproportionately impacted the United States at the expense of everyone else. I subscribe to the basic viewpoint acknowledged in Andrew Rotter's book *Hiroshima: The World's Bomb*. About Hiroshima, he stated,

> For the atomic bombing of Hiroshima was not merely a decision made by US policymakers in order to punish the Japanese, not just an issue in Japanese-US relations, but instead the product of years of scientific experimentation, ethical debate within the scientific community, and significant changes in the conduct of war—all undertaken globally. Americans alone did not decide to build the bomb, and neither did they alone actually build it…[S]o too did the bomb have implications that stretched beyond the sensibilities of Americans and Japanese…[11]

Yet, providing a global context is not just about participation in developing these new technologies or reacting to them, but in the shared optimism and pessimism about their impacts. It is about different world views on establishing strategic policies. It is the local and regional factors that determine the needs and desires for a new source of energy generation. It is the perspective of the actors and those acted upon.

The literature available on topics nuclear is vast and deep. Hopefully what distinguishes *Atomic Age America* from most of that literature is twofold. First, the book follows simultaneously the development of both nuclear weapons and nuclear power in a single, narrative over a long chronological period (from the inception of the idea of atomic theory to the present). Second, it attempts to place heavy emphasis upon the context in which these technologies emerged. Chapter 1 begins with the origins of atomic theory. While substantially out of the primary time sequence of the other chapters, it is important to understand where ideas come from and how abstractions find their way into practice. One of the extraordinary stories about atomic energy is how a question like "What is the world made of?" led to "How can we use that information?" "How can we harness a newly discovered source of power?" The leap from the first query to the others is an essential starting point in any narrative about nuclear weapons and nuclear power.

The remainder of Chapter 1 and part of Chapter 2 follow scientists through the process of learning about the atom and its properties. The achievement of splitting the atom and releasing its power in the 1930s coincided with how governments came to exploit this knowledge in pursuit of national goals. The atom slipped from the hand of scientists into those of bureaucrats and soldiers early in the twentieth century. The commitment to the Manhattan Project and the race by the United States to build an atomic bomb focuses on how government mobilized the atom for war. As described in Chapter 3 the bombings of Hiroshima and Nagasaki demonstrate most dramatically that splitting the atom revolutionized war and gave the idea of "total war" a whole new meaning.

In the post-World War II era, nuclear weapons become the centerpiece of a new kind of war, Cold War (as discussed in Chapters 3 through 5). The major point of confrontation was a relentless arms race that went on for decades, accompanied by an escalation from A-bombs to H-bombs. These weapons caused a rise in the destructive equation to an unbelievable level that was not accompanied by a commensurate recalibration of Cold War goals. Enmeshed in the Cold War, the nuclear arms race also provoked questions about loyalty and espionage in an increasingly poisonous atmosphere of recrimination. Deep concern about the spreading of atomic secrets fed suspicion of foreigners and citizens alike. The demands of the arms race, in addition, led to incessant testing of new and better weapons. Such testing produced a fallout scare that brought the risk of radioactivity to the shores of the United States for the first time. In this 1950s Age of Anxiety, fear of fallout and fear of nuclear holocaust were two sides of the same coin.

To dwell only on the destructive power of the atom was to ignore its potential as a promising source of everlasting energy, and the debate that ensued around that issue. Chapter 6 discusses attempts at finding "peaceful uses" of the atom from the sublime (nuclear medicine) to the ridiculous (*Project Plowshare*). The latter was a plan to employ atomic and thermonuclear bombs to dig canals and to release underground natural gas. The main thrust of finding peaceful uses for the atom was stimulating civilian nuclear power to generate electricity. In an era of abundant fossil fuels, such a program seemed unnecessary, but the long-range promise of an abundant source of energy, plus the propaganda value (for the United States at least) in promoting peaceful uses made commercialization logical for the time.

Chapter 7 turns back to the Cold War demonstrating how the accelerated production of nuclear arms became an essential ingredient of the military-industrial complex and a catalyst for dangerous international crises in Berlin and Cuba during the early 1960s. Subsequent chapters follow post-Cuban Missile Crisis developments to the end of the Cold War in the early 1990s. This is followed by a discussion of how a multi-polar world at the turn of the century replaced bilateral politics of the superpowers. In this new context, nuclear weapons engage a world of terrorism and a new round of nuclear proliferation. Additional chapters follow the fall and rise and fall again of the nuclear power industry still bent on reaching its promise as the twenty-first century opens. The disasters at Chernobyl and Fukushima challenge the industry's resurgence. The seeming wild card in the commercial nuclear power debate is the convergence of climate change warnings with a changing view of nuclear power as the preeminent carbon-free alternative to fossil fuels. Proliferation, terrorism, and climate change converge in recent years to fundamentally alter the discourse of the world of the atom (discussed in Chapter 11). From Hiroshima to Fukushima, nuclear weapons and nuclear power are not subtext in American history but central influences.

In exploring the military and peaceful uses of nuclear power over the course of many years, *Atomic Age America* relied primarily on the widely available and vast wealth of secondary materials in print and on contemporary documents. Historical studies on almost every aspect of atomic energy are plentiful, but rarely cover the whole topic. As a collective body, these works nonetheless provide a literal reservoir of information and interpretations essential for this book. New books and articles seem to appear every week. I also utilized commercial and documentary films as essential research and general sources of information. It is not my intention to give film synopses or to offer extensive film criticism. Instead, I use elements in several movies to illustrate key general points or to give the reader a sense of contemporary responses to a range of actions, values, or attitudes. Historians too often neglect films as documents, equally valuable to the printed page, especially in narratives like this one that truly need a sense of the times to engage the audience. I urge the readers to view these films and many others for themselves, since visual images do even more to enlighten us about visceral issues and help to put the viewer into a variety of times and places.

Has the existence of nuclear weapons and the potential of nuclear power defined the American experience since the mid-twentieth century? Taken individually, probably not. But as some observers have correctly noted, such technologies cannot be "disinvented," and thus we deal with them as the essential sinews of modern history. And it matters little if nuclear power has yet to meet its promise or how nuclear weapons have been employed as a policy tool. They both shaped the historical dialogue in so many ways that it is difficult to view them in any way other than as part of the historical fabric. Atomic energy is more than a prism through which to view our times, but is fused to the very essence of what defines us in a modern world.

Endnotes

1. In a much more modest way, the book also touches upon a variety of other nuclear technologies including nuclear medicine. Nuclear medicine is the most intimate use for atomic energy in the form of radiation. The field of nuclear medicine, about fifty years old, has become central to diagnosing and treating serious diseases. Doctors internally administer radioactive materials—radioisotopes— into the human body for a variety of purposes. Widespread use of nuclear medicine started in the 1950s to measure thyroid function and to diagnose and treat thyroid disease. In 1971 nuclear medicine was formally recognized as a medical specialty.
2. John Canaday, *The Nuclear Muse: Literature, Physics, and the First Atomic Bombs* (Madison, WI: University of Wisconsin Press, 2000), 3.
3. Jonathan Stevenson, *Thinking Beyond the Unthinkable: Harnessing Doom from the Cold War to the Age of Terror* (New York: Viking, 2008), 8.
4. Nina Tannenwald, *The Nuclear Taboo: The United States and the Non-Use of Nuclear Weapons Since 1945* (New York: Cambridge University Press, 2007), 1.
5. M. Joshua Silverman, "Nuclear Technology," in Carroll Pursell, ed., *A Companion to American Technology* (Malden, MA: Blackwell, 2008), 298.
6. Spencer R. Weart, *Nuclear Fear: A History of Images* (Cambridge, MA: Harvard University Press, 1988), 16.
7. See ibid., 35.
8. David E. Nye, *Technology Matters: Questions to Live With* (Cambridge, MA: MIT Press, 2007), 47.
9. Merritt Roe Smith and Leo Marx, eds., *Does Technology Drive History? The Dilemma of Technological Determinism* (Cambridge, MA: MIT Press, 1994), xii. For a variety of views interpreting the place of technology in the historical discourse, also see John G. Burke and Marshall C. Eakin, eds., *Technology and Change* (San Francisco, CA: Boyd & Fraser, 1979).
10. Smith and Marx, eds., *Does Technology Drive History?* xiii–xiv, 226.
11. Andrew J. Rotter, *Hiroshima: The World's Bomb* (New York: Oxford University Press, 2008), 3–4.

CHAPTER 1

A Community of Scientists
Atomic Theory over the Centuries

INTRODUCTION: ATOMIC THEORY TO ATOMIC BOMBS

Atoms and void; matter and space; being and nothingness. These extremes, or sets of apparent contradictions, have become the basis for knowing the world in which we live. These pairings also are at the heart of atomic theory, which unlocked knowledge of the material world. The subject of this chapter seems far removed in time and topic from a book focused on Atomic Age America. Indeed, there is no single, direct historical line between the hypothesizing of atomic theory centuries ago and the atomic bomb. The science behind nuclear weapons was not, in other words, first theorized in ancient times. But in several respects speculation about the physical nature of our universe, why matter and energy shape everything around us, and the way in which we come to know how things work began as a rather simplified atomic theory. A rationalist view of the universe grew up alongside philosophical and religious queries into determining the place of humans on the globe.

Sometimes the free exchange of ideas was muffled by governmental and religious institutions only to reemerge later. Intellectual cross-fertilizing was necessary for a process that led to basic changes in a variety of disciplines to help understand the physical world. It also resulted in refining atomic and subatomic theory. By the 1930s, however (as Chapter 2 demonstrates), warring powers in World War II stifled the free exchange of ideas about the atom, institutionalized science like never before, and directed the application of atomic theory to practical ends—primarily atomic bombs and nuclear reactors. Not all the queries about atoms flowed toward one inexorable conclusion, but the limits on the free exchange of scientific ideas challenged an intellectual process that had been going on for centuries. Therefore, this book needs to begin in the distant past to let us better grasp some of the most important philosophical and scientific ideas that shaped the modern world and our understanding of the atom.

A COMMUNITY OF THINKERS, A WORLD OF IDEAS

Atomism is a very old idea originating with the ancient Greeks and ancient Hindus. Science as we know it today was not practiced in antiquity. Scholars sought answers through philosophy. In fact, there was no physics community for hundreds of years to come. Yet, there has always been a community of thinkers (at least loosely constructed) who shared common interest in perplexing ideas, argued vigorously over the major and even most minor concepts and precepts, and passed along their curiosity and speculations to future generations. Transmitting ideas in ancient times was no easy task, and much was lost simply in the chaos of living. Confusion sometimes occurred in the translation of ideas and concepts from language to language, from culture to culture. Ultimately, scholarly communities survived and prospered, transcending country borders and even continents. A scientific community as we understand it today, however, was not established until the late nineteenth century.

Atomic theory, like other great ideas, depended on collaboration and on sharing thoughts and speculations without fetter. At times politics and religion made such free exchange difficult or impossible. The well-known case of Galileo Galilei (1564–1642) became a prime example, when the Catholic Church forced Galileo to renounce his notion of a sun-centered as opposed to an earth-centered universe. Even when ideas flowed without restraint, grand theories of science were not simply the result of an accumulation of suppositions, discoveries, and inventions. Most big ideas go through paradigm shifts, that is, fundamental changes in the shared common beliefs that govern the thinking of an intellectual community. Scientific revolutions arise out of paradigm shifts. Atomic theory has gone through such change from the fifth century B.C.E. to this day.[1] Twentieth-century chemist and philosopher of science Michael Polanyi identified a "republic of science" (or what I call the "democracy of science") which operated as "a highly simplified example of a free society" almost tribal in its adherence to its own conventions and practices.[2] The "community structure of science,"[3] and its predecessor scholarly communities, allowed atomic theory to blossom.

ATOMISM IN THE ANCIENT WORLD

The fifth century B.C.E. introduced a precursor to atomism borne out of simple observation as much as through abstract theory. Philosophers initially surmised that water, fire, and/or air were primary (or primordial) substances modified in various ways to produce all matter. They speculated not only about the basic substances but about their permanence and their ability to change. The philosopher Empedocles of Agrigentum in 445 B.C.E. asserted that all existing matter emerged from a combination of four primary substances: air, earth, fire, and water. This idea was most important to the development of atomic theory, especially the notion that building blocks made up nature. Empedocles's theory endured in one form or another until modern chemistry challenged it many centuries later.[4]

The four primordial elements represented the most basic way to make sense of the physical world. But it was contemplating what the naked eye could not see that revealed the substance of the universe, that is, atoms and void. In the West of the fifth century B.C.E., several Greek philosophers developed and refined the concept that matter consisted of small indivisible particles—atoms or corpuscles (*atomos* in Greek meaning "indivisible"). Many regard Leucippus (early fifth century B.C.E.) as the founder of atomism. He may have been the first to state the concept of the indivisibility of matter, and the first to assert that all events have rational explanations rather than supernatural causes. About 420 B.C.E. Democritus of Abdera, a wealthy Thracian, further developed this atomic theory (or at least left more writings about it than Leucippus) making clear that atoms were infinite in number, indivisible, made of the same substance, differed in shape and size, and were always in motion. To account for observable changes in the world, he added that things do not differ in their nature. Variations in the position or arrangement of atoms modify them. To make this possible space itself had to be understood as a vacuum (nothingness or void) where motion could take place. The universe, therefore, was composed of the two realities of atoms and void. Others refined the ideas of Democritus, incorporating them into a materialistic philosophy emphasizing a world governed by the laws of nature, rather than by the intervention of a deity or other supernatural forces. The concept of atoms and void, identified as interdependent, challenged and then replaced the idea of primordial substances.[5]

While ancient Greeks shaped the idea of atomism, it also was a topic of much speculation in Hindu philosophy in what is now modern India. Which came first? No one knows for sure. There is a distinct possibility that atomism developed independently in the West and East. What seems fairly clear is that while both cultures shared a similar concept, the details were different. For example, the atomic theory of the Nyaya-Vaisheshika movement of Brahmanism joined atomic theory with the idea of the primordial elements (nine rather than four in this case). Atomistic doctrine also appeared in philosophical accounts of Buddhism and Jainism.[6] Whether East or West, for more than fifteen centuries the atomic theory remained in limbo, fading from memory, contested, or ignored. The philosophical giants of the classical era, Plato (c.427–c.347 B.C.E.) and Aristotle (384–322 B.C.E.) and their followers failed to embrace the doctrine. Plato regarded reason, not perception, as the way to know the world. His universe was governed by an intelligent power (God), and although he believed that matter was eternal, he viewed the observable as a pale copy of the world of forms and ideas. Plato modified the theory of the four elements (placing him in line with atomist thought if not atomic theory) by devising a geographic variation relying more on abstraction than empiricism. Aristotle was clearly the premier anti-atomist of his day. The world, he argued, was as we saw it, interacted with it, and understood it. There was no gap between matter and form. They were bound together. While he accepted atoms as everlasting and supported the idea of the four elements, he rejected the notion of void.[7] Yet, in the ancient world superstition, myth, and simple observation were making way for a fairly sophisticated set of ideas about the construction of the physical universe.

ANTI-ATOMISM IN THE MIDDLE AGES AND THE COMING SCIENTIFIC REVOLUTION

The quashing of the atomic theory in the Christian sphere through the Middle Ages was virtually unconditional. The Catholic Church did not tolerate the notion of primordial substances which questioned the spiritual origins and sustaining of the universe. While comfortable with Aristotle's theory of the four elements, the Church rejected the idea of the void. Some medieval Christians helped compile historical narratives about atomic theory, but few were even sympathetic to the idea. Jewish philosophers also were staunch anti-atomists. Some Arab scholars embraced atomism, but with a decidedly religious focus. Arab atomism relied on a concept of recurring creation, but creation dependent on Allah's will and not through real-world cause-and-effect. In this interpretation the material world and the spiritual world were reconciled, unlike in Christianity, where atomism appeared to pose a threat to the notion of an omnipotent God.[8] By and large atomism faded from thought in the West for hundreds of years. In the fifteenth century, scholars unearthed and published a long-lost work on Greek atomism which helped to invigorate scientific inquiry about atomic theory in the sixteenth and early seventeenth centuries. The ancient text raised questions about atomic theory that were particularly well-received in England among such luminaries as physicist and chemist Robert Boyle (1627–1691), an early founder of modern chemistry, and Isaac Newton, who united the laws of physics. England was fertile ground for atomism at the time, since it was well beyond the powerful religious authority of the Catholic Church, which was omnipresent in Italy, France, and Spain. Even in Catholic countries, some individuals hoped to reconcile atomic doctrine with Christian views.[9]

The Scientific Revolution of the seventeenth century (really underway in the sixteenth century or earlier) built upon the rebirth of experimental science at the height of the Renaissance, and was receptive to atomism. Scientists of this era drew upon the classical and medieval roots of science in the West as well as works by Islamic scholars. The great advantages for scientific inquiry in this period were the invention of instruments of measurement, such as fine microscopes and telescopes, and improved knowledge of mathematics.[10] German physician Daniel Sennert likely was the first to apply the classical atomic theory to chemistry in 1619. His work, *De Chymicorum*, described chemical change as the interaction of atoms, distinguishing between atoms of an element and atoms making up a chemical substance. Sennert's work suggested the existence of different types of atoms.[11] However, there was no agreement among scientists and philosophers about atomism during the Scientific Revolution.[12] Any atomic theory worth embracing had to be based on empirical phenomena, which was a position that took time to develop. The Tuscan physicist and astronomer Galileo was crucial in establishing the science of motion, and along with German mathematician and astronomer, Johannes Kepler (1571–1630), helped move science toward an understanding of the unification of the terrestrial and celestial worlds as part of one universe. Galileo's open support for atomism was an element in his troubles with Rome as it had been for others.[13]

THE NEWTONIAN WORLD

As the Scientific Revolution blossomed, scientists and philosophers approached atomic theory from many directions and through several disciplines. French philosopher and mathematician René Descartes (1596–1650) tried to build a framework for the new science, and developed analytic geometry as a deductive tool to reduce physical sciences to mathematical principles. He was the first to present a mechanical explanation of the solar system, but one that dismissed atoms and the void. In his view, space could not be a container for matter; material objects implied space. It was a self-contradiction, he argued, that "nothingness" existed. Descartes surmised that to exist a void "should possess extension," and since nothingness or emptiness did not have extension, space must constitute a continuation of matter. His theories also excluded the notions that atoms were indivisible and that gravity was an intrinsic property of matter.[14]

Despite the eventual rejection of many of Descartes's scientific theories, his general aims had an important influence on the titan of the Scientific Revolution, Isaac Newton (1642–1727). The English physicist and mathematician went down in history as the man who unified the celestial and terrestrial worlds through his universal law of gravitation (used by physicists to describe the motion of all objects in the universe, and referred to as "classical mechanics"). This was a paradigm that remained unshaken until the revelations of Albert Einstein in the twentieth century. While Newton eventually became critical of Descartes, he embraced the idea of explaining the universe in terms of matter and motion.[15] Unlike Descartes, Newton was an atomist. He accepted the idea of matter as composed of corpuscles and acknowledged the existence of the void. He broke with classical atomism over its purely mechanistic worldview, ruled by chance and devoid of God. Newton conceded the responsibility for creating atoms to a supernatural power. Like others before him, he was unable to demonstrate why atoms behaved as they did. Newton's image of the atom as "something like a miniature billiard ball," one historian noted, clearly limited his

Sir Isaac Newton (1642–1727) explained the universe in terms of matter and motion. *Source:* Library of Congress, Prints & Photographs Division, [LC-USZ62-10191]

speculations.[16] An atomic theory of matter was not necessary in his mathematical treatment of motion and material bodies. Yet his support for atomism, even with its strong religious overtones, gave credibility to the idea. Particularly important was that he added the concepts of mass and force to the classical definition of atoms, which had centered primarily on size and shape.[17]

During the "Age of Newton," scientific disciplines beyond the burgeoning field of physics explored the atomistic view of matter.[18] Boyle was one of the leading advocates of the theory of atoms in England. Like Newton, he was deeply religious and believed that God created fundamental particles of matter. His "corpuscular philosophy" emphasized that these particles formed myriad shapes with specific properties. As an experimentalist, he looked for the chemical and physical properties of atoms in their various shapes. Rejecting Newton's idea of a force of mutual attraction, Boyle fell back on the notion that the movement of atoms could not be relegated to chance, but directed by a spiritual force. He surmised that the affinities of atoms for one another were due to their spatial configurations. Boyle's pneumatic experiments were particularly noteworthy. He asserted that air, like other forms of matter, had weight. He also believed that air was not composed of a singular element, but made up of several kinds of particles with different functions and characteristics. Descartes strongly influenced Boyle's ideas, especially the notion that extension was related to matter. Creating a vacuum (void) experimentally was beyond Boyle's ability at the time, and thus proving the existence of the void remained contested. Not until the early eighteenth century did someone produce the first modern atomic or molecular model of a gas.[19] Other studies led to a kinetic view of matter in which particles of all bodies were in constant motion, (with those of different masses moving at different speeds). Some made deeper queries into issues of attraction and repulsion of atoms.[20] But without a substantial body of concrete, measurable data, atomism continued to be open to fervent debate not only among scientists but among philosophers as well.[21]

Consensus on atomic theory was far from possible in the eighteenth century. The Renaissance, the Protestant Reformation, and the Scientific Revolution had produced clear indicators that men of learning (indeed philosophy and science remained provinces clearly restricted to men) were developing very different attitudes about the universe. The pace of scientific inquiry had also accelerated since the mid-seventeenth century, particularly due to the appearance of private academies and societies in Europe offering places to share and publish scientific research. Philosopher and scientist Francis Bacon (1561–1626) argued that scientific work needed to be a collective endeavor, and that information must be exchanged especially to avoid duplication. The Royal Society founded at Oxford, England, in 1662 (possibly the first permanent organization of its type) fit this model by acting as a clearing house for research that could benefit the public. It published *Philosophical Transactions*, the first professional scientific journal. Soon after—in 1666—the French organized its Royal Academy of Sciences, and similar institutes followed in Naples and Berlin by 1700. A more formal intellectual community of scholars was taking shape as science was becoming an issue for wider public discussion.[22]

THE NINETEENTH CENTURY: A TRANSITION ERA

Scientists and scholars increasingly pushed aside the philosophical musings about atomism in the nineteenth century for more pragmatic concerns about the nature and function of atoms. Antiatomism was not dead. Yet scientific atomism gained greater attention with demand for tangible data through observation and experimentation, for better research tools, and for more creative theoretical concepts. The path was not direct, and there was plenty of dissonance among researchers from a variety of scientific disciplines. At this time Newton's mechanistic view of the world largely bound together scientists, raising questions of atomic theory within that paradigm. Years passed before Newtonian physics proved limited in cracking the code of the microscopic world of the atom. Startling findings opened new doors or unravelled long-held beliefs. The nineteenth century was a transition era for atomic science, from general concept to more direct testing.[23]

Progress in developing an atomic theory of chemistry and in the physics of energy (since heat is a form of atomic motion) made the nineteenth century special. The hypothesis that atoms were the smallest particles in the universe was derived from conjecture not empirical research, and would soon be challenged. English chemist and physicist John Dalton (1766–1826), 100 years after Newton, gave "fresh meaning" to the atom.[24] The pioneer of the modern atomic theory of matter began his work with the physical structure of gases. Like many of his predecessors, he accepted the idea that atoms were indivisible and indestructible. Dalton believed that matter consisted of individual atoms, that each element was composed of a particular kind of atom, that compounds were made of molecules (combinations of atoms of two or more elements), and that all atoms of the same element had the same weight. He concluded that it should be possible to calculate the atomic weight of each element by measuring equal volumes of various elements. In 1808 he devised a system of chemical symbols describing the elements and then arranged them in a table according to atomic weight. Throughout the nineteenth century, a great deal of chemical experimentation rested on atomic weight as a key to understanding the properties of atoms.[25] Few accepted Dalton's atomic theory immediately, but the idea fared better on the Continent than in England.[26]

Establishing a periodic table of elements was an important step forward in chemistry and atomic science. At the time of Dalton's atomic theory scientists knew about 36 elements; by 1860 they identified twice that many with more being added all the time. Dimitri Ivanovich Mendeleev (1834–1907), a Russian chemist from St. Petersburg, published his version of the periodic table of elements (organized by order of atomic weights) in 1869, which essentially remains the version used today. Among the applications of the system was Mendeleev's ability to predict elements

Dr. John Dalton (1766–1826), pioneer of the modern atomic theory of matter. *Source:* Library of Congress, Prints & Photographs Division, [LC-USZ62-64936]

to fill gaps in the table, with the weights and properties he anticipated. The development of the periodic table was the culmination of approximately 60 years of investigation of and speculation on the physical properties of atoms, which was resurrected by Dalton's atomic theory.[27]

MERGING PHYSICS AND CHEMISTRY TO UNDERSTAND THE ATOM

Scientists in the nineteenth century were moving on other fronts as well to understand the atom. In the 1840s and 1850s, physicists intensely studied the conservation and transformation of energy and the increase in entropy (the measure of the disorder of a system, in this case at the atomic level), essential to the development of atomic science. By 1870 many agreed on a universal law that energy is conserved through all transformations, that is, energy is neither created nor destroyed in the processes of nature. This suggests, for example, that mechanical energy can be translated into heat energy, and heat energy has an exact mechanical equivalent. It also suggests that electrical motion and chemical reactions were forms of energy. Conservation and transformation of energy helped to unify the study of physics, as chemistry had coalesced around the periodic table, to support the belief in the atomic composition of matter.

A new kinetic model of gases (gases consisting of rapidly moving atoms and molecules) connected the two disciplines and grounded them on the use of statistics.[28] Building on the work of other scientists interested in atoms in motion, Scottish physicist James Clerk Maxwell (1831–1879) was greatly responsible for developing the kinetic theory of gases (referred to at the time as the "dynamic" or molecular theory). Assuming that gases are composed of enormous numbers of molecules, as a gas gets hotter the molecules move faster thus increasing pressure. Such a perspective offered a way to understand the elasticity of gases, the process of diffusion, how gases expand to fill space, and how they mix together. For the purposes of atomic theory, this perspective also offered a way to calculate the size of molecules. Development of the kinetic-molecular theory was a major event in nineteenth century science, not only through the work of Maxwell but through others in Austria, Holland, and the United States. The theory was significant for reinforcing the discussion of the physical atom in terms of mechanical force (for instance, heat as the motion of atoms) at least for the time being.[29] Light had its own chemistry, and scientists observed that each element had its own spectrum or its own light. In 1826 scientists applied the new discipline of spectroscopy to chemical analysis. Using modern instrumental optics, they noted that each element had a characteristic spectral series. Therefore it should be possible to explain the spectral lines with respect to the elements that produced them. Until 1896 no one developed a convincing theoretical explanation concerning any spectral regularity. Spectroscopy, however, helped significantly to systematize observable properties of light.[30]

As he had done with gases, Maxwell took the study of light in new and exciting directions in the nineteenth century, placing him among the greatest scientists of his generation. In 1864 Maxwell first presented his electromagnetic theory, including the notion that light is transmitted as an electromagnetic vibration. It would take another 25 years before experiments in electricity caught up with theory. Nonetheless, the electromagnetic theory eventually proved important for atomic scientists in demonstrating how to understand the emission and absorption of radiation.[31] Maxwell's theory also was instrumental in uniting a wide range of phenomena which had been considered unrelated. The unification he accomplished in his work led to a very complex model that would later proved unworkable. While scientists attempted to explain a range of phenomena (including light, electricity, and magnetism) in mechanical terms, and as their equations became more and more accurate in describing and predicting certain phenomena, "the physical basis of those equations was becoming less and less comprehensible."[32]

Through a variety of new institutions researchers studied, discussed, and debated the heady scientific achievements of the nineteenth century and the confounding puzzles that resulted from those very achievements. Shared experiences, especially in formal laboratories and learned societies, augmented interaction through private correspondence, visits, meetings, and scientific journals. In 1860 the first international congress of chemists was held in Karlsruhe, Germany. Attendees wanted to provide order to a confusing mix of terminology and concepts that the scientific community utilized. While the delegates did not achieve standardization, they generally reached consensus on accepting a distinction between atoms and molecules.[33] In the field of physics, employment opportunities steadily increased. Employers hoped that scientific discoveries would translate into practical uses, helping to inspire a new industrial revolution.

Equally important was greater emphasis on research teams. Aside from societies in large cities, however, collective enterprise struggled for many years. Chemical laboratories in Germany were important in setting a new standard of collaboration. Part of the reason for teamwork in chemistry was the cost and complexity of the apparatus needed, and the field's close connection to industry. Justus Liebig's laboratory in organic chemistry at the University of Giessen set an important precedent in the 1820s. From such laboratories came tangible products: fertilizers, the manufacture of cast iron, pharmaceuticals, and dyestuffs. Also, the laboratories proved essential to the military, especially in the development of explosives. From chemistry the model spread to medicine, particularly in the area of physiology. Many years before the "military/industrial complex" of the twentieth century, scholars and researchers wedded the theoretical and the practical, and the scientific and the political in some fields more than others. Chemistry was more inclined to go in this direction earlier than physics.

There was a hazy line between theoretical and applied (or experimental) physics before the twentieth century, other than what some called "pick and shovel work" to investigate the possibilities of utilizing new theory. Physics lagged behind other scientific disciplines until late in the century, largely because many people could not discern the requisite practical advantages more obvious in chemistry or medicine. Amateurs and academic scientists had dominated the field of physics, which made the scientific societies all the more important in promoting it. Physiologist Hermann Helmholtz's Institute of Physics in Berlin (1870s) took advantage of the Kaiser's desire to make the imperial capital the center of science and culture in Prussia. The institute attracted most every important German physicist for many years. In addition, German technical schools and colleges became important centers of research and the teaching of technological studies. These included the Polytechnische Schule in Karlsruhe (1826), modeled after the French École Polytechnique in Paris. Several others sprouted up in Munich, Stuttgart, and Zurich. After 1870, other countries organized technical colleges along the lines established in German schools. In 1882, the École de Physique et Chimie Industrielles appeared in Paris, surprisingly modeled after the German equivalent. By 1899 German technical colleges could grant degrees. The British did not want to be outdone by their continental rivals. Between 1867 and 1875, nine laboratories were founded in the United Kingdom. In 1875, the Cavendish Laboratory opened under the leadership of James Maxwell. This important laboratory in experimental physics became part of Cambridge University. Teaching and researching science spread throughout several colleges in the UK in the late nineteenth century, and also throughout the British Empire.[34]

THE MECHANICAL WORLDVIEW UNDER SIEGE

By the 1890s, scientists regarded as less and less progressive the mechanical worldview so central to atomic physics and chemistry in the preceding decades. A shift was underway from mechanics to thermo- and electrodynamics. The kinetic theory and an electromagnetic explanation of mass were now fundamental to understanding the nature of atoms and molecules.[35] The fields of physics and chemistry themselves also were changing. Men still dominated science, with a few very important but numerically limited exceptions. Physicists, especially in Germany, came from the middle and upper classes like their counterparts in the humanities. Chemists, especially in organic chemistry, grew out of the business community. The definition of "physicist" by 1900 was not unlike its modern meaning, that is, a professional scientist employed as a faculty member at a university or at a polytechnical college. Amateurs or secondary school teachers calling themselves physicists were a dying breed, although some engineers and technical experts engaged in broadly defined applied physics.

Despite some inroads in Japan, most academic physicists were found in Europe and North America. In 1900 the total numbers were modest, possibly 1,200 to 1,500 worldwide (only about ten percent engaged in studying what became the very essential research in radioactivity). Physicists were most highly concentrated in Germany, Great Britain, France, and the United States, with Germany holding the primary position. The United States clearly was the leader in total numbers of physicists at the turn of the century, but by measures of productivity and original research it lagged behind Europe. Within ten years, American physics made key inroads, although it remained provincial and relatively weak in theoretical physics, and much stronger on the experimental side.[36] Physics still remained an international enterprise in these years. Physicists pooled ideas and results. Contacts between physicists of different nationalities were typical through visitations and study at various labs and institutes, personal meetings and collaborations, and conferences. Large gatherings were atypical, but in summer, 1900, the first International Congress of Physics along with the International Congress of Mathematics met in Paris in connection with the World Exhibition. Between 1900 and 1914, 22 new physics laboratories were built in Germany, 19 in Great Britain, 12 in France, and 13 in the United States. Physics journals flourished.[37]

Outside Europe and North America, only Japan made strides in physics in the period. Before 1870, physics was almost nonexistent there with hardly any identifiable leaders and a few poorly equipped facilities. Tokyo University (established in 1877) taught physics, but few faculty conducted original research. Hantaro Nagaoka was a pioneer in Japanese physics. A graduate of Tokyo University in 1887, he studied in Munich, Berlin, and Vienna from 1893 to 1896. Japan's foremost physicist, he was a professor at the renamed Tokyo Imperial University in 1910. European scientists showed little interest in his atomic theory presented in 1904. After World War I atomic science became better established in Japan, especially because of links to European universities. Also important, the Japanese fought on the winning side in the war and did not suffer the humiliation, nor the loss of colonies and other resources that the Germans and other Central Powers experienced.[38]

THE ELECTRON, X-RAYS, AND MODERN PHYSICS

By the late 1890s atomic theory was almost universally accepted, being largely an academic and theoretical topic of interest. New discoveries and a shift away from the mechanistic perspective challenged atoms as indivisible and as the nominal particles of matter. Atoms would prove to be composites, functioning in much different ways than thought before.[39] The turning point came with the discovery of the electron. Like most other major discoveries, many scientists carried out the work that led to the electron over decades of painstaking effort, and, at times, with a little luck. Experiments in different areas of science, numerous debates, and theoretical speculations all contributed. A shift in emphasis from mechanics to electromagnetism was crucial, with the first concrete evidence of the interaction between electricity and magnetism in 1819–1820. The general research in electrical technology proved essential, and provided insight about the relationship between electricity and the atom. Vacuum discharges resulting in the identification of cathode rays was a promising line of inquiry leading to breakthroughs in subatomic particles. Researchers employed tubes as mini-chambers for such studies. Electric current was sent through gas in these devices, which themselves were placed between a strong magnet. Obtaining a vacuum, however, proved difficult.[40]

Scholars claim that "modern physics" was born in November 1895 when German physicist Wilhelm Conrad Roentgen (1845–1923) discovered the x-ray. This event started the search for the inner structure of the atom, leading to a new scientific pathway and major revelations. But it also pointed to the way science was taking center stage in world events involving a growing community of researchers in several fields.[41] The momentous finding was not a lucky accident, as it could have been scripted in a melodramatic science-fiction movie. Working in his darkened laboratory with a cathode tube, Roentgen noticed that a paper screen coated with barium plantiocyanide some distance away was glowing. He then undertook a series of rigorous investigations to see if various materials could block the effect (wood, rubber, tinfoil) but they did not. Placing his hand in front of the screen, the bones cast a shadow, but not the flesh. He soon learned that these "x-rays" would expose a photographic plate. The discovery, publicly announced seven weeks after the initial observations, made Roentgen a celebrity. Many hailed the possibilities for practical applications of x-rays, especially in medicine. Physicians used x-rays to find foreign objects imbedded in human bodies, and the x-ray machine became an essential aid to surgeons and other doctors. Soon, however, rash use of this new scientific miracle revealed its dark side in terms of radiation burns and worse. For many years people did not take adequate precautions, and the practical applications of the discovery clouded the potential risks.[42]

In January 1896 Roentgen's observation was the central topic at the French Academy of Science. Some proposed that the x-rays might not be electrical but related instead to the fluorescing part of the cathode tube. This point interested French physicist Antoine Henri Becquerel (1852–1908), who set out to test it later that year. Utilizing uranium salts (which produced phosphorescence) and photographic plates to capture the emissions, he observed that the uranium salts gave off invisible, penetrating rays after being subjected to chemical and physical processes. This occurred even after being isolated for a lengthy period, or also in the absence of sunlight. The new phenomenon called "uranium rays," appeared to be neither x-rays nor fluorescence. Becquerel had discovered radioactivity.[43]

The significance of the discovery remained obscure for almost two years, until a young scientist, Marie Curie (1867–1934) and her husband Pierre Curie (1859–1906), brought radiation squarely to international attention. The Polish-born Marya Sklodowska (Marie Curie) moved to Paris in 1891 to study physics and mathematics. She fought an uphill battle to gain credibility in a world dominated by men. Yet despite all odds, and because of her brilliance as

Pierre (1859–1906) and Marie (1867–1934) Curie brought radiation to international attention. *Source:* Library of Congress, Prints & Photographs Division, [LC-USZ62-73356]

an experimentalist, she became one of the most celebrated scientists of her day, receiving two Nobel Prizes. Her early work with x-rays, radium, and other radioactive elements (eventually conducted with her husband, Pierre) helped lay the groundwork for additional research into the structure of atoms. She began studying the properties of uranium in fall 1897, and experiments with pitchblende led to the discovery of polonium and radium (two new radioactive elements) in 1898 within months of each other. The discoveries of substances much more active than uranium propelled radioactivity into the public spotlight and pushed physicists to further experimentation. Before people learned about its dangers, the fascination with radioactivity led to a bevy of radioactive products, including hair tonic, chocolates, and bath salts. Fortunately such crazes proved short-lived, but not before they inflicted awful consequences.[44] After Pierre's untimely death in 1906, Marie succeeded him as professor of physics at the Sorbonne, University of Paris. She was the first woman to achieve that honor. Having worked in a shabby laboratory for years, the French government funded the Radium Institute before World War I, and Marie became the director of the Radium Laboratory. Ironically, her death in 1934 was the result of radiation poisoning (leukemia) incurred from years of tireless research.[45]

The work of Becquerel, the Curies, and others introduced radioactivity as an important physical phenomenon with concrete applications. But it also provided important evidence leading to a clearer understanding of the inner structure of the atom and its constituent parts, most importantly the electron. English physicist J.J. Thomson (1856–1940) of the Cavendish Laboratory is credited with "discovering" the electron in the 1890s. This is somewhat oversimplified, since knowledge of the electron can be traced to the early nineteenth century. It was, however, more of a hypothetical entity than Thomson's experiments demonstrated.[46] Thomson's experiments over a four-year period are nonetheless extremely important for essentially wiping away the idea of the basic structure of matter that scientists had accepted for 200 years or more. He also settled the debate about cathode rays, stating that they were electrically charged particles (electrons or minute particles of matter carrying a negative electric charge) not electromagnetic waves. Thus atoms are not indivisible; electrons are constituents of all atoms, and are of the same mass. This subatomic perspective attracted a variety of claimants, and in some cases others borrowed or tested it.[47]

A PLANETARY MODEL OF THE ATOM

Ernest Rutherford (1871–1937), New Zealand-born physicist and a former assistant under Thomson, took up the study of radiation from uranium likely near the end of 1897. Rutherford initially worked in Canada, but moved to England, where he eventually became Thomson's heir as head of the Cavendish. One of the foremost atomic scientists of the twentieth century, Rutherford is regarded as the founder of nuclear physics for developing the modern understanding of the structure of the atom. His discovery of alpha and beta radiation was a major advance in recognizing the nature of radiation itself. Along with gamma rays (discovered in 1900), alpha and beta rays were present in the process of nuclear fission (to be discussed later). There were several other claims to new rays as well, but most proved specious. Rutherford with others concluded that the number of electrons in an element was roughly proportional to its atomic weight. He also concerned himself with the role of the electron in the structure of the atom. He needed to determine the origin and the nature of the positive electrical charge in the atom that had to offset the negative charge of the electron. Rutherford and his collaborators posed a solution that the existence of a positively charged nucleus in the atom was equal to the negative charge of the electrons. The atom was essentially empty space with a small, but highly concentrated nucleus and a respective number of electrons with charges balanced for the sake of neutrality (or stability). Rutherford's "planetary model" of the atom (1911) was so-named because it concentrated most of the atom's mass into a small core in much the same way that the mass of the solar system is concentrated in the sun. All of the protons were in the nucleus with the electrons orbiting around it like planets. His model, modified several times over the years, soon became the standard for understanding the structure of atoms.[48]

With a chemist colleague, Rutherford also made a major advance with the study of radio-active decay in 1902, nine years before proposing the nuclear atom. They concluded that the disintegration of atoms produced radioactivity. This suggested that radioactivity was not a permanent phenomenon, but actually decreases over time, and that a radioactive substance transforms (or transmutes) into another substance.[49] While this was a major claim, designing atomic models that would explain radioactivity in a mechanical way proved confounding.[50] Rutherford and others had taken giant leaps in describing the structure of the atom during the first decade of the twentieth century, but atomic physics was now up against a very sturdy wall.

THE QUANTUM AND THE EINSTEINIAN WORLD

The inner structure of the atom was now a principle focus of physics, but the old mechanistic theory of the universe seemed wanting for understanding subatomic phenomena. Physicists and chemists struggled with the swirl of ideas and speculations about the nature of the atom. They also struggled with Newton's mechanical basis of physics versus Maxwell's electromagnetic one. In the background a new way of understanding that world was emerging. The "quantum," the smallest unit of energy, signaled a paradigm shift in physics by helping to reveal the subatomic world. In December 1900, German physicist Max Karl Ernst Planck (1858–1947) experimented in the area of thermal radiation, especially the interaction between radiation and matter. In a paper he read before the Physical Society of Berlin, Planck proposed that energy is emitted in small, indivisible amounts which he called "quanta." The classical view was that energy exchanges only took place in continuous amounts. Planck, a professor of physics at the University of Berlin and a specialist in thermodynamics, hoped that his work would lead to practical applications in the German heating and lighting industries. A long-time opponent of atomic theory, he believed that if atoms existed, they were not mechanical. Neither he nor his contemporaries at first believed that the idea of quanta required serious attention at least on a theoretical level. For five years few scientists discussed the quantum hypothesis.[51]

The work on electromagnetic radiation by the young German-born Albert Einstein (1879–1955) eventually brought more attention to the quantum theory. One of the few proponents of Planck's ideas at the time, Einstein was not working as a physicist but as an inspector of patents in Berne, Switzerland. In 1905, the same year that he introduced his monumental theory of relativity about the relationship of space and time[52], Einstein proposed that light itself was made up of particles which he called "photons." Einstein's theory of photoelectric effect resulted from experiments made to explain how electrons were ejected from a metal exposed to ultraviolet light. The fusion of the Planck and Einstein hypotheses resulted in a quantum theory recognizing that energy is not continuously emitted nor absorbed, and that the carrier of a quantum of energy is the photon. As in the case of Planck's hypothesis, physicists ignored or even rejected Einstein's light quanta (light-quantum hypothesis) as much too radical. He nonetheless continued to work on the problem, and in 1909 provisionally proposed a theory that fused both waves and particles in light. The view that colleagues believed to be contradictory became dogma with the emergence of quantum mechanics in the mid-1920s.[53]

Albert Einstein (1879–1955), ca 1920, developed the theory of general relativity and brought attention to the quantum theory. *Source:* Library of Congress, Prints & Photographs Division, [LC-USZ62-106038]

Exciting things were happening in the early years of the twentieth century with respect to the atom. Rutherford's planetary model invited vast speculation, as did the quantum theory pushed by Planck and Einstein. Yet quantum theory was not well understood at the time when Rutherford proposed his model in 1911, and only a few theoretical physicists were studying it.[54] There was a growing interest, or at least curiosity, in quantum theory as demonstrated by the convening of the first Solvay Congress in Brussels during November 1911. Belgian Ernest Solvay, an industrialist, philanthropist, and amateur physicist, underwrote what became an important series of international physics meetings. He invited 21 participants (all from Europe), including Planck, Einstein, Rutherford, Marie Curie, and German physicist Arnold Sommerfeld (1868–1951). The participants resolved little, but they sharply debated problems surrounding radiation and the quantum theory. A wariness over quantum theory remained, but the meeting helped to provide a common understanding about the key concerns related to it. Out of the meeting also came the establishment of the International Institute of Physics (1912), which Solvay endowed.[55]

THE BOHR-SOMMERFELD ATOMIC MODEL

While the Solvay Congress did not allay skepticism about quantum theory, the fact remained that orthodox Newtonian physics did not operate on the subatomic level and Maxwell's electromagnetism was incomplete. Danish physicist Niels Henrik David Bohr (1895–1962), who had not been much interested in the structure of the atom until his postdoctoral years, developed what one scholar called "a conceptual hybrid" in producing his own model of the atom which combined

Newton's mechanics with quantum theory. Bohr did not take Rutherford's planetary model literally, more interested in the fact that the same substance always had the same property, that is, the
stability of matter. Classical mechanics, he believed, could not explain this, nor could it explain
why spectral rays were specific for each atom. Bohr's great achievement was determining that
atomic physics was best understood and its problems best resolved through quantum theory.

In 1913, two years after Rutherford's pronouncement on atomic structure, Bohr produced
his own model. A theoretical genius, but ponderous in his explanatory style, Bohr moved in a very
radical way to look at subatomic structure. He made two propositions. First, he challenged the
notion that in a planetary model electrons plunged into an atom's nucleus because of continually
changing orbits and loss of energy. Instead, he postulated that electrons held selected orbits with
constant energy or were at fixed distances from the nucleus. Second, he stated that atoms can
change from one stationary state to another, a change that is determined by probability. In this
way, electrons can "jump" from one orbit to another or from one stationary state to another. One
could only imagine how his colleagues greeted these revolutionary, albeit bizarre, ideas. On one
level, the old cliché applied: the devil was in the details. But, on another, as one writer observed,
"Bohr opened a door into a new mansion of nature. If the furnishing proved more elaborate than
appeared to his first glance, that is only to be expected."[56] Intense criticisms of Bohr's hypotheses ultimately led to refinements. While Bohr's ideas seemed unconvincing or hard to grasp,
they were verifiable through experimentation. Confirmations followed in the work of others over
several years. Sommerfeld, among other things, substituted elliptical electron orbits for Bohr's
circular ones, improving calculation of quantum numbers. By as early as 1914, the theory was
becoming an essential way to analyze atomic structure in European and American laboratories.[57]

In this fruitful period of insight and speculation, other discoveries opened up the world of
the atom like never before. The discovery of the "neutron" in the early 1930s by English physicist
James Chadwick (1891–1974) helped deal with the measured mass of an atom and its electric
charge. Neutrons had the mass of a proton, but no electrical charge, thus they could explain
discrepancies in the number of protons and electrons in a given element. A related issue was the
identification of isotopes, which were atoms of the same element with roughly the same chemical and physical properties but with different masses. Isotopes had the same number of protons and electrons, but a different number of neutrons. Such a discovery undermined the older
view that atoms of any given element were the same. Such a discovery would be crucial to many
future applications of atomic theory.[58] While Bohr's model had proven to be somewhat fragile,
his hypotheses and those of others led the scientific world into a whole new era of discourse
about the state of nature on a cosmic and a subatomic level. Nothing revolutionized science on a
cosmic level more than Einstein's theory of relativity, in which matter and energy were no longer
two distinct entities. But the work of Bohr and others on the atom was remarkable as well. The
speed with which old ideas were leading to new ones was astonishing. A good example was the
work of French physicist Louis de Broglie (1892–1987), who developed the principle that an electron or any particle of matter could behave as a wave as well as a particle, in much the same way
that radiation could have a dual nature. His hypothesis made clear the limits of Bohr's atomic
model, especially Bohr's ability to give an adequate picture of the physical qualities of the electron. It thus stimulated new experiments and new theory.[59]

Few doubted that Bohr's quantum theory of atoms was not a finished theory, but a beginning point of research that would lead to unfamiliar ground. Well into the early 1920s, and
despite World War I, the theory was broadened and modified by Bohr himself, Germans like
Sommerfeld, and by physicists from the United Kingdom, the Netherlands, and Japan. Three
centers in Europe were central to the research, one in Copenhagen and two in Germany. So

significant were Sommerfeld's contributions that in scientific circles Bohr's theory became the Bohr-Sommerfeld quantum theory.[60] Quantum theory had forced a reconstruction of the concepts of physics. Several still questioned the validity of the Bohr-Sommerfeld theory; others wanted to link older, more established ideas in some way with the new. In a 1924 paper, German physicist Max Born (1882–1970) expressed the need for a "quantum mechanics" as a successor to the Bohr-Sommerfeld theory. Quantum mechanics, "the modern picture of matter and energy," ultimately provided a mathematical explanation for the behavior of an electron in an atom. Indeed, scientists would employ a mathematical system to describe the physics of atoms in general, but as yet no one knew what it might entail.[61]

UNCERTAINTY

Work in radiation theory and wave functions offered a pathway to the new quantum mechanics. Austrian physicist Erwin Schroedinger (1887–1961) created "wave mechanics" which laid out rules for understanding de Broglie's waves, beginning a series of responses and refinements to this approach. Werner Karl Heisenberg (1901–1976) took a different path. He was a German physicist in Gottingen, who later worked with Bohr and also became director of the Max Planck Institute. In 1925 Heisenberg called his version of the new mechanics "matrix mechanics," which was the first complete form of quantum mechanics as an approach for solving a range of atomic problems.[62] To describe the state of an electron in an atom, Heisenberg utilized matrices made up of rows and columns of numbers. Matrix mechanics and wave mechanics were equivalent in mathematical terms, but the latter is more widely used today to explore the structure of atoms and molecules. This is because the mathematics was more familiar to physicists. The two approaches developed along parallel but separate lines, and collectively became known as quantum mechanics.[63]

Heisenberg's major contribution to quantum mechanics was a conceptual rather than an empirical breakthrough known as the "uncertainty principle." In quantum physics, knowing one variable in an equation may either reduce or eliminate knowledge about another. For example, one could not determine the exact position and the precise velocity of an electron at any given moment; one could determine one or the other, but not both. In this sense, the certainty of the classical theories was replaced with probability and statistical measurement. Heisenberg's principle not only challenged classical mechanics, but also the limits of accessible data in scientific experiments and in natural phenomena in general. His conclusion was that in analyzing physical systems it was inappropriate to use concepts that could not be measured. His conclusion also was a connection between the wave theory and matrix mechanics. Uncertainty therefore is really the culmination of quantum mechanics.[64]

Such an explanation did not please everyone. It was uncomfortable to think in new ways, and to learn a new language about science. Some physicists denounced the explanation; others tried to modify it. In question was whether uncertainty applied to physics was part of an evolutionary or revolutionary process, that is, to what extent it contributed to the paradigm shift away from a Newtonian to an Einsteinian universe. Not until 1928 did scientists successfully apply quantum mechanics to such issues as alpha particle radioactivity and other questions related to the atomic nuclei.[65] In philosophical circles, debate over uncertainty raised many uncomfortable issues as well and offered no easy resolution. Heisenberg was asking: How do we gain knowledge about our world? In the physics community, and in atomic physics in particular, scientists took sides principally around Bohr, who favored the new world-view stimulated by quantum mechanics,[66] and Einstein, who championed more orthodox, and some say, conservative concepts. Einstein ultimately accepted the technical accuracy of Heisenberg's and Bohr's system, but he would not concede that it was the final word.[67]

CONCLUSION: TOWARD HARNESSING ATOMIC ENERGY

In essence, scientists and philosophers were carrying on a dialogue in the twentieth century about what made up the material universe going back to queries concerning atoms and void in the fifth century B.C.E. The issues were grand, their comprehension perplexing, and disagreement not easily put aside. Contributions came from many quarters, from many places, and from many countries. After all, these were compelling matters facing humankind as a whole, even if the average person did not have them in mind as he or she pursued the struggle of living. The debates about quantum mechanics, the inner structure of the atom, and the future of physics continued throughout the early twentieth century, preoccupying scientists and their work. However, applying knowledge gained from studying atomic theory to some practical ends seemed to await a distant future. That future, to many people's surprise, was looming just over the horizon with the discovery of nuclear fission in the 1930s. In that decade and beyond, an abrupt shift occurred in atomic science from theory to application. Along with it came the disruption of the long-standing democracy of science, and the increasing central role of government in dictating priorities in scientific inquiry and practice. Questions about the universe made way for more mundane, but nonetheless critical interests in human security, national dominance, and the practical utility of science. Questions about atomic theory turned to the potential for reactors and bombs. Harnessing atomic energy became a passion.

MySearchLab Connections: Sources Online

READ AND REVIEW

Review this chapter by using the study aids and these related documents available on MySearchLab.

✓•⌐**Study** and **Review** on **mysearchlab.com**

Chapter Test

Essay Test

📖•⌐**Read** the **Document** on **mysearchlab.com**

Aristotle, Excerpts from *Physics and Posterior Analytics* (350 B.C.E.)

Abu Hamid Al-Ghazzali, On the Separation of Mathematics and Religion (1100 C.E.)

Galileo Galilei, Third Letter on Sunspots (1612)

Galileo Galilei, Letter to Madame Christine of Lorraine, Grand Duchess of Tuscany (1615)

Isaac Newton, from *Opticks* (1704)

RESEARCH AND EXPLORE

Use the databases available within MySearchLab to find additional primary and secondary sources on the topics within this chapter.

Endnotes

1. Thomas S. Kuhn, *The Structure of Scientific Revolutions* (Chicago: University of Chicago Press, 1996; third ed.), 1–7, 176–80.
2. Richard Rhodes, *The Making of the Atomic Bomb* (New York: Simon & Schuster, 1986), 29, 31–32.
3. Coined by professor of linguistics and philosophy, Thomas S. Kuhn.
4. Henry A. Boorse and Lloyd Motz, eds., *The World of the Atom* v. 1 (New York: Basic Books, Inc., 1966), 4; Lisa Rosner, ed., *Chronology of Science: From Stonehenge to the Human Genome Project* (Santa Barbara, CA: ABC-CLIO, 2002), 2, 14; Bernard Pullman, *The Atom in the History of Human Thought* (Oxford: Oxford University Press, 1998), 13, 15–24.
5. Rosner, ed., *Chronology of Science*, 14–15; Gerald Holton and Stephen G. Brush, *Physics, The Human Adventure: From Copernicus to Einstein and Beyond* (New Brunswick: Rutgers University Press, 2005), 30–31; Boorse and Motz, eds., *The World of the Atom* v. 1, 3–5; Pullman, *The Atom in the History of Human Thought,* 31–45.
6. Pullman, *The Atom in the History of Human Thought*, 71–83.
7. Ibid., 49–60; Rosner, ed., *Chronology of Science*, 2; Holton and Stephen G. Brush, *Physics, The Human Adventure*, 265.
8. Pullman, *The Atom in the History of Human Thought*, 85–114.
9. Rosner, ed., *Chronology of Science*, 23; Holton and Stephen G. Brush, *Physics, The Human Adventure*, 31; Boorse and Motz, eds., *The World of the Atom* v. 1, 3; David Lindley, *Boltzmann's Atom: The Great Debate That Launched a Revolution in Physics* (New York: Free Press, 2001), 3–8; Edward M. MacKinnon, *Scientific Explanation and Atomic Physics* (Chicago: University of Chicago Press, 1982), 15.
10. "Scientific Revolution," online, http://www2.sunysuffolk.edu/westn/science.html; *The History Guide*, "Lectures on Early Modern European History: Lecture 12: The Scientific Revolution, 1642–1730," online, http://www.historyguide.org/earlymod/lecture12c.html; Boorse and Motz, eds., *The World of the Atom*, v.1, 22.
11. Rosner, ed., *Chronology of Science*, 63.
12. Pullman, *The Atom in the History of Human Thought*, 118–22.
13. Ibid., 122–32. See also Holton and Stephen G. Brush, *Physics, The Human Adventure*, 50–60, 77–87; MacKinnon, *Scientific Explanation and Atomic Physics*, 22.
14. Boorse and Motz, eds., *The World of the Atom*, v. 1, 22–26. See also Pullman, *The Atom in the History of Human Thought*, 157–60; Holton and Stephen G. Brush, *Physics, The Human Adventure*, 63; Rosner, ed., *Chronology of Science*, 362; MacKinnon, *Scientific Explanation and Atomic Physics*, 22–23.
15. Holton and Stephen G. Brush, *Physics, The Human Adventure*, 63; MacKinnon, *Scientific Explanation and Atomic Physics*, 23; Mark Erickson, *Science, Culture and Society: Understanding Science in the 21st Century* (Cambridge, UK: Polity, 2005), 96–97.
16. Rhodes, *The Making of the Atomic Bomb*, 29.
17. Pullman, *The Atom in the History of Human Thought*, 136–38; Lindley, *Boltzmann's Atom*, 9; MacKinnon, *Scientific Explanation and Atomic Physics*, 26–37; Boorse and Motz, eds., *The World of the Atom*, v. 1, 86, 89; Alex Keller, *The Infancy of Atomic Physics: Hercules in His Cradle* (Oxford: Clarendon Press, 1983), 9.
18. The word *physics* is derived from the Greek, meaning "nature or natural things." "As such, physics is defined as that branch of science, which studies natural phenomena in terms of basic laws and physical quantities. The study is generally structured to satisfy queries, arising from the observed events occurring around our world. In this sense, physics answers questions about universe and the way elements of universe interact to compose natural phenomena." See "What is Physics?" *Connexions*, online http://cnx.org/content/m13250/latest/.
19. Pullman, *The Atom in the History of Human Thought*, 140–142; Boorse and Motz, eds., *The World of the Atom*, v. 1, 35–41, 110–11; Holton and Stephen G. Brush, *Physics, The Human Adventure*, 234, 269–70, 272; Lindley, *Boltzmann's Atom*, 3; Rosner, ed., *Chronology of Science*, 356.

20. Boorse and Motz, eds., *The World of the Atom*, v. 1, 52–58, 117–21; Lindley, *Boltzmann's Atom*, 13–14; Pullman, *The Atom in the History of Human Thought*, 175–77; Rosner, ed., *Chronology of Science*, 104, 372.

21. Pullman, *The Atom in the History of Human Thought*, 144–45, 152–54, 179–82, 183–88.

22. "Scientific Revolution;" "Lectures on Early Modern European History: Lecture 12: The Scientific Revolution, 1642–1730."

23. Pullman, *The Atom in the History of Human Thought*, 193–96.

24. Keller, *The Infancy of Atomic Physics*, 10.

25. Ibid.; Pullman, *The Atom in the History of Human Thought*, 197–198; Holton and Brush, *Physics, The Human Adventure*, 275–79; Rosner, ed., *Chronology of Science*, 116.

26. Rosner, ed., *Chronology of Science*, 117, 367; Boorse and Motz, eds., *The World of the Atom*, v.1, 144, 157–58, 171; MacKinnon, *Scientific Explanation and Atomic Physics*, 96; Pullman, *The Atom in the History of Human Thought*, 202–03. Dalton's studies and those of a few others built upon the work of French chemist Antoine-Laurent Lavoisier (1740–1794), regarded by many as the father of modern chemistry. Lavoisier undercut the age-old idea of four elements with his experiments on the compound structure of water, and helped establish systematically the law of conservation of matter (matter is neither created nor destroyed during any physical or chemical change). He also demonstrated the importance of gravimetric studies, that is, methods of analysis in which the final step involves weighing. See Clive Buckley, "Gravimetric Analysis," online, http://www.newi.ac.uk/buckleyc/gravi. htm; MacKinnon, *Scientific Explanation and Atomic Physics*, 95; Pullman, *The Atom in the History of Human Thought*, 197–98; Boorse and Motz, *The World of the Atom*, v. 1, 129, 141, 143.

27. Holton and Brush, *Physics, The Human Adventure*, 297; Keller, *The Infancy of Atomic Physics*, 10; Rosner, ed., *Chronology of Science*, 96, 163, 379, 500.

28. Keller, *The Infancy of Atomic Physics*, 15–16; Rosner, ed., *Chronology of Science*, 486; Boorse and Motz, eds., *The World of the Atom*, v. 1, 236.

29. Rosner, ed., *Chronology of Science*, 160, 357, 367, 378, 392; Keller, *The Infancy of Atomic Physics*, 16; Boorse and Motz, eds., *The World of the Atom*, v. 1, 195–96; Holton and Brush, *Physics, The Human Adventure*, 310, 339, 352.

30. MacKinnon, *Scientific Explanation and Atomic Physics*, 97–101; Keller, *The Infancy of Atomic Physics*, 17.

31. Boorse and Motz, eds., *The World of the Atom*, v. 1, 260–61.

32. Holton and Brush, *Physics, The Human Adventure*, 339–40.

33. Pullman, *The Atom in the History of Human Thought*, 207.

34. Keller, *The Infancy of Atomic Physics*, 23–32.

35. MacKinnon, *Scientific Explanation and Atomic Physics*, 107–116; Rosner, ed., *Chronology of Science*, 160, 359.

36. Helge Kragh, *Quantum Generations: A History of Physics in the Twentieth Century* (Princeton: Princeton University Press, 1999), 9–16.

37. Ibid., 16–22. See also Diana Preston, *Before the Fallout: From Marie Curie to Hiroshima* (New York: Walker & Company, 2005), 3.

38. Kragh, *Quantum Generations*, 22–25; Preston, *Before the Fallout*, 56–57.

39. Pullman, *The Atom in the History of Human Thought*, 255–56.

40. In the late nineteenth century, English physicist and chemist William Crookes (1832–1919), using a "Crookes tube," noticed a beam of radiation originating from the negative pole of the tube (cathode), which struck the glass opposite the cathode, produced a green glow, and could be deflected by a magnetic field. He was not the first to notice such phosphorescence (cathode rays) but in 1879 he suggested that the rays were molecules that picked up a negative electric charge by the cathode and then were repelled by it. Cathode rays became a curious, but difficult-to-explain phenomenon. Crookes's intriguing and controversial particle theory (especially that particles were deflected by magnetism but not by electricity) led to many more years of experimentation with cathode rays before the seminal discovery of x-rays in 1895. Holton and Brush, *Physics, the Human Adventure*, 369–382; Keller, *The Infancy of Atomic Physics*, 37–54; Rosner, ed., *Chronology of Science*, 361, 365,

381; Boorse and Motz, eds., *The World of the Atom*, v. 1, 344–45; Pullman, *The Atom in the History of Human Thought*, 257.

41. Holton and Brush, *Physics, the Human Adventure*, 382; Stephen E. Atkins, ed., *Historical Encyclopedia of Atomic Energy* (Westport, CT: Greenwood Press), 42.

42. Keller, *The Infancy of Atomic Physics*, 55–64; Holton and Brush, *Physics, The Human Adventure*, 382–383; Boorse and Motz, eds., *The World of the Atom*, v. 1, 385–86; Kragh, *Quantum Generations*, 28–30.

43. Kragh, *Quantum Generations*, 30–31; Holton and Brush, *Physics, The Human Adventure*, 384; Atkins, ed., *Historical Encyclopedia of Atomic Energy*, 42; Rosner, ed., *Chronology of Science*, 356; Boorse and Motz, eds., *The World of the Atom*, v. 1, 403.

44. Preston, *Before the Fallout*, 58–59.

45. Rodney P. Carlisle, ed., *Encyclopedia of the Atomic Age* (New York: Facts on File, 2001), 72–73; Atkins, ed., *Historical Encyclopedia of Atomic Energy*, 102–103; Keller, *The Infancy of Atomic Physics*, 82–84; Kragh, *Quantum Generations*, 32. See also Preston, *Before the Fallout*.

46. Rosner, ed., *Chronology of Science*, 391; Kragh, *Quantum Generations*, 38, 40.

47. Holton and Brush, *Physics, The Human Adventure*, 384–386; Pullman, *The Atom in the History of Human Thought*, 257–58; Boorse and Motz, eds., *The World of the Atom*, v. 1, 408; Atkins, ed., *Historical Encyclopedia of Atomic Energy*, 364–365; Keller, *The Infancy of Atomic Physics*, 66, 72–73, 76.

48. "Rutherford's Planetary Model of the Atom," online, http://www.iun.edu/~cpanhd/C101webnotes/modern-atomic-theory/rutherford-model.html; Keller, *The Infancy of Atomic Physics*, 94; Pullman, *The Atom in the History of Human Thought*, 257–60; Carlisle, ed., *Encyclopedia of the Atomic Age*, 289–90; Rosner, ed., *Chronology of Science*, 374, 386; Atkins, ed., *Historical Encyclopedia of Atomic Energy*, 316–17, 393.

49. Radioactive decay is measured by its "half-life," which is "the time required for one-half of the radio-active nuclides present to undergo radioactive decay…Half-lives can be as short as a few seconds to as long as thousands of years. After a duration of time equivalent to one half-life has elapsed, one-half of the amount of radioactive material originally present remains. After a second half-life has elapsed, half of the half remains." Carlisle, ed., *Encyclopedia of the Atomic Age*, 131.

50. Boorse and Motz, eds., *The World of the Atom*, v. 1, 446–47, 451; Rosner, ed., *Chronology of Science*, 388; Atkins, ed., *Historical Encyclopedia of Atomic Energy*, 316–17; Keller, *The Infancy of Atomic Physics*, 95, 101.

51. Kragh, *Quantum Generations*, 58–63; Carlisle, ed., *Encyclopedia of the Atomic Age*, 247; Rosner, ed., *Chronology of Science*, 377; Pullman, *The Atom in the History of Human Thought*, 261; Keller, *The Infancy of Atomic Physics*, 115–18.

52. Einstein proposed two theories to deal with space, time, and motion (thus undermining Newton's mechanistic view of the universe): the special theory and the general theory of relativity. The special theory, focusing on bodies in motion at constant velocities, asserted that the speed of light always is the same and that there is no single frame of reference for making dependable observations. The general theory, focusing on accelerating bodies, states that it is impossible to distinguish between acceleration and the effects of a universal gravitational field. Light is bent by gravity, and space itself is curved. See Erickson, *Science, Culture and Society*, 98–99.

53. Pullman, *The Atom in the History of Human Thought*, 262; Kragh, *Quantum Generations*, 66–68; Carlisle, ed., *Encyclopedia of the Atomic Age*, 91, 271; Kevles, *The Physicists*, 85–87.

54. Pullman, *The Atom in the History of Human Thought*, 262; Kevles, *The Physicists*, 91–92; Kragh, *Quantum Generations*, 65.

55. Kragh, *Quantum Generations*, 70–72. See also Pierre Marage and Gregoire Wallenborn, eds., *The Solvay Councils and the Birth of Modern Physics* (Basel: Birkhauser Verlag, 1999).

56. Keller, *The Infancy of Atomic Physics*, 169.

57. Ibid., 170–177; Kragh, *Quantum Generations*, 53–57; Pullman, *The Atom in the History of Human Thought*, 262–69; Kevles, *The Physicists*, 92–93; MacKinnon, *Scientific Explanation and Atomic Physics*, 121–22, 158–59.

58. Pullman, *The Atom in the History of Human Thought*, 269–71; Boorse and Motz, eds., *The World of the Atom*, v. 1, 779–92.

59. Rosner, ed., *Chronology of Science*, 358; Pullman, *The Atom in the History of Human Thought*, 273–88; Henry A. Boorse and Lloyd Motz, eds., *The World of the Atom* v. 2 (New York: Basic Books, Inc., 1966), 1042, 1094.

60. Kragh, *Quantum Generations*, 155–59.

61. Boorse and Motz, eds., *The World of the Atom*, v. 2, 1223. Rosner, ed., *Chronology of Science*, 357; Kragh, *Quantum Generations*, 159; Atkins, ed., *Historical Encyclopedia of Atomic Energy*, 296.

62. This was in collaboration with Max Born and German physicist Pascual Jordon (1902–1980).

63. Atkins, ed., *Historical Encyclopedia of Atomic Energy*, 112, 296; Holton and Brush, *Physics, The Human Adventure*, 452; MacKinnon, *Scientific Explanation and Atomic Physics*, 191; Rosner, ed., *Chronology of Science*, 363; Boorse and Motz, eds., *The World of the Atom*, v. 2, 1166.

64. Kragh, *Quantum Generations*, 161–163; Pullman, *The Atom in the History of Human Thought*, 289–95; Boorse and Motz, eds., *The World of the Atom*, v. 2, 1094–095; Atkins, ed., *Historical Encyclopedia of Atomic Energy*, 162. See also David C. Cassidy, *Uncertainty: The Life and Science of Werner Heisenberg* (New York: W.H. Freeman and Co., 1992); David Lindley, *Uncertainty: Einstein, Heisenberg, Bohr, and the Struggle for the Soul of Science* (New York: Doubleday, 2007), 7.

65. Kragh, *Quantum Generations*, 177–81.

66. Bohr's theory of "correspondence" (a way to connect the quantum world of physics with the classical world), and his philosophy of "complementarity" (while wave and particle behavior may be contradictory, they are equally important), were attempts to make quantum mechanics more practical and understandable to scientists. See Lindley, *Uncertainty*, 152–53, 201.

67. Pullman, *The Atom in the History of Human Thought*, 297ff; Lindley, *Uncertainty*, 2–6, 142. See also Lewis S. Feuer, *Einstein and the Generations of Science* (New Brunswick, NJ: Transactions Publishers, 1989; 2nd ed.), 78–238.

CHAPTER 2

Government Mobilizes the Atom

War, Big Science, and the Manhattan Project

INTRODUCTION: SCIENCE AND THE STATE

Science and government forged formal associations in several countries between World I and World War II. This was not the first time that political leaders sought the aid of scientists for the ends of government, but at this time the terms of that relationship were galvanized. In the United States the bond was made through the Manhattan Project. The goal was to mobilize science for the sake of the war effort. Government did not simply fund scientific research, but made explicit what the outcome of that research was to be. Such a bargain had immediate impact on the war, and also long-term impact into the postwar years.

The era of "Big Science" began in the late 1930s. The scale of scientific research from that time forward was imposing.[1] As Alvin M. Weinberg, physicist and director of the Oak Ridge National Laboratory stated,

> When history looks at the 20th century, she will see science and technology as its theme; she will find in the monuments of Big Science—the huge rockets, the high-energy accelerators, the high-flux research reactors—symbols of our time just as surely as she finds in Notre Dame a symbol of the Middle Ages.[2]

Several important elements characterized the change in the nature of scientific research itself and the development of institutions focusing on massive or complex projects. These included large facilities and instruments, funding from governments or international agencies, and emphasis on team research.[3] The high cost of research (in this case weapons) made Big Science only affordable to the governments of industrialized nations. In capitalist countries in particular, business enterprise also played a major role. Essentially, government bureaucracies, their primary supporters, and research universities reshaped the role of science in support of national aims. These changes did not come without resistance, especially among scientists who felt the loss of individual autonomy and the growing necessity to deal with the world beyond their disciplines.[4] The advent of Big Science was not a question of governments becoming the major patrons of scientists, but an instance when they increasingly set the scientific agenda.[5]

The world of atomic physics and chemistry was turned on its head by the coming conflict in the late 1930s. Before that time, scientists strenuously debated the theory in a relatively open arena of ideas. Atomic science, however, now became subservient to applied approaches conducted behind walls of secrecy and security. The 1930s and early 1940s, unlike previous decades, found government leaders setting the scientific standards. "Aryan physics" deposed "Jewish physics" in Germany. "Proletarian science" replaced independent inquiry in Russia. A race to build an "ultimate weapon" gripped several nations, spearheaded by the United States, England, Germany, and others. In essence, the 1930s proved to be the moment in history when the practical application of atomic theory (fission) merged with a powerful desire to develop a war-ending weapon (the atomic bomb) and probably use it. The coming of World War II proved to be the caldron for this mixture.

SCIENCE AND GOVERNMENT THROUGH WORLD WAR I

Well before the 1930s there was historical precedent for governments seeking scientific expertise in peace and war. Over the years, several countries employed scientific and technical know-how in navigation and exploration, in building great monuments and edifices, and in constructing war machines. Governments and petty princes alike periodically engaged scientists to do their bidding, offered patronage, or simply utilized the knowledge of people of learning.[6] Science has always been related to society in one way or another. In more recent times, a former director of the British Museum noted, "The fact is that since science has achieved so vast an importance in the affairs of men, in times of national danger science now belongs to Caesar."[7]

In the middle of the nineteenth century in Europe, Germany set the standard in organizing scientific training at the university level. Higher education remained elitist and exclusive for decades, but it likewise became the haven for numerous scientists.[8] Also in the nineteenth century many fields of science experienced the growth of professionalism; specialization was more common and scientific societies cropped up in several countries.[9] Thomas Alva Edison is credited with developing the archetype research laboratory, his "invention factory," in Menlo Park, New Jersey, in the 1870s. By the turn of the century, scientific facilities of all types were found worldwide. In the early decades of the twentieth century, some of the biggest and most successful companies sponsored industrial research.[10] Despite these preconditions, a sustained relationship between science and government did not develop until at least World War I. At the onset of the Great War, no nation specifically mobilized scientists to develop new weapons. In the summer of 1915 British leaders agreed to enlist scientists to aid in the war effort through the Admiralty's Board of Invention and Research. The Royal Aircraft establishment at Farnborough employed some physicists, and the Cavendish Laboratory developed a new type of radio receiver. Ernest Rutherford collaborated with the French and Americans on the antisubmarine effort. In Germany, a group of scholars produced a declaration of support for their nation in the war.[11]

While civil science did not come to a halt in Europe, the war stifled cross-national scientific collaboration and consultation especially between belligerent powers. Emblematic of the rift was a mounting hostility toward German wartime practices, such as the seemingly anti-intellectual act of burning the library at Louvain, Belgium in 1914. Even after the war, there were those who wanted to isolate German scientists from the rest of the world's learned community.[12] In the

United States, industrial research attracted some scientists, especially in chemistry. This was an attempt to make American companies more competitive with or independent of their European counterparts. Competing federal bureaus, however, did not pool resources to underwrite military research. The armed services instead relied upon individual inventors and industrial firms for new weapons. Laboratories run by the army and navy simply tested various devices and materials, although aviation as the newest field of military technology drew increasing attention.[13]

The onset of war attracted Edison to government service. The iconic inventor initially concerned himself with his far-flung business empire threatened by the conflict. Yet he also favored war preparedness and the opportunity to influence the military application of his inventions. He chaired the Naval Consulting Board (NCB) that focused more on manufacturing and standardization than on science or new inventions.[14] The Wizard of Menlo Park therefore wanted "practical men" rather than pure scientists on the NCB. This stance irked people such as astronomer George Ellery Hale because it shunned the pure scientist and excluded members of the National Academy of Science (too elitist by Edison's standards) from the board. Hale's reaction resulted in the establishment of the National Research Council (NRC), which encouraged the use of pure and applied research for the sake of national security. As a group, American scientists split over wartime involvement; some did not want to see the openness of scientific exchange undermined by national (or political) objectives; others saw the value of such an approach because of growing antipathy toward the Central Powers, particularly Germany.[15]

Beyond an advisory role, American scientists during World War I participated in a number of projects related to submarine detection, aviation, and chemical warfare. The development and use of poison gas on the European Front was significant in galvanizing the relationship among science, technology, and warfare well before the atomic bomb. Indeed, several people referred to World War I as "the chemists' war." At this time, the United States as well as other nations added chemical weapons to its munitions store. (Chemist James B. Conant who would be a central player in the development of the atomic bomb, engaged in making mustard gas and other toxic agents at the time.) The NRC with the Bureau of Mines (BOM) led a massive program of research in chemical weapons in the United States. By mid-1918, the BOM supervised about 700 chemists who focused on that work. As the war progressed, the military became more uncomfortable with civilian research efforts for security and control reasons, and the entire poison gas program was transferred to the new Chemical Warfare Service under army control. The war ended before the Chemical Warfare Service's giant factory at Edgewood Arsenal, Maryland, could produce gas in any quantity.[16] Chemists and chemical engineers, not surprisingly, were in high demand because of the chemical weapons program. Physicists (a much smaller number) worked on submarine detection and aviation-related technology. The army and navy also funded war-related research at about forty colleges and universities.[17]

RUSSIA, WORLD WAR I, AND BIG SCIENCE

In Russia, the foundations for the Soviet system of planned, applied research were laid in World War I. Before then the czars tended to view science as fostering free inquiry and aspirations for modernization, and thus discouraged it.[18] The universities and other teaching institutions were the only places offering opportunities for scientists in Russia, since industrial development was very limited and prospects for applied research negligible. The War Ministry had no research facilities until the end of 1914. Russia's state-owned military industry and munitions factories were built on buying and copying non-Russian innovations and inventions. As it mobilized for war, the imperial government looked to its allies and to neutral countries for needed technology and supplies.[19] The Great War ultimately stimulated rapid expansion of Russian industrialization, including the production of weapons such as poisonous gas and technologies for protection against it. The military also provided some administrative structure for wartime development that was missing earlier. The Russian academic community changed as applied research became more widely accepted among the ranks of scientists. In 1915 the Imperial Academy of Sciences formally set aside its long-standing pure-science tradition and established the Commission for the Study of Natural Productive Forces of Russia (KEPS), which encouraged collaboration among natural and physical scientists, social scientists, and technicians.[20]

Russia was not only facing problems of a weak scientific infrastructure and military peril from without in the 1910s, but the threat of revolution from within. In the last days of Czarist Russia various groups demanded that scientists come out from behind academic walls and think more about the public value of research. In the Bolshevik's new political environment, college professors often were viewed as "counter-revolutionary." Under V.I. Lenin, the government grew wary of "bourgeois" scientists, believing that survival in the postwar world required commitment to new technologies and to modernization. Applied science was a way to meet these goals. The vast social changes triggered by the Russian Revolution broke down a system that for 100 years had been almost exclusively based on pure science within the universities.[21]

The changing role of scientists from cloistered academics to a separate profession serving the people (and the state) was the characteristic feature of the postwar Soviet Union. To a large degree, the Soviet's launched the phenomenon of Big Science in the 1920s that all of the great powers embraced later in the twentieth century. What characterized the Soviet model were large, specialized institutes organized and funded by government, mostly independent of universities. With the cessation of the Russian civil war by about 1921, the isolation of science ended but not without some constraints. The ascendancy of Joseph Stalin in the late 1920s emphatically elevated "proletarian science" over "capitalistic science." Ideology shaded science policy, but funding of applied science nonetheless was lavish. Because applied research was more highly prized than basic research, mistrust between academicians and Soviet bureaucrats deepened. In the late 1920s officials used the Communist Academy to weaken and then replace the Academy of Sciences, which undermined the autonomy of scientists and gave more control to the party.[22]

THE 1930s: THE GREAT TERROR, NAZISM, ANTI-SEMITISM, AND SCIENCE

In the 1930s, the intersection of science and government was increasingly defined by an impending world conflict, and by internal politics in almost every major industrialized country. The consolidation of state power under Stalin was manifest in the Great Purge (Great Terror) in the 1930s directed at the military, political leaders, academics, diplomats, party officials, the secret police, and scientists. Stalin's desire to quickly and aggressively transform Soviet Russia into an industrial nation had, in part at least, led to the massive and aggressive purges with millions of Russians killed or imprisoned. In 1937–1938 more than 100 physicists were arrested in the Leningrad area alone. A few fled, but some such as Pyotr Kapitza (a future Soviet Nobel laureate, who worked in England with Rutherford for more than ten years but had his passport removed on a return visit to the Soviet Union in 1934) was not allowed to leave again. The promise of a thriving scientific community in the Soviet Union reunited with the rest of the academic world shriveled. Instead focus turned to projects necessary to sustain the state, increasingly cut off from the West, and built on heightened intimidation and repression.[23]

Dr. Max Planck (1858–1947) proposed that energy is emitted in small, indivisible amounts which he called "quanta." *Source:* Library of Congress, Prints & Photographs Division, [LC-DIG-ggbain-06493]

The rise of Adolf Hitler and Nazism in Germany also brought an iron hand to the scientific community, but with unanticipated results. Internal political forces scattered key physicists, talented chemists, and a wide array of other important scientists around the world. One of Hitler's first measures after he was appointed Chancellor of Germany in January 1933, was the *Law for the Restoration of the Professional Civil Service* (April 11). This action forced Jews, socialists, and others deemed to be political enemies of the state from their jobs in civil service and universities. It also implied that other political parties were forbidden. The initial round of dismissals included more than 1,000 university professors. One of the first major figures to respond to the rising tide of anti-Semitism and state intimidation was Albert Einstein, who was in the United States at the time and chose not to return to Berlin. Einstein had been under attack for years in Germany for his "Jewish science" and his political views (he was a Zionist, supporting the self-determination of the Jewish people), and saw no future in returning to his birthplace. Prophetically Max Planck tried to convince Hitler that dismissing Jewish scientists would severely hamper German science in general, but the chancellor was unmoved and responded angrily to Planck's suggestion. Because of the lower status of theoretical work in Germany at the time, many Jewish physicists found their best opportunities in atomic science, and ironically spearheaded the successful development of the American and British atomic bomb projects.[24]

The response to Nazi policies varied among German scientists. Most non-Jewish physicists did not protest and wanted to continue working in and for their country. Others were Nazi sympathizers. Some Jewish physicists resigned in protests, and many others were dismissed from their positions. While Nazi Party membership was rare for non-Jewish scientists before 1933, afterwards membership grew sharply especially among young scientists. The German nuclear project, the *Uranverein*, employed most of Germany's nuclear physicists. Among the physicists, chemists, and engineers in the group (including 71 scientists in all), 56 percent were members of the party. Many of the scientific institutions that existed in the Weimar Republic prior to Hitler's rise in power continued to function. They often were even further developed under Nazism, but were not as politicized as the *Uranverein*. Physics in Germany, nonetheless, was seriously undermined in the Third Reich because of the dismissal policy. Key scientists left the country or were driven out. Although the total number of physicists did not greatly decline, the field failed to grow at the high rate occurring in the United States. Some estimated that Germany lost 25 percent of its 1932 physics community in all. Aside from Einstein, such luminaries as Max Born, James Franck, Edward Teller, and many other Jewish physicists were pushed aside in the name of racial purity and Nazi politics. Between 1933 and 1940, Germany lost six Nobel laureates and eight others who would receive the Nobel Prize. Many other important physicists from throughout Eastern Europe also fled the Continent because of anti-Semitism and other forms of persecution.[25]

The "Aryan Physics" movement was used not only to discredit Jewish scientists, but also to turn Germany back to traditional experimental physics and away from quantum mechanics. Its adherents were conservative scientists, anti-Semites, or right-wing nationalists. The movement began in the 1920s, supported by those opposed to Planck's quantum theory and Einstein's theory of relativity, peaking in about 1939. By that time, the group held a minority of physics positions at the German universities as it was clear that such a view of science was well out of step with cutting-edge theory and practice. While short-lived and not widely accepted among scientists, Aryan Physics served the Nazis well. In addition, anti-relativist physicist Phillip Lenard and his right-wing supporters at a conference in Leipzig (1922) leveled the derogatory epithet "Jewish Physics" (a description used to attack the theories of Albert Einstein) at some colleagues. Prior to the conference, Lenard published "A word of warning to German scientists" in which he dismissed relativity as just an hypothesis and attacked his critics with an anti-Semitic harangue. Aryan Physics (and its counterpart Jewish Physics) came under much broader usage in the Third Reich as an additional way to promote Nazi racial policies. Even some of the physicists who stayed in Germany faced criticism for their embrace of quantum physics and sympathies toward their Jewish colleagues. Planck, Sommerfeld, and even Heisenberg were branded as "white Jews," racially pure but nonetheless suspect.[26]

Anti-Semitisim was neither a product of the twentieth century nor a German invention. Although anti-Semitism took its most virulent form in Nazi Germany, particularly with the Holocaust, opponents used it aggressively in many forms to flush out Jews. For example, eugenics (the notion of improving the human species through the control of hereditary factors) was a term coined in the 1880s and an idea often used as justification for racial and class prejudice.[27] Anti-Semitism also was prevalent in various degrees of intensity in the United Kingdom, France, Sweden, the United States, and Italy. Benito Mussolini's racial laws in Italy in the late 1930s also made emigrants of Italian scientists as they had in Germany. Interestingly, during the early 1920s physics failed to maintain its once-impressive standing in the Italian scientific community. Yet between 1925 and 1938, stellar physicist Enrico Fermi and several of his colleagues reinvigorated modern physics in Fascist Italy. Mussolini himself and most Italian scientists were chary to collaborate too closely with Hitler's Third Reich. Fermi and others published their important papers in English, not German. But as Italy moved closer to Germany in their political, economic, and military objectives, Mussolini instituted racial laws on the Nazi model. Italian Jews faced a similar fate as their German counterparts. Fermi, whose wife was Jewish, came under attack. He and several other Italian physicists fled their homeland.[28]

The internal turmoil in Germany, Italy, and Eastern Europe resulted in significant migration of scientists at first to nearby countries such as France, Denmark, Switzerland, and the Netherlands. About thirty countries harbored emigrant physicists between 1933 and 1945, but most of them ultimately ended up in Great Britain and the United States. German Jews made up

Enrico Fermi (1901–1954) received the 1938 Nobel Prize in physics. He is well-known for his work that led to the development of the first nuclear reactor. *Source:* National Archives

the majority of physicists fleeing their countries, but others who were not refugees left as well. Some scientists working outside their homelands before 1933, like Einstein, chose not to return.

Once so vibrant and extensive, the international community of scientists was fragmented by the political upheavals of the 1930s. Members of that community in England, France, Denmark, the United States, and elsewhere worked, often through aid organizations, to obtain financial support and jobs for their displaced colleagues. Many scientists openly condemned the attack on academic freedom, the dismissals, and Aryan Physics. Countries like Great Britain and the United States were safe havens for Jewish physicists and others, but their presence was not universally welcomed. In the United States universities and other institutions gave priority to more senior immigrant scholars as opposed to younger ones who might compete for jobs with their American colleagues. From the late 1920s, physics in the United States was experiencing rapid growth. By the early 1930s, it was competitive with the Europeans by attracting foreign scientists to American shores, increasing the numbers of home-grown scientists, strengthening centers of higher learning, and providing opportunities in industry. The shift of the physics community (and other sciences) to the English-speaking world presaged a reorientation in atomic energy and its uses in the mid-twentieth century.[29]

While the emigrants did not face the persecution they encountered at home, anti-Semitism in American universities was of long standing. In addition, many Americans harbored grave concerns about the political baggage of émigrés, especially if they espoused socialism or communism. For years Jews, Catholics, and women in general faced difficulties in attending college and landing academic jobs. In 1922, the president of Harvard recommended imposing a quota on admission of undergraduates with Jewish backgrounds. (During the Great Depression, Jews accounted for a significant portion of undergraduates in the United States, yet anti-Semitism was intense.) Celebrated scientists such as Einstein found ready acceptance, but young newcomers were warned that Jews had little chance in the academy. Since anti-Semitism existed in many other fields, academic careers were no less attractive to Jews. In the 1930s, Jewish scholars who took their Ph.Ds in physics in American universities were as well-represented as any group (maybe more so), but this did not necessarily translate into academic posts. American physics was significantly improved by the contributions of the emigrant scientists, but assimilation into American culture took time for some, and went more quickly for others. This was not just one way. European academics had to adapt to a culture where the professoriate was not as elitist as in the Old Country, and where university and institute life was not as hierarchical. Yet, from the American standpoint, general acceptance was made much more palatable because the intellectual migration of scientists obviously benefitted the United States as war clouds gathered in Europe.[30]

FISSION: THE TURNING POINT

Experts normally greeted scientific breakthroughs with jubilation. Nuclear fission sent a mixed message. From the very moment James Chadwick discovered the neutron in the early 1930s, physicists understood that since it had no electrical charge, the neutron could be used as "an effective projectile in nuclear reactions."[31] When such a projectile fractures a nucleus, the atom "fissions," or splits into several smaller fragments (fission products) and also releases energy.[32] Such an occurrence opens up great possibilities not only for further study of the subatomic world, but for finding ways to harness energy release. What scientists may not have understood when nuclear fission was made public in January 1939 was that this new discovery could not have come at a worse time.[33]

As early as 1932, Cavendish physicists Englishman John D. Cockcroft and Irishman Ernest T.S. Walton were among the first to split the atom by bombarding lithium with protons generated

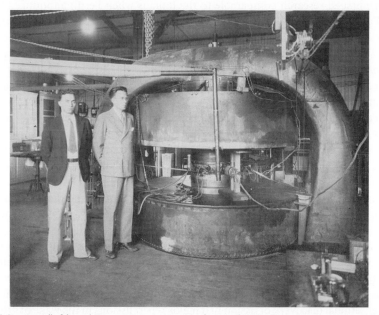

M. Stanley Livingston (left) and Ernest O. Lawrence in front of 27-inch cyclotron at the old Radiation Laboratory at the University of California, Berkeley, 1934. *Source:* National Archives

in their particle accelerator ("atom smasher" in popular parlance). In the United States, physicist Ernest O. Lawrence at U.C. Berkeley, put into operation what many believed to be the first particle accelerator, the cyclotron. Scientists built a similar device at Princeton. Both machines bombarded nuclei of various elements with the intent on splitting atoms.[34] Within a short time the bombardment technique was making light atoms radioactive (a spontaneous emission of radiation). Using proton beams and alpha particles to split atoms required great amounts of energy. Thus many in the scientific community, including Rutherford, Bohr, and Einstein, were skeptical that a breakthrough in utilizing the power of the atom could be accomplished any time soon. In an oft-repeated statement in 1933, Rutherford branded such speculation about fission as "moonshine."[35]

The strong nuclear force, or "strong force," is one of the four basic forces in nature which also includes electro-magnetic force, weak nuclear force, and gravity. The strong force holds together the subatomic particles of a nucleus, or nucleons, consisting of protons (positive charge) and neutrons (no charge). As the name suggests, it is the most powerful of the four forces, and thus splitting atoms would be a significant scientific achievement. To break down or split an atom, scientists had to determine the nuclear binding energy necessary to cause fission.[36]

Beginning in 1934, Enrico Fermi and his colleagues in Rome replaced protons with neutrons in their experiments, hoping that these projectiles would experience no resistance when they entered the nucleus. The team bombarded stable elements to produce new radioactive ones, and they slowed down the bombardment process for better results. Fermi and other scientists thought that they might have produced the first transuranic element (having a higher atomic number than uranium, and thus radioactive), but did not realize that they had caused the nucleus to be broken into radioactive fragments.[37] In Berlin the story developed differently. German radiochemists Otto Hahn and Fritz Strassmann and Austrian physicist Lise Meitner made up the team. Meitner traveled to Berlin in 1907 to work with Max Planck, but was soon drawn to the work of Hahn. Together they conducted radiation experiments for several years. Hahn and Meitner pursued research comparable to the Italians when Fermi announced his findings in 1934. In Paris physicists Irene Joliot-Curie (the daughter of Marie and Pierre Curie), her husband Frederic, and a Serbian collaborator were on a similar quest. After Hitler annexed Austria in March 1938, Meitner (an Austrian Jew) was no longer safe in Germany and ultimately secured a modest position in Stockholm. Hahn was distressed by this turn of events, but maintained contact with Meitner and continued the work with Strassmann; now two chemists carried on the work without the aid of a crucial physicist.[38]

In December 1938 Hahn and Strassmann conducted what became a revolutionary experiment. Using neutrons for bombardment, they split a uranium atom into two substances almost equal in atomic weight (one they thought was radium), but not equal to the original weight of uranium. Like Fermi they could not believe they had divided the nucleus. Upon communicating the results to Meitner, she and her nephew, physicist Otto Frisch, were able to explain what had occurred. They concluded that the fragments were barium isotopes, and the difference in weight from the original uranium was due to a substantial release of energy. Frisch, borrowing a term from cell division in biology, binary fission, christened the process "nuclear fission."[39] Hahn and Strassmann's report of the findings in January 1939 did not mention Meitner and Frisch. This was not a slight, since such an acknowledgment would have been dangerous in the current political climate. By the time Meitner and Frisch's important work was published, however, the news about fission had already been widely circulated, and their role obscured. One additional dimension to the sensational discovery of fission had major repercussions in the wake of growing world tensions, that is, the possibility of a chain reaction caused by splitting atoms. Fission emitted neutrons, which under the right circumstances, might split other nuclei and give off energy with the process continuing to repeat itself with a vengeance. A controlled chain reaction produced heat and power; an uncontrolled chain reaction could create an explosion.[40]

RELUCTANT STEPS TOWARD BOMB RESEARCH

The fission discovery spread rapidly through a surprised scientific community, and the practical implications of such a find became more obvious. Frisch was quick to inform Niels Bohr of the results of their work, and the Great Dane quickly accepted the hypothesis. Bohr soon thereafter left for the United States, and at a late-January meeting of the Washington Conference on Theoretical Physics discussed fission with Fermi and others. Within a short time researchers had widely tested and accepted the fission hypothesis as valid. By 1940, scientists had published more than 100 papers on fission, and began to speculate about the possibility of a self-sustained chain reaction being achieved.[41] If a chain reaction was possible (one that released a great amount of energy), an atomic bomb also was possible. The idea of subatomic energy per se was not a new supposition in the late 1930s. Aside from scientists, others had put forth such speculation. Novelist H.G. Wells, for example, wrote about atomic bombs in his 1914 book, *The World Set Free*.[42]

The discovery of fission was the development most responsible for advancing the idea of a new kind of bomb. This certainly was the thinking of Jewish émigré Leo Szilard. The eccentric (few of these physicists were not, but he was more than most) Hungarian physicist had worked in England and had recently moved to the United States when the announcement about fission was made public. He was an active participant in radiation research, having postulated in 1934 that a neutron chain reaction was possible and could lead to a massive explosion. By 1939, he completed his own experiments to confirm fission. After talking with Hungarian colleagues such as Eugene Wigner and Edward Teller, he came to fear that Hahn and Strassmann's work in Berlin could put the means to create a uranium bomb in the hands of the Nazis. By virtue of their discovery, Germans had a substantial lead in nuclear research. In January 25, 1939, Szilard wrote to Lewis Strauss, a government bureaucrat and financial supporter of physics research, about the fission discovery: "I see, however, in connection with this new discovery potential possibilities in another direction [other than generating power]. These might lead to a large-scale production of energy and radioactive elements, unfortunately also perhaps to atomic bombs."[43]

Later in 1939 observers learned that Hitler had banned the export of uranium from Czechoslovakia after its takeover in March. Szilard's first response, as a scientist but also as a political activist, was to implore his colleagues to squelch all information about uranium research.[44] While several scientists supported the idea, withholding data went against the tenets and long history of scientific inquiry, and was met with strong skepticism. Throughout most of 1939, publications on the subject continued apace, but with the onset of war things changed. Beginning in 1940 physicists in Great Britain and the United States agreed to curtail all publication related to atomic energy.[45] As an alternative to the secrecy campaign, Szilard turned to a more direct route in spring 1939, seeking to inform the American government of his growing concerns and the need for action. Even as war approached in early fall (England and France declared war on Germany on September 3, 1939), the general mood in the country clearly favored aloofness from the conflict. The United States did not become a belligerent until the day after the attack on Pearl Harbor (December 7, 1941). Before that time, strong isolationist sentiment kept the Franklin Roosevelt administration from raising appropriations for war-related research and development or taking any provocative actions that might embroil the country in war.[46]

Even behind the scenes, the idea of an atomic bomb remained abstract and theoretical, rather than a weapon to be readily available to Germans let alone Americans any time soon. The military potential of nuclear research was not a pipe dream, but neither was it a fervent concern. Szilard nevertheless approached the navy with no luck. He then set a loftier goal to approach the president himself. To get President Roosevelt's attention about the possible atomic weapons, development in Germany required someone better known than an émigré physicist (at least this émigré physicist). Szilard and Wigner decided to approach Einstein to elicit his support and to utilize his celebrity. The Szilard-written "Einstein letter" (August 2, 1939) was delivered to FDR through economist and New Dealer Alexander Sachs, but not until several weeks after the eruption of war in Europe on October 11.[47] In tone, the letter was firm but restrained. It began: "Sir: Some recent work by E. Fermi and L. Szilard...leads me to expect that the element uranium may be turned into a new and important source of energy in the immediate future. Certain aspects of the situation which has arisen seem to call for watchfulness and, if necessary, quick action on the part of the Administration..." The letter went on to mention the real possibility of producing a nuclear chain reaction and the construction of "extremely powerful bombs of a new type." It encouraged the need to speed up experimental work, and noted Hitler's decision to stop the sale of uranium from Czechoslovakia.[48] The president was impressed and passed the letter along to his personal secretary saying "this requires action."[49]

Little action was immediately forthcoming. The sluggishness in response to the Einstein letter in part had to do with the lack of a scientific and technical infrastructure in the federal government (or in the military for that matter). It also reflected the paucity of concrete evidence that Germany was developing a bomb or that the United States could develop a bomb quickly. Leaders in Washington needed to make substantial institutional adjustments (plus they had to be attentive to potential political fallout) to move this project beyond conjecture. After all, the United States was not yet at war, and knowledge about the possible development of a bomb by Americans or someone else rested on sparse information. FDR's creation of an ad hoc Uranium Committee in October 1939 (under Lyman J. Briggs, physicist and director of the National Bureau of Standards) was an important step toward an unprecedented crash program. The committee composed of civilian and military members studied the current state of research on uranium. Yet the action of the group was unhurried, if not glacial. In early 1940 it recommended that the government provide some modest research funds for work on uranium isotope separation (which might result in a fuel source for bombs by producing isotopes that were highly fissionable) and for the work Fermi and Szilard carried out on chain reactions at Columbia University. During 1939 and 1940, scientists conducted most of the research on isotope separation and chain reactions in university laboratories with funding primarily from private foundations.[50]

The Uranium Committee was the narrowest path toward a broad governmental commitment to scientific research in high energy physics, and it operated very cautiously under the conservative Briggs. Soon after World War II began, however, Vannevar Bush (engineer, mathematician, and president of the Carnegie Foundation) began pushing to involve the federal government more aggressively in supporting war-related scientific research. With the backing of the armed services and various science agencies, Bush convinced Roosevelt to create the National Defense Research Committee (NDRC) within the executive branch. The NDRC, established in June 1940 with Bush as its chair, reorganized the Uranium Committee into a scientific entity without military participation, and began pressing for work on isotope separation. Bush curtailed further publication of uranium research and like Briggs closed off the committee to foreign-born scientists. The NDRC also studied other potential wartime technologies, such as radar and anti-submarine devices, and coordinated scientific research related to warfare.[51] At this stage, scientists were developing a closer relationship with government policymakers than ever before.[52]

Ernest Lawrence urged the government to speed up uranium research in early 1941. Desirous of seeking more funding for his own Radiation Laboratory, he was convinced that the NDRC was making slow progress in uranium research and in mobilizing scientists in general. Lawrence enlisted Karl T. Compton and Alfred L. Loomis of Harvard University to make the case to Bush. Compton painted an optimistic picture of the possibilities for radiation research, warned about Germany's lead in this field, and asserted that even the British were ahead of the Americans. Two reports from the National Academy of Science discussed the potential for radiation research, but did not convince Bush that there was sufficient reason to believe it could aid the war effort if and when the United States entered.[53] The bureaucratic reshuffling was somewhat confusing, but little by little the scientific community and the federal government regularized linkages between them. Officials needed some indication of a practical breakthrough in uranium research to further move the process along, especially since bomb development competed for funding with other war-related research.

THE GERMAN BOMB PROGRAM

By the beginning of the war, the Germans were making what appeared to be excellent headway with their nuclear project, although they did not sustain the advantage in producing a bomb. Under Heisenberg's leadership, Germany devised a national plan for the "exploitation of nuclear fission." In addition, like the Americans and the British, the Germans placed restrictions on publications and established a Nuclear Physics Working Group. Not all of the remaining German scientists participated, but the work in Berlin was a promising step. Heisenberg submitted a report to the War Office in December outlining his own views and what he gleaned from American, British, and French journals, essentially setting a blueprint for a German-designed bomb.[54]

Uranium was an obvious choice for producing a nuclear chain reaction and for making a bomb. Uranium is one of the heaviest naturally occurring elements, almost 19 times as heavy as water, and has the capacity for producing vast amounts of energy. It can be found in many rocks in the earth's crust including tin, tungsten, and molybdenum, and in sea water. Like other elements, uranium occurs in several isotopes differing in the number of neutrons in the nucleus. Natural uranium is a mixture primarily of two isotopes, uranium-238 (U-238), which accounts for 99.3 percent, and uranium-235 (U-235), representing a much smaller 0.7 percent. U-238 is much less radioactive than U-235, which can readily be split and thus is considered "fissile," that is, it fissions when hit by a neutron. U-238 is alternatively a "fertile" material; after capturing a slow neutron (in a reactor core for example), it can become fissile (plutonium-239).[55]

Heisenberg's primary focus was to demonstrate a nuclear chain reaction, using uranium and heavy water. At this stage, the Germans had limited supplies of refined uranium (although by the middle of the war, they may have had greater supplies than the Americans). They initially tried using dry ice (a solid form of carbon dioxide) as a moderator to contain and slow down neutrons in a chain reaction to encourage fission.[56] Results, however, were discouraging. The Germans rejected using graphite as a moderator, which Enrico Fermi later used to produce the first successful chain reaction. Like the French they turned to heavy water. Chemically similar to regular water, in heavy water hydrogen atoms are replaced with deuterium. Heavy water worked well as a moderator, but its availability was limited. The only substantial quantity was located at Norsk Hydro's Vemork plant in Norway. Having gotten wind of the German interest in this supply, and wanting it for their own uses, French operatives acquired the entire stock. Although the

Germans failed to obtain the heavy water, through their conquests in early 1940, they did capture key documents, a cyclotron from Copenhagen (not used), a partially completed cyclotron from Paris, the Vemork plant, and Belgian stockpiles of uranium. The Allies attacked the plant in February 1943 and again in November. They destroyed one ton of heavy water on site and more as it left the plant.[57]

Because of these events and other resources problems, material and human, the Germans never fulfilled the early promise of developing a bomb during the war. One view was that Heisenberg and others at the *Uranverein* mistakenly believed that they were far ahead in fission research (even after the Manhattan Project began to accelerate) and discarded the uranium/graphite reactor too quickly, focusing more attention on reactors than explosives.[58] There also is an argument that Heisenberg, who was responsible for the theoretical component of the project, had been misled by inexact experimental data concerning carbon as a moderator, and thus argued to eliminate it as an option. He also purportedly overestimated the amount of enriched uranium-235 necessary to sustain a chain reaction. More generally, the German nuclear project suffered from a lack of coordination and increasing pressures of the war itself, and certainly from the loss of physicists who had fled the country before the war.[59]

Heisenberg's role in the German bomb project has long been a source of interest and speculation. He remained loyal to his country even when others had left, and had a complicated relationship with the Nazis. Michael Frayn's *Copenhagen*, an award-winning smash hit play in London and Broadway (and later adapted for PBS on film) dramatized a meeting in Denmark between Bohr and his German protégé in September 1941. Exactly what transpired at that meeting is not known, but speculation has it that Heisenberg was attempting to discern what Bohr knew about the development of an American-British bomb. Although Frayn was admired for attempting to explain to laypersons the dynamics of quantum mechanics and the tensions of the early war years, some criticized his work for historical inaccuracies and his own conjecture about Heisenberg's motives for remaining in Germany during the war. In particular, Heisenberg biographer Paul Lawrence Rose accused Frayn of writing an apologia for the German physicist. Rose believes that Heisenberg never really faced the moral dilemma of working for the Nazis and miscalculated the amount of fissionable material needed to construct a practical bomb.[60] Whatever real problems Heisenberg faced, the Americans, the British, and others feared that Germany was moving full speed on a project that could determine the outcome of the war. Without precise information about the state of the German nuclear project, they assumed the worst, and acted on it.

GREAT BRITAIN AND THE MAUD REPORT

Outside Germany, Great Britain (not the United States) made the first significant effort to contemplate developing an atomic bomb. Although several scientists and others assumed that the Germans were moving forward on the project, an embryonic program in the United States had stalled. The French concluded that producing a bomb at the time was highly implausible. When the British became frustrated with their failure to build a reactor, they turned to uranium isotope separation and bomb design. Working in England, German refugee physicists Otto Frisch and Rudolf Peirels[61] concluded in March 1940 that a uranium bomb could be built and that only one kilogram of metallic uranium-235 would be necessary to produce the equivalent of "several thousand tons of dynamite." They cautioned that such a bomb could kill large numbers of innocent civilians which would make it "unsuitable as a weapon for use by this country." The caveat was essentially ignored, and the British formed a committee, code-named MAUD (after Bohr's former governess), to work on what now appeared to be a viable weapon. The committee completed its final report by mid-1941.[62] The report stated that such a bomb was practicable, making clear that it would be very potent, could be assembled in a size small enough to be deployed by aircraft, and possibly could be ready in two years. It also concluded that building such a device required a large organization and a financial investment beyond the resources of Great Britain alone.[63]

The MAUD Report ("The Use of Uranium for a Bomb") crossed the Atlantic in July 1941, and ultimately intensified the American commitment to an atomic bomb. However, at the time, the wheels already were in motion in the United States to support some sort of bomb project, accelerated by the American entry into the war after the Pearl Harbor attack. Shortly before the British transmitted the final version of the MAUD Report to US officials, the president signed an executive order creating the Office of Scientific Research and Development (OSRD) to strengthen the role of science on the federal level. Vannevar Bush was named chairman. The NDRC (headed by James B. Conant, president of Harvard University) became an advisory body to the OSRD. By the end of 1941, the Uranium Committee (now transferred to OSRD), became the Uranium Section (or S-1), the center of nuclear weapons work. The "S-1 project" was the initial name given to the effort of building the bomb in the United States.[64]

While the Advisory Committee on Uranium had given support to Fermi and Szilard's work on producing a self-sustaining chain reaction, too little uranium and graphite (used as a moderator) was available for anything but limited experiments. One promising line of inquiry to expand the supply of radioactive material focused on plutonium. The human-made metallic element was discovered by nuclear chemist Glenn Seaborg and his team in 1940 in the U.C. Berkeley cyclotron. Plutonium, a transuruanic, was highly radioactive and its most important isotope (Pu-239) was fissionable. It could be produced by bombarding U-238 with neutrons. The area of isotope separation produced other encouraging, but limited developments. They amounted to the first cautionary steps to extract very small amounts of radioactive material for use in a fission bomb.[65] In this context, the arrival of the draft MAUD report to North American shores found a receptive hearing and seemed to push the United States over the top toward developing an atomic bomb, rather than creating inertia. Americans had been in contact with the MAUD Committee since fall 1940, but the availability of a tangible document with specific recommendations proved to be the most vital element of the report for an American program needing clearer direction. A powerful, deliverable bomb seemed not only plausible, but possible. The British were hoping for cooperation with the Americans on the project, which was problematic since the United States was not yet a belligerent in the war. The report pushed specifically for gaseous diffusion of uranium-235 on a massive scale (to the exclusion of other isotope separation methods). Its conclusions about the feasibility of a bomb, however, and concern over German fission experiments, offered additional *raison d'être* for what became the Manhattan Project.

Without waiting for a new assessment of the uranium program that he requested from the National Academy of Science, Bush met with President Roosevelt and Vice President Henry Wallace on October 9, 1941. He got the president's approval to move quickly on research and development, and to determine if a bomb could be built and at what cost. The president also established the Top Policy Group (including himself, Vice President Wallace, Secretary of War Henry Stimson, Army Chief of Staff George Marshall, Bush, and Conant) which had overall responsibility for the bomb project. Bush did not have permission to move toward a program of production at that time. Compton's NAS committee reported in November that a bomb could be produced. FDR did not give authorization to proceed until January 19, 1942, more than a month after the United States entered the war.[66] What appeared to begin as a joint venture with the British (considered and discussed) became primarily an American-directed and American-funded project. Prime Minister Winston Churchill very much wanted the British involved, not only because his country had helped to pioneer the venture but because of the promise it held for postwar defense. Great Britain became the "junior partner" by American design, resulting in little collaboration at all.[67]

On June 17, 1942, the US Army Corps of Engineers had responsibility for the construction phase of the new weapons venture, organized under the code name "Manhattan Engineer District." The S-1 Project became the Manhattan Project. As the undertaking was reaching the developmental phase, the partnership between scientists (especially the science managers and science advisors) and the federal and military bureaucracy was tilting more heavily toward the politicos and the military. Bush and Conant, the most visible scientists at the highest levels of government, consistently wanted to advance the public role of science and to collaborate in decision making about the bomb.[68] But as historian Martin Sherwin vividly asserted, "Science administrators in their Washington offices and research scientists in their Manhattan Project laboratories began fighting a two-front war: on the one hand, there were the enemies of the United States; and on the other, the potential enemies of science—men and institutions seeking to control research."[69] The government's determination to keep security and secrecy paramount in the bomb project clashed regularly with meaningful efforts at collaboration and the scientists' need to freely exchange information with their peers.

IN SEARCH OF FISSIONABLE MATERIAL

Many books and films that deal with the Manhattan Project tend to focus on the challenge of building the atomic bomb, especially on Robert Oppenheimer and his intrepid team of scientists cloistered away in Los Alamos, New Mexico. The first such movie, *The Beginning or the End*, directed by Norman Taurog (MGM, 1947), was a pseudodocumentary which has been analyzed *ad nausea*, often criticized as an official justification for using the bomb, and filled with historical inaccuracies. There is an important common thread in this film connecting it with many others. They include the fairly realistic PBS docudrama, *Day One* (1989) based on a book by Peter Wyden; the not-so-realistic Hollywood production, *Fat Man & Little Boy* (Paramount, 1989), which severely miscasts Paul Newman as General Lesley Groves; the History Channel's *The Manhattan Project* (2002); and *Trinity and Beyond* (1995) shown on the Discovery Channel and narrated by *Star Trek*'s William Shatner.[70] That common thread was the focus on bomb building with very limited discussion of the less sexy, but the highly significant infrastructure and the people necessary to produce enough fissionable material of sufficient quality and quantity to actually construct the bomb.[71] These film portrayals also tend to understate, and even undervalue, the central role of General Lesley R. Groves who spearheaded the Manhattan Project, and is responsible as any single person for completion of the mission.[72]

By late May 1942, S-1 officials decided to pursue several possible methods of isotope separation, necessary for gathering sufficient quantity and quality of fissionable material for bombs. Because of the urgency brought on by the war and no clear determination of what method would prove superior, the committee decided to move on five fronts simultaneously; three to separate U-235 from U-238 and two to create plutonium. None were sure-fire. One of the two separation methods chosen was electromagnetic separation, tested by Lawrence's group in California, which used the cyclotron to pass an electrically charged mixture of uranium through a magnetic arc to collect U-235. Gaseous diffusion, pushed hard by the MAUD Report, was a second separation method chosen, which passed a uranium mixture in gaseous form under pressure through spongy, metallic porous barriers. To produce plutonium, scientists chose a uranium reactor utilizing graphite as a moderator because good progress was being made with graphite. Electromagnetic separation, gaseous diffusion, and the "pile process" for producing plutonium, each with its own promise and problems, formed the basis upon which a massive building project was to follow.[73]

Fermi's success in producing the first self-sustaining chain reaction on December 2, 1942, validated the uranium/graphite method and moved atomic science much closer to practical application for both energy production and weaponsmaking. Project officials earlier that year gave the task of building a nuclear pile (to produce a controlled chain reaction) to the Metallurgical Laboratory (Met Lab) then located at the University of Chicago under the leadership of Arthur Compton. Compton was responsible for combining all existing pile research (at Columbia, Princeton, and Berkeley) in Chicago, although it created quite a stir among researchers at each campus. The Met Lab at its peak employed more than 5,000 people for the whole project.[74] Fermi, Szilard, and others relocated from Columbia and took up the task of planning and constructing the pile in May. By October, they acquired enough uranium and graphite to test a chain reaction. The team fashioned bricks of synthetically produced graphite of high purity to construct the pile. (One reason why the Germans had given up on graphite as a moderator was that they tried naturally mined graphite with many impurities which "poisoned" the chain reaction.) Chicago Pile-1 was located at the squash courts under Alonzo Stagg Football Stadium on the University of Chicago campus. It took more than 80,000 pounds of uranium oxide, 771,000 pounds of graphite, and 12,400 pounds of uranium metal to construct the pile. A group of about forty people witnessed the successful test on December 2, 1942, sharing a bottle of Chianti in celebration after researchers slowly removed the control rods from the pile and Fermi declared that "The pile has gone critical." After about four and a half minutes, workers reinstated the control rods, and the chain reaction ceased. Compton then sent out a coded message that "the Italian navigator has just landed in the new world."[75]

THE ARMY AND GENERAL GROVES

The race to build the atomic bomb meant commitment to a massive wartime infrastructure and the mobilization of thousands of skilled people working directly on the device. As one historian noted, "[T]he situation changed from one of too little money and no deadlines to one of a clear goal, plenty of money, but too little time."[76] S-1's production planning, which began months before the successful chain reaction, resulted in Bush transferring responsibility to the US Army Corps of Engineers in four crucial areas: process development, materials procurement, engineering design, and site selection (with a lion's share of the 1943 budget as well). This action essentially completed the transition of the bomb project from one dominated by research scientists to one largely controlled by the army and its government support network. The navy was excluded from the process, partly because it had not demonstrated a strong interest in involving civilian researchers in its own projects. But possibly more important, army resistance kept the rival service out. Bush and Vice Admiral Harold G. Bowen, Director of the Naval Research Laboratory and Technical Aide to the Secretary of the Navy, were on very poor terms.[77]

On September 23, 1942 (without consultation with Bush) the army leadership selected then—Colonel Groves to manage the Manhattan Project. This was a clear indication of where the momentum shift was headed as the project reached its operational stage. In May 1943, Stimson as Secretary of War had full responsibility for administering the entire project, which diminished the role of the other members of the Top Policy Group including Bush and Conant. Conant's role shifted from being chair of the S-1 Executive Committee to becoming Groves' chief science adviser. Little by little the army and civilian bureaucratic leadership almost exclusively took over the role of policy making with little input from the scientists or their buffers.[78]

General Groves was "an engineer by training and a field commander by temperament."[79] Among his accomplishments in the Corps of Engineers, Groves had been involved in the opening of the Port Isabel Harbor near Galveston, Texas; investigating the possibility of a new two-ocean canal in Nicaragua; and constructing the Pentagon (the largest office building in the world). Elevated to brigadier general, Groves's vast engineering experience, his high capacity for work, and his reputation for getting things done made him a good choice to lead what undoubtedly was a building program unrivaled by anything before it. Even in the military, Groves had made enemies because of his brashness, his impatience, and his lack of tact. It would remain to be seen if his style wore well with other engineers, politicians, industrialists, and especially the research scientists. Even before Groves assumed his position, many scientists expressed disdain for the army administering an enterprise they had previously spearheaded.[80] In the end, Groves proved to be what his biographer called the "indispensable man," making many crucial decisions, choosing among myriad technical alternatives, and selecting effective subordinates.[81]

In theory, the Manhattan Project should have received unlimited support from government leaders to produce a bomb that could be the turning point in the war. In reality, the project was not the highest priority in 1942 and 1943, only a very risky, very speculative venture. Groves did not see it that way. He had a job to do and he was unwilling to let red tape or anything else stand in his way. He therefore moved on several fronts. For example, the general arranged to purchase 1,250 tons of uranium ore in storage in New York for the separation processes. He obtained the government's first-priority AAA rating necessary for acquiring a whole range of supplies and materials. And he began looking into vast land purchases for the needed space to build reactors, plants, and countless structures necessary for the production of fissionable material and bomb design. As Groves stated rather dramatically in his memoirs about plutonium development: "It was a phenomenal achievement; an even greater venture into the unknown than the first voyage of Columbus."[82]

The general kept the direction of physical and chemical research for the bomb under civilian program directors: Lawrence at Berkeley for electromagnetic separation and plutonium studies; Arthur Compton at Chicago for chain reactions and weapons theory; and Harold Urey at Columbia for gaseous diffusion. By April 1943, Groves had complete control over the operations necessary to build the bomb, ending the central role of the OSRD.[83] The most colossal task he faced was developing the infrastructure to produce sufficient quantities of fissionable material. He had to do so in a secret environment where no single method had a proven track record and where time was of the essence. The general's approach was a multi-front engagement. He combined facilities for both gaseous diffusion and electromagnetic processes. In 1943 Groves arranged to use a 59,000 acre site eighteen miles from Knoxville, Tennessee, acquired by the Corps a year earlier. This would become the Oak Ridge National Laboratory. The advantage of the site was its relative isolation, its access to electrical power from the Tennessee Valley Authority, and abundant water from the Clinch River. On the site, engineers constructed the gigantic gaseous diffusion plant (K-25), the largest industrial building in the world. It had thousands of diffusion tanks; a complex for electromagnetic separation eventually with 268 buildings including a major plant (Y-12); a pilot plant for the plutonium facility (X-10); and a thermal diffusion installation (S-50).[84] While Oak Ridge was being developed, the project directors built plutonium plants in western Washington State during 1943–1944. They selected Hanford, near Pasco, also for its isolation and for its access to abundant water, this time the Columbia River. Grand Coulee Dam supplied much of the electricity. Groves secured more than six hundred square miles of land for the Hanford site, where the engineers intended to build production reactors and separation plants. All of this activity set off a major construction boom in the area.[85]

That these massive facilities were able to provide fissionable material by early 1945 is all the more amazing because of the chronic wartime problems with logistics, technology, and personnel. At Oak Ridge, breakdowns in the electromagnetic plant were common. The primary contractor, Stone and Webster, also was frustrated because Lawrence's lofty optimism about producing adequate amounts of U-235 was unrealistic. Gaseous diffusion faced the knotty problem of finding a suitable diffusion barrier. Kellex Corporation, the primary for this technology, also lacked conviction about fixing the glitch. Bush, Conant, and Groves urged accelerated efforts to produce the barrier, while Lawrence suggested scaling back on gaseous diffusion. By mid-1944 the barrier dilemma, plus problems with the necessary work on pumps, had yet to be solved. Groves also made a bold move by experimenting with thermal diffusion technology (something the navy had been studying for its submarines) to see if he could get better results but without navy cooperation. Other problems arose over tensions, here and elsewhere, between university and industrial scientists, and between research scientists and contractors (notably Urey's group at Columbia and Kellex). Even the relationship with the British was suffering, especially because the army-controlled projects meant that the flow of information to the UK was reduced to a trickle. British personnel were excluded from several key areas, including the Met Lab. K-25 finally was up and running by the end of 1944, and producing the needed U-235 by 1945.[86]

In August 1942, chemists isolated pure plutonium. DuPont was primarily responsible for running a plutonium production plant and separation facilities at Hanford. They had some 5,800 management personnel with a peak workforce of 45,000. By February 1945, shipments were being sent to the bomb construction site at Los Alamos. Here, too, problems arose. Labor turnover was high because of the arduous conditions, with approximately 140,000 workers engaged for various periods during 1944 and 1945. Some viewed Hanford as little more than a DuPont company town, and scientists at the Met Lab especially felt undermined, suspicious of big industry in general, and resentful of DuPont in particular. Technical problems also plagued Hanford, including reactor design. Researchers and operators again gave some attention to heavy water, but they did not take this route.[87]

Few would question the astounding speed and astonishing results achieved at the industrial site in Oak Ridge and Hanford. These "secret cities," new planned communities where thousands of people lived and worked, were a lasting reminder of the new atomic age with its promise and trepidation. As photographer and historian Peter Bacon Hales suggested, "The Manhattan Engineer District created a new form of American cultural landscape with one Herculean goal in mind: the manufacture of an atomic superweapon in time to use it on the Japanese."[88] Years after the war, a more grisly contribution became evident when these sites that produced fissionable material were found to have vast radioactive contamination.

OPPENHEIMER AND LOS ALAMOS

The gigantic investment in infrastructure, isotope separation, and creation of plutonium was meant to feed into the third "secret city," or instant city, at the special weapons laboratory in Los Alamos, New Mexico.[89] Between July and September 1942, a team of scientists at U.C. Berkeley, led by J. Robert Oppenheimer, were able to establish the theoretical basis for the design of a fission bomb. In October General Groves met with Oppenheimer, which led to the Berkeley professor's appointment to head up the special weapons laboratory. By this time, the American-born Oppenheimer had earned an international reputation as a theoretical physicist and had put Berkeley (and Cal Tech) on the map as a place(s) for aspiring young theoretical physicists to study.[90] Countless biographies, movies, and even an opera have delved into the psyche of this

Dr. J. Robert Oppenheimer, head of the Manhattan Project and "father of the atomic bomb," ca. 1944. *Source:* RIA Novosti/Alamy

extraordinary man. The seven-part mini-series, *Oppenheimer: The Father of the Atomic Bomb*, produced by the BBC in 1980 and aired in the United States in 1982, provides a nuanced view of Oppie (Sam Waterston) from his Berkeley days through the Cold War, but with strong emphasis on his ambition. Other portrayals in a vast string of films range from oblivious professor to earnest superstar scientist. John Adams' opera, *Doctor Atomic* displays the great stress and anxiety that Oppenheimer faced in the events surrounding the *Trinity* test (the first detonation of an atomic bomb device). At the end of the first act, Oppenheimer, in response to giving the test the name *Trinity*, begins his aria with:

> *Batter my heart, three person'd God: For you*
> *As yet but knock, breathe, knock, breathe, knock, breathe*
> *Shine, and seek to mend;*
> *Batter my heart, three person'd God;*
> *That I may rise, and stand, o'erthrow me, and bend*
> *Your force, to break, blow, break, blow, break, blow*
> *burn and make me new.*[91]

Kai Bird and Martin Sherwin's Pulitzer prize-winning biography may have summarized Oppie best: "Robert Oppenheimer was an enigma, a theoretical physicist who displayed the charismatic qualities of a great leader, an aesthete who cultivated ambiguities."[92] There is little doubt that Oppie relished the opportunity to lead the bomb project, possibly for many reasons but particularly because his ego made him aspire to the notoriety it would bring him, especially because he never received a Nobel Prize. In addition, his determination never wavered to produce a weapon that he hoped would end that war.[93] He later recalled, "The prospect of coming to Los

Alamos aroused great misgivings." It would be like a military post with restrictions on travel and on the families of the scientists and others. The location was remote. "But there was another side to it. Almost everyone realized that this was a great undertaking. Almost everyone knew that if it were completed successfully and rapidly enough, it might determine the outcome of the war."[94]

General Groves looked beyond Oppenheimer's lack of experience in directing such a project, and saw instead a leader who could bridge the gap between the scientists and himself. He considered Lawrence, Compton, and Urey for the post, but could not spare them in their various capacities. In his memoirs, Groves stated,

> My own feeling was that he [Oppenheimer] was well qualified to handle the theoretical aspects of the work, but how he would do on the practical experimentation, or how he would handle the administrative responsibilities, I had no idea. I knew, of course, that he was a man of tremendous intellectual capacity, that he had a brilliant background in theoretical physics, and that he was respected in the academic world. I thought he could do the job. In all my inquiries, I was unable to find anyone else who was available who I felt would do as well.[95]

While Groves was concerned that Oppenheimer had neither the administrative experience nor a Nobel Prize, others harbored deeper trepidation. Most significantly they worried about his history as a champion of left-wing causes and his possible membership in, or at least close association with, the Communist Party. Despite Groves' obsession with security, the general put aside this potential trouble, for the moment at least, focusing instead on the project's central objective. He would not let himself believe that he had made a poor choice.[96]

In seeking a location for the Nuclear Weapons Laboratory, Oppenheimer recommended a remote spot in New Mexico about 40 miles from Santa Fe (*Los Alamos*, Spanish for "cottonwoods") on the current site of a boys' preparatory school. This was an area that Oppenheimer knew well, and where he vacationed regularly. Like Oak Ridge and Hanford, isolation enhanced security, which was central in importance to Groves. For Oppenheimer, the chance to bring key scientists together in one place, without distractions, increased the chances to work out the complex problems associated with producing a theoretically possible but untested weapon in a very short time. The new facility opened in spring 1943.[97] What came to be known as "the Hill" also was called a "Nobel Prize Winners' Concentration Camp," a place that "absorbed physicists like a sponge," "Shangri-La," and "an unreal world, part mountain resort and part military base."[98] It was all those things.

Oppenheimer aggressively recruited scientists and technicians to populate the new think tank. What he assumed would require a few dozen people ballooned into more than 2,000 scientific and technical staff and another 600 army enlisted men in the Special Engineering Detachment (SED) by spring 1945. While the senior scientists remained civilians, the army drafted many of the younger scientists and technicians and assigned them to Los Alamos, resulting in about 42 percent of the scientific personnel belonging to the SED. In all, there were more than 5,700 people living there at the time. Scientists were poached and coaxed from several other laboratories. The Radar Laboratory (Rad Lab) at MIT proved to be an excellent source, although some were seriously concerned that the brain raiding of young and brilliant scientists might cripple radar research in the effort to recruit the best minds for bomb work. The list of luminaries who worked or passed through Los Alamos was a veritable who's who in physics and chemistry and other scientific disciplines, including Bohr; Fermi; Hungarian émigré Edward Teller; Italian immigrant Emilio Segrè; mathematician Stanislaw Ulam; Rad

Lab veteran I.I. Rabi; German émigré Hans Bethe; Hungarian émigré John von Neumann, and many more. Younger stars-to-be also populated Los Alamos in droves. The average age of the scientific community in residence was 32, which was not young for physicists but young nevertheless. In December 1943, with its own nuclear program scaled back, the British contributed a small but vital group of twenty-two. Klaus Fuchs, a German physicists and Russian spy, also was part of that contingent.[99]

While the special weapons laboratory accomplished astounding things in a very short time, life on the Hill was not without its obstacles. As Edward Teller observed, "The first thing I noticed on arriving was that we were all going to be locked up together for better or for worse…"[100] Such isolation, even in a remarkably beautiful area, was very difficult to endure for almost everyone at Los Alamos. Those men with wives and families in particular (the scientific community itself was a male bastion almost exclusively) faced hardships of living conditions and socialization.[101] General Groves's overpowering demand for security was the major point of contention between the "army way" and the requirements of science. At all of the Manhattan Project facilities, Groves sought to impose a policy of "compartmentlization," which, simply put, had people conducting their specific tasks with no clear knowledge of how they fit into to the larger project. Some functioned without knowing what that larger project might be. This meant that scientific information could only flow on a "need to know" basis. For scientists accustomed to exchanging ideas, working in teams, and viewing problem solving as a group activity, Groves' demands seemed unworkable. On one level at least, this was a stand-off between academic science and practical engineering. Oppenheimer's intervention was crucial here, since he argued that the isolation of Los Alamos provided the security essential to the project of producing the so-called "gadget." He implored Groves to loosen his more rigid demand for compartmentalization. Groves agreed, but not happily.[102]

The curious relationship between the brilliant theoretical scientist with somewhat shadowy leftist political associations and the blunt, driven, and pragmatic Army Corps of Engineers general kept the project functioning effectively despite the hardships and distractions. But Oppenheimer's successes as a mediator between the world of science and the military world sometimes camouflaged his extraordinarily difficult task. He was constantly overwhelmed by details, logistical problems, and having to wrestle with many big egos. He often failed to delegate some of his authority, at times relying on assistants but inconsistently. He was non-Nobel laureate leading Nobel laureates. When the brilliant but sensitive Teller felt slighted in the pecking order of the physicists, and when he chose to focus on his dream of a fusion bomb (discussed later) rather than give full attention to the fission bomb, Oppie could do little more than plead with him. Oppenheimer, nonetheless, showed an amazing ability to remain on point and move toward the project's awesome objective. This strength probably kept Groves on his side through their association at Los Alamos.[103]

The primary task at hand was to move from theory to practice in developing an atomic bomb, and to do it with great haste. Oppenheimer set up five divisions at first: Theory, Experimental Physics, Chemistry (and later Metallurgy), Ordnance and Engineering, and Administration. Early on he decided to include work on plutonium alongside the uranium work. Initial attention turned to determining how a critical mass of uranium-235 or plutonium behaved between the initiation of a chain reaction and the explosion, that is, critical mass behavior. Among both theorists and experimentalists, the next task was determining how to detonate the bomb. The idea that initially won favor was a gun design that would fire a subcritical mass of fissionable material into another with great force and speed. This worked well with the uranium, but not at all with plutonium. The latter emitted many more neutrons than U-235 and thus could prematurely detonate the bomb assembly before releasing a large portion of its explosive energy.

In April 1943 a young physicist from Cal Tech, Seth Nedermeyer, proposed an "implosion" approach instead of the gun method. The idea was to encircle plutonium with high explosives, which would produce a spherical shockwave traveling inward that would compress the pluto-nium to create a critical mass and thus avoid predetonation. The idea was brilliant, but the execu-tion was most difficult. In February 1944, Oppenheimer brought Russian-born physical chemist George Kistiakowsky to Los Alamos, and gave him primary responsibility to turn Nedermeyer's idea into practice. Not until mid-April 1945 did he find a way to create a symmetrical shockwave using fine mirrors to detonate a plutonium bomb.[104]

Along with the detonation devices themselves, Oppenheimer and the group focused atten-tion in mid-1944 on design and bomb fabrication, especially the development of an implosion bomb. By mid-1945, the scientists felt confident that the gun method of detonation for a ura-nium bomb did not require an explosive test prior to use, and thus *Little Boy* was ready. The implosion weapon was another matter. There was substantial pressure to produce a plutonium bomb in concert with *Little Boy*, because there was not enough U-235 available for a second uranium device. This was particularly important because part of their directive was to begin stockpiling atomic weapons rather than simply creating one lone bomb.[105] In summer of 1945, the Los Alamos group poised for a first test of the plutonium "gadget."

CONCLUSION: FROM FISSION TO BOMB

It is almost impossible to imagine that between the fission experiments in the 1930s and World War II in the early 1940s, science theory (even science fiction) had become technical reality. What is sometimes lost in this whirlwind of change in these years is how the very role of science in society had been altered. The heady success of Hahn, Meitner, Strassmann, and others with the discovery of fission made possible the remarkable feat of Oppenheimer and his team just a few years later. Yet, the achievements of both sets of scientists and technicians were produced in a world increasingly tethering scientific accomplishment not only to practical ends, but to state-mandated and state-directed objectives. World War II invigorated science, but also inhibited it. The ultimate product of the flurry of scientific activity in these years was a weapon of such magnitude and destructive power that neither scientists nor politicians alone could assess its potential impact. The atomic bomb emerged nonetheless, and would find its use, its victims, its supporters, and its critics.

MySearchLab Connections: Sources Online

READ AND REVIEW

Review this chapter by using the study aids and these related documents available on MySearchLab.

✓•⌐**Study** and **Review** on **mysearchlab.com**

Chapter Test

Essay Test

▣•⌐**Read** the **Document** on **mysearchlab.com**

Albert Einstein, Letter to Franklin D. Roosevelt (1939)

Manhattan Project Notebook (1945)

Report by MAUD Committee on the Use of Uranium for a Bomb (1943)

Bernice Brode, Tales of Los Alamos (1943)

General Groves, Address to Officers Regarding the Atomic Bomb (1945)

Emilio G. Segrè, The Discovery of Nuclear Fission (1989)

RESEARCH AND EXPLORE

Use the databases available within MySearchLab to find additional primary and secondary sources on the topics within this chapter.

Endnotes

1. Peter Galison, "The Many Faces of Big Science," in Peter Galison and Bruce Hevly, eds., *Big Science: The Growth of Large-Scale Research* (Stanford, CA: Stanford University Press, 1992), 1.
2. Alvin M. Weinberg, "Impact of Large-Scale Science on the United States," in Norman Kaplan, ed., *Science and Society* (Chicago: Rand McNally & Co., 1965), 551.
3. "Big Science," *Britannica Online Encyclopedia*, online, http://www.britannica.com.
4. Galison, "The Many Faces of Big Science," 2. See also Bruce Hevly, "Reflections on Big Science and Big History," in Galison and Hevly, eds., *Big Science*, 355–61; Norman Kaplan, ed., *Science and Society* (Chicago: Rand McNally & Co., 1965), 1–13.
5. Galison, "The Many Faces of Big Science," 2; Mark Erickson, *Science, Culture and Society: Understanding Science in the 21st Century* (Cambridge, UK: Polity, 2005), 116; Weinberg, "Impact of Large-Scale Science on the United States," 551.
6. Lawrence Badash, *Kapitza, Rutherford, and the Kremlin* (New Haven: Yale University Press, 1985), 2.
7. Sir Gavin de Beer, "The Sciences were Never at War," in Norman Kaplan, ed., *Science and Society* (Chicago: Rand McNally & Co., 1965), 18.
8. D.S.L. Cardwell, "Science and Society" and Joseph Ben-David and Awraham Zloczower, "Universities and Academic Systems in Modern Societies," in Kaplan, ed., *Science and Society*, 57, 64, 67, 75, 78–79, 83.
9. Cardwell, "The Professional Society," in ibid., 102.
10. See Martin V. Melosi, *Thomas A. Edison and the Modernization of American* (New York: Pearson, 2008), 46–61; Daniel J. Kevles, *The Physicists: The History of a Scientific Community in Modern America* (Cambridge, MA: Harvard University Press, 1977; fourth printing, 1995), 100.
11. Alex Keller, *The Infancy of Atomic Physics: Hercules in His Cradle* (Oxford: Clarendon Press, 1983), 206–08.
12. Kevles, *The Physicists*, 113, 141; David Lindley, *Uncertainty: Einstein, Heisenberg, Bohr, and the Struggle for the Soul of Science* (New York: Doubleday, 2007), 57, 84.
13. Kevles, *The Physicists*, 100–04; A. Hunter Dupree, *Science in the Federal Government: A History of Policies and Activities* (Baltimore: Johns Hopkins University Press, 1986), 302–25.
14. Melosi, *Thomas A. Edison and the Modernization of American*, 156–71.
15. Kevles, *The Physicists*, 102–54.
16. Edmund Russell, *War and Nature: Fighting Humans and Insects with Chemicals from World War I to Silent Spring* (New York: Cambridge University Press, 2001), 28–45.
17. Kevles, *The Physicists*, 117–38.
18. Andrew J. Rotter, *Hiroshima: The World's Bomb* (New York: Oxford University Press, 2008), 23.
19. Alexi B. Kojevnikov, *Stalin's Great Science: The Times and Adventures of Soviet Physicists* (London: Imperial College Press, 2004), 1–6.

20. Ibid., 18–19.
21. Rotter, *Hiroshima*, 23.
22. Kojevnikov, *Stalin's Great Science*, 7–32, 43–44. See also Badash, *Kapitza, Rutherford, and the Kremlin*, 37–50.
23. Badash, *Kapitza, Rutherford, and the Kremlin*, 48. See also Paul R. Josephson, *Red Atom: Russia's Nuclear Power Program from Stalin to Today* (Pittsburgh: University of Pittsburgh Press, 2000), viii.
24. Lindley, *Uncertainty*, 171–72; Helge Kragh, *Quantum Generations: A History of Physics in the Twentieth Century* (Princeton: Princeton University Press, 1999), 230–31, 243; David C. Cassidy, *Uncertainty: The Life and Science of Werner Heisenberg* (New York: W.H. Freeman and Co., 1992), 96–97.
25. Kragh, *Quantum Generations*, 231–35.
26. Cassidy, *Uncertainty*, 135; Stephen E. Atkins, ed., *Historical Encyclopedia of Atomic Energy* (Westport, CT: Greenwood Press), 21, 186; Kragh, *Quantum Generations*, 236–38; Lindley, *Uncertainty*, 174–75.
27. Daniel J. Kevles, *In the Name of Eugenics: Genetics and the Uses of Human Heredity* (New York: Knopf, 1985), 164. See also "Eugenics," Bioethics Research Laboratory, Georgetown University, online, http://bioethics.georgetown.edu/publications/scopenotes/sn28.htm#kevles1.
28. Kragh, *Quantum Generations*, 238–40.
29. Ibid., 245–49; Kevles, *The Physicists*, 197–204, 221; Lindley, *Uncertainty*, 177.
30. Kragh, *Quantum Generations*, 249–256; Kevles, *The Physicists*, 210–215, 276–282; Lindley, *Uncertainty*, 173.
31. Kragh, *Quantum Generations*, 257.
32. "Nuclear Fission: Basics," Atomicarchive.com, online, http://www.atomicarchive.com/Fission/Fission1.shtml.
33. Diana Preston, *Before the Fallout: From Marie Curie to Hiroshima* (New York: Walker & Company, 2005), 115.
34. Prior to World War II, the United States had more than 30 cyclotrons in operation, mostly for medical research. The Russians developed their first in 1937, and the Germans did not have one until 1944. See Atkins, ed., *Historical Encyclopedia of Atomic Energy*, 106.
35. F.G. Gosling, *The Manhattan Project: Science in the Second World War*, Energy History Series (Washington, D.C.: History Division, US Department of Energy, August 1990), 1–2.
36. The mass of a nucleus is less than the masses of the protons and neutrons in it. The difference is a measure of the "nuclear binding energy" holding together the nucleus. "Nuclear Binding Energy," *Hyperphysics*, online, http://hyperphysics.phy-astr.gsu.edu/hbase/nucene/nucbin.html. See also "The Strong Nuclear Force," online, http://aether.lbl.gov/elements/stellar/strong/strong.html "Nuclear Binding Energy," *Super Glossary*, online, http://www.superglossary.com/Definition/Chemistry/Nuclear_Binding_Energy.html.
37. Gerald Holton and Stephen G. Brush, *Physics, The Human Adventure: From Copernicus to Einstein and Beyond* (New Brunswick: Rutgers University Press, 2005), 506; Kragh, *Quantum Generations*, 257.
38. Preston, *Before the Fallout*, 49, 104–113; Kragh, *Quantum Generations*, 257–59; Holton and Brush, *Physics, The Human Adventure*, 506.
39. Keller, *The Infancy of Atomic Physics*, 217; Lisa Rosner, ed., *Chronology of Science: From Stonehenge to the Human Genome Project* (Santa Barbara, CA: ABC-CLIO, 2002), 258; Gosling, *The Manhattan Project*, 2.
40. Preston, *Before the Fallout*, 114; Kragh, *Quantum Generations*, 260; Gosling, *The Manhattan Project*, 2.
41. Kragh, *Quantum Generations*, 261, 263.
42. Ibid., 262–63.
43. Quoted in Michael Kort, *The Columbia Guide to Hiroshima and the Bomb* (New York: Columbia University Press, 2007), 171.
44. Martin J. Sherwin, *A World Destroyed: Hiroshima and the Origins of the Arms Race* (New York: Vintage Books, 1987), 22.

45. Kragh, *Quantum Generations*, 263–64; Ronald E. Powaski, *March to Armageddon: The United States and the Nuclear Arms Race, 1939 to the Present* (New York: Oxford University Press, 1987), 4; Preston, *Before the Fallout*, 119–33.

46. Kevles, *The Physicists*, 289.

47. Sherwin, *A World Destroyed*, 26–27. See also Powaski, *March to Armageddon*, 4; Atkins, ed., *Historical Encyclopedia of Atomic Energy*, 116–17.

48. Quoted in Kort, *The Columbia Guide to Hiroshima and the Bomb*, 172.

49. Sherwin, *A World Destroyed*, 27–28.

50. Gosling, *The Manhattan Project*, 5, 7; Sherwin, *A World Destroyed*, 28–30.

51. Richard G. Hewlett and Oscar E. Anderson, Jr., *The New World, 1939–1946*, volume 1, *A History of the United States Atomic Energy Commission* (University Park, PA: Penn State University Press, 1962), 29; Gosling, *The Manhattan Project*, 6–7; Powaski, *March to Armageddon*, 5; Kevles, *The Physicists*, 295–300; Sherwin, *A World Destroyed*, 29–30.

52. Sherwin, *A World Destroyed*, 40–41.

53. Kevles, *The Physicists*, 300–301; Gosling, *The Manhattan Project*, 7–9; Sherwin, *A World Destroyed*, 30–34.

54. Thomas Powers, *Heisenberg's War: the Secret History of the German Bomb* (New York: Knopf, 1993); Paul Lawrence Rose, *Heisenberg and the Nazi Atomic Bomb Project* (Berkeley, CA: University of California Press, 1998.

55. "What is Uranium? How Does it Work?" World Nuclear Association, online, http://www.world-nuclear.org/education/uran.htm.

56. A way to define the moderator: "The probability that a high energy neutron emitted in a fission reaction can directly cause fission is quite low. In order to build a system in which fast neutrons sustain the chain reaction, very highly enriched uranium is needed, which is very expensive. A more easily feasible way is to use materials that slow the neutrons down to such low energies at which the probability of causing a fission is significantly higher. These materials that slow down the neutrons are the so-called moderators. With the aid of some adequate moderator one might as well achieve [a] chain reaction using natural… uranium… [T]there are two major requirements that a moderator material should meet: it should have as low an atomic mass number as possible and its neutron absorbing ability should be as low as possible." Pure graphite and heavy water are good moderators. See "Nuclear Chain Reaction," online, http://paksnuclearpowerplant.com/download/1226/nuclear_chain_reaction.pdf.

57. Alwyn McKay, *The Making of the Atomic Age* (Oxford: Oxford University Press, 1984), 44–46; Atkins, ed., *Historical Encyclopedia of Atomic Energy*, 161–62; Carlisle, ed., *Encyclopedia of the Atomic Age*, 84, 213; Rotter, *Hiroshima*, 78.

58. McKay, *The Making of the Atomic Age*, 44–52, 101–03.

59. Preston, *Before the Fallout*, 159–60; Rotter, *Hiroshima*, 80–81.

60. Paul Lawrence Rose, "Frayn's 'Copenhagen' Plays Well, at History's Expense," *Chronicle of Higher Education* (May 5, 2000). See also John Lukacs, "The Conversation: *Copenhagen* by Michael Frayn," *Los Angeles Times* (May 21, 2000); David C. Cassidy, "A Historical Perspective on 'Copenhagen,'" online, http://www.ashp.cuny.edu/nml/copenhagen/Cassidy.htm.

61. As in the United States, émigré physicists had been closed off from research on radar and other high priority war projects, leaving them to work on activities of lower priority, such as—ironically—the uranium bomb. Sherwin, *A World Destroyed*, 34–35.

62. Klaus Fuchs, a German refugee physicist and a Soviet spy working in Great Britain, passed along the information about the British work on a bomb to Moscow in 1941. See Kragh, *Quantum Generations*, 265.

63. Ibid., 264–65; Gosling, *The Manhattan Project*, 9; McKay, *The Making of the Atomic Age*, 53–55; Cynthia C. Kelly, ed., *The Manhattan Project: The Birth of the Atomic Bomb in the Words of Its Creators, Eyewitnesses, and Historians* (New York: Black Dog & Leventhal Pubs., 2007), 50–51; Atkins, ed., *Historical Encyclopedia of Atomic Energy*, 139, 227; Rotter, *Hiroshima*, 88–91.

64. Powaski, *March to Armageddon*, 5; Gosling, *The Manhattan Project*, 7–9.

65. McKay, *The Making of the Atomic Age,* 60–62; Atkins, ed., *Historical Encyclopedia of Atomic Energy*, 289.

66. Sherwin, *A World Destroyed*, 35–39; Gosling, *The Manhattan Project*, 9–10; Kort, *The Columbia Guide to Hiroshima and the Bomb*, 18. See also Richard Rhodes, *The Making of the Atomic Bomb* (New York: Simon and Schuster, 1986), 357–78; Kelly, ed., *The Manhattan Project*, 60–66.

67. Vincent C. Jones, *Manhattan: The Army and the Atomic Bomb,* Special Studies, *United States Army in World War II*, (Washington, D.C.: Center of Military History, United States Army, 1985), 227–52; Atkins, ed., *Historical Encyclopedia of Atomic Energy*, 60–61.

68. David C. Cassidy, *J. Robert Oppenheimer and the American Century* (New York: Pi Press, 2005), 216.

69. Sherwin, *A World Destroyed*, 41.

70. See A. Costandina Titus, " The Mushroom Cloud as Kitsch," in Scott C. Zeman and Michael A. Amundson, eds., *Atomic Culture: How We Learned to Stop Worrying and Love the Bomb* (Boulder, CO: University Press of Colorado, 2004), 112–13; Jerome F. Shapiro, *Atomic Bomb Cinema: The Apocalyptic Imagination on Film* (New York: Routledge, 2002), 62–65; Joyce A. Evans, Celluloid Mushroom Clouds: Hollywood and the Atomic Bomb (Boulder, CO: Westview Press, 1998), 26–43; Michael J. Yavenditti, "Atomic Scientists and Hollywood: The Beginning or the End?" *Film and History* 8 (December, 1978): 73–88; Jack G. Shaheen and Richard Taylor, "The Beginning or the End," in Jack G. Shaheen, ed., *Nuclear War Films* (Carbondale, IL: Southern Illinois University Press, 1978), 3–10.

71. Jones, *Manhattan*, 117.

72. Robert S. Norris, *Racing for the Bomb: General Leslie R. Groves, The Manhattan Project's Indispensable Man* (South Royalton, VT: Steerforth Press, 2002).

73. "The Manhattan Project," online, http://www.me.utexas.edu/~uer/manhattan/project.html; Gosling, *The Manhattan Project*, 10–11; Carlisle, ed., *Encyclopedia of the Atomic Age*, 115; Kevles, *The Physicists*, 327; Hewlett and Anderson, Jr., *The New World, 1939–1946*, 29–31; McKay, *The Making of the Atomic Age*, 68.

74. Gosling, *The Manhattan Project*, 11.

75. Atkins, ed., *Historical Encyclopedia of Atomic Energy*, 84–85, 234–235; Carlisle, ed., *Encyclopedia of the Atomic Age*, 56–58, 127; Kort, *The Columbia Guide to Hiroshima and the Bomb*, 21–22.

76. Gosling, *The Manhattan Project*, 10. See also Kort, *The Columbia Guide to Hiroshima and the Bomb*, 19.

77. Harvey M. Sapolsky, *Science and the Navy: The History of the Office of Naval Research* (Princeton: Princeton University Press, 1990), 18–19, 25–28; Gosling, *The Manhattan Project*, 11–12; Sherwin, *A World Destroyed*, 45.

78. Sherwin, *A World Destroyed*, 42–45.

79. Ibid., 58.

80. McKay, *The Making of the Atomic Age*, 69–70; Atkins, ed., *Historical Encyclopedia of Atomic Energy*, 152–54; Kevles, *The Physicists*, 326–27; Sherwin, *A World Destroyed*, 48; Hewlett and Anderson, Jr., *The New World, 1939–1946*, 71–75.

81. Norris, *Racing for the Bomb*, x.

82. Quoted in Kelly, ed., *The Manhattan Project*, 88.

83. Kort, *The Columbia Guide to Hiroshima and the Bomb*, 19–20; Kevles, *The Physicists*, 326.

84. Atkins, ed., *Historical Encyclopedia of Atomic Energy*, 267–69; Kort, *The Columbia Guide to Hiroshima and the Bomb*, 22; McKay, *The Making of the Atomic Age*, 71–72.

85. Jeff Hughes, *The Manhattan Project: Big Science and the Atom Bomb* (New York: Columbia University Press, 2002), 59; Atkins, ed., *Historical Encyclopedia of Atomic Energy*, 157–58; Michele Stenehjem Gerber, *On the Home Front: The Cold War legacy of the Hanford Nuclear Site* (Lincoln: University of Nebraska Press, 1992), 31.

86. Kevles, *The Physicists*, 327–28; McKay, *The Making of the Atomic Age*, 70–78; Hewlett and Anderson, Jr., *The New World, 1939–1946*, 31; Sherwin, *A World Destroyed*, 48–49.

87. Hughes, *The Manhattan Project*, 59; Atkins, ed., *Historical Encyclopedia of Atomic Energy*, 158, 289; Kevles, *The Physicists*, 328–29; McKay, *The Making of the Atomic Age*, 79–89.

88. Peter Bacon Hales, *Atomic Spaces: Living on the Manhattan Project* (Urbana, IL: University of Illinois Press, 1997), 4. See also Gerber, *On the Home Front*; Bruce Hevly and John M. Findlay, eds., *The Atomic West* (Seattle: University of Washington Press, 1998.

89. Jon Hunner, *Investing Los Alamos: The Growth of an Atomic Community* (Norman, OK: University of Oklahoma Press, 2004).

90. Kai Bird and Martin J. Sherwin, *American Prometheus: The Triumph and Tragedy of J. Robert Oppenheimer* (New York: Alfred A. Knopf, 2005).

91. John Adams, *Doctor Atomic*, premiering in 2005 in San Francisco, 2005.

92. Bird and Sherwin, *American Prometheus*, 5.

93. Cathryn Carson and David Hollinger, eds., *Reappraising Oppenheimer: Centennial Studies and Reflections* (Berkeley: University of California Press, 2005), 1–9; Gregg Herken, *Brotherhood of the Bomb: The Tangled Lives and Loyalties of Robert Oppenheimer, Ernest Lawrence, and Edward Teller* (New York: Henry Holt and Co., 2002), 43.

94. Quoted in Kelly, ed., *The Manhattan Project,* 160.

95. Ibid., 112.

96. Ibid., 111–151; Herken, *Brotherhood of the Bomb*, 51–53; Cassidy, *J. Robert Oppenheimer and the American Century*, 226–227; Jeremy Bernstein, *Oppenheimer: Portrait of an Enigma* (Chicago: Ivan R. Dee, 2004), 75–76, 79–80; Rotter, *Hiroshima*, 103–08, 116–20.

97. Kort, *The Columbia Guide to Hiroshima and the Bomb*, 20–21.

98. Atkins, ed., *Historical Encyclopedia of Atomic Energy*, 221; Kevles, *The Physicists*, 329; Ferenc Morton Szasz, *The Day the Sun Rose Twice: The Story of the Trinity Site Nuclear Explosion, July 16, 1945* (Albuquerque; University of New Mexico Press, 1984), 17.

99. Kevles, *The Physicists*, 329; Szasz, *The Day the Sun Rose Twice*, 18–21; Hughes, *The Manhattan Project,* 64, 68; Atkins, ed., *Historical Encyclopedia of Atomic Energy*, 221; McKay, *The Making of the Atomic Age*, 93–94.

100. Edward Teller, *Memoirs: A Twentieth Century Journey in Science and Politics* (Cambridge, MA: Perseus Pub., 2001), 167.

101. For insight into the lives of women at Los Alamos, see Mary Anne Schofield, "Lost Almost and Caught Between the Fences: The Women of Los Alamos, 1943–1945 and Later," in Rosemary B. Mariner and G. Kurt Peihler, eds., *The Atomic Bomb and American Society: New Perspectives* (Knoxville, TN: University of Tennessee Press, 2009), 65–87.

102. Sherwin, *A World Destroyed*, 54–63; Kevles, *The Physicists*, 16.

103. Kevles, *The Physicists*, 330–331; McKay, *The Making of the Atomic Age*, 91–92. See also Hewlett and Anderson, Jr., *The New World, 1939–1946*, 227–54; Jones, *Manhattan*, 465–81.

104. Kevles, *The Physicists*, 331–32.

105. Ibid., 332; Lillian Hoddeson, "Mission Change in the Large Laboratory: the Los Alamos Implosion Program, 1943–1945," in Galison and Hevly, eds., *Big Science*, 265–66; Hughes, *The Manhattan Project,*73–83; Kort, *The Columbia Guide to Hiroshima and the Bomb*, 21–24.

CHAPTER 3

Hiroshima, Nagasaki, and the Aftermath

From Total War to Cold War

INTRODUCTION: FROM THE LAB TO THE FIELD

The atomic bomb ceased to be an idea and became a reality in 1945. It was conceived as a practical application of relentless probing into the inner workings of the atom, and forged by a collaboration of science and the state. The race to beat the Germans to the bomb united those in North America, Great Britain, and elsewhere in a singular cause. The use of the new weapon was another issue entirely. US officials made a decision in the heat of battle to up the ante in destructiveness and to quickly end the conflict that engulfed the world. Such a choice had impacts well beyond the fatal days in August when the people of Hiroshima and Nagasaki became the first casualties of a new kind of total war. The bombings and their aftermath clearly redefined the execution of war itself and the definitions of vulnerability and security for all nations.

TRINITY: THE GENIE OUT OF THE BOTTLE

Robert Oppenheimer was not sure, but he vaguely recalled some years later why he gave the first explosion of a nuclear device the code name *Trinity*. It was likely a line from a John Donne poem: "Batter my heart, three person'd God." Yet, given his regard for Hinduism, the moniker could have signified the trinity of Brahma, the Creator; Vishnu, the Preserver; and Shiva, the Destroyer. Others suggested that *Trinity* referred to three bombs being constructed at the time, or that someone else other than Oppenheimer came up with the name.[1] Whatever the reason, there was a sort of otherworldliness about this giant step in atomic history.

    Almost everything about the test of the plutonium (or implosion) bomb was imbued with portent. The desert location where the detonation took place was more than 200 miles south of Los Alamos on the Alamogordo Bombing Range. This ninety-mile site in the valley between the Rio Grande River and the Sierra Oscura mountains was called Jornada del

Muerto—*Journey of Death* or *Dead Man's Way*. The effort to ready the bomb for testing was an exhausting journey itself for the Los Alamos team. To relieve the tension of the work the scientists organized a betting pool on how big the blast might be, but its size and intensity caught everyone by surprise. General Groves in particular constantly worried about sabotage of the bomb and beefed up security before zero hour. He also feared that the very limited amounts of plutonium would be destroyed if the test failed, seriously delaying the mission of the Manhattan Project. To say the least, Washington, now in the throes of the Pacific War, anxiously awaited the results of the test.[2]

The detonation itself on July 16, 1945, came on the heels of feverish preparation. The implosion studies were the first step, followed by developing measures to determine the release of nuclear energy and to assess the potential damage resulting from the explosion, including radiation. On July 12, the plutonium core for the bomb was delivered to Alamogordo. The scientists at Los Alamos completed the bomb assembly on the fourteenth and hoisted it onto a 100-foot steel tower. On Sunday, July 15 several key dignitaries arrived at the base camp, including Groves and Conant. Tensions were high. As Groves' deputy recalled, "For some hectic two hours preceding the blast, General Groves stayed with the Director [Oppenheimer], walking with him and steadying his tense excitement. Every time the Director would be about to explode because of some untoward happening, General Groves would take him off and walk with him in the rain, counselling [sic] with him and reassuring him that everything would be all right."[3] Bad weather threatened the test, but Groves insisted that it take place on Monday, July 16 barring hopeless conditions. Violent thunderstorms at 2:00 A.M. on that day looked likely to compel the team to delay beyond the 4 A.M. target hour—possibly forcing a cancellation. Oppenheimer, test director Kenneth Bainbridge, and the chief meteorologist decided to set the new time at 5:30 A.M., and much to Groves' surprise and approbation, detonation occurred at 5:29 A.M.[4]

The "Trinity" explosion at Los Alamos, Alamogordo, New Mexico, on July 16, 1945 was the first detonation of a nuclear device. *Source:* National Archives

Light, heat, sound, and radiation. The blast produced a blinding yellow light that could be seen as far as Albuquerque to the north and El Paso to the south. The heat generated by the bomb was intense, followed by the upward thrust of an enormous blue-glowing white mushroom cloud rising 38,000 to 41,000 feet into the air. The sound of the blast was so intense that it growled in the hills surrounding Ground Zero for several minutes. The explosion itself left a crater of 1,200 feet in diameter. Sand at the detonation site turned to glass; the explosion killed animals, insects, birds, and plants within a mile of the now-vaporized tower.[5] Observer reactions were mixed. As a group the atomic scientists were thrilled and relieved by their success, but they were equally stunned by the destructive power of what had modestly been called "the gadget." Ernest Lawrence later wrote in a report to Groves, "I was enveloped with a warm brilliant yellow-white light—from darkness to brilliant sunshine in an instant, and as I remember I momentarily was stunned by the surprise."[6] One eyewitness wrote, "The whole spectacle was so tremendous and one might almost say fantastic that the immediate reaction of the watchers was one of awe rather than excitement."[7] Physicist Victor Weisskopf conceded, "Our first feeling was one of elation, then we realized we were tired, and then we were worried."[8]

The sheer size of the explosion, equal to 20,000 tons of TNT, was astonishing to all who were present. This first experience with radioactive fallout also was a lingering reminder of the uniqueness of this new weapon.[9] To Brigadier General Thomas Farrell, the roar of the explosion "Warned of doomsday and made us feel that we puny things were blasphemous to dare tamper with the forces hitherto reserved to the Almighty."[10] More matter-of-fact, Groves was concerned about keeping "this thing quiet" and concluded, "The war's over."[11] In his own cerebral manner, Oppenheimer later recalled that what flashed through his mind at the time was a line from Hindu scripture, the *Bhagavad-Gita*, "Now I am become death, the destroyer of worlds." In some respects this belied the most immediate responses of the scientists who were dancing and shaking hands. Yet Kenneth Bainbridge brought things down to an elemental level when he told Oppie at the control center, "Now we're all sons-of-bitches."[12]

Whether contemporaries realized it or not, the *Trinity* detonation was a dividing line in history. Scientific theory became physical reality. The impacts were at once local and global. For many years the National Park Service hoped to make the *Trinity* site into a national monument, but strong resistance came from government officials concerned about security or questions of health and safety. In 1965 the area was designated as a National Historic Landmark, but open to the public only twice a year. *Trinity* became one of the first "nuclear landscapes" in the late twentieth century. The immediate result of the tremendous release of energy was directed toward building bombs with crucial impact on World War II, but with much more far-reaching implications for humankind. For historian Ferenc Szasz, "What happened at *Trinity* that Monday morning must go down as one of the most significant events in the last thousand years."[13]

POTSDAM AND THE PACIFIC WAR

The work of the scientists was now done. The atomic bomb, and all of the crucial decisions about it, passed to the hands of politicians and the military. Among General Groves' immediate tasks was containing the rumors about what had happened in the New Mexico desert. The press release simply stated that an ammunition magazine had exploded.[14] His most important duty was to inform President Harry Truman about the successful detonation of the plutonium bomb (he radioed a coded message to Washington about one hour after the test) and the delivery of the weapons to the war front. On April 12, 1945, President Franklin Roosevelt had died, leaving his relatively inexperienced vice president in charge of the American war effort. Harry S Truman, the former US senator from Missouri, was FDR's third vice president, and held that position less than three months when he was elevated to the country's highest office. In fact, not until Roosevelt's death did anyone inform Truman of the exact nature of the work of the Manhattan Project.[15] Less than a month later, on May 8, German Admiral Karl Donitz signed the unconditional surrender documents ending World War II in Europe. One day after the *Trinity* test, Truman arrived in Potsdam, Germany (for the last major wartime conference of the Grand Alliance) to meet with Soviet leader Joseph Stalin and British Prime Minister Winston Churchill. The chief item on the agenda for the United States was the Pacific War against Japan.

President Truman had agreed in May to meet with the Allied leaders on July 1, and the White House was pressuring General Groves to arrange for the test of the plutonium bomb as quickly as possible. Initially, the team in New Mexico stated that the earliest achievable date would be July 13. The president had been reluctant, and even nervous, to confer with the other more seasoned leaders, who might be able to grab control of events leading up to the end of the war. Truman clearly wanted the leverage of a successful test to bolster US bargaining power before meeting with Stalin and Churchill, and thus postponed the conference until the *Trinity* test could be scheduled. Even with that strategy in place, he arrived in Germany determined to coax the Soviets into participating in the defeat of Japan.[16] In Truman's first meeting with Stalin on July 17, the Soviet leader pledged to have his country enter the Pacific War on August 15. The president was relieved, and presumed that he had achieved his primary goal. Other knotty problems were still on the table, including the postwar fate of Germany and Eastern Europe. At the time, these issues seemed to be secondary in importance to the Soviet promise, which along with the availability of the atomic bomb, would certainly secure a victory over the Japanese. Truman's view of the bomb, however, grew more ambivalent as the conference proceeded. The new president was enthusiastic about the achievement, but also noted in his diary that "It seems to be the most terrible thing ever discovered, but it can be made the most useful."[17]

When the details of the *Trinity* test arrived in Potsdam, Truman appeared more self-assured, more aggressive in dealing with his fellow leaders. Churchill later recalled that after the US president read Groves' full report on the test he was a "changed man." The prime minister received a full description of the *Trinity* test on July 22, and he enthusiastically noted that the atomic bomb was "the Second Coming in wrath." The weapon, he too believed, would surely mean an end to the war. As a savvy realist, he also viewed it as a potentially effective tool against future Soviet ambitions in the Far East and elsewhere.[18] Truman did not tell his Soviet ally about the existence of the atomic bomb until July 24. While Stalin reacted almost indifferently to the disclosure, privately he was very troubled about the American advances in atomic energy and the possibility that its use of the bomb would limit his country's options in Asia. Stalin's lack of surprise was a result of effective Soviet spying within the Manhattan Project and related projects in England and Canada well in advance of Potsdam. Also, at least since 1940, Stalin had pushed the Soviets to develop their own bomb, and at the time of the German invasion of the Soviet Union in June 1941, they may have been ahead of the Americans at least in some areas of atomic energy.[19]

Soviet agents and collaborators had been gathering intelligence information in the United States and Canada since the 1930s. The NKVD (People's Commissariat of Internal Affairs) conducted most of the spying. The security program in the United States was considerable not only because agents attempted to gather sensitive information but also because many of them monitored thousands of the Soviets' own citizens working in the country on Lend Lease activities.[20] According to one study, Soviet operatives ranged from "highly sophisticated practitioners of tradecraft to bumbling amateurs."[21] Wartime spying focused on manufacturing firms, industrial espionage, and military enterprises in areas as wide ranging as avionics, radar, proximity fuses, and the atom bomb. Much of it went undetected, but even in the early days of the war the FBI and other agencies began paying greater attention to Soviet covert activities. Ironically, the FBI only learned about the Manhattan Project when agents somewhat inadvertently discovered information about the secret program while uncovering documents concerning Soviet efforts to infiltrate it.[22]

Beginning in 1940, Soviet espionage focused some of its most intense efforts on gaining information about "Enormoz" (Enormous), which was the code name for nuclear research and production in the United States, Canada, and England, but especially the Manhattan Project. Agents first began gathering information on scientists, among other places, in Cambridge University and the Cavendish Laboratory in England and at Columbia and MIT in the United States. Some of the secrets obtained allowed Soviet scientists to bypass cumbersome steps in bomb development. The Soviets succeeded in enlisting Klaus Fuchs who ultimately sent extensive information to the USSR about the scale of the bomb project, details about the plutonium bomb itself, and other pertinent data.[23] Fuchs had fled Nazi Germany for England in the 1930s largely because of his German Communist Party affiliation, and became a naturalized citizen in the UK. He traveled to the United States in late 1943 with a British team to work on gaseous diffusion for the Kellex Corporation and at Columbia University. In August 1944 Fuchs went to Los Alamos to actively participate in the implosion program. He had begun spying for the Soviet Union possibly as early as 1939, taking a brief hiatus in 1945–1946, and resuming the covert activities in 1947. Authorities finally arrested him on February 2, 1950.[24]

Unaware that the Soviets knew more about the bomb project than they could have imagined, some key advisers within the Truman administration did not want to disclose the existence of the bomb to their wartime ally at all. They wanted to assume a harder line with the Soviets, and continually prodded Truman to do so. The hawkish, but independent Secretary of State (former US senator from South Carolina, Supreme Court justice, and close personal friend of the president) James F. Byrnes, was most influential in constructing diplomatic policy for the United States at the time. He believed that the bomb provided a unique opportunity to check Soviet control of Eastern Europe and Asia in the postwar years, and he very much wanted to delay or avert the entry of the Soviet Union into the war with Japan. This change in strategy from encouraging Soviet entry was predicated on the belief that such a move would increase a potential competitor's influence in East Asia. In this instance, top military advisers agreed with Byrnes. General George C. Marshall, the Army Chief of Staff, argued that the bomb made Soviet participation in the war with Japan unnecessary. Truman came around to this view, but still believed that Stalin needed to be told to avoid needlessly threatening the wartime alliance. More than anything, Truman wanted a quick end to the war.

The chess game between and Americans and the Soviets (an analogy in their on-going relations) continued beyond the end of the war. With much remaining unresolved at Potsdam, the Americans nevertheless left the conference with Stalin's promise still intact to enter the Pacific War (whether they needed it or not), and a joint ultimatum from the United States, Great Britain, and China to Japan. The somewhat vague Potsdam Declaration on July 26, which was made without the Soviet Union (who Byrnes feared would ask too much in return for its signature), called for unconditional surrender of the "Japanese armed forces" and the establishment of a new government. If they refused, the Japanese faced the prospect of "prompt and utter destruction." The declaration did not hold to a strict definition of "unconditional surrender" that included the stepping down of the Japanese emperor, who his people regarded as a deified leader. Nor did it mention the entry of the Soviet army into the war or the bomb. It did make clear that diplomatic negotiations were not on the table, and that the United States and its allies were prepared to bring World War II to a quick end.[25]

TOTAL WAR: CONTEXT FOR THE USE OF THE BOMB

Since the end of World War II there has been much speculation as to why President Truman decided to drop the atomic bomb on Japan. There is much less debate about whether he would do so.[26] Context is everything. Simply put, the dropping of the atomic bomb was the final act in a new and deadly total war. The war aims of the Allied forces, whether finally realized or not, were unconditional surrender of the enemy and their total military defeat. The use of the atomic bomb was not inevitable, but hardly unthinkable under the circumstances.[27] World War II has been characterized in many ways, as a global conflict, a revolution, a war of extremes, mass warfare, a fully mechanized war, and even a "good war." There is little doubt that it also was a total war. As historians Roger Chickering and Stig Forster aptly argued, "The history of total war was driven by material and ideological forces that culminated respectively in Hiroshima and Auschwitz—in weapons that did not discriminate and policies that did so with a vengeance." After the war, they added, the concept of total war lived on as an ideal type, "now as the nuclear nightmare that would succeed the Second World War, set the ultimate parameters of 'totality,' and capped the master narrative in apocalypse."[28] General Erich Ludendorff, the German First Quartermaster General in 1918, popularized the term "total war" with reference to World War I. He believed that conventional warfare fought in short campaigns between opposing armed forces was being replaced by war waged between entire populations, civilian and military. In such a case, whole countries mobilized and any person, resource, or speck of land became a legitimate objective for attack and annihilation.[29] Depending on the definition, the idea of total war has been applied to conflicts going back to the Peloponnesian Wars in Ancient Greece. But in a practical sense and in terms of scale, World War II was the first truly total war.

There were limits to the full impact of warfare on noncombatant populations during World War II. While all countries together mobilized more than 70 million people for military service, and combat took place on every continent but Antarctica (if you count coastal waters), fighting concentrated in fewer locations, civilian destruction was confined, economic mobilization was uneven, and surrender conditions were not uniform.[30] This being said, in those places where the war was most aggressively taken to civilians (USSR, the UK, Germany, Eastern Europe, China, Japan) devastation was vast. Overall casualty figures were numbing. World War II was the deadliest war of all time with from 50 to 70 million fatalities or more. This figure includes not only battlefield deaths, but Holocaust victims, those killed by aerial bombing of civilian sites, and war-related famine deaths in China, Indonesia, Vietnam, the Philippines, India, and Bangladesh. Civilian losses represented at least half of the total casualties, with about 70 countries involved in the war in some way. The death toll excludes other horrors including cruelty and torture of prisoners of all kinds, slave labor and the inhumanity of concentration camps, vast homelessness, malnutrition and disease, and permanent disappearance of thousands. Property loss was almost incalculable, along with the destruction of whole cities such as Warsaw, Hamburg, Dresden, Tokyo, Hiroshima, and Nagasaki.[31]

What made World War II the ultimate total war for its time was air power, a true revolution in warfare. Beginning at the end of World War I, but flourishing in the 1930s and 1940s, the airplane unalterably changed battle on land and at sea. While aircraft entered in a supporting role, bombing became "the supreme instrument of total war." As one expert noted, "Bombing strategy was deliberately aimed not at forces in the field but at the war-willingness and productive capacity of the society behind them." Bombing became a weapon to destroy economic targets and/or "domestic morale."[32] The shift from systematic strategic bombing of civilian targets in Germany and Japan with conventional bombs and incendiaries (as well as rocket attacks like the German V-1s and V-2s on England and Belgium) was but a step away from justifying the use of atomic weapons.[33] Among the complex issues surrounding the momentum of war, the technological means to exact total war led to Hiroshima and Nagasaki.

TARGETING JAPAN

The primary justification for the Manhattan Project was to develop the atomic bomb before the Germans. More than six months prior to the surrender of Germany on May 8, 1945, the Allies knew that the Third Reich had not developed a bomb. General Groves sent a special civilian-military unit to Italy in 1943, code name *Alsos* (Greek for "grove"), to learn the status of the German bomb project, but with few immediate results. After gathering data from documents captured in November 1944, it determined that the Germans had accomplished nothing beyond the experimental stage. In addition, Groves gave the *Alsos* mission the task of rounding up German nuclear physicists to keep them away from the Soviets. (The mission also seized several hundred tons of Belgian uranium in an area that later fell to the USSR.) In 1945 President Truman brazenly authorized "Project Paperclip" to capture and/or recruit more German scientists to help with the United States' postwar military programs. These scientists proved crucial to US projects during the Cold War, as did scientists apprehended by the Soviets become crucial to their projects.[34]

Attention quickly turned to the Pacific War, and soon, diplomatic wrangling at Potsdam. Well before the meeting of the Allied leaders, plans were underway to shift not only the war effort in general to the Asian Front, but the possible use of the American "ultimate weapon" against the final foe. That Japan had never been in the equation concerning the atomic bomb required some recalibration of war strategy. Several of the atomic scientists signed on to the Manhattan Project under the assumption that Germany posed the threat that necessitated the crash bomb project; Japan was a different story. Additionally Japan never mounted a serious effort to produce its own bomb, and thus the race against German scientists was not an issue in the case of the Pacific rival.[35] In reality, there is scant evidence suggesting that after developing the bomb the United States would not consider using it, let alone spare Japan from that fate. Fighting in the Pacific itself conditioned the military approach to end the war. The conflict in the Pacific was long and brutal for both sides. After the initial shock and success of the Japanese at Pearl Harbor, it took until spring 1942 for the war to shift in the United States' favor. The Battle of Guadalcanal, from August 1942 until February 1943, was the first offensive campaign of the Allied forces. The United States began to "close the ring" on the Japanese by late 1943 by taking the Gilbert Islands and then the Marshalls. A major battle on the island of Saipan suggested the intensity and the brutality of the fighting. The marine commander there proclaimed, "Saipan was war such as nobody had fought before: a campaign in which men crawled, clubbed, shot, burned, and bayoneted each other to death."[36] Unfortunately, more such battles were to follow, especially in Iwo Jima and Okinawa in early 1945. The former was the largest amphibious marine assault ever mounted, lasted five weeks, and clearly was the bloodiest. The latter was the largest amphibious assault of the war, lasted five weeks, and exacted heavy casualties on both sides (more than 100,000 Japanese killed, captured or committing suicide; more than 50,000 American troops killed or wounded; tens of thousands of non-battle casualties including thousand of Okinawans).[37]

Near the end of the Okinawa campaign in June 1945, President Truman instructed the Joint Chiefs of Staff to assess the key issues in achieving victory against the Japanese. On June 18, they presented their case to the president for an invasion of Kyushu (the southmost island of Japan) in November, followed by an invasion of Honshu (where Tokyo is located). Everyone agreed that Soviet participation was very desirable, but would not replace the land invasion.

On the central issue of casualty figures, however, the Joint Chiefs were not able to predict what they believed would be an accurate figure. (The probable human cost of the invasion is a sticking point among many scholars and commentators to this day.) On the surface of it, the meeting seemed to suggest that there was unanimity of opinion among high-level advisers about an invasion, agreement on casualties, and the use of army, navy, and air forces. But a wider range of opinions was expressed privately or unofficially.

The civilian and military leadership considered four options other than invasion, but did not scrutinize them thoroughly at the June 18 meeting. The first was to intensify the bombing and blockade of Japan to force surrender without an invasion. A second, which received little debate, was to wait for the Soviet Union to enter the war against Japan, hoping it might produce surrender. The third, receiving the most attention and support within the administration, was to modify the unconditional surrender policy to allow the Japanese to retain the emperor. (Truman's reaction to this possibility was unclear.) And fourth, was to use the atomic bomb without relying on a costly invasion. The use of the bomb, however, was not viewed as an alternative to the other options, but clearly discussed within the context of speeding up the end of the war.[38] Not lost on Washington was the investment in dollars and labor that went into building the bomb (approximately $2 billion, or much more than $20 billion in today's dollars) which suggested a high-priority commitment.[39] Along with wartime violence and the desire to end a long and destructive conflict, hatred of the enemy played a significant, if not easily defined role in the Pacific War. American racism directed at the Japanese (and other Asians) over many years, and a reciprocal hatred of the West by the Japanese, cannot be dismissed as part of the context for the intensity of a conflict called by some a "war without mercy."[40]

In May 1945, Secretary of War Henry Stimson set up the Interim Committee under the president's authority. The members were Stimson as chair, Vannevar Bush, James Conant, Karl Compton, and James Byrnes, plus the undersecretaries of state and the navy. A Scientific Panel included Robert Oppenheimer, Arthur Compton, Ernest Lawrence, and Enrico Fermi. The committee was heavily tipped toward government officials, which was typical of the new facts of life about the influence of scientists in decision making on the threshold of the Atomic Age. The charge to the Interim Committee was to evaluate the wartime use of the atomic bomb. It did not deliberate over whether the United States should use the bombs. The committee considered future weapons development and research into civilian nuclear power. It also devoted a great deal of time to considering postwar relations with the Soviet Union. Byrnes in particular brushed aside a suggestion from Oppenheimer and Compton that information about atomic energy should be shared with the Soviets. The most important immediate recommendation from the committee was agreement to use atomic bombs against Japan without warning. A discussion about staging a non-military demonstration of the weapon for the Japanese, as a deterrent to using it, received incidental attention. Some non-member atomic scientists pushed for a demonstration, but this was not reflected in the Scientific Panel's support for use of the bomb against Japan. Truman approved the committee's recommendations on July 5, several days before Potsdam.[41]

A clear line was drawn between the value of scientists in developing the bomb and their ability to influence its use. The satisfaction of turning theory into reality with the completion of the bomb came up against some sobering implications. For several scientists, the magnitude of the test explosion, the shift to Japan as a prime military target, and the uncertain future about the place of nuclear weapons in the postwar world kindled anxieties and produced reservations. Niels Bohr was among the first to propose sharing data about the development of the bomb with the Soviets to encourage an open exchange of information, something he hoped would return after the tight controls on science during the war years. Leo Szilard was as strident in his opposition to the use of the bomb on Japan as he had been in encouraging the United States to beat the Nazis to this awesome technology. He told Secretary Byrnes in May 1945 that using the bomb against Japan would likely push the USSR more quickly toward developing its own bomb. Byrnes was not convinced.[42]

Returning to the Met Lab in Chicago, Szilard learned that he was stirring up a controversy, one that Groves and others did not appreciate. In order to keep the dispute from boiling over, Arthur Compton organized several committees to study the potential consequences of using the bomb; the most important was the Committee on Social and Political Implications. James Franck, a German-born physicist and associate director of the chemistry section of the Met Lab, chaired that committee. Franck, who held serious qualms about the bomb, crafted a report with Szilard in June 1945. The report read in part, "We feel compelled to take a more active stand now because the success which we have achieved in the development of nuclear power is fraught with infinitely greater dangers than were all the inventions of the past."[43] Its conclusions did not favor the use of the bomb against Japan without a demonstration. A surprise attack, it argued, could ultimately lead to postwar proliferation of nuclear weapons and set off a potential arms race, creating "grave political and economic problems for the future of the country."[44] The group presented the Franck Report to Secretary Stimson, who referred it to the Scientific Panel of the Interim Committee. Its primary recommendation was either rejected or ignored. (There is no evidence that the president ever saw the report.) The dissent did not end, but even among the scientists there was no unanimous agreement as made clear by the formal position of the Scientific Panel. Oppenheimer himself seemed to waiver in his views. After *Trinity* physicist Robert Wilson, a former student of Lawrence and Oppie, had serious regrets about "a terrible thing that we made." He noticed that Oppenheimer himself was expressing regret during discussions over target selection, and repeatedly said "Those poor little people, those poor little people," referring to the Japanese. While Oppenheimer was important to the preparations to do so, he also hoped that the use of the bomb against Japan would not set off a postwar arms race.[45] Most of the talk about the bomb at the time was confined to political and military implications. But in some cases, as in a petition prepared by Szilard, "moral responsibility" surfaced. It was a difficult idea to reflect upon in the midst of total war.[46]

Targets in Japan were determined in the context of total war. Some argued that President Truman and his advisers rationalized the targets they chose as essentially "military," rather than facing the fact that the bulk of casualties would be civilians.[47] The Targeting Committee selected Kokura (an arsenal), Hiroshima (army depot and embarkation port), and Niigata (seaport) as primary targets because they were manufacturing centers regarded as part of the Japanese war machine. They chose to drop the bomb(s) at city centers rather than aim for industrial zones for maximum effect. Japan's population was more concentrated in cities than Germany's, and industry was closely connected to residential areas. Because of the building materials used in Japan, the cities were highly inflammable and thus "temptingly vulnerable to attack."[48] Secretary Stimson removed Kyoto, originally on the list of primary targets, because he believed the ancient capital and religious shrine should be spared to avoid future anti-American sentiment.[49] Stimson actually had been troubled by all of the choices because he was hoping for a quick surrender of Japan. He also did not want the United States to appear to be committing wartime atrocities or inflaming anti-American sentiment after the war for the sake of victory. But the decision to use the new bomb had already been settled, and Stimson fell in line with the target committee's view that only "virgin" cities (that is, Japanese cities which as yet had not suffered war damage) should be bombed with atomic weapons for psychological effect and to graphically demonstrate US power.[50]

Truman gave the order to drop the bombs as they became available; no sign-off by the president was needed for deploying individual bombs. While the Washington leadership was aware of the postwar political implications of the decision to use the bomb against Japan, especially in its future dealing with the Soviets, it seems clear that military considerations took precedent in Truman's decision. He wanted to end the war as quickly as possible with a weapon that fit within the context of the total war in which the United States was engaged.[51]

HIROSHIMA AND NAGASAKI

Hiroshima, Japan's seventh largest city, was the first mission. On early Sunday morning, August 6, 1945, a US Air Force specially modified *Flying Fortress* B-29 bomber, under the command of veteran pilot Colonel Paul W. Tibbetts, Jr., headed for their primary target. Two support planes with scientific and photographic equipment accompanied the bomber when it left its base at Tinian Island in the Marianas. The *Enola Gay*, named for Tibbett's mother, was carrying a U-235 bomb, *Little Boy*. At approximately 8:15 A.M., the bombardier released the weapon and it detonated 1,890 ft. above the city. The explosion immediately produced temperatures exceeding 1.8 million degrees Fahrenheit in the air and 5,400 degrees on the ground. Five square miles of the city were virtually leveled. Within a radius of 1.5 miles all wooden houses collapsed. About 62,000 of the 90,000 buildings in the city were destroyed. Only three of the city's 55 hospitals and first-aid centers remained. The intensity of the fire grew stronger less than an hour after the explosion, and then heavy rain (black rain) saturated everything with soot, dust, debris, and radioactive material. By December 1945, the death toll reached about 140,000 out of a population of 350,000. (This included American POWs and 10,000 Korean forced laborers.) Many people died immediately; many more died soon after from radiation burns and radiation sickness. Civilians and soldiers at the epicenter of the blast were vaporized.

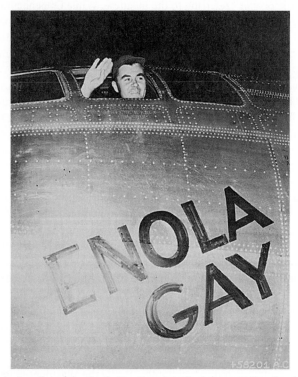

Colonel Paul W. Tibbets, Jr., pilot of the *Enola Gay*, the plane that dropped the atomic bomb on Hiroshima, waves from his cockpit before the takeoff on August 6, 1945. *Source:* National Archives

The LB (*Little Boy*) unit on trailer cradle in pit. Note the bomb bay door in upper right-hand corner. The uranium bomb was dropped on Hiroshima, Japan, on August 6, 1945. *Source:* National Archives

Most of the deaths took place in a four square mile area around the Aioi Bridge where the city's major residential, commercial, and military quarters were located.[52] Akira Onogi was 16 years old on the fateful day of the bombing: "When the blast came, my friend and I were blown into another room. I was unconscious for a while, and when I came to, I found myself in the dark. Thinking my house was directly hit by a bomb, I removed red soil and roof tiles covering me by hand and for the first time I saw the sky. I managed to go out to open space and I looked around wondering what my family were (sic) doing. I found that all the houses around there had collapsed for as far as I could see." Isao Kita was the chief weatherman for the Hiroshima District Weather Bureau, which was located about 2 miles from the epicenter. He kept on working even after being exposed to radiation. When the black rain started falling, he observed,

> It was a black and sticky rain. It stuck [to] everything. When it fell on trees and leaves, it stayed and turned everything black. When it fell on people's clothing, the clothing turned black. It also stuck on people's hands and feet. And it couldn't be washed off… I couldn't see what was taking place inside the burning area. But I was able to see the extent of the area which was on fire…The atomic bomb does not discriminate. Of course, those who were fighting may have to suffer. But the atomic bomb kills everyone from little babies to old people. And it's not an easy death. It's a very cruel and very painful way to die.[53]

The news of the bombing reached Groves late Sunday evening, and he immediately reported it to Secretary Stimson and Chief of Staff Marshall. President Truman, who was sailing

A general panoramic view of Hiroshima after the bombing which shows the devastation at about 0.4 miles. Official US Army photo. *Source:* Library of Congress, Prints & Photographs Division Washington, DC, [LC-USZ62-134192]

home from Potsdam aboard the cruiser *USS Augusta* heard the news and told a group of sailors, "This is the greatest thing in history."[54] His expectations for bringing the war to a quick end seemed to become real. For their part, the Japanese government did not hear of the bombing until midday because of the loss of local communication. But the chance of an immediate surrender was not likely. Despite the Hiroshima blast and American warnings that the Japanese must capitulate or suffer more attacks, the militarist faction continued to resist the inevitable. The conditions they proposed were totally unacceptable to the United States, but for the moment at least they had the support of the emperor.[55]

A second bombing scheduled for August 11 left the Japanese little time to absorb the horrors of Hiroshima. Due to a foreboding weather forecast, officials moved up the mission to August 9. *Bock's Car,* a B-29 named for its captain Frederick Bock, was carrying a plutonium bomb, *Fat Man,* headed for its primary target of Kokura. Because of heavy overcast, the bomber low on fuel was diverted to Nagasaki. The city had 270,000 inhabitants living in a river valley surrounded by mountains in southern Japan. The blast destroyed about four square miles in the urban area, home to Mitsubishi Shipyards, Electrical Equipment Works, and an arms plant. By the end of 1945, Nagasaki suffered almost 74,000 fatalities and another 75,000 injured.[56] While *Fat Man* delivered a bigger yield than *Little Boy*, the topography, the distribution of the buildings, and an off-target release of the bomb limited the physical destruction compared to Hiroshima.[57] But the human toll was just as grisly and overwhelming. Michie Hattori Bernstein was a fifteen-year-old schoolgirl at that time. When the air raid warning sounded in Nagasaki, she ran to a cave dug into the side of a hill for use as a shelter for students:

> When the bomb exploded, it caught me standing in the entrance to the shelter, motioning for the pokey girls to come in. First came the light—the brightest light I have ever seen. It was an overcast day, and in an instant every object lost all color and blanched a brilliant white. My eyes couldn't cope, and for a little while I went blind.

The FM (*Fat Man*) unit being placed on a trailer cradle in front of Assembly Building #2. The plutonium bomb was dropped on Nagasaki, Japan, on August 11, 1945. *Source:* National Archives

A searing hot flash accompanied the light that blasted me. For a second I dimly saw it burn the girls standing in front of the cave. They appeared as bowling pins, falling in all directions, screaming and slapping at their burning school uniforms…

Immediately, a powerful wind struck me. It propelled me farther into the cave; then in an instant it threw me out the front entrance…What a terrible feeling! I could see nothing. My hands and face singed, intense pain gripped my body. I tried to walk a little and stumbled over a fallen tree. I lay there, not knowing for sure where I was or whether something else might happen to me.

When my senses, including my sight, began returning, I heard crying from the girls in front of the shelter. All, except one, were now standing and blowing on their skin. Looking at the one lying down, I saw her leg twisted at a crazy angle…The face and hands of the other girls quickly turned bright red…[58]

Two atomic bombs dropped within a matter of days were shocking events. On his return from Potsdam, Truman made a radio address in which he stated:

Having found the bomb we have used it. We have used it against those who attacked us without warning at Pearl Harbor, against those who have starved and beaten and executed American prisoners of war, against those who have abandoned all pretense of obeying international laws of warfare. We have used it in order to shorten the agony of war, in order to save the lives of thousands and thousands of young Americans.[59]

His stern response was a reflection not only of his relief in bringing the war to an end, but a commitment to the "war without mercy" in a new era of total war. The president, upon the recommendation of Secretary Stimson, gave the order to suspend further bombing on August 10.[60]

Another blow to Japan occurred before the Nagasaki attack, immediately after midnight on August 9. Desirous of protecting its Asian interests as soon as possible, Soviet forces as promised entered the War in the Pacific, invading Manchuria and quickly subduing the Japanese army there. Despite these immeasurable pressures on the Japanese government, officials crafted no surrender message as yet. The major obstacles were persistent demands by the Japanese to retain their emperor and the resolve of the Allies for unconditional surrender. (On several occasions, elements in Japan had sought Soviet aid in bringing the war to an end.)[61]

While full disclosure of the grim details were not made available for years, American troops on the battlefield received the news of the Hiroshima and then Nagasaki bombing with relief and elation. At home, the reaction was the same. The *New York Times* headline for August 7 read: FIRST ATOMIC BOMB DROPPED ON JAPAN; MISSILE IS EQUAL TO 20,000 TONS OF TNT; TRUMAN WARNS FOE OF A "RAIN OF RUIN."[62] The first polls showed about 85 percent of Americans approved of the bombings. Of nearly 600 editorials in American newspapers less than 2 percent opposed the use of the atomic bombs against Japan. Of those few dissenters, most were ill at ease about the potential future use of the bomb rather than the morality of the wartime actions. Similar favorable reports flooded British newspapers, but not necessarily so throughout the world. Stalin had serious political if not military concerns about the bomb. The attack on Hiroshima prompted him to protect Soviet interests in Asia and to move troops quickly into Manchuria. He also (and quite significantly) established a committee to make the USSR's atomic projects a top priority, and purportedly stated, "The balance has been destroyed."[63]

Finally, on August 14, Japan surrendered on Emperor Hirohito's command. Debate continues to rage over the exact impact of the Soviet entry into the Pacific War. Some argue that it was more important than the Hiroshima and Nagasaki attacks in eventually bringing the war to an end; others are chary to dismiss the central role of the atomic bombs. Evidence is not persuasive for simply one or the other argument. It appears most plausible that Hiroshima, Nagasaki, and Manchuria collectively sealed the decision.[64] Nagasaki undermined the case of the Japanese military who had held firm for some time on three issues: preserving of the emperor, balking at occupation after the war, and demanding that "war criminals" be tried in Japan. By that point, there was no doubt that an atomic explosion had destroyed the two cities and that the United States may have even more such deadly weapons. The Soviets were an additional threat. Even given all of these circumstances it still took several days for Hirohito to break the deadlock. Japan accepted the Potsdam Declaration as the best option for avoiding an unconditional surrender that toppled the position of the emperor (although the Allies already had blunted some of the hardest edges of unconditional surrender). The American response did not offer a guarantee about the future role of the emperor, but stated that his authority would be subject to the Supreme Commander of the Allied powers who would occupy Japan after the war.[65]

On September 2, the War in the Pacific officially ended with the signing of documents aboard the *USS Missouri* moored in Tokyo Bay. The war concluded without a costly land invasion of Japan, although the debate continued for years as to whether such an invasion could have avoided the use of the atomic bomb, what the scale of casualties of such an invasion might have been, and how the war might have ended differently. Such speculations ignore the reality of events as they played out. The bombs were dropped with little reservation in the context of total war. As military historian Russell Weigley concluded, "Yet if employing the first atomic bombs could seem at the time a mere extension of a strategy in use, it soon became evident that by carrying a strategy of annihilation to the literalness of absurdity, the atomic bomb also represented a strategic revolution."[66]

THE BOMB IN OUR LIVES: POSTWAR REACTIONS

The immediate response to the use of the bomb against Japan was favorably received in the United States and Great Britain, if not in the Soviet Union. Very soon a more complex array of responses, emotions, and attitudes emerged. The wide variety of reactions must be understood against a backdrop of readjusting to life in the postwar world and were also the result of different vantage points. Obviously, if one lived in America as opposed to Japan "vantage point" meant something quite different. In addition, contemporaries knew as much about the bomb and its impact as governments were willing to tell them. There was relief that the war was coming to an end, but little celebration took place in the United States over the news and there was some real uneasiness.[67] Gruesome atomic humor, an "atomic cocktail" at local bars, and atomic-themed songs were seemingly acceptable at the time, when hatred of the Japanese and war weariness had yet to recede. But as historian Paul Boyer aptly noted, "Despite the outpouring of post-Hiroshima ephemera, it would be wrong to conclude that Americans took the bomb casually or that its impact quickly faded. Just below the surface, powerful currents of anxiety and apprehension surged through the culture."[68] Over the years, anxiety (even primal fear) and indifference ebbed and flowed throughout the United States and elsewhere. Several polls taken late in 1945 and into 1946 suggested real ambiguities in how Americans felt about the possibility of future nuclear war and their own fears about the existence of the bomb.[69]

Hiroshima and Nagasaki sparked a movement against the bomb, with Japan as a focal point, and revitalized interest in pacifism around the world. The long years of World War II had devastated the peace movement. In Germany and in other parts of the fascist world, pacifists and anti-war advocates were outlawed, imprisoned, and executed.[70] During the postwar years three major pacifist organizations gained renewed momentum after being decimated during the war: the secular War Resisters' International, the religious International Fellowship of Reconciliation, and the Women's International League for Peace and Freedom.[71] Attention quickly turned to Japan. Amazingly in some instances the Japanese people were more inclined to blame their military leaders for the destruction of their homeland than to criticize the United States' use of the atomic bomb against them. Or they simply invoked hatred of warfare itself. The bombings caused horrible physical and psychological damage in Japan, but the US occupation forces and their censorship policy seriously constrained public discussion after the war. Criticism of the bomb did surface through the *hibakusha,* those people affected by the attacks on Hiroshima and Nagasaki. On the first-year anniversary of the attack, the Hiroshima branch of Japanese Association of Religious Organizations sponsored a Memorial Day. By 1948, the Hiroshima Peace Association conducted demonstrations regularly in the two ravaged cities. In Japanese political circles, resistance to rearming the country and favorable disposition to maintaining neutrality overtook American pressure to the contrary.[72] As for the victims, both American and Japanese officials long ignored them. Physicist Henry DeWolf Smyth wrote the Smyth Report (August 12, 1945), which served as the official account and justification of the Manhattan Project, providing theory and technical details of the building of the bombs, but did not shed light on the impact of the attacks. For their part, the Japanese government did not extend special assistance to the victims until 1952 when the occupation ended.[73]

Official US reports of the attacks tended to focus on physical damage to the cities and less on humans, clearly as a way of underplaying the horror of the bombings. Early in the occupation, a Japanese film crew directed by Akira Iwasaki went to Hiroshima and Nagasaki to record the human impacts of the blasts. About halfway through the project American military police arrested the cameramen in Nagasaki. Occupation officials banned all filming of the destroyed cities. The US Strategic Bombing Survey, however, authorized and directed an American film project under its auspices. Lt. Daniel McGovern, in charge of the film unit, urged that the Japanese footage be retained because of its matchless value. He then hired Iwasaki himself to shoot thousands more additional feet of film. The result was an almost three-hour documentary, *The Effects of the Atomic Bombs on Hiroshima and Nagasaki* (1945), the earliest film record of the bombings. Classified "secret" and shipped off to the United States in May 1946, it "disappeared" and some fragments did not resurface until the 1960s. Columbia University Professor Erik Barnouw and his staff found some of the obscurely archived footage in the National Archives in Washington, D.C. They put together a powerful sixteen-minute film entitled *Hiroshima-Nagasaki: August 1945*. Sparse in dialog with some haunting music, the film is a powerful visual rendering of the attacks, but not one that contemporaries in the 1940s or 1950s ever saw. As one review noted, "The film provides a sense of being an eyewitness to the bomb's aftermath, literally walking through the rubble and hospitals jammed with dying people."[74] Since the 1960s, *The Atom Strikes!* (1945), a documentary commissioned by the US Signal Corps Pictorial Division, is often paired with Barnouw's film in showings. In contrast with *Hiroshima-Nagasaki: August 1945*, it focused almost exclusively on physical damage to the cities as if no people ever died there. *The Atom Strikes!* clearly sanitized the events, playing down the human tragedy so well-presented in the Barnouw film.[75]

American scientists ignored or discounted the first reactions to stories of radiation injuries and death in Hiroshima and Nagasaki. General Groves believed the stories to be a hoax or the product of Japanese propaganda. He did, however, send a small survey team to Japan in September/October 1945, but it became clear that a more thorough study needed to be undertaken. Subsequently the Atomic Energy Commission organized the civilian Atomic Bomb Casualty Commission under the operation of the National Academy of Science and the National Research Council. The commission discovered elevated forms of several kinds of cancer, but the group was unpopular locally because physicians were only allowed to conduct examinations not treat patients. Policy changed in the 1950s, and commission doctors could administer to the needs of the suffering people.[76]

Over time, tracking the exact effects of radiation on bomb survivors proved very difficult. Health physicians dealt with imprecise statistical evidence and the fact that nuclear fission created many radioactive isotopes not found in nature. The immediate public reaction to the bombings focused on the blasts rather than on the effects of radiation, which were much less understood or publicized. No new studies reexamined original claims and speculations until the 1980s and 1990s, but even these sometimes proved contradictory on long-term impacts. The study of the survivors of Hiroshima and Nagasaki was the most important source of data on the dangers of radiation to large population groups. A 1995 report stated that the inhabitants of the two cities were exposed to high doses of radiation immediately after the bombings, but that the "average survivor" received an "intermediate dose," and a fewer number of people a "high dose." The first indication of radiation effects were increased rates of leukemia that showed up in the early 1950s, although the absolute numbers were not large. Newer studies indicated that there were no drastic increases in birth defects or infant mortality, or no "excess deaths" in cancer above established norms.[77]

While the overall implications of illness, deaths, and genetic mutations due to the bombings did not result in a worst case, concern and fear over the potential risk of radioactivity from atomic explosions escalated in the coming years. No telling how world opinion would have reacted to a frank disclosure of radioactivity on humans following the blasts at Hiroshima and Nagasaki. Possibly for that reason and no other, such information was kept under wraps. John Hersey's *Hiroshima*, however, gave Americans and others a horrifying glimpse of the personal tragedies of the atomic attacks on Japan, and the inhumanity of using such a weapon. First published in the August 31, 1946 issue of the *New Yorker*, it presented accounts of six people who experienced those deadly hours in Hiroshima in straight-forward and detailed prose. As Hersey writes,

> A hundred thousand people were killed by the atomic bomb, and these six were among the survivors. They still wonder why they lived when so many others died. Each of them counts many small items of chance or volition—a step taken in time, a decision to go indoors, catching one streetcar instead of the next—that spared him. And now each knows that in the act of survival he lived a dozen lives and saw more death than he ever thought he would see. At the time, none of them knew anything.[78]

The issue in which Hersey's remarkable narrative was published sold out quickly. Newspapers wanted reprint rights. The whole article was read word-for-word over the ABC radio network. Most everyone who read *Hiroshima* was appalled, but it did not necessarily change the minds of those who viewed the weapon as a powerful American advantage. It did put a human face on what had been an abstraction, a distant event.[79] Yet *Hiroshima* was not only a story of individual experiences; it raised questions about the bombings as justified or rationalized by war.[80] One of the great ironies of the devastation in Japan was, as one writer put it, the "biological anomaly" of plants and flowers emerging out of the devastation and destruction at Hiroshima, a situation replicated at the site of the great nuclear reactor disaster at Chernobyl in Ukraine in 1986.[81]

Underway before the end of the war, an incipient scientist movement added to the rising tide of dissent over atomic weapons. The success of the bomb made heroes of the Manhattan Project scientists, although many did not feel like heroes. The efforts of Bohr, Szilard, and the Franck Group raised many serious questions about the use of the bomb, postwar threats, and nuclear proliferation. On November 1, 1945, several groups organized the Federation of Atomic Scientists (renamed the Federation of American Scientists, FAS). Within a few months there were seventeen local chapters with almost 3,000 members focusing on civilian control of atomic energy, the dangers of radioactivity, and the issue of nuclear testing. FAS organized the National Committee on Atomic Information to distribute materials on atomic energy, and two members started what became the *Bulletin of the Atomic Scientists* with a circulation of about 20,000 by mid-1947.[82] Motives for participation in FAS and other such groups varied. Some had a serious change of heart once the idea of the bomb became a reality; others looked into the future and feared the worst; still more were hoping for a carefully thought out decision-making process on atomic energy.[83]

Leading atomic scientists who continued to work intimately within government circles took a murky stance on nuclear weapons in the postwar years. Oppenheimer, who remained in government service, had become a celebrity and was comfortable with the almost universal approbation. In several public appearances, he dwelled on his reservations about "a most terrible weapon," but the more time he spent in Washington the more he became an insider. Despite the fact that he told a disgruntled Truman in private, "Mr. President, I feel I have blood on my hands," he still worked within the system which often meant compromise and in some cases acquiescence.[84] The renowned scientist was not the only one who had mixed emotions about the bomb. Truman, Stimson, and other political leaders, while supporting the Hiroshima and Nagasaki decisions, were not oblivious to the larger issues. Fleet Admiral William Leahy saw the "cruelty" in the bomb. Supreme Allied Commander General Dwight D. Eisenhower after the war recalled having "grave misgivings" about using the bomb on Japan, but apparently never conveyed them to the president. At the least, the decision to drop the atomic bombs was a sobering experience for those who had to defend it.[85]

Beyond the reorganization of pacifist groups and the emergence of the atomic scientist organizations, a short-lived, but dramatic movement for world government arose after the war. Among many books, pamphlets, and other writings, Wendell Wilkie's *One World* (1943) caught public attention. Willkie was a corporate lawyer who ran for president in 1940 on the Republican ticket against FDR. Although not explicitly calling for world government, *One World* rejected "narrow nationalism" and "international imperialism." Norman Cousins, editor of the *Saturday Review of Literature* and deeply active in the peace movement, believed that the "need for world government was clear long before August 6, 1945."[86] The bomb plainly was a catalyst for world government advocacy in Europe as well. Welsh-born philosopher and mathematician Bertrand Russell became one of the best known anti-war activists internationally, and an influential early backer of world government in Britain. But not all advocates were cut from the same political cloth. The idea attracted religious groups, scientists, those of various political persuasions, journalists, and others, and had its serious detractors.[87] Yet the growing debate over internationalizing the bomb, the distraction of other issues, and the rise of the Cold War drowned out the early hopes and idealism imbedded in the world government movement. In reality it was never a viable alternative to nationalism in the postwar years, but clearly sometimes a visceral and at other times a cerebral reaction to an uncertain future.

ESTABLISHING INSTITUTIONAL CONTROL: ORIGINS OF THE AEC AND JCAE

It is astonishing, although not surprising, how quickly the bomb inspired a policy of absolute control of atomic energy. Soon it also became the central instrument of deterrence and aggression in the Cold War. Maintaining such a monopoly ("nuclear nationalism") was an American postwar objective with two dimensions. Domestically the fundamental question was how rather than if the federal government, and only the federal government, would manage this new military and civilian resource. On the world scene, the basic issue was how the United States could maintain its atomic monopoly without undermining good relations with its friends or unduly aggravating its foes.[88]

The military significance of the bomb afforded little discussion over the necessity for government control. Those who believed that government should not be in the weapons or energy businesses provided few answers on how the government could avoid it.[89] Prior to the war's end, military planners developed postwar strategies for a peacetime agency to manage atomic energy and to protect atomic secrets. Representative Andrew Jackson May (Dem., KY) and Senator Edwin C. Johnson (Dem., CO), the respective heads of the congressional military affairs committees, presented a bill in 1945 that would create a board to oversee the atomic energy program. Justified on the grounds of national defense, the new law intended to give the military control in future development of atomic weapons. The May-Johnson bill was ill-conceived, and confused the need for security with the need for haste. When news of the bill reached the scientists at the Chicago and Los Alamos laboratories, there was a storm of protest. Several scientists, with fresh memories of General Groves' insistence on secrecy and security, objected to the central role of the military. Interestingly, Oppenheimer, Fermi, and Lawrence supported May-Johnson. The protesting scientists quickly called for public hearings and lobbied hard to defeat the bill. Their efforts, the continuing debate in the Senate, and the withdrawal of President Truman's support killed the bill.[90]

An alternative piece of legislation came from freshman Democratic senator from Connecticut, Brien McMahon. He was chair of the Senate Special Committee on Atomic Energy and a political rival of General Groves. The McMahon bill deemphasized the military role in regulation and stressed the potential for civilian applications of nuclear power. On August 1, 1946, after almost one year of congressional debate, Truman signed the Atomic Energy Act. It gave the government a monopoly over atomic energy under civilian control, and prohibited the exchange of information on the use of atomic energy with other nations. The Act established two new bodies: the Atomic Energy Commission (AEC) and the Joint Committee on Atomic Energy (JCAE). The AEC was unique, holding a virtual monopoly over atomic energy development, including exclusive control and ownership of fissionable materials. Attached to this body was an elaborate administrative system including the General Advisory Committee (chaired by Oppenheimer), which provided guidance in scientific and technical matters, and a Military Liaison Committee, which reviewed AEC decisions. The latter wielded the real power on the commission. The British, Canadians, and French set up similar commissions in the late 1940s as well.[91]

After a two-month search, Truman chose four Republicans and one Independent to serve on the AEC. David E. Lilienthal became chair. His appointment, in particular, set off a political flare-up. The former chair of the Tennessee Valley Authority (TVA) and coauthor of a controversial plan for internationalizing atomic energy had strong credentials as an administrator. Yet prominent individuals attract powerful enemies, and Lilienthal was subjected to a scathing and relentless attack by Senator Kenneth McKellar (Dem., TN) during the confirmation hearings. Lilienthal had apparently deflected McKellar's efforts to exercise his patronage power in TVA several years before, and the peevish, old senator claimed that the nominee had been part of a Communist cell within the authority. Prolonging the confirmation hearings only raised bad feelings, and delayed the work of AEC projects.[92]

The significance of the AEC was not simply in its oversight powers, but in its control of an extensive atomic-era infrastructure. In 1947 it assumed responsibility for US army installations and personnel linked to atomic energy, including 37 installations located in 19 states and Canada, almost 2,000 army personnel, 3,950 government workers, about 38,000 contractor employees, and an array of additional properties and facilities. In that year the AEC's budget was $300 million, reaching more than $4 billion by 1953. AEC's inventory included the Argonne National Laboratory (1946) near Chicago, which conducted research on radiation and radioactive materials. Soon thereafter a number of universities lobbied to have research facilities situated near them, including Harvard, MIT, and Columbia. In 1946 a consortium of nine universities formed Associated Universities, Inc., for that purpose. The institutional relationship between universities, the federal government, and private industry with respect to atomic energy and other practical scientific and technical research grew rapidly in the postwar years. In 1947 Brookhaven National Laboratory on Long Island, New York, became the third in a series of national laboratories in the United States to conduct advanced research in atomic energy. The Sandia National Laboratories (1945) near Albuquerque, New Mexico, was the largest nuclear weapons facility in the country. The Nevada Test Site, established in 1950, was the only area in the United States used for testing atomic weapons. In 1949 the federal government acquired a reactor testing site in Arco, Idaho, renamed the National Reactor Testing Station.[93]

The Atomic Energy Act did not leave regulatory authority solely in the hands of AEC. Congress made sure that it was not excluded from assuming a pivotal role in national defense and national security policy vis-a-vis atomic energy. It was always alert to protect its interests within the system of checks and balances. In the sense that AEC was an executive creation, the Joint Committee on Atomic Energy was a legislative innovation. The eighteen-member body had substantial jurisdiction in the area of regulation, and it had power as a regular standing committee in Congress. The JCAE's primary functions were maintaining legislative power on all atomic energy bills; acting as a watchdog over matters of secrecy and security; and maintaining policy and review functions on related programs. Through its acquisition of data on atomic energy, the JCAE became an indispensable source of information with great influence and authority.[94]

THE DANCE: TALK OF INTERNATIONALIZING THE BOMB

The Cold War was, among other things, a power struggle between rising superpowers emerging at the end of World War II. The development, control, and potential use of the atomic bomb and its progeny shaped the rivalry. As experts on international relations Campbell Craig and Sergey Radchenko argued, this ultimate weapon "was an object of statecraft, a grim means of pursuing national ends."[95] The Cold War came close at times to turning hot but did not; the old maxim that the atomic bomb kept the Cold War cold is fairly accurate. In hindsight it seems obvious that the global rivals never quite understood the other's postwar intentions. To many American leaders (starting with Truman and the hardliners among his advisers), the Soviets clearly had expansion in mind despite being deeply wounded in World War II. To Soviet leaders (particularly Stalin), the United States had postwar ambitions of its own. These perceptions led to the mutual assumption that the differences of the two countries were irreconcilable. The existence of the bomb made compromise precarious.[96]

In the immediate postwar world, the sides were not evenly matched to pursue a shared strategy of deterrence. The United States had the bomb, not the Soviets. The game plan, therefore, seemed relatively obvious. The Americans wanted to monopolize atomic weapons; the Soviets wanted them destroyed or appeared to support internationalizing the bomb until they had their own. International control of atomic weapons was preconditioned by US wartime experiences and shaped by the onset of the Cold War. Those weapons also helped define power relationships between the two countries. Americans and their leaders wanted World War II brought to a quick end, but neither FDR nor Truman was naïve enough to assume that wartime alliances would remain exactly the same after the defeat of Germany and Japan. In Truman's case, the world he and other American leaders envisioned after the war did not include accommodation to Soviet interests. The global stature of the United States depended on political stability, acceptance of America's own brand of exceptionalism, and economic growth unimpeded by a major challenger.

Most everyone recognized that the atomic bomb gave the United States a military advantage after the war (save possibly Stalin and some others who were not convinced of that fact, having more faith in troops on the ground). Both FDR and Truman realized that the bomb also had value as a political and diplomatic tool. Stalin's unvarnished perspective was that the Americans utilized the bomb against the Japanese more to forestall Soviet advances in Asia than to save American lives. While Truman's top priority was to end the war quickly, he and his advisers well appreciated the bomb as a diplomatic lever against the Soviets.[97] The diplomacy of the bomb began to shape FDR's overall atomic energy policy as early as 1943. The president, especially at Churchill's urging, saw an Anglo-American monopoly of the bomb as an essential counterweight to the postwar ambitions of other countries, particularly the Soviet Union. Truman was more suspicious of the Soviets than FDR, and much more adamant that the possession of the new weapon could make Stalin more cooperative.[98]

A true desire for internationalization of the bomb did not exist at the war's end. Truman and several key advisers, however, believed that the Soviets would make significant concessions in exchange for neutralization of atomic weapons. Truman's "atomic diplomacy" also was built on the false assumption that the Soviets were many years away from developing their own device. The president realized that some type of international control was needed, if for no other reason than to avoid the United States being labeled as an atomic bully. In November 1945, Truman,

British Prime Minister Clement Attlee, and Canadian Prime Minister Mackenzie King issued the Washington Declaration calling for international control over atomic energy, that is, a willingness to share information for peacetime uses. They also recommended that the newly established United Nations create a commission on atomic energy, which became the United Nations Atomic Energy Commission (UNAEC) in 1946. It included members of the Security Council and Canada. This body was authorized to examine and make proposals on the exchange of scientific information, peaceful uses of atomic energy, and elimination of atomic weapons. A month before the birth of UNAEC the Council of Foreign Ministers called for a commission to investigate international control.[99]

Soon thereafter Secretary of State Byrnes assembled a committee to examine similar questions, which he would use to make recommendations to UNAEC. The Acheson Board, chaired by the Under Secretary of State Dean Acheson, represented political insiders, business interests, and the leaders of the Manhattan Project. For technical support Acheson appointed a Board of Consultants including AEC director Lilienthal as chair and Oppenheimer. Not surprisingly the group rejected the idea of outlawing atomic weapons or curtailing commercial development. It also rejected a system of international inspection as a singular solution to control. Oppenheimer then proposed an international agency, the Atomic Development Authority (ADA), to focus upon developing peaceful uses of atomic energy. While the plan received applause by many of the dissenting scientists, such a strategy did not hold out much benefit for the Soviets who believed it would impede their own efforts.[100]

Enter Bernard M. Baruch. In March 1946 President Truman appointed the businessman as ambassador to the UNAEC, and one writer noted that Baruch was an anticommunist with a "twenty-kiloton ego."[101] Baruch had the task of presenting the Acheson-Lilienthal Plan, but Truman gave him assurances that he would be able to shape the final proposal. Committee members were appalled by his appointment, especially Lilienthal and Oppenheimer. Baruch had no technical qualifications, was hardly young and hearty at 75, and was chary to giving in to the Soviets on almost anything. It was that last point that made him attractive to the president. The "Baruch Plan" as presented to UNAEC on June 12, 1946, looked quite different from what the Acheson Board had crafted. Baruch's proposal denied ADA control of mining and fissionable material, presented a rather threatening set of punishments for violators, and stipulated that negotiations include all weapons not just atomic bombs. The revisions split the Acheson Board, with Groves supporting a punishment clause and Lilienthal and Oppenheimer opposed to most if not all of the changes. The FAS supported the plan, either because it was naïve about its intentions, or the FAS was overly hopeful about any chance for international control. Many atomic scientists ultimately withdrew from the political battlefield, coming to the conclusion that their idealism would not produce immediate results. Truman appeared willing to moderate some of Baruch's more strident provisions, but backed off when he threatened to resign.[102]

Not surprisingly the Soviets balked particularly because the United States would maintain its nuclear monopoly during the transition period to internationalization, and the plan generated unprecedented powers of inspection. Atomic weapons testing off the Bikini Atoll in the Marshall Islands at the time of the negotiations starkly brought into question the Americans' willingness to cooperate. In response, the Soviets presented a counterproposal that simply called for outlawing nuclear weapons. The sincerity of both sides at this point was questionable.[103]

The American Nuclear Testing Program (beyond *Trinity*) began in earnest in 1946 and lasted, with brief breaks, until the early 1990s. In 1946 the United States launched the *Crossroads* series at the Bikini Atoll.[104] The United States had delayed testing since Nagasaki largely for political reasons. Secretary Byrnes and President Truman understood the "unsettled conditions" of international relations after the Japanese surrender, plus the president was concerned about the high financial cost of a testing program. The timing gave the Soviets a perfect opportunity to question the seriousness of the interest of the United States in disarmament.[105] The decision to test atomic weapons also must be understood within the context of inter-service rivalry between the army and navy. The army's view was that the scale of the blast at *Trinity* clearly indicated the ability to use the bomb against large industrial targets, and questioned a detonation on or under water. The navy's goal at the Bikini tests was to demonstrate that the A-bomb was not effective against a fleet, thus showing the sustaining value of the service. The navy got its wish. On July 1, a bomber dropped *Able* into a harbor against a mothballed fleet as part of a full navy exercise. The detonation was off target with inconclusive results about the blast, but officials took little immediate account of the damage caused by radiation to the sailors on the ships or to the displaced natives in the atoll.[106] Despite this major oversight, both the army and navy urged acceleration of the next test, *Baker*, which took place in late July at a greater depth than *Able* and with considerably better blast results.[107]

In 1948, in a period when the United States was producing a greater number of bombs than earlier, the *Sandstone* series began. The tests utilized newly designed bombs that scientists believed would be more efficient than those dropped on Hiroshima and Nagasaki. Personnel involved in the operation topped 10,000. Three detonations took place in April and May on the islands of Enjebi, Aomon, and Runit in the Eniwetok Atoll in the Marshall Islands, 3,000 miles west of Hawaii. Scientists warned that the site was unsafe for testing for meteorological reasons, but the series at Eniwetok went forward and ended the first phase of tests in the South Pacific. The *Sandstone* series took place at a time when concern about its impact on Soviet-American relations was low, and the determination of the United States to beef up its nuclear arsenal was quite high. For a decade 43 detonations in all took place at the atoll, mostly in the atmosphere. Not until 1977 did a cleanup of radioactive material commence. And very little was written or told about the inhabitants of the South Pacific Islands who were displaced or suffered radiation poisoning from the chronic testing.[108]

THE SOVIET BOMB: THE END OF PRETENSE

On September 29, 1949, the White House announced: "We have evidence that within recent weeks an explosion occurred in the USSR." The Soviets built a bomb and detonated it on August 29 on the steppes of Kazakhstan. This occurred two years before American intelligence sources had predicted. The United States' monopoly was over, the Cold War deepened, and the opportunity for international control slipped away. Espionage (discussed more fully in Chapter 4) along with advances in Soviet science and effective planning led to the dramatic turn of events resulting in a successful test. The American leadership was badly caught off guard largely because they had such low regard for the possibility of such an achievement in Stalinist USSR.[109]

The Soviets' path to the bomb had a decidedly different quality than the American counterpart. By the late 1930s, party and government control of science weakened Soviet physics, since government pressured scientists to contribute to practical ends in industry. But even on that score opportunities were very few given the economic system in place. What kept nuclear physics alive in the 1930s were the scientists themselves. From 1923 to 1950, Abram Fedorovich Ioffe was the director of the Leningrad Physico-Technical Institute, which was the center of Soviet nuclear research in the 1930s. Others pushed for research into radioactivity and the search for uranium deposits. As a result, Soviet nuclear physics finally reached a high level, but this was little known to many outside the USSR's closed society. In addition, American and British scientists were more keenly interested in what was transpiring in Germany than in the USSR. As world war loomed, the international exchange of scientific information was everywhere drying up beyond national borders.[110]

Stalin's decision to support nuclear research more aggressively emerged in the wake of the USSR's most perilous moments. On June 22, 1941, Nazi Germany launched an attack on its former ally. The Soviets were now fighting for their lives. The day after the beginning of the German assault, several scientists associated with the Academy of Sciences expressed interest in turning their expertise to the war effort. Georgii Flerov, a young physicist, urged that nuclear research be continued, but it was not a high priority. (As the British atomic research began to gear up, however, the Soviets were receiving intelligence information about elements of the MAUD Report.) War conditions were too desperate from late 1941 through the summer of 1942 to influence a change in Soviet policy on atomic power. Early in 1943, coinciding roughly with the start of the Soviet counteroffensive against the German advance on Stalingrad, nuclear research resumed in the USSR. Physicist Igor Kurchatov was appointed as the scientific director of the nuclear project, a position he held until 1960. The Soviets regarded him as the father of the Soviet atom bomb.[111]

In 1943 Kurchatov began the task of assembling scientists and engineers to design a weapon. The timing of the activity suggests that the Soviets were moving on this project well before *Trinity* and Potsdam, not in response to them. The larger body of information that developed over the years on nuclear fission informed the work of the group. However, it was substantially transformed by intelligence data coming from the United States in early 1945, and coincided with the arrival of Klaus Fuchs at Los Alamos in August 1944, months before *Trinity*. Not only did Kurchatov and his colleagues have a good general sense of the Manhattan Project, but also the Anglo-Canadian project as well. The capture of German scientists on the Red army's push westward did not produce much in the way of technical value, although the occupation of what became East Germany did net valuable uranium deposits.[112]

Like the Americans, the Soviets desperately needed sufficient fissionable materials to produce bombs. The advance of the Soviet army into Central Europe provided access to Czechoslovakian uranium. Ultimately, the USSR got uranium from several sources. Without sufficient uranium or good plutonium-generating technology, progress for the bomb project lumbered along on a small scale during the remainder of the war.[113] Hiroshima made the difference in pushing Stalin toward accelerating the atomic bomb project and making it a priority.[114] The argument that sharing information with Stalin early on may have avoided a weapons race seems to have little historical validity. Stalin wanted his own bomb, not only because it was a powerful weapon but also because it would influence the balance of power and symbolized international status for his country. Building the necessary industrial infrastructure, along with obtaining sufficient fissionable material and supporting the scientific research, became paramount, just as it had been in the United States a few years before.[115]

The man chosen to head up the massive project, Lavrentiy Beria, was quite unlike general Leslie Groves. Beria had been long active in the work of the secret police, first in the Cheka in the 1920s, and then in its successor organization the NKVD. Stalin brought him to Moscow in 1938 during the Great Purge. Beria was an excellent organizer, but also cruel and ruthless. He used his various connections and the NKVD to handle intelligence information and to oversee all project activities. On matters related to atomic research and bomb design, Beria did not interfere with the work of the scientists. He did place great pressure on everyone through a very rigid security system. Among the scientists a substantive issue was whether to utilize what they had learned about the American bomb or, as physicist Pyotr Kapitza recommended, seek a cheaper and quicker approach. Stalin eventually made it clear that he wanted the bomb constructed as quickly as possible, and placed no strictures on expense.[116]

The development of the industrial infrastructure posed many of the same problems of scale that the Americans faced, but with decided advantages and disadvantages. The Soviet's highly centralized economic system allowed directing resources to specific ends, not unlike what Groves accomplished in the US wartime setting. The search for sufficient quantities of specific materials, particularly uranium, was extremely difficult. But as Soviet expert David Holloway stated, "The project was a curious combination of the best and the worst of Soviet society—of enthusiastic scientists and engineers produced by the expansion of education under Soviet rule, and of prisoners who lived in the inhuman conditions of the labor camps."[117] Central Intelligence Agency figures (although only estimates) suggested that between 330,000 and 460,000 people worked on the project at some time. Most of the number toiled in mines in the Soviet Union and Eastern Europe (about 255,000–361,000); about 70,000–90,000 in construction and production; and 5,000–8,000 in research. Many of those doing manual work, especially in the mines, were prison laborers. Working conditions, although difficult in the United States, were not nearly as brutal as on the Soviet project. However, radiation sickness, other public health issues, and the impact of the construction on the natural environment occurred in both projects.

The equivalent to Los Alamos was Arzamas-16 near Sarov east of Moscow. In early 1949 it became the major weapons production site. The Soviets built the first production reactor in the Ural Mountains near Cheliabinsk, and began operations to produce plutonium in June/July, 1948. It became the most important plutonium production center in the USSR.[118] The Soviets also explored all of the various separation methods attempted in the United States, including electromagnetic, thermal, and gaseous diffusion. The crash program produced a weapon for testing in summer 1949. On August 29, "the article" was detonated on a 100-foot tower, much like *Trinity*, at a test site in Kazakhstan; it yielded approximately 20-kilotons. The game had changed.[119]

CONCLUSION: FROM TOTAL WAR TO DETERRENCE

By 1949, atomic energy was tangled up in a new war context—this time not total war but Cold War. Deterrence rather than destruction was the objective. While avoidance of total war was paramount, the bomb's specter loomed just the same. Below the surface of postwar nuclear policy making, but not much below, was an eerie subtext. This was the threat of radioactivity on a worldwide scale in peacetime and in war. This unseen danger often was suppressed, but quite real nonetheless. In the looming arms race mutual destruction or mutual survivability were the concerns that remained center stage. The fate of nations was in the hands of scientists and technicians who designed the weapons and the delivery systems, the spies who ferreted out well-kept secrets, and especially the military and civilian leaders who squared off against each other across hardened borders. The bomb influenced the emergence of the Cold War, and the contest between the United States and the Soviet Union, and thus changed the world.

MySearchLab Connections: Sources Online

READ AND REVIEW

Review this chapter by using the study aids and these related documents available on MySearchLab.

✓•—[**Study** and **Review** on **mysearchlab.com**

Chapter Test

Essay Test

⬚●─ **Read** the **Document** on **mysearchlab.com**

Eyewitness Accounts of the Trinity Test (1945)

Potsdam Declaration (1945)

Harry S Truman, Draft Statement on the Dropping of the Bomb (1945)

The Effects of Atomic Bombs on Hiroshima and Nagasaki (1946)

RESEARCH AND EXPLORE

Use the databases available within MySearchLab to find additional primary and secondary sources on the topics within this chapter.

◉─ **Watch** the **Video** on **mysearchlab.com**

Atomic Bomb at Hiroshima

Effect of Atomic Bomb on Hiroshima and Nagasaki

Robert Oppenheimer, Now I Have become Death the Destroyer of World

Endnotes

1. Kai Bird and Martin J. Sherwin, *American Prometheus: The Triumph and Tragedy of J. Robert Oppenheimer* (New York: Alfred A. Knopf, 2005), 304. See also David C. Cassidy, *J. Robert Oppenheimer and the American Century* (New York: Pi Press, 2005), 243; Ferenc Morton Szasz, *The Day the Sun Rose Twice: The Story of the Trinity Site Nuclear Explosion, July 16, 1945* (Albuquerque, NM: University of New Mexico Press, 1984), 40–41.
2. Bird and Sherwin, *American Prometheus*, 303–07; Robert S. Norris, *Racing for the Bomb: General Leslie R. Groves, the Manhattan Project's Indispensable Man* (South Royalton, VT: Steerforth Press, 2002), 395–98, 404; "The Manhattan Project," online, http://www.me.utexas.edu/~uer/manhattan/project.html; Diana Preston, *Before the Fallout: From Marie Curie to Hiroshima* (New York: Walker & Company, 2005), 283.
3. Quote by Brigadier General Thomas F. Farrell in Cynthia C. Kelly, ed., *The Manhattan Project: The Birth of the Atomic Bomb in the Words of Its Creators, Eyewitnesses, and Historians* (New York: Black Dog & Leventhal Pubs., 2007), 294.
4. Preston, *Before the Fallout*, 283–84; Richard G. Hewlett and Oscar E. Anderson, Jr., *The New World, 1939–1946*, volume 1, *A History of the United States Atomic Energy Commission* (University Park, PA: Penn State University Press, 1962), 377–79; Vincent C. Jones, *Manhattan: The Army and the Atomic Bomb,* Special Studies, *United States Army in World War II,* (Washington, D.C.: Center of Military History, United States Army, 1985), 511–15.
5. Jones, *Manhattan*, 516; Gerard J. DeGroot, *The Bomb: A Life* (Cambridge, MA: Harvard University Press, 2005), 59–62; Rodney P. Carlisle, ed., *Encyclopedia of the Atomic Age* (New York: Facts on File, 2001), 335–36; Stephen E. Atkins, ed., *Historical Encyclopedia of Atomic Energy* (Westport, CT: Greenwood Press), 374; Richard Rhodes, *The Making of the Atomic Bomb* (New York: Simon and Schuster, 1986), 617–78.
6. Quoted in Gregg Herken, *Brotherhood of the Bomb: The Tangled Lives and Loyalties of Robert Oppenheimer, Ernest Lawrence, and Edward Teller* (New York: Henry Holt and Co., 2002), 137.
7. Quoted in Kelly, ed., *The Manhattan Project*, 308.
8. Quoted in DeGroot, *The Bomb*, 62–63.

9. Preston, *Before the Fallout*, 287.

10. Quoted in DeGroot, *The Bomb*, 62–63.

11. Quoted from Ronald E. Powaski, *March to Armageddon: The United States and the Nuclear Arms Race, 1939 to the Present* (New York: Oxford University Press, 1987), 22.

12. Quotes from Bird and Sherwin, *American Prometheus*, 309.

13. Szasz, *The Day the Sun Rose Twice*, 3–4. See also Peter Bacon Hales, *Atomic Spaces: Living on the Manhattan Project* (Urbana, IL: University of Illinois Press, 1997), 319–24.

14. Norris, *Racing for the Bomb*, 406–08.

15. "The Manhattan Project"; Powaski, *March to Armageddon*, 13; J. Samuel Walker, *Prompt and Utter Destruction: Truman and the Use of Atomic Bombs Against Japan* (Chapel Hill, NC: University of North Carolina Press, 2004; rev. ed.), 13–14.

16. Norris, *Racing for the Bomb*, 406; Andrew J. Rotter, *Hiroshima: The World's Bomb* (New York: Oxford University Press, 2008), 161–62.

17. Quoted in Walker, *Prompt and Utter Destruction*, 60.

18. Ibid., 56–59, 62–63. See also Powaski, *March to Armageddon*, 23.

19. Powaski, *March to Armageddon*, 23; Richard Rhodes, *Dark Sun: The Making of the Hydrogen Bomb* (New York: Simon and Schuster, 1995), 51; Herken, *Brotherhood of the Bomb*, 129–30; Walker, *Prompt and Utter Destruction*, 66–67.

20. Lend-lease was a US program that provided war material to its Allies between 1941 and 1945.

21. Allen Weinstein and Alexander Vassiliev, *The Haunted Wood: Soviet Espionage in America—The Stalin Era* (New York: The Modern Library, 2000), xviii. See also John Earl Haynes and Harvey Klehr, *Venona: Decoding Soviet Espionage in America* (New Haven: Yale University Press, 1999), 21.

22. Katherine A. S. Sibley, *Red Spies in America: Stolen Secrets and the Dawn of the Cold War* (Lawrence: University Press of Kansas, 2004), 1–2,5; John Earl Haynes and Harvey Klehr, *Early Cold War Spies: The Espionage Trials That Shaped American Politics* (New York: Cambridge University Press, 2006), 24; Amy Knight, *How the Cold War Began: The Gouzenko Affair and the Hunt for Soviet Spies* (Toronto: McClelland & Stewart, Ltd., 2005), 1; Weinstein and Vassiliev, *The Haunted Wood*, 205.

23. Weinstein and Vassiliev, *The Haunted Wood*, 172–78, 183–98, 211–12; Rhodes, *Dark Sun*, 24, 54, 76, 80–82, 91–151, 165–77, 187–98, 215, 244–48; Herken, *Brotherhood of the Bomb*, 87–102, 129–30.

24. Richard C.S. Trahair and Robert L. Miller, *Encyclopedia of Cold War Espionage, Spies, and Secret Operations* (New York: Enigma Books, 2009), 105–07; Herken, *Brotherhood of the Bomb*, 118–20, 129–30; Atkins, ed., *Historical Encyclopedia of Atomic Energy*, 140–41; Carlisle, ed., *Encyclopedia of the Atomic Age*, 112–13. See also Robert Caldwell Williams, *Klaus Fuchs, Atomic Spy* (Cambridge, MA: Harvard University Press, 1987).

25. Walker, *Prompt and Utter Destruction*, 60–72. See also Rotter, *Hiroshima*, 161–65; Powaski, *March to Armageddon*, 14.

26. For a good overview on the debate, see J. Samuel Walker, "Recent Literature on Truman's Atomic Bomb Decision: A Search for Middle Ground," *Diplomatic History* 29 (April 2005): 311–34.

27. Robert L. Messer, "'Accidental Judgments, Casual Slaughters,' Hiroshima, Nagasaki, and Total War," in Roger Chickering, Stig Forster, and Bernd Greiner, eds., *A World at Total War: Global Conflict and the Politics of Destruction, 1937–1945* (Cambridge, MA: Cambridge University Press, 2005), 297.

28. Roger Chickering and Stig Forster, "Are We There Yet? World War II and the Theory of Total War," in Chickering, Forster, and Greiner, eds., *A World at Total War*, 13.

29. Richard Overy, "Total War II: The Second World War," in Charles Townshend, ed., *The Oxford History of Modern War* (New York: Oxford University Press, 2000), 140.

30. Chickering and Stig Forster, "Are We There Yet? World War II and the Theory of Total War," 2, 7–8. See also Frederick M. Sallagar, *The Road to Total War* (New York: Van Nostrand Reinhold Co., 1969), 1–4, 41.

31. See "Source List and Detailed Death Tolls for the Primary Megadeaths of the 20th Century," online, http://necrometrics.com/20c5m.htm#Second; "Statistics of World War II," the History Place, online, http://www.historyplace.com/worldwar2/timeline/stati-stics.htm; "World War II: Combatants and Casualties (1937–45)," online, http://web.jjay.cuny/~jobrien/reference/ob62.html; Digital History,

online, www.digitalhistory.uh.edu; Messer, "Accidental Judgments, Casual Slaughters,' Hiroshima, Nagasaki, and Total War," 298.

32. Overy, "Total War II: The Second World War," 148–49. See also Russell F. Weigley, *The American Way of War: A History of United States Military Strategy and Policy* (New York: Macmillan Pub., 1973), 363–64.

33. Chickering and Forster, "Are We There Yet? World War II and the Theory of Total War," 12–13; Hew Strachan, "Total War: The Conduct of War 1939–1945," in Chickering, Forster, and Greiner, eds., *A World at Total War*, 51–52; Richard Overy, "Allied Bombing and the Destruction of German Cities," in Chickering, Stig Forster, and Bernd Greiner, eds., *A World at Total War*, 277, 287.

34. Clarence G. Lasby, *Project Paperclip: German Scientists and the Cold War* (New York: Antheneum, 1971–1975); Atkins, ed., *Historical Encyclopedia of Atomic Energy*, 8–9; Powaski, *March to Armageddon*, 12; Carlisle, ed., *Encyclopedia of the Atomic Age*, 7–8; Rotter, *Hiroshima, 271.*

35. Japan never provided sufficient resources to develop an atomic bomb program, but some Japanese officials nevertheless seriously considered building one in World War II. They gave tentative approval to start such a program in October 1940, and in April 1941 the Imperial Army Air Force authorized research on an atomic bomb. The Imperial Navy also committed to exploring nuclear power for ship propulsion. By 1943, military defeats turned more attention to radar and aircraft. Some believed developing a bomb still was possible, but would require probably ten years. They also assumed that neither Germany nor the United States could produce a bomb during the current war. Atkins, ed., *Historical Encyclopedia of Atomic Energy*, 183–84.

36. Quoted in Michael Kort, *The Columbia Guide to Hiroshima and the Bomb* (New York: Columbia University Press, 2007), 37.

37. Ibid., 43–44.

38. Walker, *Prompt and Utter Destruction,* 3–52; Kort, *The Columbia Guide to Hiroshima and the Bomb*, 104–06.

39. Messer, "'Accidental Judgments, Casual Slaughters,' Hiroshima, Nagasaki, and Total War," 300; Weigley, *The American Way of War*, 363–64; Rotter, *Hiroshima*, 133–48.

40. John W. Dower, *War Without Mercy: Race and Power in the Pacific War* (New York: Pantheon, 1987). See also Ronald Takaki, *Hiroshima: Why America Dropped the Atomic Bomb* (Boston: Little, Brown and Co., 1995).

41. Powaski, *March to Armageddon*, 15–16; Walker, *Prompt and Utter Destruction,* 14–15; Atkins, ed., *Historical Encyclopedia of Atomic Energy*, 173–74; Rotter, *Hiroshima*, 152–55; Kort, *The Columbia Guide to Hiroshima and the Bomb*, 49–51.

42. Rotter, *Hiroshima*, 148–50.

43. Quoted in Jeffrey Porro, ed., *The Nuclear Age Reader* (New York: Alfred A. Knopf, 1989), 11.

44. Quoted in Michael B. Stoff, Jonathan F. Fanton, and R. Hal Williams, eds., *The Manhattan Project: A Documentary Introduction to the Atomic Age* (Philadelphia: Temple University Press, 1991), 146.

45. Quoted in Bird and Martin J. Sherwin, *American Prometheus*, 313–14.

46. Carlisle, ed., *Encyclopedia of the Atomic Age,* 110–11; Stoff, Fanton, and Williams, eds., *The Manhattan Project,* 136–37; Atkins, ed., *Historical Encyclopedia of Atomic Energy*, 134–36; Rotter, *Hiroshima*, 150–51; Hewlett and Anderson, Jr., *The New World, 1939–1946*, 365–67.

47. Rotter, *Hiroshima*, 156–57.

48. Weigley, *The American Way of War*, 364.

49. Hewlett and Anderson, Jr., *The New World, 1939–1946*, 365; Cassidy, *J. Robert Oppenheimer and the American Century,* 243; Jones, *Manhattan*, 528–30; Kelly, ed., *The Manhattan Project*, 319–21.

50. Messer, "'Accidental Judgments, Casual Slaughters,' Hiroshima, Nagasaki, and Total War," 307; Cassidy, *J. Robert Oppenheimer and the American Century*, 243–44. See also Sean L. Malloy, *Atomic Tragedy: Henry L. Stimson and the Decision to Use the Bomb Against Japan* (Ithaca, NY: Cornell University Press, 2008).

51. For a balanced treatment of Truman's decision, see Walker, *Prompt and Utter Destruction*. For a more assertive assessment of the political justifications for the bombings, see Gar Alperovitz, *Atomic*

Diplomacy: Hiroshima and Potsdam: The Use of the Atomic Bomb and the American Confrontation with Soviet Power (New York: Pluto Press, 1994); Martin J. Sherwin, *A World Destroyed: Hiroshima and the Origins of the Arms Race* (New York: Vintage,1987).

52. Preston, *Before the Fallout*, 290–303, 305, 309; DeGroot, *The Bomb*, 84–85; Walker, *Prompt and Utter Destruction*, 75–78; Powaski, *March to Armageddon*, 26–27; Atkins, ed., *Historical Encyclopedia of Atomic Energy*, 163–64; Kort, *The Columbia Guide to Hiroshima and the Bomb*, 3–4; Gordon Thomas and Max Morgan Witts, *Enola Gay: The Bombing of Hiroshima* (Old Saybrook, CT: Konecky & Konecky, 1977), 261–62.

53. Both quotes from *Voice of Hibakusha, Eye-witness accounts of the bombing of Hiroshima, from the video Hiroshima Witness produced by Hiroshima Peace Cultural Center and NHK*, online, http://www.inicom.com/hibakusha/.

54. Quoted in Powaski, *March to Armageddon*, 26.

55. Walker, *Prompt and Utter Destruction*, 78; Preston, *Before the Fallout*, 309.

56. Over 14,000 of 52,000 homes were destroyed.

57. Atkins, ed., *Historical Encyclopedia of Atomic Energy*, 49–50, 240; Carlisle, ed., *Encyclopedia of the Atomic Age,* 137, 200; Walker, *Prompt and Utter Destruction*, 78–80; DeGroot, *The Bomb*, 99–102; Preston, *Before the Fallout*, 310–11.

58. Quoted from "Michie Hattori: Eyewitness to the Nagasaki Atomic Bomb Blast," June 12, 2006, *Historynet.com*, online, http://www.historynet.com/michie-hattori-eyewitness-to-the-nagasaki-atomic-bomb-blast.htm.

59. Quoted in Preston, *Before the Fallout*, 311.

60. Robert Jay Lifton and Greg Mitchell, *Hiroshima in America: Fifty Years of Denial* (New York: G.P. Putnam's Sons, 1995), 29.

61. Powaski, *March to Armageddon*, 27–28; Stoff, Fanton, and Williams, eds., *The Manhattan Project*, 220; Preston, *Before the Fallout*, 311.

62. Printed in Stoff, Fanton, and Williams, eds., *The Manhattan Project*, 226.

63. Quoted in Walker, *Prompt and Utter Destruction*, 81. See also Preston, *Before the Fallout*, 311–14; Powaski, *March to Armageddon*, 103–04; Messer, "'Accidental Judgments, Casual Slaughters,' Hiroshima, Nagasaki, and Total War," 300–03; Kelly, ed., *The Manhattan Project*, 336–38; Lifton and Mitchell, *Hiroshima in America*, 23–39.

64. See Tsuyoshi Hasegawa in *Racing the Enemy: Stalin, Truman, and the Surrender of Japan* (Cambridge, MA: Harvard University Press, 2005); Tsuyoshi Hasegawa, ed., *The End of the Pacific War: Reappraisals* (Stanford, CA: Stanford University Press, 2007. See also Walker, *Prompt and Utter Destruction*, 82–97; Kort, *The Columbia Guide to Hiroshima and the Bomb*, 107–09.

65. Walker, *Prompt and Utter Destruction*, 82–92; Preston, *Before the Fallout*, 311; Powaski, *March to Armageddon*, 28; Kort, *The Columbia Guide to Hiroshima and the Bomb*, 67–74; Jones, *Manhattan*, 534–50; Hewlett and Anderson, Jr., *The New World, 1939–1946*, 403–06.

66. Weigley, *The American Way of War*, 365. For more detail and analysis on Hiroshima and Nagasaki, see Rhodes, *The Making of the Atomic Bomb*, 679–747.

67. Paul Boyer, *Fallout: A Historian Reflects on America's Half-Century Encounter with Nuclear Weapons* (Columbus, OH: Ohio State University Press, 1998), 7–8; Allan M. Winkler, *Life Under a Cloud: American Anxiety About the Atom* (Urbana, Il: University of Illinois Press, 1999), 26; Nina Tannenwald, *The Nuclear Taboo: The United States and the Non-Use of Nuclear Weapons Since 1945* (New York: Cambridge University Press, 2007), 89–91.

68. Paul Boyer, *By the Bomb's Early Light: American Thought and Culture at the Dawn of the Atomic Age* (Chapel Hill, NC: University of North Carolina Press, 1994), 12.

69. Ibid., 22–25. See also Winkler, *Life Under a Cloud*, 25–30; Stoff, Fanton, and Williams, eds., *The Manhattan Project*, 252–57; DeGroot, *The Bomb*, 113–15.

70. Lawrence S. Wittner, *One World or None: A History of the World Nuclear Disarmament Movement Through 1953*, v. 1 *The Struggle Against the Bomb* (Stanford, CA: Stanford University Press, 1993), 80–154.

71. Ibid., 40–43.

72. Ibid., 45–54.

73. John W. Dower, "The Bombed: Hirsohimas and Nagasakis in Japanese Memory," in Michael J. Hogan, ed., *Hiroshima in History and Memory* (Cambridge, MA: Cambridge University Press, 1996), 124–26; Atkins, ed., *Historical Encyclopedia of Atomic Energy*, 332–33; Kelly, *The Manhattan Project*, 376; Jones, *Manhattan*, 543–62; Rotter, *Hiroshima*, 201, 204, 224.

74. "Hiroshima Nagasaki August 1945," *The Video Project*, online, http://www.videoproject.com/hir-042-v.html.

75. Lifton and Mitchell, *Hiroshima in America*, 57–61; Robert W. Duncan, "Hiroshima-Nagasaki: August 1945," in Jack G. Shaheen, *Nuclear War Films* (Carbondale, IL: Southern Illinois University Press, 1978), 132–37.

76. See Atkins, ed., *Historical Encyclopedia of Atomic Energy*, 25–26.

77. J. Samuel Walker, *Permissible Dose: A History of Radiation Protection in the Twentieth Century* (Berkeley, CA: University of California Press, 2000), 141. See also 10, 18, 114, 129–33. The cancer cases linked to the bombings were attributed to gamma radiation.

78. John Hersey, *Hiroshima* (New York: Knopf, 1946), 4.

79. Kelly, ed., *The Manhattan Project*, 377–81; Lifton and Mitchell, *Hiroshima in America*, 86–92; Winkler, *Life Under a Cloud*, 30–31. The Japanese translation was not released in Japan until after occupation censorship ended in 1949. It became a best seller there in 1950. See Dower, "The Bombed," 135.

80. Margot A. Henriksen, *Dr. Strangelove's America: Society and Culture in the Atomic Age* (Berkeley, CA: University of California Press, 1997), 41–42; Tannenwald, *The Nuclear Taboo*, 92–93; Wittner, *One World or None*, 58–59; Boyer, *By the Bomb's Early Light*, 203–10.

81. Rotter, Hiroshima, 205.

82. In the June 1947 issue, the editors printed a picture of a "Doomsday Clock" indicating imminent danger of a nuclear catastrophe.

83. Wittner, *One World or None*, 59–66; Boyer, *By the Bomb's Early Light*, 49–64, 74–75; Atkins, ed., *Historical Encyclopedia of Atomic Energy*, 35, 118–19, 127–28; Winkler, *Life Under a Cloud*, 34–56; Daniel J. Kevles, *The Physicists: The History of a Scientific Community in Modern America* (Cambridge, MA: Harvard University Press, 1977; fourth printing, 1995), 333–58; Arthur Steiner, "Scientists, Statesmen, and Politicians: The Competing Influences on American Atomic Energy Policy, 1945–46," *Minerva* 12 (October 1974): 469–509; Jessica Wang, *American Science in an Age of Anxiety: Scientists, Anticommunismm, and the Cold War* (Chapel Hill, NC: University of North Carolina Press, 1999), 10–43.

84. Quotes from Bird and Martin J. Sherwin, *American Prometheus*, 315–54.

85. Rotter, *Hiroshima*, 128–29, 151, 232–34; Barton J. Bernstein, "Truman and the A-Bomb: Targeting Noncombatants, using the Bomb, and his Defending the 'Decision,'" *Journal of Military History* 62 (July 1998): 561–62.

86. Quoted in Wittner, *One World or None*, 45, 66.

87. Boyer, *By the Bomb's Early Light*, 29–45.

88. Martin V. Melosi, *Coping with Abundance: Energy and Environment in Industrial America* (New York: Knopf, 1984), 223–26; Shane J. Maddock, "Ideology and US Nuclear Nonproliferation Policy since 1945," in Mariner and Piehler, eds., *The Atomic Bomb and American Society*, 125.

89. S. David Aviel, *The Politics of Nuclear Energy* (Washington, D.C.: University Press of America, 1982), 15.

90. Melosi, *Coping with Abundance*, 223; Atkins, ed., *Historical Encyclopedia of Atomic Energy*, 228–29; DeGroot, *The Bomb*, 116–17.

91. Carlisle, ed., *Encyclopedia of the Atomic Age*, 20–21, 23; Atkins, ed., *Historical Encyclopedia of Atomic Energy*, 27, 29, 33, 95–96,144.

92. Melosi, *Coping with Abundance*, 225–26.

93. Atkins, ed., *Historical Encyclopedia of Atomic Energy*, 19–20, 29–30, 63–64, 242, 247–49, 319–20.

94. Melosi, *Coping with Abundance*, 226.

95. Campbell Craig and Sergey Radchenko, *The Atomic Bomb and the Origins of the Cold War* (New Haven: Yale University Press, 2008), ix.

96. Ibid., x. Wilson D. Miscamble, *From Roosevelt to Truman: Potsdam, Hiroshima, and the Cold War* (New York: Cambridge University Press, 2007) does not attribute clear or coherent motives to Truman's wartime atomic strategy and sees the United States as more defensive than proactive after the war. See 262ff.

97. The most forceful case for the use of the bomb as a diplomatic weapon against the Soviets was initially raised in Gar Alperovitz, *Atomic Diplomacy: Hiroshima and Potsdam* (New York: Vintage Books, 1965) and modified in his later works.

98. Sherwin, *A World Destroyed*, 67–115, 143–238.

99. Atkins, ed., *Historical Encyclopedia of Atomic Energy, 31,* 398.

100. Powaski, *March to Armageddon*, 41, as well as 40, 42.

101. DeGroot, *The Bomb*, 122.

102. Powaski, *March to Armageddon*, 42–43; Wittner, *One World or None*, 62–64; Herken, *Brotherhood of the Bomb*, 162–66; Bird and Sherwin, *American Prometheus*, 339–50; Boyer, *By the Bomb's Early Light*, 93–101.

103. Powaski, *March to Armageddon*, 44–45; Craig and Radchenko, *The Atomic Bomb and the Origins of the Cold War*, 111–61.

104. A French designer used the name 'bikini' for his new revealing swimsuits.

105. Herken, *The Winning Weapon,"* 175–76; Atkins, ed., *Historical Encyclopedia of Atomic Energy*, 13, 48–49.

106. Boyer, *By the Bomb's Early Light*, 90. *Radio Bikini*, a documentary originally aired in 1988 on PBS' "The American Experience" graphically and chillingly tells the story of American sailors and natives exposed to radioactive fallout during the *Crossroads* tests.

107. Herken, *The Winning Weapon,"* 224–26; Rhodes, *Dark Sun*, 261–63; Powaski, *March to Armageddon*, 44; Lifton and Mitchell, *Hiroshima in America*, 83–86.

108. Atkins, ed., *Historical Encyclopedia of Atomic Energy*, 120–21; 320; Herken, *The Winning Weapon,"* 252; Powaski, *March to Armageddon*, 51–52; Rhodes, *Dark Sun*, 320–21; Carlisle, ed., *Encyclopedia of the Atomic Age,* 184–86.

109. David Holloway, *Stalin and the Bomb* (New Haven, CT: Yale University Press, 1994), 5.

110. Ibid., 21, 24, 28, 46–48, 70–71.

111. Ibid., 72–90, 449; Carlisle, ed., *Encyclopedia of the Atomic Age,* 167–68. For more on Soviet physicists, see Alexei B. Kojevnikov, *Stalin's Great Science: The Times and Adventures of Soviet Physicists* (London: Imperial College Press, 2004), 99–185. See also Thomas B. Cochran, Robert S. Norris, and Oleg A. Bukharin, *Making the Russian Bomb: From Stalin to Yelsin* (Boulder, CO: Westview Press, 1995), 7–8.

112. Holloway, *Stalin and the Bomb*, 96–108, 110–11. For revelations on atomic espionage related to bomb secrets passed along to the Soviets, see Joseph Albright and Marcia Kunstel, *Bombshell: The Secret Story of America's Unknown Atomic Spy Conspiracy* (New York: Times Books, 1997).

113. Cochran, Norris, and Bukharin, *Making the Russian Bomb,* 9; Holloway, *Stalin and the Bomb*, 108–10, 177. Sources for uranium included (all estimates): East Germany (45 percent), Soviet Union (33 percent), Czechoslovakia (15 percent), Bulgaria (4 percent), and Poland (3 percent).

114. Craig and Radchenko, *The Atomic Bomb and the Origins of the Cold War*, 93.

115. Holloway, *Stalin and the Bomb*, 133.

116. Ibid., 134–49, 447–48; DeGroot, *The Bomb*, 135–43; Rhodes, *Dark Sun*, 1–381. After Stalin died in 1953, Beria was executed for "crimes against the state."

117. Holloway, *Stalin and the Bomb*, 172. This is not to suggest that all Soviet scientists rushed to join in the project. Scientists often were recruited for the task; they did not necessarily volunteer. See DeGroot, *The Bomb*, 139.

118. The Obninsk Atomic Energy Station south of Moscow became the center for research on nuclear reactors beginning in 1950.

119. Holloway, *Stalin and the Bomb*, 172–95, 213–15; DeGroot, *The Bomb*, 144–45; Atkins, ed., *Historical Encyclopedia of Atomic Energy*, 23, 79, 269; Cochran, Norris, and Bukharin, *Making the Russian Bomb*, 11–12.

CHAPTER 4

The Cold War and Atomic Diplomacy

Deterrence, Espionage, and the Superbomb

INTRODUCTION: HOPE, PROMISE, AND RISK

The story of the atom in postwar America was grounded in hope, promise, and risk. There was hope that atomic weapons would insure national security. There was the promise of nuclear power to deliver an abundant source of energy. And there was the duel risk of war and environmental peril. Hope, promise, and risk weaved through history for the remainder of the twentieth century and into the twenty-first. All three ideas never really marched hand in hand, but one or two seemed to take precedence over the others at different times and in different places. Yet we have never been without this trinity in Atomic Age America. From the end of World War II until the early 1950s, the Truman administration faced the question of how to integrate the atomic bomb, and then its hydrogen successor, into national security policy. The danger of war loomed large, despite the recent end to the most tortuous conflict in human history. Entwined in the security debate were dramatic revelations about atomic espionage that people believed directly produced Soviet successes in developing its own atomic weapons. Testing new atomic and hydrogen devices in the Pacific offered the first public glimpse of the threat of fallout. The "promise" of nuclear power wistfully discussed in the debates on whether to internationalize the atom remained largely an abstraction in the first postwar decade. In this period hope and risk trumped promise.

THE BOMB AND A "GET TOUGH" POLICY IN THE COLD WAR

Understanding the nature and intensity of the Cold War is essential to grasping the role of the bomb in the mid-1940s and beyond. Animus between the United States and the Soviet Union was clearly exposed with the fracturing of the wartime Grand Alliance soon after the Red Army turned the tide against Nazi Germany in 1943. Before that time, Joseph Stalin had little chance to look beyond his country's survival, and American forces were preoccupied fighting a two-front conflict in Europe and Asia. After the war the Soviets controlled Eastern Europe with an iron fist. This circumstance convinced President Harry Truman and several of his advisers that the USSR had further designs in Europe and Asia, and could not be trusted. Stalin and other Soviet officials believed that American and British leaders did not recognize the long-standing threat to their security coming from the West through Eastern Europe; did not appreciate how much of the war against Adolf Hitler the Soviet peoples had born; and failed to understand their need for a rigid territorial buffer. Thus the Soviets charged Western powers with expansionist designs of their own. Meant as a temporary military and diplomatic action after the war, the establishment of zones of occupation in Germany[1] (decided at the Potsdam meeting) turned into a permanent political reality. This was an important step in admitting the unbreachable chasm between Western and Eastern Europe. Truman now believed that the Soviets only understood force, while Stalin increasingly became wary of the Americans' unwillingness to compromise.[2]

As tensions mounted between the United States, its allies, and the Soviet Union, the role of the bomb became more critical. It emboldened the American government to protect Western Europe and still carry on strategic interests elsewhere. It increased Soviet suspicions that the United States might actually use the weapon against them. And it sparked a gargantuan arms race between the two emerging superpowers. Beyond its game-changing potential, the initial US monopoly over the bomb was a symbol of American power, scientific and technical achievement, and global status.[3] There was an alternative set of issues as well. Such bombs were not conventional weapons. Truman himself, especially after reflecting on Hiroshima and Nagasaki, viewed them as "weapons of last resort." Speaking for many around the world who viewed the bomb with great trepidation, the United Nations (UN) wanted to "delegitimize" the bomb by seeking nuclear disarmament. The UN also created the category, "weapons of mass destruction," to distinguish nuclear weapons from other types. The language was coined by US officials initially to describe biological weapons, but the UN quickly applied it to the A-bomb and its successors.[4] During the Cold War, therefore, nuclear weapons were at once central to the prosecution and strategy of international affairs (first for the United States, then the USSR, and much later for other nations) and also viewed by many as justifiably out of bounds. Such a tension provided a subtext for the new arms race throughout much of the twentieth century.

An emerging "get tough" policy against the USSR grew out of the rise of the United States as a great world power after the war. Germany, Italy, and Japan were defeated, and the once powerful imperial nations of Great Britain and France were major players no more. Left standing was a potent United States and a wounded but potentially forceful USSR pitted against each other in a new bipolar environment. The prestige of the United States was never higher—its economic potential staggering; and its monopoly over atomic weapons sobering. These realities led American leaders to believe that their country was capable of bringing an order to the world based on its faith in a democratic capitalist model. The hardened policy also was built upon Truman's belief that Stalin would not do the right thing in Eastern Europe. Such an assumption perplexed the Soviets, who never quite understood how Eastern Europe could be considered within the American sphere of influence. And they could not accept that Western-style democracy and capitalism was superior to communist ideology and governance. For both countries, however, learning how to translate power into genuine influence proved difficult.[5]

The unilateral control of the bomb was a fundamental principle influencing Truman's decision to take a hard line with the Soviet Union. Since the *Trinity* test, the bomb became a persistent feature in negotiating with (or posturing against) the USSR. Controlling weapons of mass destruction seemed to offer multiple benefits. Since the United States and its European allies could not match the size and strength of the gigantic Red Army (which remained intact after the war), atomic bombs offered a deterrence factor or a military offset, as long as the United States held a monopoly (or overwhelming numbers of bombs). Constructing bombs cost far less than maintaining a huge standing army, thus accommodating Republicans and fiscally minded Democrats at home who wanted to see the postwar federal budget scaled down. Faith in the deterrence capability of atomic weapons, however, had built-in limits. Could one threaten to use the bomb and be assured that the opponent respected the threat? What if the Red Army moved rapidly west? Would the United States and its allies be willing to destroy Berlin or Paris to counteract a Soviet attack there? Still, Truman and his successors were unwilling to give up on the notion that the atomic arsenal of the United States provided significant diplomatic and military advantage over its enemies.[6]

President Truman's foreign policy perspective often led him, as historian George Herring suggested, to see "a complex world in black and white terms." This was not simply a product of his insecurities and inexperience as a new world leader. Indeed, he was typical of many Americans who were raised to believe that their country's way of doing things was best, that is, American exceptionalism. Lacking exposure to much of the world, Truman was guilty of the rough stereotyping of other peoples and other nations.[7] The president also relied on the most hawkish members of his administration, albeit they had a more nuanced understanding of foreign policy than he, who were no less partisan. A growing "Cold War consensus" in the United States left little room for alternative views. Stalin, while proving over time to be more pragmatic than ideological, was opportunistic and ruthless. He also was cruel with his own people let alone his enemies. In dealing with such a leader, Americans had a difficult time determining when Stalin's word could be trusted and when it could not. They easily demonized him as they had demonized Hitler, often with good reason.[8]

IMPLEMENTING CONTAINMENT

In a speech delivered in Fulton, Missouri, on March 5, 1946, (with Truman's approval) former British Prime Minister Winston Churchill hardened the language of the Cold War declaring, "From Stettin in the Baltic to Trieste in the Adriatic, an iron curtain has descended across the [European] continent."[9] In his mind, the time was right to liberate Eastern Europe and to resist the forward movement of the Soviets into places like Turkey and Iran. At this stage, the United States and the USSR did little more than exchange accusations. The Soviets had a country to put back together, and the American Congress was not prepared to finance new international adventurism. By 1947 the United States had almost completely demobilized its military forces. Many Americans were hopeful that tensions would ease with the Soviet Union, but several government officials were wary about what the Soviet Union would do next. Rumors and undercurrents of espionage and possible internal subversion fueled the anxiety.

In northern Europe the primary focus was Germany. Americans encouraged industrialization in the western sectors and even flirted with rearming Germany, which was disquieting for its former enemies. A German revival deeply troubled Britain and France (as it did the Soviet Union), who preferred a weak neighbor, not a stout competitor. Stalin's concerns about the eastern sector of Germany ranged from the desire to maintain rigid political control to competing with the burgeoning economy across the border. A growing concern for the Truman administration was Great Britain's decision to roll back its presence in the Eastern Mediterranean, making the area vulnerable to Soviet interests and local rebellion. Access from the Black Sea to the Mediterranean via the straits controlled by Turkey also was a sore spot for the USSR. When the British announced in February 1947 that they would no longer provide aid to Greece and

Turkey, attention turned to a possible collapse of a delicate house of cards. Civil war breaking out in Greece provided an opportunity for the Soviet Union to gain more influence close to home.

Truman was determined to solicit congressional support to blunt growing isolationist instincts to deter new Soviet advances. He received strong bi-partisan support from Senator Arthur Vandenberg (Rep., Michigan), who was lobbying his colleagues to consider a new aid program for Greece. The president took a daring step, making a "hard sell" to the public for a great crusade against communism and a global commitment to American-style democracy. This transcended the Greek matter, where the United States had put blame on the USSR rather than internal strife for creating instability. The so-called Truman Doctrine (announced March 12, 1947) without specifically mentioning names painted the struggles between the United States and the USSR as good versus evil, democracy versus communism. Determined to "scare hell out of the American people" (allegedly in Vandenberg's words)[10] Truman stated, "I believe that it must be the policy of the United States to support free peoples who are resisting attempted sub-jugation by armed minorities or by outside pressures." And aiming directly at the Soviet Union, he added, "The seeds of totalitarian regimes are numbered by misery and want. They spread and grow in the evil soil of poverty and strife."[11] This was not a call to arms *per se* (although Truman was surely emboldened by America's atomic monopoly) but a clear statement that American foreign policy would keep Soviet influence and communism in check throughout the world.

The official American Cold War strategy became "containment," or more precisely, a strategy to limit and prevent Soviet expansionism. Such a daunting task equated freedom with political stability and revolution with communism. Containment became virtual dogma in American foreign policy for several decades. The Truman Doctrine sought the status quo rather than calling for the liberation of countries under Soviet control or influence, but clearly drew a political and ideological line between the emerging super powers. Most immediately the new direction in American foreign policy opened the door for a permanent state of military readiness and massive aid to any part of the globe that acknowledged it was resisting communism.[12] US chargé d'affaires in Moscow George F. Kennan articulated the idea of containment in late 1946, and publicly elaborated it in an article published in *Foreign Affairs* in July 1947(authored by "X").

Diplomat George F. Kennan articulated the idea of "containment" in late 1946, which framed the early US Cold War policy. *Source:* Library of Congress, Prints & Photographs Division, NYWT&S Collection, [LC-USZ62-131268]

The article stated, "In these circumstances it is clear that the main element of any United States policy toward the Soviet Union must be that of a long-term, patient but firm and vigilant containment of Russian Expansive tendencies."[13] Kennan took a hard line in devising the concept, but the threat as he saw it was political and economic, not essentially military. He clearly disapproved of the crusading tone of the Truman Doctrine, yet by its very tone containment left little room for a conciliatory approach for dealing with the Soviet Union.

The subsequent Marshall Plan (the European Recovery Program, 1948), for rebuilding the economies of Western Europe, was an attempt to protect the region from Soviet influence or control through political and economic means. The plan was a sublime expression of Kennan's ideas. Economic aid to Europe after World War II was a humanitarian gesture, but it also reestablished markets essential to America's own growth and provided a political bulwark against an aggressive USSR or home-grown pro-communist elements. New Secretary of State George C. Marshall (who replaced the increasingly troublesome and independently minded James Byrnes) announced the plan at Harvard University on June 5, 1947. Through its provisions, the United States agreed to provide financial aid to Europe (including Germany), which had to act collectively rather than as individual nations, to devise a rebuilding program.

The United States invited the Soviet Union and Eastern European countries to participate, but as much as Stalin and other officials desperately wanted (and needed) the support, they realized that the offer was politically motivated with strings attached. In a speech before the UN General Assembly in September 1947 a Soviet representative asserted, "The so-called Truman Doctrine and the Marshall Plan are particularly glaring examples of the manner in which the principles of the United Nations are violated, of the way in which the Organization is ignored."[14] Stalin forbade participation by any of the countries under Soviet control, and announced his own alternative program. Although the funding promised under the Marshall Plan was substantially smaller than the Europeans hoped, the influx of US dollars strengthened the Western European economy more quickly than otherwise from the destruction of war. The Marshall Plan, temporarily at least, gave credibility to the containment strategy.[15]

NEW GOVERNMENT APPARATUS AND CHANGING MILITARY STRATEGY

Important structural changes were underway in the civilian government and the military in the postwar years. These changes oriented the United States to the Cold War environment and integrated the bomb into its military policy. While scientists continued to participate in the discussion about the atom, government officials and military leaders held a tight grip on questions of nuclear strategy.[16] Congress passed the National Security Act (July 1947) which fashioned a single Department of Defense out of the formerly independent services; granted statutory power to the Joint Chiefs of Staff (JSC) which had been a wartime entity only; established the National Security Council (NSC) as an advisory body to the president to coordinate foreign policy; and created the Central Intelligence Agency (CIA). These changes significantly increased the role of the military in making foreign policy, and enhanced the use of covert activities.[17]

The first Secretary of Defense, James Forrestal, did not quell service rivalries nor produce the kind of integration that the president hoped for. The air force became a stand-alone service with an increasing role as an offensive arm of the military. As the importance of atomic weapons grew in the American policy mix, so did its stature.[18] The United States Air Force (USAF), which grew out of the army air forces, became an independent branch of the armed services in September 1947. This action clearly demonstrated the rising importance of air power since

World War II. The USAF was divided into three units: Strategic Air Command (SAC, formerly the Continental Air Forces), Tactical Air Command (TAC), and Air Defense Command (ADC). SAC was given the bombardment function, with TAC handling ground support, and ADC responsible for defending the skies over US territory from enemy bombers.[19]

Slowly, the Truman administration established a framework for nuclear strategy. As Commander in Chief the president had the ultimate responsibility for nuclear weapons. During these early postwar years, the United States conducted war planning on a regular basis, a practice unparalleled in peacetime. On a level below the president, military planners attempted to interpret official national policy for strategic plans. In the late 1940s the Joint Chiefs engaged in strategic planning (its foremost function) to determine the primary military threats to the United States and how to combat them. On the operational level, SAC prepared its own war plans and also had to carry out strikes if mandated by the JCS. Other units in the army, navy, and air force with nuclear capability were under the planning and operational control of JCS commanders in Europe and in the Atlantic and Pacific.[20]

The path to developing a nuclear weapons strategy was complicated. For the first few years of his presidency, Truman was not much engaged in the process. For example, he had no official knowledge of the size of the nuclear stockpile between fall 1945 and spring 1947. In addition, historian David Rosenberg argued, Truman was unwilling or unable "to conceive of the atomic bomb as anything other than an apocalyptic terror weapon, a weapon of last resort."[21] Through 1948 he focused on initiatives related to civilian control over nuclear assets, and by that time the JSC had just prepared its first operational target list in the previous summer of 1947. The president remained concerned about using the bomb for offensive purposes, and planning for such an eventuality started and stopped and started again.[22] Without clear policy guidance, attention turned to the capability of the current American nuclear force. Between 1945 and 1948, the stockpile was low and the delivery system for bombs limited. By one estimate, in July 1946 there were nine bombs and by July 1948, 50. None of them were assembled, and they were so large (*Fat Man* implosion bombs) that they were difficult to load into bombers. In addition, data about Soviet territory was limited and intelligence gathering poor. Even an expanded stockpile could not resolve those problems. And after 1949 planners had to begin thinking about defense against potential enemy atomic weapons.[23]

Within the armed services, both the newly formed air force and the navy wanted control of the strategic bombing mission. The former won out and SAC was to be its instrument. The stated role of SAC could not be clearer: if necessary, its long-range bombers would carry out missions against the USSR. SAC also became the dominant force in operational planning for nuclear war.[24] SAC was not well-prepared for its mission, however, and was further constrained by an ambiguous war strategy and the small atomic stockpile.[25] Between late 1948 and 1957, General Curtis E. LeMay became the prime architect of SAC's development. He was a vigorous advocate of strategic bombing since World War II and knew more about it than anyone else. LeMay was clearly in camp with those who believed that strategic bombing had been vital to defeating Germany and Japan. Typical of his thinking was this statement: "I don't mind being called tough, because in this racket it's tough guys who lead the survivors."[26]

LeMay headed the 20th Bomber Command in the China-India-Burma Theater in August 1944 at the age of 37, and in 1945 took charge of the 21st Bomber Command involved in strategic bombing of Japan. He also helped plan the attacks on Hiroshima and Nagasaki. Declining appointment to complete a term as senator from Ohio, he assumed leadership of the USAF in Europe after the war. In 1948 he directed formation of the Berlin airlift. Later that year his superiors placed him in charge of SAC. At that time long-range bombers were the only means of delivering nuclear weapons, and this was a prestigious assignment.[27] A hawk among hawks with an

General Curtis LeMay was the prime architect of the Air Force's modern Strategic Air Command (SAC).
Source: Library of Congress, Prints & Photographs Division, [LC-USZ62-90022]

uncompromising nature, LeMay played a decisive role in building SAC into a well-functioning organization. He successfully turned an ill-equipped and poorly trained group into an awesome force. In turn SAC transformed the atomic (and hydrogen) bomb from a deterrent into a fearsome offensive weapon. In October 1947 SAC had 18 B-29s modified for atomic weapons with 30 combat crews. By 1953 SAC had more than 140,000 personnel, and over 1,600 aircraft (almost all atomic-capable) organized into 39 wings and 23 separate squadrons.[28] With a staff of strong supporters, General LeMay faced many challenges in moving toward his goals for SAC, including equipment shortage, funding restrictions, and poor intelligence about potential Soviet targets.[29] He nevertheless built SAC into a formidable body that, while controversial, would strongly influence atomic policy in the United States for many years.[30]

Applauded by supporters and appalled by critics, LeMay favored delivering "a decisive atomic attack in the minimum possible time interval." He believed that in the postwar world the only workable approach was to apply overwhelming force as quickly as possible and in a preemptive fashion. Such a plan would get immediate results and save lives that would have been lost in a protracted battle. In his war plans he had to prepare, among other things, a strategy to protect the United States from attack of enemy long-range bombers. The Japanese had no such capability in World War II, but the Soviets might at some point. Deterrence was an early goal, but there was no guarantee that an enemy would be deterred simply because the United States had an impressive atomic arsenal, especially at a time when the American nuclear stockpile was so low. [31] The "first strike" preventive war was necessary, LeMay argued. This meant destroying 70 Soviet cities within a 30-day period utilizing 133 atomic bombs with almost seven million casualties.

"If you are going to use military force, then you ought to use overwhelming military force. Use too much and deliberately use too much; you'll save lives, not only your own, but the enemy's too."[32] The air force high command might not always appreciate his bombast, but they agreed to his plan in December 1948 and gave SAC top budget priority.[33]

Advocates viewed preventive war, especially in a period of American nuclear monopoly, as a worthwhile plan. They were not all conservatives, such as General Groves and some commentators, but included Truman's science advisor William Golden, Winston Churchill, and even philosopher Bertrand Russell. To others, a first-strike approach in anticipation of an enemy attack might not work, or was unreservedly aggressive. Plenty of US citizens and several political leaders regarded preventive war as morally reprehensible and inconsistent with American democratic values.[34] The First Strike Doctrine, although accepted as a workable scenario in the 1945 JSC document, and supported by the air force, was never formally adopted. It nevertheless remained embedded in discussions of the country's nuclear strategy.[35] The buildup of nuclear weapons capability accelerated as American policy makers accepted the bomb as the heart of American defense policy.[36]

CONTAINMENT EVOLVES

Containment, at least in its initial form, did not remain a strictly political and economic strategy for long. The Cold War intensified globally in 1947–1948. The United States provided military aid to the Greek government, but guerillas fought on. Mao Zedong and communists forces made great strides against the Nationalist government of Chiang Kai-shek in China's epic civil war. Stalin tightened his grip over Eastern Europe when in February 1948 the Soviets clamped down hard on Czechoslovakia. The one bright spot was the US Senate's overwhelming endorsement of the Marshall Plan. Between 1948 and 1952, the program provided $13 billion in economic aid to Western Europe, which played a key role in its economy and helped to promote the integration of a European-wide financial system. With war scares ever present, Truman authorized the CIA to begin covert operations against the communists in the USSR and elsewhere. An early triumph in Italy (by successfully supporting the Christian Democrats against the Communist party) encouraged the CIA to broaden its activities.[37]

Almost inevitably the real test of containment came in Berlin. The odd circumstance of having western zones of occupation in Berlin (which was deep within the Soviet's East German sector) invited conflict. In June 1948 the western powers agreed to establish a West German government. At the same time, the Joint Chiefs of Staff proposed a military alliance headed by an American supreme commander. Containment was rapidly being transformed. The new alliance itself required many more divisions than were available, which necessitated rearming Germany. Such a commitment entangled the United States in European affairs well beyond its historic traditions. Then again, Western European countries were clamoring for protection from the USSR. Between the Marshall Plan and a pending military pact, the Soviets felt extreme pressure. Reindustrialization and possible militarization of Germany under American and Western European control was anathema. Stalin also was facing internal strife from a recalcitrant Marshal Josip Broz Tito who was seeking an independent course for Yugoslavia within rigidly managed Eastern Europe.

Since he had concluded that German reunification was off the table, the Soviet leader questioned why Berlin should be considered as the future capital of Germany. He took a high stakes gamble by implementing a blockade of ground and water traffic into the Prussian city. The Americans responded by placing a counter blockade of goods coming into West Germany. Western powers could abandon Berlin completely, but Truman would not hear of it. Because of the numbers, the United States was unable to face the blockade squarely on the ground or by water. Instead Army Chief of Staff Omar Bradley suggested what became the Berlin airlift. This ingenious plan (during its peak planes landed every 45 seconds) avoided direct contact with the Soviet Union but kept the flow of goods moving non-stop into the western sector of the city. In addition, SAC sent two groups of B-29 bombers to Great Britain, a less-than-subtle reminder of the United States' atomic strength. (This was a bluff because none of the planes carried atomic bombs, nor had crews been trained in bomb assembly.) The blockade stood until May 1949, and much to the Soviet's dismay likely accelerated a unified West German state. It also gave the United States a powerful propaganda tool.[38]

Kennan's non-military approach to containment ended. In April 1949 Truman signed the North Atlantic Treaty. Aside from the United States, the North Atlantic Treaty Organization (NATO) included Great Britain, France, Belgium, the Netherlands, Italy, Portugal, Denmark, Iceland, Norway, and Canada. A new era began as the United States and its allies now situated military assets along an arbitrary line drawn on the European continent between East and West. Depending on one's point of view, NATO was the high point or low point of containment policy.[39] NATO did not only draw a proverbial line in the sand, but it came with an implicit, if not overtly stated, promise to use the bomb if necessary to deter the Soviets. The United States sought to aggressively seek bases for bombers, and to move toward remilitarizing Germany. The USSR upped the ante when it tested an atomic weapon of its own in September 1949. This action, along with the threat of the Red Army, met action with intense reaction. Truman's decision to proceed with the development of the hydrogen bomb continued the dizzying pace of escalation and an unbridled arms race. Mao's stunning successes in China and the domestic nightmare of McCarthyism further intensified the situation.[40]

All of these events, particularly America's loss of its nuclear monopoly, funneled down into a momentous decision in the form of a policy paper called NSC-68 (April 1950). This document encapsulated an utter restatement of national security policy for the United States.[41] Written by National Security Council staff, NSC-68 grew out of a request from the president for a broad review of American foreign policy. Taking a global approach, it incorporated a "worse case view" of Soviet abilities and goals. The policy paper asserted bluntly that the United States unilaterally should prevent any and all Soviet expansion. Among its proposals, NSC-68 advocated a massive buildup of conventional and nuclear arms, strengthening Western European defenses, extending containment into East Asia, and supporting an educational (publicity or propaganda) program to garner popular support at home. This grandiose and expensive commitment to fulfilling the spirit of the Truman Doctrine was meant to achieve lasting military superiority over the Soviet Union. In many respects, NSC-68 may have been the logical extension of the nation's burgeoning Cold War policy, but could its goals be accomplished or, most importantly, should they be attempted at all? For the moment, Truman put the report on hold, but the administration resurrected it quickly as a shooting war became a reality in Korea.[42]

"A WEAPON OF GENOCIDE"

The development and use of the atomic bomb in World War II and the continuing American monopoly afterwards framed US Cold War policy. The development of the hydrogen bomb not only raised the stakes, but elevated the potential level of annihilation beyond imagination. The United States' first detonation of a thermonuclear device took place on October 31, 1952, at Eniwetok Atoll in the South Pacific.[43] The new weapons were "city killers." Revelations about the magnitude and potential impact of the H-bomb were not immediately obvious to the public because of the secrecy surrounding the development of the weapon. Nor did moral qualms and potential health risks of the H-bomb become pronounced until after atmospheric testing, fallout, and open criticisms by some scientists and others.[44]

Edward Teller pushed for the idea of a superbomb while working on the Manhattan Project.[45] The moody and strong-willed Hungarian refugee, gifted physicist, and fervent anti-Nazi and anti-communist viewed the hydrogen bomb as his life's work, and also clearly a personal obsession.[46] The initial idea about a fusion weapon did not begin with Teller. As Richard Rhodes stated, "The idea of a superbomb exploiting the fusion of light elements as well as the fission of heavy elements was a logical extension of basic ideas in nuclear physics known to physicists throughout the world."[47] As early as 1934, Ernest Rutherford and colleagues at Cambridge discovered a hydrogen fusion reaction. In 1941 Tokutaro Hagiwara, a physicist from the University of Kyoto, was the first on record to observe that a fission chain reaction might produce enough energy to allow hydrogen to fuse to helium, and thus create a much larger explosion than fission alone.[48] Almost exclusively in the United States, where sufficient resources were available and where substantial fission research was underway in the 1940s, could speculation about a hydrogen bomb translate into reality in the short term.

Enrico Fermi shared the possibility of nuclear fusion with Teller in September 1941, which motivated the Hungarian physicist to take the building of atomic weapons in an entirely new direction. Fermi speculated on whether a fission bomb could heat a mass of deuterium (a stable isotope of hydrogen) to set off a thermonuclear reaction. If this was possible, deuterium would be easily distilled from seawater and added to much more costly U-235 or plutonium. He calculated that a gram of deuterium converted to helium could release energy eight times that of an equivalent amount of U-235. One fission bomb igniting large amounts of cheap deuterium could produce an unbelievably massive explosion. A hydrogen bomb, therefore, is really a fission-fusion-fission bomb, working through these three stages before detonation.[49]

Teller's first calculations seemed to suggest that fusion would not work, but further efforts were more promising. An important problem to solve was how to get the fission reaction to create enough pressure to allow fusion to occur. At the Berkeley Summer Colloquium in 1942, Teller met with Robert Oppenheimer and other physicists during which time the possibility of a thermonuclear device was among the issues they discussed. One participant suggested using tritium (an isotope of hydrogen) instead of deuterium. Although it was less easily available, tritium could produce thermonuclear reactions at a much lower temperature. By fall 1943 Teller conducted serious theoretical studies of thermonuclear weapons at Los Alamos, although Oppenheimer would have much preferred for him to devote major time to the atomic bomb. (The tension between the two never really abated.) In 1945, with Oppenheimer's reluctant approval, Teller turned full attention to the hydrogen bomb. He enlisted a Polish mathematician, Stanislaw Ulam, to aid him in his work. It was important first to understand the fission explosion in detail, and then turn to the extremely complex thermonuclear calculations. Oak Ridge produced a small amount of tritium for their use, and they utilized the Manhattan Project's heavy-water production not only for reactor research but for thermonuclear studies as well.[50] The thermonuclear computations were so complex that mechanical calculating machines and IBM punch-card sorters that were available could not carry the load. Interestingly, the history of the super and the development of modern electronic digital computers go hand in hand. The hydrogen bomb was the first problem fed to a working electronic digital computer. The crude equipment took six weeks to run a draft of the thermonuclear calculations beginning in December 1945.[51]

A secret conference held in mid-April 1946 at Los Alamos assessed the work on the superbomb and discussed the viable designs. (Klaus Fuchs was in attendance at this meeting.) One salient point was that unlike a fission blast, a thermonuclear explosion had no particular physical limits. As long as a thermonuclear explosion could be ignited and sustained, it would continue while fuel was available. The final report of the meeting stated, "It is likely that a super-bomb can be constructed and will work." However, Washington took no immediate action, especially in the midst of efforts at international control of atomic weapons and with no urgent military demand for such a bomb. Soviet scientists explored the possibility of a super of their own in 1946, and were aware of the American program because of Fuchs. Like the United States, the Soviet Union was preoccupied with other matters, most particularly development of an atomic bomb. In 1947, however, the Soviet government authorized thermonuclear weapons research at the Physics Institute of the Soviet Academy of Sciences.[52]

Two years after World War II, the work on thermonuclear explosions at Los Alamos remained theoretical. Various designs were still under consideration including the super and Teller's so-called Alarm Clock—a spherical, layered design with a more limited yield that, as he said, could "wake up the world." There were still design problems and also the need to develop detailed calculations of a fission explosion to produce the best type of a fission device for the various thermonuclear designs. In 1948 the *Sandstone* test series added a crucial piece to the puzzle by demonstrating that small amounts of fissionable material could produce large yields (useful in the design of a super or other new weapons), and could effectively divide the existing stock of plutonium and U-235 into more usable units. The tests also proved that the implosion method, considered as a possible source of detonation in the first fission phase for an H-bomb, was more efficient than the gun-type mechanism. All of this work did not yet add up to a practicable hydrogen bomb or wide support among Teller's peers.[53]

In the fall of 1949, soon after the unsettling news that the Soviets had detonated an atomic device, Truman requested that the General Advisory Committee of the AEC take up the question of whether or not the United States should develop the new weapon.[54] Teller and his supporters did not like its response. Despite the fact that in the Cold War era concerns over national security were paramount and military requirements for research and development of new weapons were strong, rifts existed within the civilian AEC. In the debate over the super, Lilienthal was much opposed, while former naval officer, businessman, and conservative Republican Lewis Strauss was very supportive. In some instances, moral and ethical reservations along with economic and military considerations surfaced within AEC and among the larger community of scientists, such as Szilard and Fermi. Oppenheimer, who was chair of the GAC, was torn between the sides.

The GAC report completed on October 30 opposed a crash program. Many of the scientists had little trouble supporting the development of the atomic bomb in a race against Nazi Germany, yet the issues were different in this case. What was the military value of such a weapon, especially in peacetime?[55] The report conceded that such a bomb could be constructed and that it could generate limitless explosive power. They drew the line, however, at initiating an "all out" effort to develop a superbomb at that time. Their technical recommendations in the main report called for building more reactors and more isotope-separation plants, utilizing low-grade ores, devising tactical atomic weapons and radiological warfare agents, and even producing tritium for boosting fission bombs. The GAC majority made its recommendations on the actual need for such a weapon, technical issues, the work required to turn theory into a practical device, and the stark impact of its use.[56] In the Majority Annex, the committee strongly warned about the "the extreme dangers to mankind" that "wholly outweigh any military advantage," adding:

> Let it be clearly realized that this is a super weapon; it is in a totally different category from an atomic bomb…Its use would involve a decision to slaughter a vast number of civilians. We are alarmed as to the possible global effects of the radioactivity generated by

the explosion of a few superbombs of conceivable magnitude. If superbombs will work at all, there is no inherent limit in the destructive power that may be attained with them. Therefore, a superbomb might become a weapon of genocide.[57]

The Minority Annex signed by Enrico Fermi and Isidor Rabi went even farther, stating that a weapon of this type could not be justified on ethical grounds: "Necessarily such a weapon goes far beyond any military objective and enters the range of very great natural catastrophes. By its very nature it cannot be confined to a military objective but becomes a weapon which in practical effect is almost one of genocide." It added, as in the majority report, that the United States had in its atomic stockpile "the means for adequate 'military' retaliation for the production or use of a 'super.'"[58]

Teller, Strauss, Senator Brien McMahon (chair of the JCAE), and other advocates of developing the super could not have disagreed more. They perceived the threat from the Soviets in much starker terms, and viewed the H-bomb's extraordinary destructive power not only as an asset for America's nuclear arsenal but a terrible weapon in the hands of an enemy. These were, in their mind, sufficient justifications for moving forward with the program. In addition, Teller was convinced that Oppenheimer in particular led the opposition, and could not forgive him for that alleged transgression. Lobbying the president began. The AEC recommendation on accelerating the super program was a split vote, with three commissioners opposed and two in favor. The Commission sent the recommendation and the GAC report to Truman, who in turn sought advice from a special committee of the NSC created specifically to deal with the H-bomb. Despite some opposition, it reported that the scientists should continue to determine the feasibility of such a weapon. Strong support, especially from the JCS and the JCAE, turned the tide in favor of a crash program. On January 31, 1950, Truman met with the Special Committee for a very brief seven minutes to hear their recommendation, although it was clear that his mind already was made up. The president asked, "Can the Russians do it?" Upon the affirmative reply he curtly stated, "In that case, we have no choice. We'll go ahead."[59]

In the context of the Cold War and Truman's assessment of the Soviets, the decision was hardly surprising. It was not that the president had intentions of using the H-bomb, but he saw the need unerringly to maintain nuclear superiority irrespective of the specific characteristics of the weapon itself.[60] Whether the super was an added deterrent or a grand new offensive weapon seemed not as important as the arms race itself. Also significant was that Truman had not allowed the debate over the hydrogen bomb to become a public issue. The announcement of his decision after the meeting was simply a pronouncement.

Teller began recruiting a team to pursue the development of a workable H-bomb under the code name *Greenhouse*. Given the controversy surrounding it, he had difficulty persuading several of the senior scientists to participate (although Ernest Lawrence and Luis Alvarez were quite sympathetic to his work and lobbied for it). Oppenheimer, obviously, was out, and so were several others. Ultimately, he was able to assemble a mix of men from the Los Alamos staff and younger physicists.[61] After several missteps, Teller and Ulam developed the design for starting thermonuclear explosions. Even Oppenheimer found the new research promising and the designs "technically sweet." The stage was set to push forward the crash program. Teller, however, received a blow when experimental physicist Marshall Holloway, director of weapons development at Los Alamos, was appointed to lead the thermonuclear program in September 1951. Not only was Teller's ego bruised, but he believed that Holloway had regularly opposed the building of the super. The "Father of the H-bomb" resigned immediately, and left "in a huff." Teller had urged the AEC with air force support to develop another weapons laboratory, but to no avail. He had been furtively recruiting scientists for such an eventuality as early as 1950. In 1952 he joined a newly established lab in Livermore, California, under the directorship of Herbert York, where

The *Mike* shot, part of the *Operation Ivy*, took place in the Eniwetok atoll in the Pacific, November 1, 1952. This was the first US test detonation of a thermonuclear device. *Source:* National Archives/NARA via CNP/Newscom

the first assignment was thermonuclear diagnostic studies. The Livermore lab immediately started competing with Los Alamos, focusing on developing innovative bomb designs.[62]

Holloway assembled the Theoretical Megaton Group, the Panda Committee, in October 1951 to build the first thermonuclear weapon. After evaluating several alternatives, it settled on liquid deuterium as the thermonuclear fuel. This was not an easy substance to work with, but the team believed it would give good results. The device itself (*Mike*) was very large and shaped like a thick-walled capsule or a refrigerator, weighing in at 65 tons. The size was necessary to accommodate a fission bomb at one end and the deuterium hanging (with a plutonium "sparkplug") in the middle. At the same time project managers and engineers oversaw production, storage, and transport of liquid deuterium.[63] The project committee estimated that the "Mike device" would produce anywhere from one to ten megatons, or even more. A five megaton blast yielded the equivalent in TNT equal to all the bombs in World War II. As Rhodes asserted, "Steel, lead, waxy polyethylene, purple-black uranium, gold leaf, copper, stainless steel, plutonium, a breath of tritium, silvery deuterium effervescent as sea-wake: Mike was a temple, tragically Solomonic, evoking the powers that fire the sun."[64]

The atoll of Eniwetok was once again the site for another test on American nuclear weapons. This time the results would be more dramatic. The *Mike* test shot, part of the *Operation Ivy* series, took place on Elugelab at the north end of the atoll, while scores of military personnel looked on and more than 500 scientific stations monitored the blast. Assembly began in September 1952, with detonation occurring on November 1. Less than one month earlier on October 3, the British had set off their first fission bomb (*Hurricane*) in the Monte Bello Islands off Australia. But *Mike* was something very different and awe-inspiring. The white fireball was more than three miles across (30 times bigger than *Little Boy*) morphing into a gigantic mushroom cloud. Elugelab disappeared into a crater 200 feet deep and one mile in diameter. Measurements fixed the yield at 10.4 megatons (1,000 times greater than the Hiroshima bomb) but in this case at least 75 percent of it came from the fission reactions, not fusion. That would change in time. Teller was not present to view his ominous creation.[65]

ANTI-COMMUNISM, ESPIONAGE, AND THE FEAR
OF INTERNAL SUBVERSION

There were several types or layers of anti-communist sentiment in the United States before and after World War II, focusing on ultraconservative activists, liberal critics of communism, anti-Stalinist radicals, and partisan politicians.[66] The Cold War fed anti-communist antagonisms, especially because of the Soviets were threatening America's atomic superiority. Public fears about nuclear war itself also increased tensions. Unfortunately, actual threats to American security rapidly got tangled up in a hysteria that blamed spying and infiltration for everything from the Soviet atomic bomb to the disloyalty of high-ranking American officials. That the United States response to pre-war and wartime spying (the so-called "golden age" of Soviet espionage) was largely unsuccessful added frustration and vulnerability to the frenzy.[67]

Evidence now available makes it clear that an extensive Soviet espionage network existed in the United States and that possibly hundreds of Americans spied for or collaborated with the Soviet Union.[68] More effective counterintelligence, expansion of the nation's security apparatus, public exposure of the clandestine activity, and the decline of the Communist Party of the United States (CPUSA) substantially weakened the Soviet espionage efforts by the 1960s. In effect, the postwar anti-communist fury concentrated much more on previous acts of espionage from the 1930s and 1940s than on the current state of affairs. As such, those involved in espionage against the United States often had motives linked to that past rather than to the ongoing Cold War controversies. Soviet spies provided all manner of information to a country bent on survival in the 1930s and 1940s. American collaborators often were sympathetic to the expressed social goals of communism or to a Soviet ally taking on the evils of Nazi Germany. None of the reasons justified their behavior if it included spying, but the sins of the past were being judged by the events of the present.[69]

Anti-communism per se was not a product of the Cold War. One would have to go back to the Bolshevik Revolution in 1917 to discover the American aversion to communism, often reflected in a suspicion of collectivist ideology and simultaneously, if not always logically, in a distaste for authoritarianism. (The latter reflected suspicion of any and all dictatorial power no matter how it was justified or no matter what the political philosophy behind it.) The emergence of the Cold War put a special spin on anti-communism. Despite the fact that the United States was the singularly most powerful nation in the world after World War II, diplomatic setbacks in general and the bomb in particular exaggerated American suspicion and vulnerability, feelings that only got stronger as the Soviets became more of a legitimate rival.[70]

Late in 1945 and 1946, the Truman White House responded somewhat indifferently about FBI reports concerning Soviet espionage. New evidence, information from defectors, and political pressure from conservative Republicans began to change the president's mind. In March 1947 he established the Federal Employees Loyalty Program to root out possible risks. (This was a more rigorous version of a program in place during the wartime Roosevelt administration.) In the same year, the House Un-American Activities Committee (HUAC) opened public hearings on alleged communist infiltration in Hollywood, and other investigations and trials turned to possible sedition among leftist groups. A jaundiced eye was directed at labor unions, universities, and other institutions where subversive activities were thought to be found. Not only conservatives, but some liberal politicians, professionals, and labor leaders, who were part of the growing Cold War consensus, also had little tolerance for communists and their supporters.[71] In addition, the so-called "lavender scare" saw the rise of fear and persecution of homosexuals alongside the anti-communist campaign in the 1950s.[72]

The year 1949 proved to be a dismal one for American diplomacy: The Soviets tested their first atomic device and mainland China fell to the Communist Chinese under Mao Zedong. The loss of the American monopoly over the bomb had come as a shock to those who saw a Soviet test as something likely in the distant future. The United States still had a sizable advantage in the area of atomic weapons, but the gap now began to close slightly. This was an event more psychological and emotional than strategically threatening. Many also saw the "loss" of China as a comedown for the United States. There had been great hope that the Nationalist government under Chiang Kai-shek would lead to the emergence of a new democratic nation in America's own image, but now the momentum moved toward a seeming expansion of the Soviet bloc. Concerns about a shifting global balance of power were real enough, but political hay was to be made and finger pointing began with a vengeance. Right-wing Republicans and some like-minded Democrats were not pleased with the administration's Europe-centered foreign policy that had neglected Asia. More revelations about Soviet espionage also raised the specter that internal subversion had much to do with Mao's victory.[73] International events appeared to be undermining our national interests and espionage at home was tarnishing mythic American invincibility and security.

Observers labeled the period of anti-communist hysteria in the late 1940s and early 1950s "McCarthyism" after the junior Republican senator from Wisconsin. Elected to the Senate in 1946 in a striking off-year victory for Republicans, Joseph McCarthy was not initially the standard bearer of the anti-communist cause, and there is little evidence that his strong espousal of this issue was part of a plan to initiate some kind of national movement. His motives, however, were not pure; anti-communism propelled him into the spotlight for approximately five years, making him as well known as any national figure of his time. McCarthy's public coming-out was a speech he made in Wheeling, West Virginia, on February 9, 1950, in which he asserted that there were 205 communists in the State Department. He had no evidence for such an assertion, and the number changed with successive public utterances. Whether he actually believed the claim or not, McCarthy's charges relied on a "big lie" tactic in which he made accusation after accusation with little proof and a great deal of innuendo. When the accused or his political opponents fought to repute his charges, he simply moved on to a new and different allegation. From a modest committee base in Congress he went after subversives wherever he could find them, from the Voice of America to the US army. The 1954 Army-McCarthy hearings ultimately led to his downfall and censure in the Senate, but not before destroying many careers and breeding deep public suspicion and intolerance. He was not the root cause or the impetus behind anti-communism in the postwar years, but as biographer David Oshinsky persuasively argued, "Above all the senator provided a simple explanation for America's 'decline' in the world."[74]

The outbreak of the Korean War in 1950 only added more tension and more fuel to the anti-communist fire. Out of various attempts to pass some sort of federal anti-communist legislation, an extremist version became law—the Internal Security Act or the McCarran Act. (It carried the name of the author, Democratic senator from Nevada, Pat McCarran.)[75] Although Truman vetoed the bill because he found it personally abhorrent and potentially ineffective, both houses of Congress overrode the veto. The new law declared that communist organizations were "a clear and present danger" to national security (although party membership itself was not a crime); members of communist-action or front groups were required to register with the Attorney General; such members were barred from government or defense jobs or from holding passports; espionage and immigration laws were tightened; and certain free speech rights curtailed.[76] In the poisonous atmosphere of McCarthyism there was little room for political dissent.

ANTI-COMMUNISM AND ATOMIC ESPIONAGE

McCarthy's supporters liked to believe that his actions were justified because Soviets spies actually infiltrated the United States and that some Americans collaborated with them. But much of what McCarthy railed about occurred before his time, did little to effectively root out continuing Soviet espionage activities, heightened the public hysteria, exaggerated the threat of internal subversion, and helped to lay bare civil liberties. The price for his celebrity was just too high. At the heart of the anti-communist frenzy of the early 1950s were public disclosures about several celebrated spy cases, atomic espionage, and the downfall of J. Robert Oppenheimer. The cases spanned the United States, Great Britain, and Canada. While fear of atomic spying per se is not equitable with the anti-communist hysteria in the United States, it clearly heightened the stakes and fueled more passionate reprisals against those who seemingly threatened American security or enhanced Soviet power. Coinciding with McCarthy's rise were some of the most celebrated atomic spy cases in history, several of which came cascading down in a flurry between 1950 and 1953. With the possible exception of Klaus Fuchs, bigger fish got away than those who made the headlines in the early 1950s.[77]

In January 1950 Fuchs confessed to British authorities that he had been a Soviet spy and had passed along top-secret information about the atomic bomb project to his contacts. He had been under suspicion for years, but largely allowed to remain free because of his value as chief theoretical atomic scientist for the British. At the end of a long career as a spy Fuchs had doubts about how Soviet scientists were using the information he was providing. Tipped off by the FBI, British counterintelligence worked for months to slowly pressure Fuchs to confess. Authorities arrested him for espionage in February, and sentenced him to 14 years in an English prison, of which he served nine. American authorities were enraged to learn that the British had known about Fuch's past history as a communist and never informed the FBI. The handling of Fuchs clearly increased ongoing tensions between the two countries over issues related to the bomb.[78] Much to the Americans displeasure, Fuchs was merely stripped of his British citizenship in 1951 and took up residence in East Germany, where he became deputy director of the Institute of Nuclear Research.[79]

After Fuchs's confession, the house of cards began to tumble. On June 15 the FBI picked up David Greenglass who had been identified by informant Harry Gold. Greenglass was a machinist at Los Alamos during the Manhattan Project, communist activist, and the brother of Ethel Rosenberg. Gold, a chemist and the son of Russian Jews, became part of the spy ring with Greenglass, Julius Rosenberg, and others. After his arrest, Greenglass signed a confession admitting that he was a contact for Gold. Because he cooperated with authorities in the Rosenberg case (he lied to protect his wife, but essentially set up his own sister for conviction), Greenglass received a 15-year sentence for passing on information. Two days after Greenglass was arrested, Julius Rosenberg, a Jewish-American electrical engineer who had been involved in the Communist Party, also was arrested on suspicion of espionage. In July Gold confessed that he was a courier for atomic energy information received from Fuchs, and transmitted information to other contacts including Greenglass. Although not a CPUSA member, Gold was a Soviet sympathizer who admired the USSR's public condemnation of anti-Semitism. He pleaded guilty to the charge of conspiracy to transmit documents to a foreign power, and was given a 30-year prison sentence in a plea bargain. He named Greenglass, Morton Sobell, and the Rosenbergs (whom he never met) as atom spies.

On August 11 authorities arrested Ethel Rosenberg, formerly Ethel Greenglass and Julius's wife, followed soon after by Morton Sobell, an American engineer-turned-spy. Some historians have argued that the there were no allegations that Sobell had anything to do with atomic

Julius and Ethel Rosenberg were executed as "atomic spies" on June 19, 1953. *Source:* Library of Congress, Prints & Photographs Division, NYWT&S Collection, [LC-USZ62-127232]

espionage, and that he was really a "sideshow" in this investigation. Ethel was arrested largely to put pressure on Julius to talk. On March 6, 1951, the controversial trial of Julius and Ethel Rosenberg and Morton Sobell began. The Rosenbergs were presumed to be the leaders of the American spy ring that included Greenglass, Sobell, and Gold. In April the Rosenbergs and Sobell were indicted for conspiring to commit espionage, but federal prosecutors decided to treat the case as if it were treason, which was a much more serious crime than other alleged spies had faced in England or in the United States before them. Since Greenglass and Gold turned state's evidence, the Rosenberg's were left in the spotlight. They continually proclaimed their innocence and refused to cooperate with authorities (which proved to be a major mistake). They received strong support from a variety of well-known figures, including Albert Einstein, and from average people who felt sympathy for the parents of two young sons facing a fight for their lives. The trial itself was a fiasco. The defense handled the Rosenberg case poorly, and critics wondered out loud to what degree being Jewish and American Communist Party members had to do with Julius and Ethel's predicament. The Rosenbergs were sentenced to death, while Sobell received a 30-year sentence (of which he served less than 18 years). On June 19, 1953, after appeals by the defense and calls for leniency, Julius and Ethel Rosenberg were executed in the electric chair at Sing Sing Prison, New York.[80]

The *Venona* cables and other government files show that Julius was definitely a spy for the Soviets, but debate continues as to whether he actually headed the spy ring and if he ever handled atomic secrets. Ethel was at best a minor operative, and the evidence presented and material now available do not prove her direct involvement other than knowledge of her husband's activities. It appears that Fuchs, Gold, Greenglass and others may have played much bigger roles in the wartime espionage activity than the Rosenbergs, but they paid the higher price.

The prosecution's focus on treason rather than a lesser crime and the harshness of the sentence indicated, among other things, the rising intensity of McCarthyism. It also suggests how powerfully the anxiety over the bomb had come to the shores of the United States, even before the "fallout scare" that was soon to sweep the nation as a result of H-bomb testing in the Pacific. Giving away bomb secrets to the Soviets, superimposed over anti-communist sentiment, was a dangerous mixture. Were the crimes considered heinous because atomic espionage was the height of treason? Or were Americans embarrassed because Soviet agents and their confederates had appeared to undermine US nuclear superiority? The real threat to security was most likely intermixed with a symbolic one.[81]

THE OPPENHEIMER CASE

Scientists, even those who had become virtual "rockstars" after the success of *Little Boy* and *Fat Man*, were not immune from the anti-communist paranoia in the early Cold War and the anxiety over the arms race. Robert Oppenheimer himself became a favorite target of speculation. Given his public notoriety, political background, leftwing associations, growing unpopularity with officials and scientists bent on developing the H-bomb, and the poisonous anti-communist atmosphere revisiting his past should not have been a surprise, but it was. Oppenheimer had always been tepid about the development of the H-bomb; not as negative as some like Fermi, Rabi, and James Conant, but certainly not as enthusiastic as Teller. For his part, Edward Teller harbored resentment against Oppie going back to their days at Los Alamos. Oppenheimer also gathered staunch enemies in the air force and in the federal government, none more critical than AEC member and then chair, Lewis Strauss; William Liscum Borden, former staff director of the JCAE; and head of the FBI, J. Edgar Hoover himself. Strauss was particularly riled about Oppenheimer's opposition to the superbomb, convinced that the former director of the Los Alamos facility was the chief culprit in that opposition. Borden tried to convince Hoover that Oppie was "more probably than not" an agent of the Soviet Union. And Hoover was quick to believe the worst. Lurking in the background was Joe McCarthy. While Truman was no great fan of Oppenheimer, incoming president Dwight D. Eisenhower gave Strauss authorization to proceed against Oppenheimer by putting a "blank wall" between him and the AEC. The new president did not want to be accused of harboring communists, give McCarthy new fuel to exploit, or allow the Wisconsin senator to get too close to the atomic energy program.

In 1953 Oppenheimer was a scientific consultant to the AEC (an important "nuclear insider" as some called him) with a contract ready to expire in June 1954. Strauss's first action well before that deadline was to inform the physicist that he no longer could have access to secret documents. This essentially would strip Oppenheimer of his ability to participate in national security matters. In response Oppie requested that Strauss's decision be brought to an inquiry committee. The new general manager of the AEC (another nemesis), General Kenneth D. Nichols, began preparations for that hearing. A Personal Security Board (the Gray Committee named after the chair Gordon Gray who had been president of the University of North Carolina and former assistant secretary of the army) was set up, and was clearly unsympathetic to Oppenheimer. The committee also included Thomas Alfred Morgan, the retired board chair of Sperry Gyroscope, and Ward V. Evans, a professor of chemistry from Loyola University in Chicago. Hearings were to begin in April 1954—an event more like a trial than a hearing. Holding the inquiry in secret had several advantages for Oppenheimer's antagonists, not the least of which were avoiding a public spectacle, protecting important state secrets, and avoiding normal legal rules and procedures found in the courts. (The Gray Committee was in sessions during the so-called Army-McCarthy hearings that proved to be the Wisconsin senator's undoing.)

Ironically, but maybe not surprisingly, most of the AEC charges brought against Oppenheimer were for dealings taking place before 1947, of which the AEC already had cleared him. During the war and soon after, when Oppie's value to the country was most critical, government officials (especially General Groves) overlooked or rationalized past indiscretions and behaviors. Times and circumstances changed by 1953. Some of the same officials deeply examined and scrutinized Oppenheimer's past associations and actions. As with other loyalty and security investigations during the era, those associations and actions going back to the 1930s became fair game. The renowned physicist had not only associated with communists, several among his family and friends, but he had been active for a time in leftist Popular Front causes from anti-Nazism to the plight of migrant workers in California. Playing the role of unofficial prosecutor, AEC's Roger Robb had at his disposal surveillance tapes of Oppenheimer's discussions with his own lawyer, and worked aggressively to destroy the physicist's credibility. Oppie's three-man team of lawyers, headed by prominent civil attorney Lloyd Garrison, was not prepared for the rough and tumble nature of the hearing (Garrison was not a trial lawyer), and was denied access to information necessary to protect their client. This included all classified documents and bursting FBI files and surveillance records. Such materials, however, were available to the other side and to members of the board.

The Father of the A-bomb never stood much of a chance. On June 1, the Gray Committee voted two-to-one against Oppenheimer's security clearance, and the AEC by a vote of four-to-one sustained that decision one day before his clearance was due to expire. Gray's report of May 27, 1954 (with the concurrence of Morgan) concluded that Oppenheimer's "continuing conduct and associations have reflected a serious disregard for the requirements of the security system." It also found that he had "a susceptibility to influence," that his conduct in the hydrogen-bomb program was "sufficiently disturbing" to raise doubt about his future participation, and that he "has been less than candid" in several instances in this testimony. (Evans in a minority report, however, favored reinstatement of Oppie's clearance although he had some reservations.) Strauss and the majority of the AEC in making their decision stressed Oppenheimer's connections to communists and other left-wing groups, his "obstruction and disregard of security," and his lack of "veracity." Strauss himself raised questions about the physicist's "fundamental defects of character and imprudent dangerous associations." [82]

Oppenheimer indeed had changed some of his stories about his associations, and continued to maintain contact with his communist friend, Haakon Chevalier. In his opening statement to the Gray committee, he concluded, "In preparing this letter, I have reviewed two decades of my life. I have recalled instances where I acted unwisely. What I have hoped was, not that I could wholly avoid error, but that I might learn from it. What I have learned has, I think, made me more fit to serve my country." But it was never proven nor was it evident in any record that Oppenheimer was a real security risk. Certainly his past associations, but also his personal style, his sometimes prickly behavior, his ambition, and his resistance to support the H-bomb probably carried more weight in those anxious times than any concrete evidence against him. Soon after the hearing, Oppenheimer lost his position as a government adviser, and returned to private life at Princeton University, where he had previously directed the Institute for Advanced Studies, and where FBI surveillance continued at his home. Not until almost 10 years later in 1963 did the Los Alamos director receive the Enrico Fermi Award for his pioneering work in nuclear physics from then President Lyndon Johnson. Whether this was a hollow gesture or atonement of some sort from a grateful government remains unclear. But Oppenheimer was never the same once in exile from the inner circle of atomic science and nuclear policy formation. Biographers Kai Bird and Martin Sherwin stated, "In assaulting his political and his professional judgments—his life and his values really—Oppenheimer's critics in 1954 exposed many aspects of his character: his ambitions and insecurities, his brilliance and naïveté, his determination and fearfulness, his stoicism and his bewilderment."[83]

There were immediate and longer-range repercussions of the decision. Exponents of the McCarthy inquisitions applauded the results of the Oppenheimer hearings. Among many colleagues and supporters, Oppie had been martyred, and soon the discrediting of McCarthy and his tactics reconfirmed that sentiment. (The National Academy of Sciences refused to issue a strong letter of support for Oppenheimer.) Teller, while influential within the halls of government and the military, became a pariah among many of his scientific colleagues. In one bit of poetic justice, political columnists Joseph and Stewart Alsop vilified Strauss in the press for denying that he taped AEC meetings and tapped the phones of some commissioners. After his term ended as AEC chair in 1958, Strauss declined reappointment by Eisenhower, believing that the Senate would not confirm him. Democratic senatorial opponents cut off another path for Strauss when they orchestrated his defeat as Secretary of Commerce the next year.[84] The Oppenheimer case had plenty of consequences to go around, but it also demonstrated in real terms how those waving the anti-communist banner abused power in these years and how they threatened civil liberties. Possible atomic espionage played a very small part in this case. It was more a clash of egos, a modest referendum over nuclear policy, and an effort to settle many scores.

SCIENCE ON THE DEFENSIVE

Well before the Oppenheimer case, scientists had fallen under scrutiny and some under suspicion in the early Cold War. While a few thrived in their roles as "scientist-administrators," many others faced pointed questioning about their role in the postwar world and, in some cases, about their views. Vannevar Bush, Conant, Lawrence, and (until his "trial") Oppenheimer, played important roles in managing or overseeing large projects, serving as top advisers to civilian and military leaders, and mediating interactions among scientists, the military, and government bureaucrats and policy makers.[85] For research scientists in various laboratories throughout the country, World War II produced a "powerful patron" in the form of the military, building what appeared to be a potential alliance between science and the state. A significant product of the wartime alliance, the atom bomb, created ambivalence and confusion in the scientific community on many levels that quickly spilled over into the Cold War. In this context, the response of scientists to anti-communism was mixed and often inconsistent.

Those who worked or continued to work in government labs and on projects linked to national defense found themselves trapped in an increasingly repressive environment, in an altered political setting with many obstacles and even risks. That a substantial number of them were foreign-born, several of them Jewish, and in large measure the product of differing social and political cultures, made the scientists particularly susceptible to the nativist bent of anti-communism in postwar America.[86] The secrecy and security that the Los Alamos scientists came to expect and tolerate was extended rather than relaxed in the Cold War years, especially with the revelations of Soviet espionage. Historian Daniel Kevles observed, "To the chagrin of most physicists, and the apprehension of some, the Cold War not only produced an escalation of the arms race; it also put barbed wire and guarded gates around the Radiation Laboratory at Berkeley."[87] The age-old problem among scientists about the need for open discussion was constantly stymied by the demands of secrecy, and some argued that such strictures hindered the development of atomic power. With the public exposure of Klaus Fuchs, in particular, security measures grew particularly rigorous and extended beyond the nuclear facilities. Security clearances, for example, were now required for AEC fellowships. In several cases the State Department refused to extend visas to foreign scientists even if they were not engaged in nuclear physics.[88]

Most overt were direct attacks on the loyalty of some scientists. In 1948 HUAC branded Edward U. Condon, of the National Bureau of Standards and president of the American Physical Society, as "one of the weakest links in our atomic security" for allegedly entertaining or associating with Soviet agents. Condon responded to the allegations, "I have nothing to report. If it is true that I am one of the weakest links in atomic security that is very gratifying and the country can feel absolutely safe for I am completely reliable, loyal, conscientious and devoted to the interests of my country, as my whole career and life clearly reveal."[89] He was cleared fully, and President Truman made a public gesture on his behalf. Yet several more scientists, including theoretical physicists, suffered public rebukes and worse for their politics, religion, or "un-American" views—even Albert Einstein. The security issues and constant suspicions drove many scientists to quit the government laboratories in the late 1940s and early 1950s, and made it very difficult for recruiters to solicit physicists to work at Los Alamos or other similar facilities. The Condon incident, and most especially the Oppenheimer case, severely chilled the scientific community and led many physicists and others to question what role in the government they might play or want to play in the future.[90]

KOREA AND AMERICA'S NUCLEAR STRATEGY

The Korean War was an early test of the Cold War operating in Asia, and an early check on our nuclear strategy in the face of a "hot war."[91] Japan annexed Korea in 1910, and by the end of World War II it had yet to be recognized as an independent nation. After the war the Allies hastily created occupation zones in Korea believing that the same approach used in Germany might provide some immediate stability until a national election could be held. Below the 38th parallel the United States backed the conservative government of Syngman Rhee of South Korea (Republic of Korea) mostly through wealthy landholders. In what became North Korea (Democratic People's Republic of Korea), the Soviets supported the leftist government of Kim Il-sung. Both regimes wanted unification, and both sides engaged in intermittent armed conflict across the artificially drawn border beginning in 1948.

The internal conflict quickly enveloped the Cold War antagonists. Initially the United States backtracked on its involvement in Korea. In 1949 Truman withdrew American combat troops believing that South Korea was outside the "Asian Defense Perimeter." The Soviets also withdrew their troops from North Korea. After the fall of China, Korea regained importance geographically because of its proximity to the People's Republic and to the Soviet Union. In addition, its potential value as a market for Japanese goods was strong. Meanwhile, Kim relentlessly put pressure on the Soviets (and the Red Chinese) to support an invasion of South Korea. After initial resistance, Stalin gave a qualified commitment in April 1950. He was emboldened by the Soviet's test of its own atomic bomb in 1949, and by the belief that United States involvement in Korea (even a lukewarm commitment) would drive Communist China closer to the Soviet Union. With Stalin's approval, Kim believed that nothing stood in the way of the invasion. He sent 100,000 troops and other military assets across the 38th parallel beginning on June 25.[92]

The action was a complete surprise to Truman, but the quick American response was equally stunning to Stalin. What prompted the president's swift action was the belief that the USSR orchestrated the attack as part of a belligerent outward-looking strategy. From a political perspective, Truman was not willing to give the Republicans cause to brand him as weak in the face of communist aggression. Rather than seeking a war declaration, Truman exacted backing from the United Nations Security Council for a military commitment; he was able to do so because the Soviets were boycotting the UN for refusal to seat the People's Republic. The president branded the North Koreans as aggressors, thus skillfully avoiding the linkage to the Soviets in explicit terms. With the UN pledge in hand, and declaring Korea "a policing action," Truman moved ahead swiftly and assertively.[93] A week before sending ground troops to Korea, the United States, in a reversal of policy, placed the Seventh Fleet between Taiwan and mainland China, thus openly supporting Chiang Kai-shek's continued resistance to the Maoist government. The Truman administration also increased aid to the French in Indochina and counterrevolutionaries in the Philippines. These actions clearly indicated that the emerging Korean conflict was situated within the broader Cold War policy framework as outlined in NSC-68.[94]

Initially the North Korean forces overran South Korea. But General Douglas MacArthur's bold gamble of an amphibious assault on the port of Inchon caught the North Korean troops in a vice grip that trapped them below Seoul (the South Korean capital). The victory drove the enemy back across the 38th parallel. MacArthur and Washington officials, flushed with success, chose not to return to the *status quo ante bellum*, and to liberate all of Korea. In doing so, they ignored the warning of the Chinese not to approach their border along the Yalu River. Chinese leaders were aware that the United States might use atomic weapons, but in late November 1950 more than 200,000 soldiers from the People's Republic crossed the river and attacked the UN forces. In 1951 the war settled into a high-attrition standoff between the sides.[95]

Throughout the war, conventional bombing was tremendously destructive. Despite his misgivings, General LeMay employed SAC airpower in support of General McArthur's forces. LeMay had been resistant to participating in the conflict because he wanted to keep his units intact and wanted more control in the decision making over strategic bombing strategy. Nevertheless, in their first four months of engagement, SAC dropped approximately 4,000 tons of bombs supporting UN ground operations.[96] There was no decision to use nuclear weapons in the war, although the Truman and Eisenhower administrations seriously contemplated it. According to some sources, the American atomic stockpile at the beginning of the Korean War stood at 298 bombs; the Soviet Union possessed about a dozen. By the end of the war in 1953 the United States had 1,161. Accompanying this significant increase was a parallel growth in nuclear-capable aircraft from 250 in 1950 to more than 1,000 in 1953.[97] From the perspective of the Soviet Union an understanding of America's nuclear-weapons strength and its policy in the Cold War was guided by two principles. First, Stalin believed that in the short term the United States would use the bomb to intimidate his country and try to extract concessions, rather than start a war. Second, the Soviet Union did not want war, nor was it ready for it. Both principles led the Soviet leader to conclude that his country should expose no weaknesses, but also avoid inciting any provocative actions. This led Stalin to pursue a "war of nerves."[98]

Just how close Truman came to giving the order to use A-bombs is debatable, but the subject itself arose on several occasions. At his first meeting with the Joint Chiefs and Secretary of State Acheson, the president asked for a study of the use of atomic weapons should the Soviet Union enter the war. Soon after, McArthur requested some bombs be transferred to the Far East Theater with the possibility of using them to block possible direct participation by the Chinese or Soviets. The United States ordered 10 B-29s with unarmed atomic bombs to Guam in August 1950. The nine that arrived (one had crashed en route) were placed under LeMay's control because of his strenuous insistence, but were recalled in September. The bombs themselves remained in Guam. The Pentagon's view was that atomic weapons could not be effectively deployed in

Korea. Truman, in a November 30 press conference, confirmed after previous denials that he considered using atomic bombs in Korea, and that McArthur would have had control of them had such an authorization been given. This was alarming news particularly for the British and French who saw the war as possibly widening to include China.[99]

In December 1950 the war continued to escalate, and Truman declared that he would increase the armed forces by 3.5 million men. Neither the JSC nor the State Department was prepared to support the use of atomic weapons at that time. As the war dragged on, Truman considered seeking a cease fire in the costly and increasingly unpopular war. McArthur and the Joint Chiefs opposed the idea, and the field commander in particular was becoming openly critical about the president's position. Again McArthur called for the possible use of atomic weapons in spring 1951. In an act of public insubordination, he wrote a letter to House minority leader Joseph Martin, a letter soon to appear in the *Congressional Record*, that there was "no substitute for victory." The UN Commander essentially was calling for all-out war against China.[100]

In April the Soviets moved approximately 200 bombers to Manchuria. In addition patrols of submarines indicated that the Soviets were preparing themselves if the United States decided to attack beyond North Korea. This threat led the JCS to order retaliation through the use of atomic weapons if the Soviets staged an assault on UN forces. Truman saw this as a chance to remove McArthur, since he told the chiefs that he could not entrust nuclear weapons to the uncontrollable commander. Officials transferred nine nuclear capsules to the air force on April 10, and the next day Truman sacked McArthur. There was much public outrage directed at the president for firing a wildly popular military hero. At about this time, the Truman administration gave serious thought to ending the war through negotiation, although Stalin was content to see the United States mired in the conflict which dragged on beyond the end of Truman's term in office.[101]

What was not known, except among a relatively small circle of officials, was that the first US military deployment of atomic weapons since the bombings of Hiroshima and Nagasaki occurred at that time. The bomber squadron and the bomb components remained in Guam until the end of the year, when a nuclear option in Korea seemed increasingly unnecessary. The nine bombs sent to Guam were not returned to the AEC, but remained in air force hands. Truman was chastened by the potential deployment of atomic weapons again in Asia, even though he privately mused whether such weapons could end the war quickly. The larger lesson for civilian and military officials was that despite their destructive power, atomic weapons possessed specialized or even limited military and diplomatic usefulness. This apparently was the case in Korea, where good targets were few and where use of the weapons against the Chinese or the Soviets presented grave consequences.[102]

CULTURAL ARTIFACTS IN THE EARLY ATOMIC AGE

The early Cold War and the bomb produced a wide array of cultural artifacts on radio, and through songs, comic books, and science fiction novels.[103] Film in particular captured moods and themes through multiple senses, providing insights that are often difficult to retrieve from historical documents. While an outpouring of films with atomic themes awaited the mid-1950s and beyond, the few produced and released by 1953 were suggestive of the perceptions, fears, anxieties, and pressures of the burgeoning Cold War and atomic age. Invasion, subversion, infiltration, espionage, and apocalypse found their way into what are now valuable documents from that time, documents that provide at least some visceral understanding of the encounter with the bomb in an early Cold War context.

In science fiction movies Martians and other aliens substituted for the Soviets, possessing the aggressiveness and wiliness of an odious rival. George Pal's *War of the Worlds* (1952) bore little resemblance to H.G. Wells's novel. However, images of the merciless attack of the Earth by waves of flying saucers graphically depicted naked, godless aggression in vivid terms. In an early scene, an alien ship turns its heat ray on Pastor Matthew Collins (Lewis Martin), a minister who offered his hand in peace, a shocking assault on Christianity without remorse. The hero of the movie, physicist Clayton Forester (Gene Barry playing something of an Oppenheimer clone) also is aggressively pursued by the aliens. When counterattack by conventional forces proved useless, military leaders turned to atomic weapons, but they too were utterly ineffectual. What finally dooms the Martians is not military prowess but an earthbound virus, which has the saucers crashing en masse with at least one plowing into the side of a church as if in memory of the good pastor Collins. The symbolism of all this is muddled, but the optimistic ending affirms a second chance for humanity.[104]

One of the best science fiction films of any generation depicting infiltration and internal subversion is *Invaders from Mars* (1953). The story is told through the eyes of young David MacLean (Jimmy Hunt), who witnesses the landing of Martians quite near his home, and their relentless efforts to colonize Earth by turning humans into a society of collectivist automatons. The nightmarish film has everyone around David, including his scientist father, his homemaker mother, and every other authority figure, surgically altered for the sake of control. The theme is repeated with great effect in *The Invasion of the Body Snatchers* (1956), but it lacked the device that made *Invaders from Mars* so powerful and terrifying—the use of a young boy as the central figure. Wise beyond his years, David (with the help of a male scientist and a female doctor) convinces the military that the Martian ship must be destroyed. As an explosion to destroy the Martian spaceship occurs, David wakes up from what seems to be a terrible dream, only to discover that events in the dream repeat themselves. The takeover, as a result, becomes all too real, and much more unsettling than *War of the Worlds*.[105] While atomic imagery plays no role in this film, it captures Cold War anxieties in a very personal way.

Straight-forward films about atomic espionage were in short supply in the early 1950s, other than through the deep camouflage of science fiction. This is what makes the low-budget *The Atomic City* (1952) such an interesting film. It takes espionage head on, but in a patriotic and somewhat modest way. Frank Addison is a nuclear scientist (again played by Gene Barry) who works and lives with his family behind the protective walls of a compound in Los Alamos. In the beginning of the movie, director Jerry Hopper utilizes documentary-like footage of actual New Mexico locations to convey a sense of authenticity. The lives of Frank and Martha Addison (Lydia Clarke) are turned upside down when "agents" (with American or non-descript accents) kidnap their son Tommy (Lee Aaker) and hold him for ransom in exchange for secrets to the H-bomb. Addison, with the aid of the FBI, track the kidnappers to Los Angeles where they eventually are captured and Tommy is released unharmed. The story quickly devolves into a rather halfhearted action film. What is unique is not the story line, but the espionage setup and the Addison family's life behind the barbed wire of Los Alamos. In one scene, Tommy tells his friend in a passing conversation, "If I grow up" instead of "When I grow up," a phrase Martha quickly picks up on and shares with Frank. The implication is clear that Dr. Addison's work, the secrecy around it, the family's life style, and the bomb have affected Tommy's anticipation of the future. Also insightful is the portrayal of the FBI as heroic steely-eyed professionals (although there is a little bumbling now and again). Throughout the 1950s, authority figures (police, federal agents, military officers, doctors, scientists, and even a few politicians) were generally portrayed with reverence and respect, befitting the conventional acceptance of authority in that period.[106]

Apocalypse has been portrayed in many ways and in many periods. Fear of the bomb and fear of human annihilation because of it, crept into the American cinema in the early 1950s. The distant horrors of Hiroshima and Nagasaki came home with the increasing number of atomic tests in the Pacific and the pressures of the Cold War. One pertinent theme in this genre was science running amuck. *The Beast from 20,000 Fathoms* (1953) was among the first of such films. Based loosely on Ray Bradbury's "The Foghorn" and brought to life by the pioneering special effects of Ray Harryhausen, the storyline is archetypal: A nuclear bomb test awakens a giant prehistoric dinosaur, who returns to its place of birth (in this case Manhattan Island) and wreaks havoc. Rather than social commentary *per se*, this was obvious exploitation of an issue of grave public concern at the time, raising questions about human responsibility for tampering with nature. Also typical was the use of an almost knee-jerk military reaction as a way to bring resolution. In this case, a scientist not the army finds a way to subdue the monster, and in an ironic turn harnesses atomic energy (a radioactive isotope) as the weapon of choice. Pitting military force against science becomes a common theme in many science fiction films of the 1950s. In this case, the story focuses, albeit indirectly or even unintentionally, on the paradoxical attitudes toward atomic energy and atomic weapons, that is, on a power with contradictory properties.[107]

More broadly apocalyptic is Arch Oboler's preachy morality play, *Five* (1951), which also spawned a genre (or subgenre) of its own. Based on Oboler's radio drama, "The Word," it was the first movie to depict survivors of a nuclear holocaust. The group is a cross-section of society (only two of the five remain by the end of the film) setting off to restart "a new Earth." The survivors meet at a mountaintop lodge along the California coast, where the domineering Michael (William Phipps) is trying to reestablish a small community. Joining him is Mr. Barnstaple (Earl Lee), an elderly white man who dies presumably of radiation poisoning; Charles (Charles Lampkin), an African American bank attendant, who is obviously on the outside of this group looking in; Eric (James Anderson), the sportsman and evil element who kills Charles; and a pregnant Roseanne (Susan Douglas Rubes), who ultimately loses her baby but becomes the protagonist's mate. Utopia this is not. What is predictable for a 1950s film is that the surviving two of five become a couple, a new Adam and Eve who hope to build a Garden of Eden cleansed of the problems of the past.[108]

The Day the Earth Stood Still (1951), directed by Robert Wise, is probably the most remarkable cautionary tale in the period. In fact the film received the 1952 Golden Globe as the "Best Film Promoting International Understanding." It was not equally applauded by the Department of Defense, however, which disapproved of the message and would not let Wise use their equipment. (He turned instead to the National Guard.)[109] Still highly regarded today, the film was based on a 1940 story written by Harry Bates for *Astounding Magazine*. An alien, Klaatu (Michael Rennie), lands his gigantic spaceship on the Mall in Washington, D.C., accompanied by his towering robot, Gort (Lock Martin). The spaceship is quickly surrounded by military men and equipment, legions of press, and throngs of passersby. As Klaatu emerges from his ship, he announces, "We have come to visit you in peace, and with good will." But as he attempts to remove an object from his spacesuit (we later discover it is a gift for the president) a nervous soldier shoots the visitor. In response, Gort unleashes a ray that kills or otherwise incapacitates the assailants. The alien is then removed to Walter Reed Hospital for treatment, where he meets Mr. Harley (Frank Conroy), the president's representative. Klaatu makes it clear that his mission is not to meet with any single government leader, but representatives from all countries. The American official does not come off very well in this exchange, expressing the wariness of every Cold War leader at the time. Nor does the military shower itself with glory because of its hostile reaction to Klaatu.

Possessing great healing powers, Klaatu is able to escape the hospital and hides out in a boarding house. Suspicions abound about this stranger who is hiding amongst them, very much in tune with the anxieties of the day linked to espionage and infiltration. Klaatu befriends a widow (Patricia Neal) and her young son Bobby (Billy Gray). Klaatu and Bobby seek out Professor Jacob Barnhardt (Sam Jaffe), an Einstein-like physicist, hoping to rally the world's scientific community to hear his message. In the course of the story, Klaatu is discovered and appears to be killed. The messianic alien, however, is resurrected through the advanced technology aboard his ship and finally makes his speech to the world's community of scientists. He appeals to them for an end to the nuclear arms race and calls for global cooperation. Klaatu makes it clear that his own people were once a violent race, and found the only solution was placing their fate in the hands of a robot police force which would insure peace—or else. In a reversal of roles, Gort becomes the master and Klaatu his subject. The Earth, therefore, must find a way to unify and seek peace or be annihilated. "It is no concern of ours how you run your own planet," Klaatu states, "but if you threaten to extend your violence, this earth of yours will be reduced to a burned-out cinder."[110]

The story is deceptively pacifist in tone. While the film criticized the Cold War consensus of the late 1940s and early 1950s, Klaatu is saying that humans do not have the capacity to control their own actions and must rely on their technology (or some third-party technology) to keep them from their worst instincts. That message appears to get lost in the seeming appeal to a single world order built on mutual respect and peace. The cautionary element in the story, therefore, is a rather pessimistic message about the limitations of human behavior and action, more than a morality play condemning the Cold War standoff. Nuclear weapons are at the root of potential global destruction, but the message is much more of an indictment of human nature. All of these films speak to colliding feelings and emotions of the day. While not always explicitly asserted, the bomb hung uneasily over the global events of the early Cold War and the suspicions and intolerance brewing at home.

CONCLUSION: THE BOMB AS COLD WAR EMBLEM

Hope, promise, and risk? Many questions remained about America's hope for security despite its overwhelming advantage in military weapons in the first decade of the Cold War. The promise of nuclear power was still at issue for the future. And risk seemed everywhere. The confrontation between the United States and the USSR was framed and defined by A-bombs and H-bombs and all of their trappings. Containment as a general policy and NATO as a military bulwark of West against East were directly linked to nuclear weapons. These weapons were, at the very least, an offset to the potentially menacing Red Army. Changes in the American military structure demanded a place for nuclear weapons, and the rise of SAC was a good indicator of how central those weapons became. Nuclear policy making per se was a different question. The atomic bombs of Hiroshima and Nagasaki seemed to fit the thrust of total war in World War II, but in a new postwar environment, one defined as much by the maintenance of peace as the prosecution of war, how to utilize these new weapons was problematic.

With the H-bomb so recently on the scene in this period, speculation about the leap from A-bombs to city killers was only beginning. The new fearsome weapon had the capacity to annihilate civilizations, and thus its use was potentially apocalyptic. But it was not only the appearance of the H-bomb, but the whole nuclear weapons mix (tactical and strategic, atomic and hydrogen) that complicated the strategy in nuclear deployment. The experience in Korea was an early test of possibly bringing the bomb into a new conflict. And the arms race that had just begun was quickly becoming the central feature of bilateral decision making for the United States and the Soviet Union. Nuclear weapons were the queen in the game of chess, with multiple moves and great leverage in capturing the king. The pressures of the Cold War, and the ominous role of nuclear weapons in it, resulted in internal convulsions in the United States manifest in the anti-communist crusades. Were the witch hunts serious attempts to root out subversives? Were they a form of self doubt about American preeminence and the country's ability to protect its own territory, its own military secrets? Did they reflect the frustration of losing the atomic monopoly or getting bested in global diplomacy? The anti-communist crusades were all of these things, and they consistently looped back to the reality that ending World War II did not, in and of itself, lead to peace and stability. In many respects, the transitioning out of World War II into the Cold War was leading deeper into an Age of Anxiety.

MySearchLab Connections: Sources Online

READ AND REVIEW

Review this chapter by using the study aids and these related documents available on MySearchLab.

✓•⎯⎡**Study** and **Review** on **mysearchlab.com**

Chapter Test

Essay Test

▭•⎯⎡**Read** the **Document** on **mysearchlab.com**

George F. Kennan, The Long Telegram (1946)

George C. Marshall, The Marshall Plan (1947)

The Truman Doctrine (1947)

FBI Report on Klaus Fuchs (1951)

Decision and Opinions of the United States Atomic Energy Commission in the Matter of Dr. J. Robert Oppenheimer (1954)

Dissenting Opinion of Henry De Wolf Smyth in Decision on J. Robert Oppenheimer (1954)

RESEARCH AND EXPLORE

Use the databases available within MySearchLab to find additional primary and secondary sources on the topics within this chapter.

◉⎯⎡**Watch** the **Video** on **mysearchlab.com**

Berlin Airlift

Operation Ivy

McCarthyism and the Politics of Fear

Endnotes

1. The Allies created three zones in the western half of Germany (and in the western half of Berlin) held by the United States, Great Britain, and France, and they established one zone in eastern Germany (and also in the eastern half of Berlin) held by the Soviet Union.

2. See George C. Herring, *From Colony to Superpower: U.S. Foreign Relations Since 1776* (New York: Oxford University Press, 2008), 595-601. Stephen E. Ambrose, *Rise to Globalism: American Foreign Policy Since 1939* (New York: Penguin Books, 1983; 3rd rev. ed.), 92–103, 106–11; Wilson D. Miscamble, *From Roosevelt to Truman: Potsdam, Hiroshima, and the Cold War* (New York: Cambridge University Press, 2007), 307–332.

3. Andrew J. Rotter, *Hiroshima: The World's Bomb* (New York: Oxford University Press, 2008), 248, 253.

4. Apparently the term was first used in a communiqué by President Truman, British Prime Minister Clement Atlee, and Canadian Prime Minister Mackenzie King in late 1945 after the bombing of Hiroshima. The leaders recommended setting up an international commission to make proposals for eliminating atomic weapons and "all other major weapons adaptable to mass destruction." Nina Tannenwald, *The Nuclear Taboo: The United States and the Non-Use of Nuclear Weapons Since 1945* (New York: Cambridge University Press, 2007), 102. See also 98–101, 103–04.

5. Ambrose, *Rise to Globalism*, 105.

6. Martin J. Sherwin, "The Atomic Bomb and the Origins of the Cold War," in Melvyn P. Leffler and David S. Painter, eds., *Origins of the Cold War: An International History* (New York: Routledge, 2005; 2nd ed.), 58–71; Rotter, *Hiroshima*, 247–55.

7. Herring, *From Colony to Superpower*, 598.

8. Melvyn P. Leffler, *For the Soul of Mankind: The United States, the Soviet Union and the Cold War* (New York: Hill and Wang, 2007), 79–83.

9. Quoted in Jeffrey Porro, ed., *The Nuclear Age Reader* (New York: Alfred A. Knopf, 1989), 33.

10. Quoted in "Truman: The Presidents," *American Experience*, online, http://www.pbs.org/wgbh/amex/presidents/video/truman_21_qt.html#v178.

11. Quoted in Porro, ed., *The Nuclear Age Reader*, 42.

12. Powaski, *March to Armageddon,* 46–47; Herring, *From Colony to Superpower*, 608–11, 614–17; Ambrose, *Rise to Globalism*, 117– 35.

13. Quoted in Porro, ed., *The Nuclear Age Reader*, 45.

14. Ibid., 43.

15. Herring, *From Colony to Superpower*, 604–05, 618–20; Leffler, *For the Soul of Mankind*, 11–83; Powaski, *March to Armageddon,* 37–38, 47–48; Robert E. Wood, "From the Marshall Plan to the Third World," in Leffler and Painter, eds., *Origins of the Cold War*, 239–44.

16. Allan M. Winkler, *Life Under a Cloud: American Anxiety About the Atom* (Urban, IL: University of Illinois Press, 1999), 57–60.

17. Herring, *From Colony to Superpower*, 614.

18. Ambrose, *Rise to Globalism*, 140–41. For more on the development of the national security system and the long-term assessment of Soviet goals, see Melvyn P. Leffler, "National Security and US Foreign Policy," in Leffler and Painter, eds., *Origins of the Cold War*, 15–41.

19. William S. Borgiasz, *The Strategic Air Command: Evolution and Consolidation of Nuclear Forces, 1945-1955* (Westport, CT: Praeger, 1996), 1–3, 12; Gerard J. DeGroot, *The Bomb: A Life* (Cambridge, MA: Harvard University Press, 2005), 150.

20. David Alan Rosenberg, "The Origins of Overkill: Nuclear Weapons and American Strategy, 1945–1960," *International Security* 7 (Spring 1983): 9–10, 25.

21. Ibid., 11.

22. Ibid., 13–14; Winkler, *Life Under a Cloud*, 63; Jonathan Stevenson, *Thinking Beyond the Unthinkable: Harnessing Doom from the Cold War to the Age of Terror* (New York: Viking, 2008), 15.

23. Rosenberg, "The Origins of Overkill," 12–17.

24. Borgiasz, *The Strategic Air Command*, 7–10; Winkler, *Life Under a Cloud*, 65; Rosenberg, "The Origins of Overkill," 19.

25. Harry R. Borowski, *A Hollow Threat: Strategic Air Power and Containment Before Korea* (Westport, CT: Greenwood Press, 1982), 4–5; Winkler, *Life Under a Cloud*, 60–63.

26. Quoted in *World War II Database*, online, http://ww2db.com/person_bio.php?person_id=509.

27. Rodney P. Carlisle, ed., *Encyclopedia of the Atomic Age* (New York: Facts on File, 2001), 172–73.

28. Samuel R. Williamson, Jr. and Steven Rearden, *The Origins of U.S. Nuclear Strategy, 1945–1953* (New York: St. Martin's Press, 1993), 162–63; Borgiasz, *The Strategic Air Command*, 12.

29. Williamson, Jr. and Rearden, *The Origins of U.S. Nuclear Strategy,* 163–65; Borgiasz, *The Strategic Air Command*, 13–20.

30. Richard Rhodes, *Dark Sun: The Making of the Hydrogen Bomb* (New York: Simon and Schuster, 1995), 346–49; DeGroot, *The Bomb,* 153–54; Stephen E. Atkins, ed., *Historical Encyclopedia of Atomic Energy* (Westport, CT: Greenwood Press), 133.

31. Rhodes, *Dark Sun,* 346–49; DeGroot, *The Bomb,* 153–54; Atkins, ed., *Historical Encyclopedia of Atomic Energy,* 133.

32. Quoted in *World War II Database.*

33. Rhodes, *Dark Sun,* 346–49; DeGroot, *The Bomb,* 153–54; Atkins, ed., *Historical Encyclopedia of Atomic Energy,* 133.

34. Tannenwald, *The Nuclear Taboo,* 105–06.

35. Atkins, ed., *Historical Encyclopedia of Atomic Energy,* 133; DeGroot, *The Bomb,* 153–54; Rhodes, *Dark Sun,* 346–49.

36. Rosenberg, "The Origins of Overkill," 22. See also David Alan Rosenberg, "'A Smoking Radiating Ruin at the End of Two Hours,': Documents on American Plans for Nuclear War with the Soviet Union, 1954–1955," *International Security* 6 (Winter 1981/82): 3–38.

37. Herring, *From Colony to Superpower,* 617–20; Ambrose, *Rise to Globalism,* 141–45.

38. Herring, *From Colony to Superpower,* 623–24; Rhodes, *Dark Sun,* 326. At the time, the Atomic Energy Commission held control over the bombs, not the air force. The president could transfer them if necessary, but since those weapons lacked locking mechanisms, anyone in possession could also detonate them. See Rhodes, *Dark Sun,* 326–27.

39. Powaski, *March to Armageddon,* 49–51; Herring, *From Colony to Superpower,* 625–26.

40. Powaski, *March to Armageddon,* 48–50; Ambrose, *Rise to Globalism,* 145–60.

41. Herring, *From Colony to Superpower,* 638.

42. Ibid., 638–39; Powaski, *March to Armageddon,* 58–59.

43. A thermonuclear device, also called a hydrogen bomb or an H-bomb, is a "weapon whose enormous explosive power results from an uncontrolled, self-sustaining chain reaction in which isotopes of hydrogen combine under extremely high temperatures to form helium in a process known as *nuclear fusion.* The high temperatures that are required for the reaction are produced by the detonation of an atomic bomb. A thermonuclear bomb differs fundamentally from an atomic bomb in that it utilizes the energy released when two light atomic nuclei combine, or fuse, to form a heavier nucleus. An atomic bomb, by contrast, uses the energy released when a heavy atomic nucleus splits, or fissions, into two lighter nuclei." See "Thermonuclear Bomb," *Encyclopedia Britannica,* online, http://www.britannica.com/EBchecked/topic/591670/thermonuclear-bomb.

44. Margot A. Henriksen, *Dr. Strangelove's America: Society and Culture in the Atomic Age* (Berkeley, CA: University of California Press, 1997), 43.

45. Confusion sometimes occurs when referring to the development of the hydrogen bomb or a thermonuclear bomb as the "super." In its strictest usage, the super represented a specific bomb design, but people often use the term as a generic description of all such weapons. See Carlisle, ed., *Encyclopedia of the Atomic Age,* 139.

46. Powaski, *March to Armageddon,* 53. See also Daniel J. Kevles, *The Physicists: The History of a Scientific Community in Modern America* (Cambridge, MA: Harvard University Press, 1977; fourth printing, 1995), 330; Winkler, *Life Under a Cloud,* 68–69. There is a growing list of Teller biographies, especially since his life and accomplishments spanned so much of the Atomic Age. For example, see Peter Goodchild, *Edward Teller: The Real Dr. Strangelove* (Cambridge, MA: Harvard University Press, 2004); Stanley A. Blumberg and Louis G. Panos, *Edward Teller: Giant of the Golden Age of Physics* (New York: Charles Scribner's Sons, 1990); Istvan Hargittai, *Judging Edward Teller: A Closer Look at One of the Most Influential Scientists of the Twentieth Century* (Amherst, NY: Promethues Books, 2010); and Gregg Herken, *Brotherhood of the Bomb: The Tangled Lives and Loyalites of Robert Oppenheimer, Ernest Lawrence, and Edward Teller* (New York: Henry Holt and Co., 2002), and his

own Edward Teller, *Memoirs: A Twentieth-Century Journey in Science and Politics* (Cambridge, MA: Perseus Publishing, 2001).

47. Rhodes, *Dark Sun*, 246.
48. Ibid., 247.
49. Atkins, ed., *Historical Encyclopedia of Atomic Energy*, 167–68.
50. Rhodes, *Dark Sun*, 248–49; Atkins, ed., *Historical Encyclopedia of Atomic Energy*, 167; DeGroot, *The Bomb,* 162–63.
51. Rhodes, *Dark Sun*, 250–51.
52. Ibid., 252–57, 332.
53. Ibid., 256, 304–07, 320–21. See also Powaski, *March to Armageddon,* 54; Herken, *Brotherhood of the Bomb*, 173–74, 186–87.
54. Powaski, *March to Armageddon,* 55; Atkins, ed., *Historical Encyclopedia of Atomic Energy*, 167; DeGroot, *The Bomb,* 163–65; Rhodes, *Dark Sun*, 381.
55. Rhodes, *Dark Sun*, 399.
56. Ibid.
57. "General Advisory Committee's Majority and Minority Reports on Building the H-bomb, October 30, 1949," online, http://www.atomicarchive.com/Docs/Hydrogen/GACReport.shtml.
58. Ibid. See also Rhodes, *Dark Sun*, 399; Powaski, *March to Armageddon,* 54–55; Kevles, *The Physicists*, 359, 377–78; Herken, *Brotherhood of the Bomb*, 209–10.
59. Rhodes, *Dark Sun*, 402–07; Powaski, *March to Armageddon,* 55–56; Winkler, *Life Under a Cloud*, 70–75; Herken, *Brotherhood of the Bomb,* 215–16; S. David Broscious, "Longing for International Control, Banking on American Security: Harry S. Truman's Approach to Nuclear Weapons," in John Lewis Gaddis, et al., eds., *Cold War Statesmen Confront the Bomb: Nuclear Diplomacy Since 1945* (New York: Oxford University Press, 1999), 35–38.
60. Stevenson, *Thinking Beyond the Unthinkable*, 18.
61. Rhodes, *Dark Sun*, 416–17; Atkins, ed., *Historical Encyclopedia of Atomic Energy*, 167; Herken, *Brotherhood of the Bomb*, 201–08.
62. Rhodes, *Dark Sun*, 478; See also 496, 542; Atkins, ed., *Historical Encyclopedia of Atomic Energy*, 167; Herken, *Brotherhood of the Bomb*, 241, 245, 261. Teller insisted that the labs cooperate on the development of the H-bomb to avoid direct competition. See Teller, *Memoirs*, 367–68.
63. Atkins, ed., *Historical Encyclopedia of Atomic Energy*, 235; Rhodes, *Dark Sun*, 482–93.
64. Rhodes, *Dark Sun*, 494.
65. Ibid., 483–512. See also Winkler, *Life Under a Cloud*, 75–76.
66. Ellen Schrecker, *Many are the Crimes: McCarthyism in America* (Boston: Little, Brown and Company, 1998), xii.
67. A good starting point about the atomic espionage issue in this period is Rhodes, *Dark Sun.*
68. Recently opened files in Russia, particularly those of the KGB, provide rare insight into Soviet espionage activity in this period. Also crucial were the top-secret *Venona* records, amounting to almost 3,000 telegraphic cables between Soviet spies in the United States and their contacts in Moscow. These cables, intercepted and decoded by the American government demonstrate that the Soviets had conducted widespread espionage, and recruited spies in most every major government agency (and among members of the Communist Party of the United States). The CIA, however, was not made an active partner in *Venona* until 1952, and the information was kept from even President Truman for a time. See John Earl Haynes and Harvey Klehr, *Venona: Decoding Soviet Espionage in America* (New Haven: Yale University Press, 1999), 1, 9, 13–15; Allen Weinstein and Alexander Vassiliev, *The Haunted Wood: Soviet Espionage in America—The Stalin Era* (New York: The Modern Library, 2000), xvii; John Earl Haynes and Harvey Klehr, *Early Cold War Spies: The Espionage Trials That Shaped American Politics* (New York: Cambridge University Press, 2006), 3; David M. Oshinsky, *A Conspiracy So Immense: The World of Joe McCarthy* (New York: Oxford University Press, 2005), ix–x.
69. Katherine A. S. Sibley, *Red Spies in America: Stolen Secrets and the Dawn of the Cold War* (Lawrence: University Press of Kansas, 2004), 1–2; Amy Knight, *How the Cold War Began: The Gouzenko*

Affair and the Hunt for Soviet Spies (Toronto: McClelland & Stewart, Ltd., 2005), 12; Oshinsky, *A Conspiracy So Immense*, xi.

70. Richard M. Fried, *Nightmare in Red: The McCarthy Era in Perspective* (New York: Oxford University Press, 1990), 6, 8–9, 47–48, 53; Ellen Schrecker, *The Age of McCarthyism: A Brief History with Documents* (Boston: Bedford Books, 1994), 9–10, 41–45, 50–53; Haynes and Klehr, *Early Cold War Spies*, 48–66; Richard C.S. Trahair and Robert L. Miller, *Encyclopedia of Cold War Espionage, Spies, and Secret Operations* (New York: Enigma Books, 2009), 10–11, 124–26; Schrecker, *The Age of McCarthyism*, 27–29.

71. Albert Fried, ed., *McCarthyism: The Great American Red Scare* (New York: Oxford University Press, 1997), 25–26, 54–55; Donald T. Critchlow, "Series Editor's Foreword, in Haynes and Klehr, *Early Cold War Spies*, xi–xii; Schrecker, *The Age of McCarthyism*, 37–40; Haynes and Klehr, *Venona*, 13–14.

72. See David K. Johnson, *The Lavender Scare: The Cold War Persecution of Gays and Lesbians in the Federal Government* (Chicago: University of Chicago Press, 2004); Elaine Tyler May, *Homeward Bound: American Families in the Cold War Era* (New York: basic Books, 2008), 91–92.

73. Herring, *From Colony to Superpower*, 636–37.

74. Oshinsky, *A Conspiracy So Immense*, 507. See also Fried, *Nightmare in Red*, 120–43; Fried, ed., *McCarthyism*, 3–7, 23–25, 70–75; Schrecker, *The Age of McCarthyism*, 62–65.

75. Julian E. Zelizer, *Arsenal of Democracy: The Politics of National Security-From World War II to the War on Terrorism* (New York: Basic Books, 2010), 104.

76. Fried, *Nightmare in Red*, 113–19; Fried, ed., *McCarthyism*, 85; Oshinsky, *A Conspiracy So Immense*, 173–74; Zelizer, *Arsenal of Democracy*, 104.

77. Many of the spies remained free because American counterintelligence was lacking, efforts at prosecuting them were unsuccessful, the Soviets were able to protect their sources, or they were allowed to continue operating to gain other information. See Haynes and Klehr, *Early Cold War Spies*, 4–5, 7–8, 11, 15–16, 23.

78. The United States abandoned the British as a partner in its nuclear program in 1945 by cutting off their physicists at Los Alamos and ordering them to leave all classified documents when they departed, even though the British provided many of the initial plans. In 1947 the British decided to develop their own atomic program, separating civilian from military applications. In the same year the Americans agreed to share some information on designing power reactors, and in exchange, the British returned much of its ore stockpile to the United States and made concessions with respect to uranium ore in the Belgian Congo. Also, the British gave up veto power over its ally's use of the bomb. (All of these provisions were labeled the *modus vivendi*, or an agreement between those with differing opinions.) Many British participants felt betrayed by the American actions that set Britain's atomic program put back at least two years. See Rhodes, *Dark Sun*, 285–301; Atkins, ed., *Historical Encyclopedia of Atomic Energy*, 60–61.

79. Robert Chadwell Williams, *Klaus Fuchs, Atom Spy* (Cambridge, MA: Harvard University Press, 1987), 2–3, 121–26; Haynes and Klehr, *Early Cold War Spies*, 152–53; Atkins, ed., *Historical Encyclopedia of Atomic Energy*, 141; Trahair and Miller, *Encyclopedia of Cold War Espionage, Spies, and Secret Operations*, 105–07.

80. Haynes and Klehr, *Early Cold War Spies*, 154–56; Schrecker, *The Age of McCarthyism*, 33, 139, 142–43; Atkins, ed., *Historical Encyclopedia of Atomic Energy*, 315; Trahair and Miller, *Encyclopedia of Cold War Espionage, Spies, and Secret Operations*, 114–16, 127–29; Carlisle, ed., *Encyclopedia of the Atomic Age*, 124–25.

81. Atkins, ed., *Historical Encyclopedia of Atomic Energy*, 314–15, Carlisle, ed., *Encyclopedia of the Atomic Age*, 285–86; Fried, *Nightmare in Red*, 115; Haynes and Klehr, *Early Cold War Spies*, 157–76; Trahair and Miller, *Encyclopedia of Cold War Espionage, Spies, and Secret Operations*, 359–62. See also Sam Roberts, *The Brother: The Untold Story of the Rosenberg Case* (New York: Random House, 2001).

82. The proceedings of the hearings were recorded in almost 1,000 pages where the direct quotes can be found. See United States Atomic Energy Commission, In *the Matter of J. Robert Oppenheimer:*

Transcript of Hearing before Personnel Security Board and Texts of Principal Documents and Letters (Cambridge, MA: MIT Press, 1971).

83. Kai Bird and Martin J. Sherwin, *American Prometheus: The Triumph and Tragedy of J. Robert Oppenheimer* (New York: Alfred A. Knopf, 2005), xi. This Pulitzer Prize winning book gives a well-analyzed, very human portrayal of Oppenheimer and the case against him. It is a good place to start.

84. For more on the Oppenheimer case, see Fried, *Nightmare in Red*, 179–80, 197; David C. Cassidy, *J. Robert Oppenheimer and the American Century* (New York: Pi Press, 2005), 305–07, 310–19; Atkins, ed., *Historical Encyclopedia of Atomic Energy*, 274–75; Sibley, *Red Spies in America*, 199–203; Priscilla J. McMillan, *The Ruin of J. Robert Oppenheimer and the Birth of the Modern Arms Race* (New York: Viking, 2005), 1–14; Trahair and Miller, *Encyclopedia of Cold War Espionage, Spies, and Secret Operations*, 311–12; Carlisle, ed., *Encyclopedia of the Atomic Age*, 239; Herken, *Brotherhood of the Bomb*, 267–69; 279–99; Schrecker, *The Age of McCarthyism*, 34–36; Bird and Sherwin, *American Prometheus*, xi–xii, 488, 546–50; Richard Polenberg, "The Fortunate Fox," in Cathryn Carson and David A. Hollinger, eds., *Reappraising Oppenheimer: Centennial Studies and Reflections* (Berkeley, CA: Office of History and Technology, University of California, 2005), 267–72. See also Jeremy Bernstein, *Oppenheimer: Portrait of an Enigma* (Chicago: Ivan R. Dee, 2004).

85. Jessica Wang, *American Science in an Age of Anxiety: Scientists, Anticommunism, and the Cold War* (Chapel Hill, NC: University of North Carolina Press, 1999), 7.

86. Ibid., 1–9.

87. Daniel J. Kevles, *The Physicists: The History of a Scientific Community in Modern America* (Cambridge, MA: Harvard University Press, 1977; fourth printing, 1995), 378.

88. Ibid., 378–79.

89. Quoted in Wang, *American Science in an Age of Anxiety*, 138.

90. Ibid., 379–80; David Kaiser, "The Atomic Secret in Red Hands? American Suspicions of Theoretical Physicists During the Early Cold War," in Carson and Hollinger, eds., *Reappraising Oppenheimer*, 185–89, 213–14; Powaski, *March to Armageddon*, 56–58; Bird and Sherwin, *American Prometheus*, 549.

91. An additional theme introduced most emphatically in Matthew Jones, *After Hiroshima: the United States, Race and Nuclear Weapons in Asia, 1945-1965* (Cambridge, MA: Cambridge University Press, 2010) is the degree to which race played a central part in the United States' use of the atomic bomb against Japan and the link between the role of racial sensitivities and the conduct of American foreign policy (especially reliance on nuclear weapons) in Asia through the Vietnam War.

92. Rhodes, *Dark Sun*, 433–35; Haynes and Klehr, *Venona*, 11; David Holloway, *Stalin and the Bomb* (New Haven, CT: Yale University Press, 1994), 276–78; Zelizer, *Arsenal of Democracy*, 97–99; Kathryn Waetherby, "Stalin and the Korean War," in Leffler and Painter, eds., *Origins of the Cold War*, 265–81.

93. Garry Wills, *Bomb Power: The Modern Presidency and the National Security State* (New York: Penguin Press, 2010), 106–07; Zelizer, *Arsenal of Democracy*, 99–101.

94. Herring, *From Colony to Superpower*, 639–41; Oshinsky, *A Conspiracy So Immense*, 165–67.

95. Herring, *From Colony to Superpower*, 641–42; Ambrose, *Rise to Globalism*, 173–79; Holloway, *Stalin and the Bomb*, 279–83. For an in-depth look at the military background to the Korean conflict and the first two years of the war, see Allan R. Millett, *The War for Korea, 1945-1950: A House Burning* (Lawrence, KS: University Press of Kansas, 2005) and Millett, *The War for Korea, 1950-1951: They Came from the North* (Lawrence, KS: University Press of Kansas, 2010).

96. Rhodes, *Dark Sun*, 440–44; Borgiasz, *The Strategic Air Command*, 27.

97. Holloway, *Stalin and the Bomb*, 230; Borgiasz, *The Strategic Air Command*, 12.

98. David Holloway, "Stalin and the Bomb," in Leffler and Painter, eds., *Origins of the Cold War*, 86–87. See also Vladislav M. Zubok, "Stalin and the Nuclear Age," in Gaddis, et al., eds., *Cold War Statesmen Confront the Bomb*, 57–60.

99. Rhodes, *Dark Sun*, 445–48; Holloway, *Stalin and the Bomb*, 283–84.

100. Quoted in Rhodes, *Dark Sun,* 449.

101. Holloway, *Stalin and the Bomb*, 288.

102. Rhodes, *Dark Sun*, 449–54; Powaski, *March to Armageddon*, 58–59; Paul Boyer, *Fallout: A Historian Reflects on America's Half-Century Encounter with Nuclear Weapons* (Columbus, OH: Ohio State University Press, 1998), 35–38; Herring, *From Colony to Superpower*, 642–45; Williamson, Jr. Rearden, *The Origins of U.S. Nuclear Strategy, 1945–1953*, 139–40; Tannenwald, *The Nuclear Taboo*, 115–40.

103. Paul Boyer, *By Bomb's Early Light: American Thought and Culture at the Dawn of the Atomic Age* (Chapel Hill, NC: University of North Carolina Press, 1994); Henriksen, *Dr. Strangelove's America*; Spencer R. Weart, *Nuclear Fear: A History of Images* (Cambridge, MA: Harvard University Press, 1988).

104. Jerome Shapiro, *Atomic Bomb Cinema* (New York: Routledge, 2002), 84–89, 132; Kim Newman, *Apocalypse Movies: End of the World Cinema* (New York: St. Martin's Griffin, 2000), 137; Errol Vieth, *Screening Science: Contexts, Texts, and Science in Fifties Science Fiction Film* (Lanham, MD: The Scarecrow Press, 2001), 174; David Wingrove, ed., *Science Fiction Film Source Book* (London: Longman, 1985), 257–58.

105. *Henriksen, Dr. Strangelove's America, 139–41, 144, 149; Vieth, Screening Science, 72, 195;* Wingrove, ed., *Science Fiction Film Source Book*, 123.

106. Ronnie D. Lipschutz, *Cold War Fantasies: Film, Fiction, and Foreign Policy* (Lanham, MD: Rowman & Littlefield, 2001), 218; Bosley Crowther, "The Atomic City," *New York Times*, May 2, 1952.

107. Joyce A. Evans, *Celluloid Mushroom Clouds: Hollywood and the Atomic Bomb* (Boulder, CO: Westview Press, 1998), 66–68, 98; Henriksen, *Dr. Strangelove's America*, 57–58.

108. Evans, *Celluloid Mushroom Clouds*, 23, 28, 70–71, 137; Newman, *Apocalypse Movies*, 99–101; Shapiro, *Atomic Bomb Cinema*, 74–77; Toni A. Perrine, *Film and the Nuclear Age: Representing Cultural Anxiety* (New York: Garland Pub., 1998), 20; Ernest F. Martin, "Five," in Jack G. Shaheen, ed., *Nuclear War Films* (Carbondale, IL: Southern Illinois University Press, 1978), 11–16.

109. Elizabeth Burton, "The Classics: The Day the Earth Stood Still," April 20, 1999, online, http://suite101.com/article.cfm/sf_and_fantasy_on_fil/18642; "The Day the Earth Stood Still," January 30, 2001, online, http://www.afionline.org/wise/films/day_the_earth_stood_still/dtess.html.

110. Henriksen, *Dr. Strangelove's America*, 50–54; Evans, *Celluloid Mushroom Clouds*, 23, 28, 126; Vieth, *Screening Science,* 59, 63; Wingrove, ed., *Science Fiction Film Source Book*, 66; Shapiro, *Atomic Bomb Cinema*, 80–82: Paul Boyer, "Nuclear Themes in American Culture, 1945 to the Present," in Rosemary B. Mariner and G. Kurt Piehler, ed., *The Atomic Bomb and American Society: New Perspectives* (Knoxville, TN: University of Tennessee Press, 2009), 4; Boyer, *By Bomb's Early Light*, 104.

CHAPTER 5

Invincible to Vulnerable in the Age of Anxiety

Massive Retaliation, Fallout, and the Sputnik Crisis

INTRODUCTION

Popular stereotypes picture the 1950s as deeply conformist and sometimes anxiety ridden—simply a prelude to the more glitzy and tumultuous 1960s. Such generalizations are easy to grasp, but hardly do justice to the time period. Like any decade, the 1950s was rife with contradictions and complexity. As such, thermonuclear weapons and global atomic policy were major contributing factors. A new era of Cold War leadership emerged in 1953 after the death of Joseph Stalin in the USSR and because of the election of Dwight Eisenhower in the United States. With new leaders came new approaches and fresh problems, but the dynamics of a world dominated by two super-powers locked in an arms race grew more intense. By late in the decade, atmospheric testing and the *Sputnik* crisis brought home the realization that the bomb had become a local threat. A new atomic reality defined the Age of Anxiety. The 1950s era did not officially begin with the election of Dwight Eisenhower, but a new phase of America's atomic age certainly did. The texture and tone of the Cold War changed, especially as it extended deeper into Asia and the Middle East. People felt the vast grip of the bomb and the uncomfortable sense of insecurity linked to it as deterrence made way to brinksmanship. The fear of fallout became less an abstraction than an insidious peril. Hope was in short supply, and risk grew deeper.

"I LIKE IKE"

President Eisenhower's stock has risen over the years. Often characterized as dull and avuncular, his views and actions belied that image. Eisenhower came to office in 1953 on a wave of per-sonal popularity, and a wish for respite from the Roosevelt-Truman era. On the domestic front the new president espoused the Republican Party line as a fiscal and social conservative with a strong anti-communist bent, critical of the New Deal legacy of his immediate predecessors.

In foreign affairs, he proved to be knowledgeable, firm, but cautious. The new president was a Cold Warrior, however, every bit as much as Truman. Like the majority of political leaders on both sides of the aisle, Eisenhower accepted the Cold War consensus of the day, which viewed the USSR as an aggressor and the United States as the bastion of liberty and democracy. He and his Secretary of State, John Foster Dulles, did not accept the existing brand of containment, but their overall foreign policy goals did not stray far from Truman's. They upheld the commitment to collective security and modified rather than junked the principles initially espoused by George F. Kennan. And, not surprisingly, their views fluctuated at times depending on the current crisis, including their perspectives on nuclear weapons.[1]

Ike's Secretary of State was his alter ego. Dulles and he shared many fundamental notions about conducting foreign policy, complemented one another, and had mutual respect. But, while the president believed that Dulles had the skill to conduct diplomacy, he trusted his own knowledge of the world and did not simply defer to his new appointee.[2] Eisenhower deliberated over the selection of secretary of state for several weeks, not quickly and unreservedly turning to Dulles from the start. Throughout his presidency, he retained a group of foreign policy advisers close to him that were independent of Dulles. In addition, as a celebrated military leader, the new president was not likely to be intimidated or unduly swayed by the Pentagon.[3]

John Foster Dulles came from a family with a long tradition in government service. A lawyer and financier, he was the son of a Presbyterian minister with the zeal of a man of the cloth. His passion often took the form of public outbursts against communism and oratory rife with allusions to "liberation" for those under its yoke, but the rhetoric never became a reality. Such talk was more for the ears of US citizens than anyone else. Serious and often gruff, few people worked harder or more aggressively than Dulles. While more pragmatic than his opponents assumed, he proved to be Eisenhower's attack dog—reined in when necessary, but often leading the charge.[4] Dulles operated under the assumption that containment as practiced was not working, and he tended to be more pessimistic than Eisenhower about where the Cold War was headed. "There has been a very definite shift in the balance of power in the world," he stated, "and that shift has been in favor of Soviet communism."[5] Dulles wanted to right the ship.

Eisenhower kept tight control over the conduct of foreign affairs. He met almost daily with Dulles and the secretary's brother, CIA director Allen W. Dulles. The National Security Council met regularly as well and its structure and role expanded under Eisenhower. The president's colleagues on Capitol Hill were less easy to deal with, squabbling with him on numerous issues including the defense budget. In 1954, the Democrats regained control of Congress, but even before then Eisenhower faced problems with members of his own party. Until his censure in 1954, Joseph McCarthy remained an albatross. While the president privately loathed him, he was not inclined to denounce the Wisconsin senator publicly; he preferred to keep McCarthy in check.[6]

On the international scene, the death of Stalin in 1953 created a strange kind of instability. The United States, as it turned out, did not take advantage of the internal power struggle in the USSR. Most immediately, loyal Stalinists Georgi Malenkov and Lavrentiy Beria succeeded the long-time dictator. Malenkov was a pragmatic technocrat, while the notoriously cruel Beria had been in charge of the nuclear program and had a long history with the secret police. In this time of transition, neither man wanted to aggravate conditions in the Cold War, but hoped to mend fences with some key countries. They even urged Communist China to help end the Korean War. They talked about "peaceful coexistence" with the United States, but American leaders chose not to regard this as an opportunity to reduce tensions with the USSR. Of course, mistrust between the parties was mutual.[7] Eisenhower deflected calls for a summit until 1955, seemingly boxed in by his party's relentless criticism of the Democrats as "too soft" on communism. Distrust of the

USSR remained American policy if and until the Soviets demonstrated with actions, not words, that they were willing to put aside their militant posture.[8]

FORMULATING THE NEW LOOK

The bilateral nature of the Cold War remained a constant. The two sides chose to use propaganda, military and economic posturing, and diplomacy to engage each other.[9] Espionage and covert operations were the staple of each. The emerging superpowers squared off on a larger global stage than before, but often through client states. An arms race built upon stockpiling thermonuclear weapons and developing new delivery systems intensified. If the H-bomb as a deterrent kept the Cold War cold, the possibility of the heat rising always was present. Several weeks after Eisenhower assumed the presidency, he sought to stake out a foreign and military policy position that distanced his administration from his predecessor and gave it the initiative in facing adversaries. For example, he quickly reversed Truman's policy of civilian control of the atomic stockpile. On June 20, 1953, he transferred a major portion of completed atomic weapons to the military. By 1961 less than 10 percent of the stockpile remained under civilian control.[10]

Eisenhower was willing to consider some form of international controls on nuclear weapons (Chapter 6 will discuss the Atoms for Peace program regarding international management of nuclear energy), but like Truman he wanted the United States to maintain a sufficient arsenal capable of upholding the nation's security.[11] Eisenhower came to the presidency with good knowledge about nuclear weapons, and as historian David Rosenberg stated, Ike "knew that nuclear weapons could not be 'disinvented,' and that they had permanently altered the character of warfare."[12] This meant that Eisenhower not only viewed the bomb as an integral part of American defense, but as a weapon of first resort.[13] He once told Congressional leaders that the United States must be "willing to 'push its whole stack of chips into the pot' when such becomes necessary."[14] Some felt that the new president saw efforts to deal with the USSR as something that could be managed, with nuclear weapons as a centerpiece, rather than as a series of isolated challenges or crises. On the European front, especially, Eisenhower believed that American allies could not be defended by conventional military methods, and wanted troop strengths reduced substantially. While the Joint Chiefs disagreed, he pushed on.[15]

In summer 1953, the president called for a full examination of national security policy. He wanted, among other things, the various branches of the armed services to share a unified military stratagem. In his State of the Union message on February 2, he stated, "Our problem is to achieve adequate military strength within the limits of endurable strain upon our economy. To amass military power without regards to our economic capacity would be to defend ourselves against one kind of disaster by inviting another."[16] The National Security Council (NSC), through an exercise referred to as *Project Solarium*, produced three position papers by three teams of experts. Two principles bounded the work: the end product needed to move away from the Democrats' position and also be economically constrained. Eisenhower attempted to merge the best of all three reports. He accepted the notion that if non-communist countries resisted the communist challenge, the former would prevail (essentially containment), and he viewed "liberation" as a longer-term objective. In addition, the president regarded nuclear weapons as an inexpensive means of maintaining military superiority, while the United States moved to reduce dependence on more expensive conventional forces. He shared with Truman the assumption that nuclear weapons always would be part of the American arsenal, but elevated them in importance. In addition, along with developing strategic weapons to deploy against Soviet targets, the administration considered the use of tactical nuclear weapons in localized conflicts.[17]

In October Eisenhower formalized the policy. It stated, "In the event of hostilities, the United States will consider nuclear weapons to be available for use as other munitions."[18] The "New Look" strategy was a bold shift in military policy, driven partly by a desire for economic restraint. It strove to meet Soviet challenges by presenting an offensive weapons posture, and also utilizing covert operations, intelligence gathering, and building new alliances.[19] The administration's *Basic National Security Policy* consisted of three priorities: massive retaliatory striking power in which the air force played a central role; tactical weapons for defense against local aggression, especially in Western Europe; and defense against nuclear attack requiring the development of a system of continental defense.[20] The New Look recognized to some extent that the United States would have to pick its fights. Furthermore, the strategy depended heavily on its credibility. The idea of "preventive war" was implicit in discussions of US objectives in a general conflict. But in an updated *Basic National Security Policy* paper in the fall of 1954, the administration rejected preventive war. Nevertheless, the idea of preemption (attacking with nuclear weapons based on irrefutable confirmation of an imminent enemy attack) was not rejected. The Strategic Air Command (SAC) prepared for preemptive strikes, although they were not implemented in JCS policy.[21]

The New Look's built-in inflexibility with strong reliance on nuclear weapons came back to haunt Eisenhower and Dulles almost before they could initiate it. Some additional reflection and reconsideration forced them to qualify their plan and even to backtrack to a traditional approach in many cases, that is, relying more heavily on conventional forces. (The Soviets followed US strategic policy and developments in NATO closely, including the deployment of tactical nuclear weapons. As a result, by 1953 the Soviets seriously attempted to adjust their own tactics and operations to the changes in US policy.[22]) In theory, harmonizing military strength and economic prudence suited the Republican president quite well. Describing the New Look as a "bigger bang for the buck," is a little simplistic. Eisenhower clearly sought steadiness in the nation's security system by striking a balance between a sufficient military force and controlled spending. He feared that unlimited military budgets would disrupt the economy, thus creating instability at home and abroad. In 1953, the defense budget was more than $85 billion, or 12 percent of the gross national product, with about 3.5 million people in the armed services. The United States operated more than 800 bases in 100 countries, and averaged $5 billion per year in foreign aid between 1948 and 1953. Eisenhower actively moved to reduce military spending and to further postwar efforts at unification of the armed services. *Reorganization Plan Six* was to be that mechanism.[23]

In practice such a balancing act proved difficult. It was a political hot potato as Democratic leaders especially feared that the United States would fall behind the Soviets in crucial areas of the arms race, notably intercontinental ballistic missiles (ICBMs). In a January 1954 speech, Secretary Dulles attempted to make clear how the administration's New Look would prevent future unwinnable wars. "We want for ourselves, and the other free nations," he stated, "a maximum deterrent at a bearable cost."[24] He added,

> Local defense will always be important. But there is no local defense which alone will contain the mighty land power of the Communist world. Local defense must be reinforced by the further deterrent of massive retaliatory power. A potential aggressor must know that he cannot always prescribe battle conditions that suit him.[25]

The declaration caused an uproar by publicly reducing the New Look to the idea that when the United States responded to aggressive acts it would depend on nuclear weapons exclusively, possibly through preventive action. "Massive retaliation" appeared to set up the United States to wave the atomic saber when necessary and to be willing to use it. Dulles denied that the administration would make a blanket judgment about responding in such a way no matter what the circumstances, but the sense of what he said seemed to indicate the contrary.[26] On so many levels massive retaliation as a policy was unworkable, raising serious questions within government and military circles and without.[27] Even as fodder for public consumption it had an unsettling ring to it. Did the United States actually intend to use A-bombs and H-bombs and under what circumstances? As policy, it faced several harsh realities. The Soviets themselves had not stood still in developing their own nuclear arsenal. Gone were the days of a face-off between American bombers and the Red army. In August 1953 the USSR tested a hydrogen bomb. While the United States held a vast lead in the nuclear arms race for several years, the gap was closing. In a very short time, delivery systems expanded beyond bombers to include longer-range bombers, submarines, and intercontinental ballistic missiles.[28]

Within American military circles, the New Look was greeted with mixed reactions. SAC and General LeMay were ecstatic with a security policy clearly favoring the air force. SAC's expanding capability influenced its growing independence, especially with its almost total control over target selection by 1955.[29] The New Look affirmed SAC's central role in the "atomic mission," providing the air force with 46 percent of the military budget and increasing the number of wings from 110 to 137 by July 1957, and increasing personnel from 913,000 to 975,000. In 1954, 90 percent of SAC's bombers were equipped for nuclear warfare. In 1953 the air force began ordering the first intercontinental jet bombers, the B-52s. It also placed major emphasis on ballistic missiles, which carried lightweight nuclear warheads. In 1955 the government approved the development of the *Atlas* missile, the first American ICBM, and *Thor*, the nation's first intermediate-range ballistic missile (IRBM).[30]

Despite the focus on SAC, several air force officers were unhappy that in committing to the New Look the administration did not give sufficient attention to potential problems in Africa, Asia, Latin America, and the Middle East. Others were concerned about the decision to create the National Aeronautical and Space Administration (NASA), which set roadblocks for militarizing outer space.[31] The army and navy were not pleased about the New Look either. Army Chief of Staff Matthew Ridgway wanted to make clear that the new strategy did not eliminate the need for ground troops, but only heightened it. He asserted that land warfare would become more complex, not less, and human decision making would always be necessary to win an engagement. Ridgway and other army generals believed that the New Look seriously threatened the ability of the army to prevent localized conventional wars, or guerilla wars, for that matter. Overall, the Pentagon adamantly opposed budget cuts of any kind implicit in the New Look, and all three branches wanted their share of the resources. To placate the army and navy, the administration gave them their own missile programs. The army developed the *Jupiter*, a carbon copy of the *Thor,* while the navy concentrated on submarine-launched ballistic missiles (SLBM), first with the *Polaris.* Tactical nuclear weapons also found their way into the arsenals of the army and navy. Despite efforts at drawing the army and navy into the New Look, their resistance to the administration's efforts at unification and budget frugality persisted into the late 1950s.[32]

Aside from its missile program, the navy entered the atomic age with the launch of the first nuclear powered submarine, *Nautilus,* in January 1954. The implications of this event went beyond its military uses to the heart of propulsion technology and also to commercial reactor development. The role of Admiral Hyman Rickover was central, and as one writer noted, his leadership left "an imprint on both the nuclear navy and the broader nuclear industry in the United States."[33] (Rickover's contributions to nuclear power are dealt with more fully in Chapter 7.)

USS Nautilus (SS-571), the Navy's first atomic powered submarine, on its initial sea trials, January 20, 1955. *Source:* National Archives

THE NEW LOOK IN ACTION

Before long Eisenhower tested the New Look policy, using it most immediately in East and Southeast Asia. In the case of Korea, the administration claimed that its implied threat to use nuclear weapons to quickly end the conflict brought the adversaries to the peace table. In reality all of the parties had tired of the war. The Soviets were in the midst of a change in power with Stalin's death in March 1953, and had much to deal with in Eastern Europe. Mao Zedong saw little benefit to China in extending the war. The United States assured South Koreans of a mutual security pact. In July 1953 the sides declared a truce along a border that continued to prolong tensions for years to come.[34]

In 1954 crisis reemerged in Indochina, when France's eight-year conflict against the communist forces guiding the Vietminh (or League for the Independence of Vietnam) was coming to a climax. Twelve thousand French troops manned Dien Bien Phu, a stronghold in northwest Vietnam, under siege by a much larger enemy force. The French appealed for US military help, and the president and his secretary of state seriously considered utilizing air and naval forces (possibly even nuclear weapons) to save its European ally. The United States already had invested heavily there to continue to prop up French control. (As early as 1950, the United States had made a commitment to aid the French in Indochina.) American leaders viewed the defeat of the French in Vietnam as a clear indication of a domino effect in Southeast Asia, which was a concern growing out of Vietnam's strategic location and its wealth of natural resources. Eisenhower stated in an April 7, 1954 press conference that if the communists captured Vietnam, surely it would be the first in a series of defeats for the West, an idea that gained considerable credibility as the United States justified its presence there. Congress rebuffed American intervention at that time when Great Britain refused to participate, and because the French were unwilling to support an independent Vietnam. While a conference in Geneva convened to consider the future of Indochina, the fortress at Dien Bien Phu fell on May 7, 1954.[35]

American participation at Geneva, and its hint of possible military involvement in ending the conflict, did not turn the tide in favor of US policy goals, but likely softened the settlement. Replaying history, the sides agreed to temporarily divide Vietnam at the 17th parallel and to set elections to be held in 1956. Not a signatory, the United States immediately violated the Geneva Accords. It established the Southeast Asia Treaty Organization (SEATO) to carry out a role similar to that of NATO for defending the region against communism. It also backed the anti-communist Ngo Dinh Diem as the leader of South Vietnam and brought the new "country" into SEATO. The United States was fully aware that Vietminh leader Ho Chi Minh would win a national election if held, and he would unify the divided country which it presumed would ally with its adversaries. The debacle that resulted in the Vietnam War was set in motion in the mid-1950s with the United States assuming the previous role of France in Southeast Asia. The threat of a nuclear response in Vietnam proved to be a bluff at best or no more than a whisper at worst (although Air Force Chief of Staff Nathan Twining apparently regretted that the United States did not use tactical A-bombs at Dien Bien Phu). Diplomatic and political maneuvering took center stage in the volatile environment of Southeast Asia.[36]

A crisis in the Taiwan (Formosa) Straits in 1954–1955 raised serious concerns about the Eisenhower administration's new Cold War policy, especially massive retaliation, and long-term relations with Taiwan and China. In September 1954 the Communist Chinese shelled the islands of Quemoy (now Jinmen) and Matsu (now Mazu), then under the control of the Nationalists. There had been a series of small-scale bombing raids against mainland shipping and ports, but Mao's deeper interest was to promote domestic mobilization in his on-going efforts at continuous revolution.[37] For Eisenhower and Dulles the islands were essentially worthless. They were neither strategically important nor crucial to American interests (although two American soldiers died in that attack). The administration did not want war, but was unwilling to display weakness in the face of continued shelling of Quemoy and Little Quemoy, and by the January 1955 seizure of one of the small Dachen (Tachen) Islands (held by a division of Nationalist troops).[38] The cumulative effect of these actions, however, produced a war scare. The "Red Chinese," the president asserted, "appear to be completely reckless, arrogant, possibly overconfident, and completely indifferent to human life."[39] In fact United States came very close to launching a nuclear attack. So jaundiced were Dulles and Eisenhower over potential menace of the People's Republic in the region that such a thought seemed chillingly possible. To reduce the tensions that were spinning out of control, the Beijing government was willing to negotiate with Washington to bring peace. The more conciliatory tone of the Communist Chinese resulted in the United States backing off from its potentially hostile response. The unfortunate misunderstanding between the parties, especially about the actual intentions of the Communist Chinese in the Taiwan Straits, had produced the dangerous crisis.[40]

After tensions declined, Quemoy became an issue again in 1958 when the Communist Chinese resumed shelling the island and eventually blockading it. Mao persisted in telling his people that the islands had been lost by "imperialist aggression," first by the Japanese and then by the Americans. This played directly into, what Chinese–American Relations Professor Chen Jian called, the Chinese people's "victim mentality" that Mao relied upon constantly.[41] Once again Eisenhower invoked massive retaliation tactics. While the crisis proved briefer than the first, it was even more intense. Once the blockade was broken and the Communist Chinese were convinced that continual shelling of the island would arouse a strong reaction from the United States, the mainland leaders sought peaceful solutions to their ongoing struggle with Chiang.[42] Eisenhower and Dulles may have come to believe that their actions were a positive use of massive retaliation tactics, but in the long run the action likely increased the resolve of the Communist

Chinese to have their own nuclear arsenal. The rattling of nuclear weapons, in addition, did not attract additional support for the New Look at home or among American allies.[43]

The mid-1950s also saw other intensifying "hot spots" in Europe, the original battle ground of the Cold War. In May 1955 West Germany formally became a member of NATO, creating a divisive issue among the Americans and their European allies. Economic hardship in East Berlin led to protests and rioting, although the United States received some credit for a relief program to help them. Dulles moved away from the old containment notion by encouraging internal dissent through the use of "psywar" (psychological warfare) techniques such as leaflet dropping by balloons and American propaganda via Radio Free Europe and Radio Liberation. Soviet leaders viewed the aid and the psywar tactics as provocative, which gave Dulles some pause about developing programs that might incite a Soviet response.[44] The use of psychological warfare and propaganda were common tools in the Cold War for both sides. Some regarded them as essential elements in non-shooting combat, as an expansion of the concept of total war, and even as a kind of "fourth weapon." American psywar tactics and the battle for "hearts and minds" was not simply the brainchild of the US Information Agency or an avenue promoted by the aggressive Secretary Dulles, but a powerful tool supported by the president himself to wage war "by other means." While these nonmilitary modes of combat were global in their reach, they also were turned inward to help shape domestic public opinion, be it making a distinction between the "free world" and the "communist menace" or simply raising morale.[45]

In 1955 Soviet leadership was in new hands with the clever political chameleon Nikita Khrushchev taking control. He quickly denounced Stalinism within the government's inner circles, and his position soon became public knowledge. Khrushchev was both general secretary of the Communist Party of the Soviet Union (1953–1964) and chairman of the Council of Ministers or premier (1958 to 1964). In the former post he headed the Presidium (or Politburo) of the party's Central Committee, where he made the most important decisions in the Kremlin. He was born of humble circumstance in Ukraine. On reflection late in life he frankly stated,

> I had no education and not enough culture. To govern a country like Russia, you have to have the equivalent of two academies of science in your head. But all I had was four classes in a church school and then, instead of high school, just a smattering of higher education. So I acted inconsistently…I offended many good people…I shouted and swore…at the intelligentsia, which actually supported my anti-Stalinist policies. They supported me, and look how I treated them in return.[46]

Yet Khrushchev entered Stalin's inner circle, and outlasted several rivals. He was not only a survivor of the Stalin years, but in some ways complicit in them by loyally serving the Soviet dictator. As premier he tried to "de-Stalinize" the Soviet system by modernizing programs in agriculture, industry, and other areas.[47] As the new leader, he also crushed the Hungarian Revolution in 1956, arrested opponents and rivals, and failed to achieve many of the reforms he hoped to institute. Khrushchev's behavior was never easy to predict.[48] The hope for vast reforms in the USSR, including a change in Cold War policy, was premature at this time. In May 1955 the leaders in the Soviet Union created the Warsaw Pact, which was an alliance counterpoint to NATO with seven Eastern European signatories coming on the heels of West Germany joining NATO. Internally, the USSR allowed Poland to implement modest reforms, but took harsh action in Hungary where discord had grown into rebellion. The ferocious crackdown led Dulles to reconsider efforts to actively encourage uprisings against the Soviets. The tepid Voice of America, using music as its primary tool of "subversion," now replaced Radio Free Europe and Radio

Liberation. The Democrats, for their part, used the American inaction in Hungary to denounce the inadequacy of the New Look.[49]

The Geneva summit held in July 1955 was the first to bring together the new leaders of the USSR and the more seasoned US president. The world of that year, with both sides in possession of thermonuclear weapons, was a different place than in 1945. By the time of the meeting, the USSR had about 300–400 atomic and thermonuclear weapons, although some leaders in the United States grossly exaggerated the closing of the so-called (and largely bogus) "bomber gap" between the two countries. The Soviets had 200–300 bombers capable of reaching North America, while the United States had more than five times that number. The United States still held a significant edge in nuclear fire power, but there was no clear assurance of victory (whatever that meant). Eisenhower understood that such a total commitment to atomic war threatened the whole world. Indeed, revisions of the New Look revealed in 1956 placed emphasis on "sufficient deterrence" rather than preventive war. Both sides were wary of the coming meeting, but the summit's historic significance was that it actually took place. German reunification and the fate of Eastern Europe were on everyone's mind. In this latest poker game, the Soviets presented an ambitious disarmament proposal. Eisenhower offered in return "Open Skies," a plan for mutual aerial surveillance, which the Soviets regarded as simply allowing the United States to spy on them. The summit ended with no concrete agreements.[50] Khrushchev left the Geneva meeting "encouraged, realizing that our enemies probably feared us as much as we feared them."[51]

Cold War ambitions for the parties remained global, but nuclear brinksmanship played no significant role in the Middle East, the Near East, or Latin America at this time. The role of the United States in the Middle East had grown since World War II, especially with American strategic military and oil interests and support for the new State of Israel. (Oil policy was complicated since the United States gave private companies the green light to conduct business with a minimum of government interference.) Arms and trade agreements with Egypt in 1955 propelled the USSR into the Middle East and raised serious concern from the White House. Eisenhower was more convinced than ever that the United States had to secure friendly governments in the region, authorizing more covert operations, and building an alliance structure to match NATO and SEATO, called CENTO (Central Treaty Organization).[52]

The most destructive conflict was the Suez Crisis of 1956. Dulles and Eisenhower's decision to support Gamal Abdel Nasser, a nationalist who had overthrown British-supported King Farouk of Egypt in 1952, met with serious opposition at home from cotton-growing Southerners, supporters of Israel, and anti-communists. When Dulles rescinded an offer to assist in the building of the Aswan Dam on the Nile, Nasser nationalized the Suez Canal. This was a serious blow to the British and French corporation operating the canal, and also threatened access to British oil supplies. Under a secret military plan with the British and French, the Israelis, also threatened by Nasser's provocation, seized the Sinai Peninsula and the Gaza Strip on October 29, 1956. The British and French followed with attacks by air and land against Egypt. The United States ultimately secured a cease fire and the agreement of the attacking forces to withdraw. The cost to its relations with its European and Middle East allies was severe, however, and the crisis allowed the Soviets to engage in a little saber-rattling of its own against Britain and France.[53] Rather than backing away in the Middle East, the Suez Crisis strengthened Eisenhower's resolve to press for friendly governments there and to provide economic aid to those who supported support United States actions. The "Eisenhower Doctrine" (1957) acknowledged the authority of the United States to intervene militarily in the Middle East when and if necessary. The New Look now had some new wrinkles that looked eerily like conventional interventionism. But unlike in Southeast Asia and Europe, the "nuclear option" was more a subtext than a publicly expressed strategy.[54]

"DUCK AND COVER": CIVIL DEFENSE IN THE 1950s

The Cold War produced public and private points of tension for the United States, the USSR, and everyone caught between. The ubiquitous bomb heightened almost everyone's sense of vulnerability, even though in this period the arms race was yet to reach its zenith. In many respects the efforts to prepare the United States for possible nuclear attack in the 1950s focused on the emotional and psychological. The reality of A-bombs and H-bombs raining down on cities offered few practical options to deflect them, destroy them, or survive them. The fear of fallout (discussed below) also made evident that rural areas were not immune from the impact of nuclear attacks.[55] Depending on who you believe, civil defense practices in the United States became either a way to dupe Americans into thinking their lives could be made safe or a genuine effort to stave off panic. Even before the existence of Soviet ICBMs, the unfathomable power of the bomb and the dangers of radioactivity forced national leaders and others to think about how to protect the nation. Beyond immediate objectives, civil defense was the manifestation of a larger understanding of national security in the Cold War. As one scholar persuasively argued, civil defense was "not quite military, but not quite civilian either. It was a *paramilitary* program, situated between the priorities of the defense establishment and the cultural ideas of the postwar home front."[56] And despite state and federal programs for civil defense, preparedness in the 1950s relied heavily on a doctrine of "self-help." Millard Caldwell, director of the Federal Civil Defense Administration (FCDA), asserted "Civil defense, like charity, is something that begins at home."[57]

In December 1950 President Truman merged older agencies into the FCDA. Its mission was to deal with disaster "on a scale neither this country nor any other country has before dreamed possible."[58] In reality the FCDA was mainly a supervisory body, leaving implementation and funding of civil defense programs to the states. In the late 1940s, while several in the scientists' movement and other organizations believed there was no real defense against a nuclear attack, social scientists began to talk seriously about urban dispersal. Although theoretically possible, the idea of relocating people by breaking American cities into thousands of smaller communities was not a realistic response to nuclear attack. To others, suburbanization itself offered a powerful means of dispersal and an additional rationale for flight from core cities. In civil defense planning, it appeared that the FDCA essentially "wrote off" the large cities by promoting an illusion that people could survive nuclear attack in the (predominantly white, middle-class) suburbs.[59]

As government-run activities, few deemed practical urban dispersal, large-scale evacuation programs, construction of extensive public shelters, or even more mind-boggling schemes. A modest preparedness program became the primary focus.[60] Civil defense was so decentralized that the main burden for preparedness fell on the individual ("self-help defense"). This meant everything from recognizing the danger signs of a would-be attack to seeking food and shelter. The federal and state governments dealt with civil defense, but were not always central players. They often provided promotional material and other communications to strike a note that nuclear attack was survivable. Breeding fear was the last thing they wanted to do.[61] In late 1952 Los Alamos was one of the first communities to stage a mock evacuation, with a national program of civil defense drills underway two years later. In 1954 the Eisenhower government staged *Operation Alert*, which was an exercise drill in more than 50 cities. It also prepared *Survival under Atomic Attack*, a pamphlet used in grammar schools. In the early 1960s focus shifted to the construction of fallout (or bomb) shelters.[62] The practical impact of all of these activities was questionable at best. A Washington, D.C. civil defense official called *Operation Alert* "so inadequate it couldn't cope with a brush fire threatening a doghouse in the backyard."[63]

The public schools were a central medium through which the government conveyed the civil defense message in the 1950s and into the 1960s. They were possibly the most important institutions that went beyond self-help defense. The FCDA produced educational and propaganda materials to mobilize individuals and local and state agencies. Among the important targeted groups were the public schools, which were the means to reach students, teachers, administrators, and parents. FDCA material was not designed expressly for the classroom; those materials came largely from the AEC. New pedagogy and curriculum focused on peaceful uses of the atom as well as civil defense, sending somewhat mixed messages about atomic energy. The new curriculum also included substantial doses of anticommunism.[64] The FDCA was careful not to paint the horrors of an atomic attack too graphically. *Survival under Atomic Attack*, for example, was bolder than later material. This was the case because the government designed the 1950 pamphlet for adults and not children, and thus it contained more detailed information about an atomic attack than most classroom materials, but no mention of the Soviet Union. The pamphlet stressed the common theme to avoid panic if an attack did occur: "YOU CAN SURVIVE. You can live through an atom bomb raid and you won't have to have a Geiger counter, protective clothing, or special training in order to do it. The secrets of survival are: KNOW THE BOMB'S TRUE DANGERS. KNOW THE STEPS YOU CAN TAKE TO ESCAPE THEM."[65] The federal government also sponsored animated films for civil defense purposes. Probably the most famous was *Duck and Cover*, sporting a catchy little jingle and introducing Bert the Turtle in a civil defense helmet (also employed in a comic book). His message was that the atomic bomb was a new danger, not unlike fire or crossing the street. "You must be ready to protect yourself." Such an approach offered little more than a false impression of survival. Some also believe that the school air raid drills in these years caused lasting trauma to many children.[66]

Aside from utilizing prepared materials, educators conducted air-raid drills and began personal identification programs. The first air-raid drills were initiated in the schools of "targeted cities," such as New York, Los Angeles, Chicago, and San Francisco between August 1950 and August 1951. These included "duck and cover" drills, "sneak attack" drills, safety education, and classroom evacuations. As Bert the Turtle guided us in the 1951 pamphlet, you must "*duck* to avoid the things flying through the air…" and you must "*cover* to keep from being cut or even badly burned." In 1951 the school budget for New York City included funds for identification tags (like military dog tags) to aid civil defense workers in identifying lost or dead children. Many other cities followed suit.[67] Even school architecture came under scrutiny as a consideration for civil defense planning. In many cases schools could serve as bomb shelters and evacuation centers. This added a sense of security to the community that there was a safe place to go if you were caught outside of your home during an attack. Especially in suburbs, where there were few hospitals, people viewed the schools as islands of security. The idea that new schools could serve double duty as places of learning and also as emergency shelters was a justification for building programs in the late 1950s and 1960s. Educators, however, were concerned that using schools as safe bases in an emergency would weaken their control over their own facilities.[68] In the end, civil defense became almost routine in American schools and children were taught that it was a necessary precaution as if it were just another everyday risk.[69] In the adult population it was not necessary to "domesticate" the bomb, but to counteract what some imparted as unwarranted fear. One cartoon, filmed by civil defense, warned about "nuclearosis," an obsession with the bomb and its horrible impacts. The doctor in the film cured the patient with a suggestion of a fallout shelter to give him a sense of well-being and safety. It is amazing that real psychological trouble or emotional anxiety of this type were often treated so carelessly.[70]

Ogden School kindergarten teacher Eleanor Keenan coaches a group of crouched children in a national air raid drill, Chicago, Illinois, circa 1954. *Source:* Library of Congress, Prints & Photographs Division, NYWT&S Collection, [LC-USZ62-125985]

TO TEST OR NOT TO TEST

Repeated testing of atomic and thermonuclear devices by the United States and the USSR was testament to the unrelenting nature of the arms race in the 1950s and early 1960s. The scores of tests also provoked serious questions about their impact on human health and the environment. Testing was done for many reasons: to discover greater explosive capabilities; to construct lighter bombs for more efficiency; to develop tactical devices; to evaluate radiation and blast effects; to find new means of detonation; to conduct explosions from various altitudes and in numerous conditions; and to help determine civil defense emergency preparedness plans. In the end, testing seemed to take place because it could be done rather than if it should be done. Except for *Trinity*, all of the early American testing of atomic and hydrogen weapons took place at sites in the Pacific. Such distant locations were inconvenient and created logistical problems in transporting people and equipment, coordinating tests, and maintaining secrecy, but kept the tests far away from the US mainland. In 1947–1948, however, the Armed Forces Special Weapons Project sponsored *Project Nutmeg* to select a test site in the continental United States. The report concluded that finding a location for conducting tests "without harm" was viable, but depended on "the elements of public relations, public opinion, logistics, and security."[71]

Designating such a site in North America would take "a national emergency," as one commissioner noted. The onset of the Korean War offered such an opportunity, since it speeded up weapons development and buttressed the nuclear stockpile. On December 18, 1950 (despite serious reservations from scientists about the potential safety risks), President Truman approved a facility primarily for testing tactical nuclear weapons to be located in southern Nevada, 65 miles north of Las Vegas. The Las Vegas-Tonopah Bombing and Gunnery Range spanned 1,350 square miles of Nevada desert. Until 1955 it was known as the Nevada Proving Grounds, and then became the Nevada Test Site (NTS). A very large area near necessary operational facilities, it was well isolated and had favorable weather conditions for atmospheric tests. The authorities did not take the geology of the area into account at the time, because the focus was on atmospheric testing. Later, when underground testing commenced, several accidents and some cave-ins took place, mainly from venting of radioactive material.[72] Under AEC auspices, the NTS became the only area in the country used for testing nuclear devices. In most cases, scientists from Los Alamos Scientific Laboratory conducted the work. Testing of hydrogen bombs would continue at the Pacific test sites.[73] The first detonation at the NTS took place on January 27, 1951, as part of *Operation Ranger*. Between January 1951 and October 1958 there were seven major series: *Ranger, Buster-Jangle, Tumbler-Snapper, Upshot-Knothole, Teapot, Plumbbob,* and *Hardtack II*. Data varies on the actual number of detonations, but between 1951 and 1963 there were roughly 100 atmospheric tests at the site. After the Limited Test Ban Treaty (1963), Americans conducted underground tests exclusively (more than 400), and effectively ended the South Sea program. Few if any other places became as radioactive as the Nevada Test Site, which the government closed in the early 1990s.[74]

The goals of the test series conducted in Nevada often varied. For example, the *Teapot* series (1955) was designed for developing lighter, more efficient tactical nuclear weapons. Scientists experimented with miniaturization and efforts to reduce the intensity of radiation. Another test was a demonstration of an ICBM warhead. The *Plumbbob* series (1957) was the most ambitious test series at the time with 25 detonations that underwent a variety of scenarios.

Operation Teapot, the *Met Shot*, a tower burst weapons effects test April 15, 1955 at the Nevada Test Site. *Source:* Photo courtesy of National Nuclear Security Administration/Nevada Site Office

In the Pacific, the *Hardtack I* tests followed in the next year, employing 33 devices in all. The military also staged powerful high altitude and underground tests there. One of the high altitude tests produced an electromagnetic pulse capable of destroying computers, communication links, and other electronic devices.[75]

American soldiers also were used as guinea pigs to, among others things, reduce concern over radiation. In a June 27, 1957 memo, the chair of the Defense Department's Armed Forces Medical Policy Council stated, "Fear of radiation is almost universal among the uninitiated and unless it is overcome in the military forces it could present a most serious problem if atomic weapons are used for tactical or strategic purposes."[76] In July the AEC's Military Liaison Committee sought permission (using a substantial amount of pressure) to have soldiers participate in atomic weapons testing at the Yucca Flat Nevada Test Site. The rationale was to indoctrinate troops to battlefield methods and strategies of atomic warfare, and to monitor their physical and psychological reactions. AEC chairman Gordon Dean approved the request, but disclaimed any mishaps that might occur and insisted that the Pentagon accept responsibility for safety compliance. Uncomfortable with the request in general, the AEC favored placing the troops several miles from the blast site. The military disagreed, wanting the 5,000 soldiers assigned to be placed closer to the detonation site, with movement to ground zero after the blast. Initially they positioned the troops approximately seven miles away, but many soon found themselves with little or no protective gear at a location a little less than four miles (and eventually one mile) from ground zero. In some cases volunteer officers could be as close as 2,000 yards from the explosion. In October the army deployed the first troops for *Operation Buster-Jangle*, with seven tests in all. Almost 900 troops conducted tactical maneuvers in several of the early tests, with almost 2,800 in the troop-observer program who did not participate directly. Over a 10-year period almost one quarter of a million soldiers were involved in the program.[77]

Great disparity existed between the research teams' initial positive reports of the psychological impact of the bomb on the soldiers, and the top-secret reports that were eventually declassified. The latter interpreted the physiological evidence to suggest much more anxiety. For future tests, the AEC relinquished all safety and health responsibilities to the military. At this juncture, it was clear that the well-being of the troops had a lower priority than the utility of test results. For example, amounts of acceptable radiation exposure doubled for the troops, and movement into blast zones was speeded up.[78] The 1954 film *The Atomic Kid* reflected the cavalier attitude that seemed to embody the government's public stance on radiation at the time. Based on a Blake Edward's story, the comedy placed Blix Waterberry (Mickey Rooney) at ground zero on an atomic testing site just before a planned detonation. Blix became lost in the desert and did not know what to make of the completely furnished middle-class cottage inhabited by a family of mannequins. By the time he figured out what was going to happen, the bomb exploded and poor Blix was not killed but irradiated. We follow this wonder of science through his hospital stay with bemused doctors and nurses, into a budding romance, and ultimately confrontation by enemy agents who want to learn the secret of his survival. From today's perspective, *The Atomic Kid* is less the light-hearted comedy it was intended to be, than a sad comment on taking radiation and atomic testing so lightly.[79]

Until 1962 hydrogen bomb testing for the Soviet Union (and dumping of hazardous nuclear waste and reactors in the adjacent waters) took place at Novaya Zemlya, a harsh, largely snow-covered archipelago in the Arctic Ocean far removed from human settlements. The USSR conducted a series of H-bomb tests between 1955 and 1959, the most important taking place in September and October 1961. The Soviets tested approximately 132 devices at Novaya Zemlya, 87 of which were in the atmosphere, another 42 underground, and three under water. On October

30, 1961, a gigantic detonation took place there, registering between 50 and 58 megatons.[80] The Soviets closed the site in 1991 after the testing moratorium went into effect in October 1990.[81] The British began conducting aboveground testing in 1952. Their hydrogen bomb detonations commenced in 1957–1958 on Malden Island and Christmas Island in the Pacific (1,000 miles from Hawaii). They conducted earlier tests of fission devices in Australia, but by agreement thermonuclear testing was prohibited there. Between 1952 and 1958, the British detonated 21 devices. The first British H-bomb test, *Operation Grapple,* took place on Malden Island in May 1957. Like the Americans, the British initially deflected criticism about the risk of its tests, but eventually compensated some aboriginal peoples in Australia for disrupting their homelands. Participants in the tests, some American, later filed suit for damages against the British for radiation exposure.[82]

THE FALLOUT CONTROVERSY

Even before the official announcement of the opening of the NTS, the US government tried very forcefully to justify the testing there and in the Pacific. Officials stressed the critical need for the program in strong Cold War language. They also emphasized the peaceful applications of atomic energy derived from the research. There was every effort to assure citizens in the vicinity of the blasts and beyond that the tests were safe and the risks were minimal. This latter point became critical when testing moved from the Pacific to the United States mainland.[83] Nevertheless, by the mid-1950s a "fallout controversy" gripped the world. Atmospheric testing produced less concern about nuclear war or nuclear proliferation and more about this new source of contamination cascading down on the earth in peacetime.[84] Simply put, fallout is "the often radioactive particles stirred up by or resulting from a nuclear explosion and descending through the atmosphere."[85] It is greatest in atmospheric testing, and considerably less in underground or very high atmospheric tests. The material in fallout is lifted into the air by a detonation of a nuclear device and then slowly deposited on the ground, on vegetation, on buildings, on water, or on anything in its path. The amount of the fallout and where it settles depends of weather, the nature and yield of the device being tested, and where the explosion takes place. Fission weapons produce "fission products" that make up most fallout, while fusion weapons produce less radioactive material. Since H-bombs depend on fission triggers, they generate fallout as do A-bombs.[86]

The *Upshot-Knothole* series at Yucca Flat (March to June 1953) produced the first public safety crisis of the American testing program. The federal government did not make public for years the physical impact of the testing on soldiers at the NTS or the damage caused by exposure to nearby civilian populations. What immediately raised red flags among the citizenry were the apparent impact on livestock and unanticipated shifts in the winds blanketing nearby towns with fallout. Most Americans initially applauded the testing as important to national security. People in the region recognized the program's economic value in terms of new jobs and additional federal largesse, and prided themselves on their patriotic duty in support of their country's defense. People living closest and downwind from the test site in Nevada, Arizona, and Utah, however, soon became wary of possible exposure to radiation. They also worried, contrarily, that negative publicity could shut down their facility and also scar the reputation of their communities.[87]

Before the *Upshot-Knothole* series the AEC assembled a study group to examine the issue more closely, but it made no hard and fast decisions. The new series, while not appearing to pose any special problems, included 11 tests over almost two and one half months. When the dust cleared (literally) the study group deemed the future of the NTS uncompromised by the tests, believing that the fallout problem was manageable. In several cases, changing winds, inaccurate weather forecasting, and human error accounted for much radioactive debris falling on the small communities. There was fallout on roads and nearby towns, to be sure, and the AEC instructed citizens to stay indoors during the tests and their immediate aftermath. But no extraordinary precautions seemed to be necessary. The AEC followed what had become a typical response to the tests—simply to reassure people rather than to inform them. Sometimes defense of testing became even more strident.[88] An article in the *Bulletin of the Atomic Scientists* claimed that the government alerted film manufacturer Eastman Kodak about areas of potentially heavy fallout in the 1950s in order to protect unexposed film. The downwinders never received comparable forewarnings.[89]

At first, the public seemed to be mollified by the official response. But the mounting evidence of damage to livestock turned reassurance into circumspection and then into downright criticism. Horses grazing in the proving ground area showed evidence of burns and cattle deaths mounted.[90] The AEC's response was to settle the claims quickly as it had done several times before. Soon thousands of sheep exposed to fallout died. And although a January 1954 internal report cleared the AEC of blame, no alternative to fallout was put forth as a reason for the deaths. Ranchers sued for damages, but lost in 1956, only to have the case reopened several decades later when a federal judge ruled that the AEC had suppressed important evidence in the case. Among the most notorious incidents affecting humans took place with the *Harry* shot (later called "Dirty Harry") in May 1953. It showered the people of St. George, Utah, with irradiated dust, and 4,200 sheep died of "mysterious causes."[91] As typical, the AEC warned the citizens of the small town east of the site to stay indoors, and yet people in a location even further east became ill. The immediate impacts of the livestock deaths and the St. George case were mixed. The AEC's credibility was tarnished in many people's eyes. While radiation exposure at St. George was short-lived and likely not extremely high, sentiment that the agency was misleading the nearby residents about the potential fallout risk was growing.[92] Utah Representative Douglas Stringfellow called for a complete end to testing in Nevada. But the *Las Vegas Review-Journal* backed the AEC. This was, in the editor's mind, Nevada's business not Utah's (and by "business" he meant jobs). By limiting the sharing of information and research, the AEC kept fallout from causing widespread alarm although it did cause suspicion.[93]

The Winter 1954 *Bravo* test off Namu Island in the Bikini Atoll brought international attention to the dangers of fallout and atomic testing. This particular device was the first important demonstration of Edward Teller and Stanislaw Ulam's research on thermonuclear weapons. They designed the device to be dropped from an airplane or used as a warhead on a missile. The detonation on the morning of March 1 exceeded expectations by almost a factor of three. The bomb yielded approximately fifteen megatons, which was the biggest thermonuclear explosion produced by the United States.[94] Scientists monitoring the blast 20 miles away on the island of Enya had to be quickly evacuated because the radiation levels were extraordinarily high. A large "danger zone" was staked out beforehand, but inhabitants of the islands of Rongelap, Alinginae, and Rongerik in the Marshall Islands and the crew of the Japanese tuna fishing boat, *Lucky Dragon* (which were both outside of the zone), were not so lucky.[95]

On Rongelap, a small island 100 miles east of Bikini with 86 residents, radioactive powder was more than one inch deep, and radioactivity in the Bikini testing range was so intense that future tests had to be moved elsewhere. Studies revealed that fallout had extended over 7,000 square miles in the Pacific, but because of weather it was not contained in the danger zone. The inhabitants of Rongelap received high doses of radiation and exhibited symptoms of severe radiation sickness. They were evacuated to Kwajalein atoll (an additional 150 Marshallese were evacuated from other Marshall Islands including Utirik), but with no acknowledgment that these people had been placed at serious risk.[96] In AEC's March 31 press release, Chairman Lewis Strauss struck a more positive note, "[N]one of the 28 weather personnel have burns. The 236 natives also appeared to me to be well and happy…"[97] The government did establish the Marshall Islands Medical Program (MIMP) within the Brookhaven National Laboratory in 1955 to monitor and treat radiation-related diseases. Groups from the laboratory visited the islands to provide medical care on several occasions. The most significant complication of the exposure to fallout was thyroid disease, which some experts attributed to the testing (questions also arose over long-term genetic effects). Researchers at Brookhaven, who later conducted studies, suggested that there were no significant differences in the incidences of thyroid disease between the Rongelap inhabitants and unexposed populations.[98] Debates of this kind continued throughout the Atomic Age.

Privately, scientists at Los Alamos and Livermore were shaken by the miscalculation of the *Bravo* test's brute force and intensity. As customary, the AEC tried to downplay the serious problems caused by the blast, releasing only the most perfunctory information. But the fate of the *Lucky Dragon* (*Fukuryu Maru*) and publicity around it was beyond their control. The ill-fated trawler had drifted into the general test site area, some 80 to 100 miles east of Bikini at the time of the explosion. Three hours later, as the boat was pulling up its nets and preparing to head home, ash from the mushroom cloud rained on the crew. There had been an exclusion zone north of Bikini, but a change of wind spread fallout due east. By the time the boat returned to its port in Yaizu all 23 members of the crew suffered radiation sickness, and the radio operator, Aikichi Kuboyama, died six or seven months later of complications from blood, heart, lung, and liver disorders. The catch itself was so radioactive that it had to be disposed of. The story of the crew's encounter became public on March 16. Other fishing boats in the area also carried radioactive catches, and soon the contamination or fear of contamination from *Bravo* (and also Soviet tests in Siberia) affected the whole industry.[99]

A few American experts resident in Japan offered to help with the crew's condition. Local Japanese physicians and hospital workers were skeptical, believing that the Americans were more interested in studying the effects of radiation than providing therapy for the patients. Japanese doctors estimated that crewmembers received high doses of radiation, about 130 to 450 Roentgens.[100] Some incredibly insensitive American officials initially viewed the tragedy of the *Lucky Dragon* as an event orchestrated by communists from Japan to embarrass the United States, and others even accused the Japanese of being overly sensitive. The boorish AEC chairman Strauss announced that the *Lucky Dragon* must have been in the exclusion zone, believing that the Japanese were themselves responsible for the accident. He also suspected that the crew on the tuna boat was a "Red spy outfit" sent by the Soviets.[101] A few American doctors also questioned the findings of their Japanese counterparts. When Japanese-American relations seemed to be threatened by such talk, the US government issued a halfhearted apology and offered some monetary compensation. The paltry sum of $2,500 to $3,000 went to Kuboyama's widow; another $151,000 to surviving crew members; and about $2 million to the Japanese government for losses to the tuna industry. No admission of liability accompanied the payments. Most immediately in Japan, and then elsewhere, the incident created great unease, then panic, and finally outrage. It was not lost on people that the first casualties of a thermonuclear detonation were Japanese as were the first casualties of the atomic bomb.[102]

The increasing number of atmospheric tests conducted by the United States, the USSR, and even Great Britain made radioactivity and the effects of fallout a genuine public issue worldwide by the late 1950s. Researchers tried to determine the extent of the immediate risk of fallout or how to define that risk. Some prominent scientists were not so ambivalent. A. H. Sturtevant, a well-known geneticist at Cal Tech, argued that both those exposed and their descendants would be harmed by fallout. Others, including Nobel Laureate in genetics Hermann Muller, concurred. The AEC stated that the risk (which they argued publicly was minimal) had to be balanced against the value of staying ahead of the Soviets in the arms race. It used this line of argument to deflect serious concerns about the impact of fallout on health. Physicist and Manhattan Project participant Ralph E. Lapp charged that the AEC was making "reckless and unsubstantiated" statements about fallout, and questioned its efforts to obfuscate the issue.[103] Concern over the effects of fallout did not go away, despite AEC's assurances of safety. Some scientists tried to determine the health impact of low-level radiation exposure, since effects of massive exposure were obvious. A central question was how or whether a "threshold" could be determined for somatic (cellular) radiation injury. (Experts agreed that there was no threshold for genetic effects.) Queries continued as the fear of fallout in America became less of a distant problem and raised questions about radiation risks closer to home.[104]

The radioactive isotope strontium-90, a by-product of atomic testing, found its way into the stratosphere and then settled back on earth over many months or years. Since it chemically resembles calcium, the body could not distinguish between the two. Children were highly susceptible because strontium-90 concentrated in newly formed bones and teeth, and could cause bone tumors. Milk was an obvious pathway for the isotope, especially because cows could ingest it while eating grass upon which fallout settled. As an important source of calcium, and a staple of the diet of children, milk became an insidious delivery vehicle for strontium-90. Knowledge about this health threat finally reached the public because of bomb testing, with scientists and even some popular magazines warning about the dangers of fallout as "the silent killer."[105] While scientists had been aware of the properties of strontium-90 for some time, the AEC and other groups were slow to test the milk supply or other possible sources of the deadly isotope. Two studies contributed to the public unease that persisted in the decade. The Committee for Nuclear Information at Washington University in St. Louis conducted one study in 1958, which found rising concentrations of strontium-90 in children's teeth (biologist Barry Commoner organized the "Baby Tooth Survey"). The Consumers Union conducted the other study in the same year, and found rising but not overt amounts in milk samples.[106]

Official denials and efforts to counter the criticism about fallout were as far as the AEC took the matter at the time. The same was true for the Eisenhower administration in general. In June 1956, on a request from the AEC to study the effects of low-level radiation, the National Academy of Sciences (NAS) released a report suggesting that as yet fallout from testing did not add significantly to the cancer risk, but any amount of radiation could cause genetic damage. The Genetics Committee of the NAS, which included Sturtevant and Muller, concluded that the fallout produced less radiation than some medical procedures, but exposure still needed to be minimized because of its long-term effects. The revelation about radioactivity's genetic impact was certainly the most riveting. The Pathology Committee, including several members with ties to the AEC, regarded the risk as less serious. Lapp and others, however, remained skeptical of the Pathology Committee's conclusions. But the door had been opened, and the sharp denials of the AEC about the dangers of fallout seemed weak and less credible, and to some people outright negligent.[107] Not until 1959 did the Eisenhower administration respond to the radiation issue in any constructive way. In that year the president created the Federal Radiation Council to offer advice on radiological safety programs. He also increased the role of the Department of Health, Education and Welfare in analyzing data on radiation risks. Protests against fallout and testing accelerated.[108]

That so many films of the 1950s and early 1960s either used the fallout scare as a plot device or as a lightly veiled warning says a great deal about its public impact. Alongside Cold War atomic spy films and infiltration thrillers, movies about "Nature Gone Amuck" were everywhere in the period. Among the first, best, and probably the most interesting of the "creature" films was *Them!* (1954). It is noteworthy for its effective treatment of fallout, its apocalyptic message, and its portrayal of key social themes. Not far from the *Trinity* Site in New Mexico, giant mutant ants were terrorizing the locals. The police find a young girl in shock wandering the desert. The ants had killed her parents. Ultimately a swarm of ants fly west to Los Angeles, where the military, police, and FBI (demonstrating great heroism) stage a final showdown with the monsters and prevail. Scientists too had been called to the scene to lend their expertise to subduing the creatures. Dr. Harold Medford (Edmund Gwenn) from the Department of Agriculture has eradication on his mind, which was not uncommon in an era when scientists were developing and applying pesticides like DDT. *Them!* pits twentieth-century expertise and firepower against the nightmare of genetic mutation through radiation. In the final scene, Dr. Medford intones, "When man entered the atomic age, he opened a door into a new world…what he will eventually find in that new world, nobody can predict."[109]

In this golden age of science fiction, the theme of *Them!* is repeated many times over. Inoshiro Honda's *Gojira*, remade for American audiences as *Godzilla* (1954), launched a monster movie franchise. The radioactive dinosaur-like creature begins as a malevolent force who destroys Tokyo, but devolves in later movies into a caricatured hero-monster. The image of the original Gojira/Godzilla rampaging through Tokyo is an all-too-obvious symbol of Japan's fated experience with the bomb.[110] Poignant on another level is *The Incredible Shrinking Man* (1957). Robert Scott Carey (Grant Williams) is exposed to a strange glowing cloud while vacationing on his brother's boat. Six months later Carey begins to shrink. His doctor concludes that this is the result of radiation from the cloud and accidental exposure to pesticides while working in the yard of his suburban home (double trouble). He shrinks to the point where his wife cannot find him, and he must stave off attacks from the family cat and a menacing spider in his basement. Shrinking actually and metaphorically, Carey is losing touch with the world he knew. As he shrinks to a speck, he accepts his fate of becoming one with the universe. There are obvious mixed messages in the movie, but it is clear that this new era is increasingly beyond human control. An alternative image is *The Amazing Colossal Man* (1957) in which radiation reverses the growth trend for Colonel Glenn Manning (Glenn Langan). He becomes a 60-foot giant, having been exposed in a trench on the NTS. This movie, although much less cerebral than *Shrinking Man*, also plays upon the disastrous impact of mutation through radiation.[111]

BAN TESTING, BAN THE BOMB

With *Bravo* and the *Lucky Dragon* as grim omens in 1954, the spate of thermonuclear atmospheric tests ignited protests all over the world. Demand for an end to testing occurred alongside calls to "ban the bomb" and to embark on nuclear disarmament.[112] Among them the United States, the USSR, and Great Britain detonated more than 300 nuclear weapons by the end of 1958. In 1957 and 1958, more bombs were tested than in all previous years combined. France and China also entered the nuclear club in these years. The Hiroshima and Nagasaki attacks originally inspired anti-nuclear responses from Japanese survivors as well as from scientists and pacifists in other countries. Yet, by the late 1940s the public disquiet about nuclear weapons and the call among some for nuclear disarmament had faded because of war exhaustion and Cold War tensions. By 1954 the uneasiness had returned with the development of the H-bomb, but also with a counterweight—the rise of the nuclear disarmament movement. The movement itself was broad-based, spanning a variety of individuals and groups with objectives ranging from political goals to uncompromising pacifism. However, they all shared opposition to testing.[113]

Overwhelming popular sentiment did not favor disarmament in this decade. For many people the bomb represented security, especially if you possessed it and your enemies did not (or if you had more). Pro-nuclear groups were not easily deterred by the rising tide of concern and criticism, yet neither could they silence the mounting anti-nuclear sentiment. The combination of unyielding testing and the nuclear arms buildup reinvigorated international peace organizations such as War Resisters' International, the Women's International League for Peace and Freedom, and the International Fellowship of Reconciliation. International religious groups, such as the World Council of Churches, called for a prohibition of nuclear arms. Pope Pius XII spoke out against thermonuclear weapons in particular as "new, destructive armaments, unheard of in their capacity of violence" and urged that every nation avoid nuclear war "by all means." He did not, however, disavow the value of nuclear deterrence in general, nor soften his strong anticommunist sentiments.[114] In England, mathematician, philosopher, and world–government supporter, Bertrand Russell, led the way in efforts to renew the fight against the bomb. He eloquently and soberly delivered an address on the BBC in December 1954 called "Man's Peril," warning about the possible annihilation of life on earth because of nuclear weapons. Albert Einstein endorsed the idea right before his death in April 1955, as did French nuclear physicist and communist, Frederic Joliot-Curie. Cold War politics weakened the effort, since several non-communist scientists refused to sign the statement, and Joliot-Curie argued over its wording before signing it. The offshoot of Russell's efforts was the so-called Russell-Einstein Manifesto supported by several well-known scientists. It asked, "What steps can be taken to prevent a military contest of which the issue must be disastrous to all parties?" Other appeals followed.[115]

Throughout the world, calls to end testing gained momentum. Japan was in the forefront for obvious reasons. On August 6, 1955, the First World Conference against Atomic and Hydrogen Bombs convened in Hiroshima. It resulted in the establishment of an important mass-movement organization, the Japanese Council against Atomic and Hydrogen Bombs (*Gensuikyo*). Broadly based at first, it ultimately became more left wing with support from the Japanese Socialist Party, its trade union groups, and the smaller Communist Party. Despite the *Gensuikyo*'s political leanings, a clear majority of Japanese people favored banning nuclear weapons and disapproved of American testing. Next to Japan, Europe probably represented the area with the strongest concerns about nuclear testing and disarmament. Pacifist groups had been active in a variety of protests prior to *Bravo*. Even the British Council of Churches, which had been silent on the issue of nuclear weapons, publicly expressed their concern at the fallout from the Pacific tests in 1954. The Labour Party pressured the Conservative government unsuccessfully to work toward ending nuclear testing. Polls conducted at the time showed a majority of British citizens in favor of the manufacture of H-bombs opposed to employing nuclear weapons. (In a secret US government study, banning atomic weapons received overwhelming support in Great Britain.) Public debate and growing anti-nuclear sentiment emerged in other parts of Europe as well. Anti-nuclear sentiment was on the rise in the Soviet Union, although the extent of such feelings was almost impossible to gauge. Outside of Europe, anti-nuclear reactions were strong in Australia and New Zealand, particularly because these countries were in close proximity to the test sites in the Pacific.[116]

The reaction in the United States was somewhat tepid until the fallout controversy intensified. Some news commentators and American pacifists made public their opposition to nuclear weapons. The *Nation*, a well-known liberal organ, urged "hard political thinking" on the issue, and even *Scientific American* showed signs of wariness about thermonuclear weapons. Atomic scientists from the days of the Manhattan Project were not as cohesive or well-organized in the

1950s, but stalwarts like Leo Szilard raised the issue of disarmament in talks with government officials. The Federation of American Scientists (FAS) seemed to be more engaged on the issue than Szilard, whose focus was directed more toward broader Cold War issues than to a testing ban. Religious groups in the United States followed the lead of those of other countries in calling for an end to nuclear weapons. One World advocates, such as Russell and Einstein, had begun to focus on banning the bomb instead of on the more grandiose plans for world government. This was particularly true for Norman Cousins, editor of the *Saturday Review of Literature*. His objections to the bomb began with Hiroshima, and his magazine soon reflected abhorrence to nuclear weapons. In a celebrated effort, Cousins teamed up with Reverend Kiyoshi Tanimoto, a *hibakusha* himself, to bring the "Hiroshima Maidens" to the United States. In May 1955, they brought 25 Japanese girls disfigured with scars from the atomic attack to Mt. Sinai Hospital in New York to obtain plastic surgery. The event, reasonably well covered in the press, was meant to be an object lesson about nuclear war.[117]

The American public at large, as expressed through opinion polls, was showing a greater uneasiness with nuclear weapons, a growing concern that the United States could be attacked, more frequent interest in disarmament, and a modest increase in sentiment against testing. The fallout controversy heightened all of these concerns. Nuclear testing also became an issue in the 1956 presidential campaign. The news about strontium-90 led Democratic candidate Adlai Stevenson to raise the issue of contamination as a result of his promise to halt the testing of hydrogen bombs. Surrogates for President Eisenhower, including Vice-President Richard Nixon and Secretary of State Dulles, dismissed the Democratic challenger's claims. Nixon called it "catastrophic nonsense." Stevenson's attempts to raise the test-ban issue, however, were not decisive in the election and did not keep him from losing in a landslide to the incumbent. The president remained relatively silent on a test ban, and had clearly demonstrated by his actions that he was a proponent of weapons testing.[118]

As the detonations increased, so did the protests. Cousins had recruited Alsatian physician, philosopher, and humanitarian Albert Schweitzer into the disarmament movement. Schweitzer broadcast his "Declaration of Conscience" on April 23, 1957, into at least 50 countries with reports in innumerable newspapers. He called for a halt to testing, stating emphatically that radiation was a health hazard to current and future generations, "a danger not to be underrated." This was just the start for Schweitzer who spread his message far and wide, receiving enthusiastic support for his call to action.[119] The international scientists' movement also became involved, overcoming divisions that occurred as a result of the early Cold War. The first Pugwash Conference on Science and World Affairs met in July 1957 in Pugwash, Nova Scotia. Mainly scientists, participants came from 10 Western, Eastern, and non-aligned countries. Despite the fact that the meeting attracted many who had spoken in favor of nuclear disarmament, the real story was the presence of scientists from across the "Iron Curtain." As a group they did not agree on the nature of the hazards nor the need for control, but 20 of the 22 participants signed a general accord nonetheless. It stated that all countries must abolish war, curtail the arms race, seek peace, and end testing. While the Pugwash Conference received limited attention in the United States and Great Britain, it was widely publicized elsewhere. The conference became a regular event, with 21 such meetings held through 1971. It was, at the very least, an important gathering place for scientists from East and West.[120]

Nobel laureate in chemistry, Linus Pauling from Cal Tech, was a leading voice in the campaign against fallout and the development and production of nuclear weapons. He became very active in mobilizing scientists against the bomb. In 1962 Pauling won the Nobel Peace prize for his efforts to ban nuclear testing, and was one of very few Nobel laureates to garner both

Linus Pauling, Nobel Laureate in chemistry and recipient of the Nobel Peace Prize (1962), was a leading voice in the campaign against fallout and the development and production of nuclear weapons. *Source:* Library of Congress, Prints & Photographs Division, [LC-USZ62-76925]

the peace and scientific prizes. His anti-testing petition distributed among scientists in 1957 ("Appeal by American Scientists to the Government and People of the World") had an important international impact. More than 10,000 scientists signed the petition, which was presented to United Nations Secretary-General Dag Hammerskjold in January 1958. Although not all scientists agreed on the issue, the fact that the protest was led by a prestigious scientist and many of his peers did not go unnoticed.[121]

In 1957 and 1958, individual or small-group protests were making way for mass movements in the United States, Great Britain, and Japan. Banning the bomb became a rallying cry. In Japan *Gensuikyo* was the primary instrument for generating anti-nuclear sentiment. In Great Britain calls for the government to cancel the H-bomb tests were closely associated with the first British H-bomb detonation scheduled for Christmas Island. The Emergency Committee for Direct Action against Nuclear War planned to live up to its name and disrupt the event if necessary, but it was not able to mobilize in time. Other groups focused on non-violent resistance.[122] In the United States nuclear issues quickly became the focus of pacifist groups, and protest in the scientific community resurfaced as well. In June 1957 FAS called for a ban on testing large nuclear weapons, and a few months later pushed for an international agreement to ban nuclear testing altogether. The National Committee for a Sane Nuclear Policy (SANE) under the leadership of Norman Cousins also pushed for the cessation of testing. Its advertisement in the *New York Times* (primarily written by Cousins and signed by almost 50 scientists, clerics, labor leaders, public figures, and business people) intertwined the issue of arms control with the concern over radioactive contamination. The ad struck a chord, propelling SANE into a national organization by 1958 with 130 chapters and about 25,000 members. SANE was weakened temporarily when it became ensnared in charges of communist infiltration. It sought a way out by bowing to pressure to adopt a resolution restricting membership to those who rejected "communist and other totalitarian doctrine[s]."[123] SANE returned to prominence with the celebrated pediatrician and activist Dr. Benjamin Spock as a national sponsor. Leaders in SANE played an active role in the passage of the Nuclear Test Ban Treaty in 1963.[124]

Even with the renewed vigor of pacifist groups, direct-action groups, and anti-testing protests, the impulse to ban the bomb was not widespread in the United States during the 1950s. Membership in groups like SANE tended to be regionally bound, often robust in the Northeast and Pacific Coast, weaker in the South and much of the West. Anti-nuclear views were predictably stronger among liberals and Democrats as opposed to conservatives and Republicans. Those against the test-ban activists could always play the "Red Scare" card (as in the case of SANE), even though it had lost some of its steam by the late 1950s. Nevertheless, contemporary opinion polls heavily favored a test ban and even nuclear disarmament as long as neither was unilateral. Calls for bans spread more widely in Western and into Eastern Europe, the Pacific region, and parts of the Third World because of the proliferation of nuclear weapons and the increase in the number of locations where countries conducted tests.[125] At the same time, debates over nuclear disarmament overall were at serious loggerheads. Many pacifists were nonpartisan, repulsed by nuclear weapons in general, and critical of both Cold War camps. Others were more clearly partisan, such as communist-led groups like the World Peace Council. They tried to score points by condemning the Western arms buildup and asserting that communist nations clearly favored peace. As early as 1950 the council appealed for a ban on nuclear weapons in their Stockholm "ban the bomb" petition (which they claimed was signed by 473 million people throughout the world). The Soviets used the petition as a way to cast doubt on the legitimacy of the United States arms program, while American leaders rejected it as propaganda.[126]

On the Beach (1959) is one of the best-known films about the bomb and amazing for the time. It was a powerful warning about the failure in choosing nuclear warfare over peace. Set primarily in Australia, the film focused on a star-crossed love affair between Commander Dwight Towers (Gregory Peck) and Moira Davidson (Ava Gardner) caught in the last days of life on earth as a radioactive shroud from a nuclear exchange makes its way Down Under. Based on Neil Shute's book, director Stanley Kramer created an evocative piece full of uncomfortable messages. Towers is an American submarine commander stationed in Australia, but dispatched to San Francisco on the *U.S.S. Sawfish* to visually observe what many fear. When he returns with the bad news, his life focuses on Moira, and the tragedy of the inevitable doom becomes very personal. A scene in which citizens line up to obtain suicide pills powerfully communicates the theme of resignation in the film. There are no wild bacchanalias, no violent outbursts, and little hope. *On the Beach* did not earn much in the way of American government approbation, largely because it so clearly displayed the unthinkable. There is a cautionary moment at the end of the film when the camera zooms in on a sign from an evangelist's gathering: THERE IS STILL TIME, BROTHER. The political message was loud and clear, and coincided with many of the fears of the anti-nuclear protests.[127]

THE POLITICS OF DISARMAMENT

One scholar referred to disarmament as the "second pathway of stigmatization" of nuclear weapons as opposed to the "first pathway" of the anti-testing and ban the bomb effort.[128] The United Nations in particular had called for the "elimination and prohibition" of weapons of mass destruction since 1946. The General Assembly proposed an international disarmament convention in 1954 in the wake of *Bravo*, and in 1955 recommended that a scientific committee investigate radioactivity from nuclear testing. By 1958 the UN's annual resolution against nuclear weapons focused on the issues of testing and surprise attacks. Third World participants wanted funds otherwise used for the production of nuclear weapons to be diverted to the needs of developing countries. Leaders of the United States took an active role in the disarmament debates and voted in favor of the resolutions, knowing full well that they were non-binding. The Soviets were playing a political game of their own (the "peace offensive"), which was a "third pathway of stigmatization."[129] Consumed with their inferior status as a nuclear power, Khrushchev hoped to seek peaceful coexistence with the West, while branding the United States and its allies as aggressor nations. The new Soviet leader came to realize that the USSR could no longer adhere to Stalin's public claims denigrating the significance of nuclear weapons when compared with the Red army. Khrushchev acknowledged the nuclear threat to both sides and the need to contain that threat. The Soviets also criticized Western efforts in disarmament asserting that they always came with reservations. He believed that early Soviet proposals were more concrete. Those proposals included phased reductions in weapons, cutbacks in Soviet troop strength in Eastern Europe, a test ban, and American exit from Germany. Such alternatives, of course, had their own political price.[130]

Early disarmament negotiations were simply part of the mutual propaganda battle and little more. When the Soviets invaded Hungary in November 1956, proposals for disarmament and a test ban clearly went off the table.[131] The UN, however, did not curb its efforts to find a route to disarmament, and negotiations between the United States and the USSR began afresh at the London disarmament talks in 1957. US delegate Harold Stassen presented an informal proposal to seek common ground on a formula for "use" of nuclear weapons, and the possibility of tying a test ban to the ending of nuclear weapons production. This was a promising start, but was doomed when NATO allies disavowed imposing limits on the potential use of nuclear weapons to deter a Soviet offensive in Europe. The Soviets, for their part, were willing to consider a temporary test ban with no inspection or no weapons production halt. The proposal from the United States, which showed Dulles' fingerprints, called for a temporary test suspension, contingent on a weapons production cutoff and effective inspection. Although the Soviets had moved closer on inspection and verification, they could not agree to any plan that called to limit the production of nuclear weapons. To do so was to harden their inferior status as a nuclear power. The talks were aborted.[132]

Facing criticism of its testing policies from around the world, the Eisenhower administration still was willing to consider some type of disarmament proposal. This was made difficult because of strident support for testing coming from the AEC, the JCS, and the weapons labs. The Soviets stole the propaganda advantage by announcing a unilateral test moratorium in March 1958 to which the Americans had to respond. In late August Eisenhower announced that the United States would stop testing for one year, unless Soviet Union testing was discovered. The US moratorium went into effect on October 31, after a furious number of last-minute blasts. Also on that date the Conference on the Discontinuance of Nuclear Weapons Tests convened in Geneva. Edward Teller discredited some of the technical findings and was party in efforts to obstruct the proceedings. The one positive step at the meetings was the passage of the Antarctic

Treaty, which demilitarized the Antarctic and prohibited dumping of radioactive waste there.[133] In 1960 another effort at a ban on testing ended badly. What appeared to be promising negotiations was derailed when Khrushchev made a startling announcement on May 7. He declared that six days earlier the Soviets had shot down an American U-2 high-altitude surveillance airplane in their territory and recovered the pilot, Francis Gary Powers. These spy planes had been flying over the USSR since 1956 to uncover its military capabilities. The Soviets had been aware of the flights before the Power's incident, but were unable to prevent them. Until now Khrushchev offered Eisenhower the opportunity to plead ignorance on the event, but the president chose to take responsibility and refused to apologize for the American action. Both sides were infuriated, and not surprisingly the test ban negotiations were doomed to failure as US-Soviet relations further deteriorated.[134]

In September 1961 the USSR resumed testing and the United States followed in March 1962. The end of the moratorium provoked a new round of public protests, and this reaction plus the sobering impact of the Cuban Missile Crisis in October 1962, prompted renewed negotiations. This led to an agreement on atmospheric testing in 1963 (to be discussed further in Chapter 7). Cold War tensions had long obstructed even the smallest step on the path to the control and management of nuclear weapons. Test ban talks and disarmament discussions were as difficult as baling the ocean, with almost the same results. In mid-1956 the United States stockpile of nuclear weapons stood at more than 3,000, with another 2,000 added the next year. It reached 18,000 by the end of 1960. The Soviet numbers were substantially less, but nonetheless impressive by world standards. The unanswerable question was: Could anything reverse this ominous trend, this unstoppable arms race?[135]

BEEP-BEEP-BEEP

In 1957 another scientific achievement redefined the arms race. On October 4 the Soviets put *Sputnik* (a small, 184 pound silver-colored orb) into space on the nose of an *R-7* intercontinental ballistic missile. *Fellow Traveler*, as it translated into English, was the first artificial satellite to orbit the earth. On that night, an NBC radio announcer told his audience: "Listen now for the sound that forevermore separates the old from the new." That sound was a simple "beep-beep" every few seconds, which was radio noise with no objective other than to announce its existence.[136] Some officials wanted to believe that the event was nothing more than a publicity stunt. But on November 3 the Soviets launched the half-ton *Sputnik II* that also carried a living thing into space (a little dog named Laika unfortunately died before reaching orbit). American scientists and politicians alike were stunned, unprepared for their rival to accomplish such feats before they did. It was bad enough that Soviet space hardware circled the globe, but it was much more ominous that a powerful missile had put it there. If rockets could carry satellites into space, they certainly could be redirected to deliver nuclear weapons to the United States. As one writer noted, "The missile gap implied a technological gap, which in turn implied a research gap, and thus an education gap and so on."[137] Partisan politics intensified. Democrats claimed that the United States was falling behind in the arms race, while Republicans tried to downplay the importance of *Sputnik*. American scientists came under attack. Not since Pearl Harbor had the United States seemed so vulnerable.[138]

From a strategic perspective, the introduction of ICBMs and other missiles into the arms race undermined the principle of massive retaliation and America's advantage with it. A couple of years prior to the *Sputnik* launch, American intelligence gathering suggested that the USSR was making progress in offensive military technologies, and the AEC was aware of Soviet advances in the miniaturization of warheads and in rocketry. Indeed, two months before *Fellow Traveler*, the USSR successfully fired the first ICBM. Their investment in rocketry started in the early 1950s, and work on a new air defense system to protect Moscow began in 1954.[139] The muscle of the Strategic Air Command was severely weakened by the prospect of middle- and long-range missiles tipped with nuclear warheads aimed at Western Europe and the United States. A Soviet surprise attack would make it difficult to get SAC bombers in the air before missiles reached their targets, and the weapons would be difficult to locate and destroy in order to neutralize Soviet military capabilities.[140]

Eisenhower administration officials had been ill-prepared for *Sputnik*. Like the Roosevelt and Truman administrations before them, they underestimated the Soviet's technical abilities and resolve. The US satellite program had been given a low priority. There was a super-secret project (*Corona*) to develop surveillance satellites, but nothing matching the Soviet program in offensive weapons. The army, navy, and air force had been working on the developing missiles since World War II. Progress was slow through 1950 because of budgetary limits and doubts that missiles could be constructed to carry the heavy nuclear weapons in production at the time. The Korean War offered an opportunity for the army and navy to develop air defense missiles, atomic tactical ballistic missiles, and cruise missiles, but not ICBMs. Under the New Look, a 1954 report of the air force's Strategic Missiles Evaluation Committee stated that the United States could deploy an ICBM by about 1960 if given high priority. In 1955 Eisenhower approved the *Vanguard* satellite project using a new rocket built by the navy rather than the more advanced *Redstone* rocket developed by the army. The early December test of the *Vanguard* failed when the rocket and its load exploded while still on the launch pad. This was insult on top of injury. The United States did not achieve a successful placement of a satellite in space until January 31, 1958, with *Explorer 1*. By 1960, the United States deployed 18 *Atlas* ICBMs, and also was beginning to produce *Titan* missiles. Both had greater lifting capacity than previous missiles, and thus more utility as military delivery systems. President Eisenhower supported the development program, but had reservations about missiles as offense weapons. He and SAC also were bent on preserving the role of the bomber as a strategic weapon.[141]

Despite the almost hysterical reaction to *Sputnik* and its military implications, the president remained relatively calm. The photographs from secret U-2 reconnaissance flights over the USSR suggested that the Soviets were exaggerating their missile strength and distance capability. Unfortunately Ike could not make this information public. In a series of speeches, he attempted to reassure the American people that the nation's defenses could discourage any Soviet attack. For public effect, he also responded by slightly increasing military spending, created NASA to promote space exploration, and supported the establishment of the National Defense Education Act to accelerate science, mathematics, engineering, and foreign language instruction. In secret he ordered the construction of an underground bunker complex in West Virginia where the government could conduct business in case of nuclear attack—just in case.[142] Eisenhower did not bend to the emotional call for numerous crash programs and massive military spending. Despite having easily won reelection in 1956, he still was recovering from a serious heart attack in the previous year which added an additional impediment to the *Sputnik* crisis before him. These were trying political times. Members of both parties criticized the president for responding too slowly; American allies hoped that the United States would stay the course in the arms race.[143] Ironically, the *Sputnik* launch actually may have set back the Soviet ICBM program. The *R-7* missile that put the satellite in orbit was much too big and bulky for warfare. As Khrushchev himself later stated, "It represented only a symbolic counterthreat to the United States" and was "reliable neither as a defensive nor offensive weapon."[144] The Soviets were forced to develop a new ICBM, and in reality lost whatever advantage they seemed to have gained in the arms space race.[145] Perceptions and reality, however, often were difficult to distinguish in these years of intense competition.

Like Cold War crises before it, the launch of *Sputnik* led many to overreact. None could fault the American people for becoming increasingly restive since it appeared that the USSR was quickly gaining power and influence in weapons development, and now in space. Particularly in Washington the mood of uneasiness was palpable. Case in point was the Gaither Report. Led by H. Rowan Gaither, Jr. of the Ford Foundation, the blue-ribbon committee of scientists, engineers, economists, and military experts presented its secret findings to Eisenhower in November 1957. The committee exceeded its mandate to assess a Civil Defense Administration's proposal to build a system of fallout shelters, and looked more broadly at basic questions of defense policy. The report's tone and message were vintage Cold War, that is, Soviet goals and objectives had not changed but intensified, its economy was growing, and its potential threat grave. In no uncertain terms the Gaither Report declared that US defenses designed to protect the public from attack were gravely inadequate and that the nation's strategic nuclear forces were vulnerable. The government needed to develop a defensive missile system to protect the American mainland and build bomb shelters for the citizenry. The United States also had to enhance its strike force to successfully annihilate Soviet targets, including improved ballistic missile capability and renewed effort to engage in limited military operations. The price tag would be about $5 billion added to the projected federal budget in 1959 of $38 billion, with about $12 billion more by 1961 and possibly more later.[146]

While Eisenhower's initial responses to the *Sputnik* crisis had been measured and his willingness to expand the defense budget cautious, the Gaither Report affected his national security policy throughout the remainder of his second term and into the John F. Kennedy administration and beyond. Eisenhower tried to suppress the report, but parts of it leaked out to the press creating an intense public reaction fueled by *Sputnik*.[147] Eisenhower did not favor the development of fallout shelters or expanding the capacity of the military to fight limited wars, but he somewhat reluctantly supported many of the other recommendations. The president called for acceleration in the development of ICBMs, and also in medium-range ballistic missiles (IRBM). He deployed the *Polaris* submarine capable of launching missiles. He ordered the construction of an early warning radar system to protect SAC assets. And he directed that SAC bombers be dispersed over several airfields. The spirit of the New Look was dead.[148]

The Gaither Report was a key Cold War document, possibly not as significant as George Kennan's containment article or NSC 68, but important nonetheless. Despite evidence to the contrary, the United States was not losing the nuclear arms race, but had a healthy lead in strategic weapons. The prospect of a vulnerable United States had been unthinkable until the late 1950s. ICBMs in particular (with thermonuclear warheads) had the capacity to reduce defensive preparation time to minutes, not hours or days. The prosecution of war most recently experienced in Korea and World War II seemed as distant and arcane as cavalry charges and hand-to-hand combat. Not surprisingly the postwar generation was trying to make rational the irrational, make comprehensible the incomprehensible. Policy formation for the superpowers, and other countries of the world, was now incredibly more challenging.

CONCLUSION: THIS ISN'T THE NOSTALGIA DECADE

The 1950s seemed to hold little hope or promise, and considerable risk for Atomic Age America. Many years later in 1982 Kevin Rafferty, Jayne Loader, and Pierce Rafferty produced a documentary called *Atomic Café*. It included a great deal of stock footage, newsreel clips, television scenes, and training films primarily from the 1950s (although it began with images of Hiroshima and its aftermath) constructed around civil defense and military training themes. It offered glimpses of atomic cocktails and other kitsch, and had a soundtrack replete with contemporary rock, blues, and country and western "bomb" songs, such as "Atomic Power" (the Buchanan Brothers), "Uranium" (the Commodores), and "Atomic Love" (Little Caesar). The 88-minute film became a big underground hit, criticizing government attempts to pacify the citizenry, poking fun at the atomic mania in an earlier time, and exposing the almost absurd decisions that were made in dealing with the bomb, radioactivity, civil defense, and the Cold War menace. One would want to believe that the filmmakers saw nothing funny at all with their film. This was black comedy sprinkled with the naiveté and even contemporary government subterfuge. Such visual and audio images speak volumes about the swirling issues engulfing the Age of Anxiety and the bomb.

The Eisenhower administration had tried to refine the nation's national security policy through the New Look, and had flirted with brinksmanship. The bomb testers and the bomb protesters raised awareness about real threats of radioactivity and the need to debate disarmament. But no one felt completely safe. There is little doubt that the H-bomb was a weapon too difficult to grasp by policy makers and citizens alike. Far from being a decade of nostalgia, the 1950s brought the bomb directly to the American public, to the Soviet public, and to every other country potentially in harm's way. Could the bomb keep the Cold War cold? A serious crisis in 1962 would challenge that hope.

MySearchLab Connections: Sources Online

READ AND REVIEW

Review this chapter by using the study aids and these related documents available on MySearchLab.

✔●—[**Study** and **Review** on **mysearchlab.com**

Chapter Test

Essay Test

📖●—[**Read** the **Document** on **mysearchlab.com**

Dwight D. Eisenhower, Special Message to the Congress on Reorganization Plan 6 (1953)

Dwight D. Eisenhower, Dien Bien Phu (1954)

American Officials' Response to *Lucky Dragon* Incident (1954)

Senate Resolution 301: Censure of Senator Joseph McCarthy (1954)

The Russell-Einstein Manifesto (1955)

The Gaither Report (1957)

RESEARCH AND EXPLORE

Use the databases available within MySearchLab to find additional primary and secondary sources on the topics within this chapter.

📖●—[**Read** the **Document** on **mysearchlab.com**

Profile: John Foster Dulles

🔍—View the Image on mysearchlab.com

Closer Look: Surviving an Atomic Bomb Blast

👁—Watch the Video on mysearchlab.com

Ike for President; Eisenhower Campaign

Duck and Cover

Endnotes

1. George C. Herring, *From Colony to Superpower: U.S. Foreign Relations Since 1776* (New York: Oxford University Press, 2008), 656, 659; Neal Rosendorf, "John Foster Dulles' Nuclear Schizophrenia," in John Lewis Gaddis, et al., eds., *Cold War Statesmen Confront the Bomb: Nuclear Diplomacy Since 1945* (New York: Oxford University Press, 1999), 63; Andrew P.N. Erdmann, "'War No Longer Has Any Logic Whatever': Dwight D. Eisenhower and the Thermonuclear Revolution," in Gaddis, et al., eds., *Cold War Statesmen Confront the Bomb*, 89, 119.
2. Robert A. Divine, *Eisenhower and the Cold War* (New York: Oxford University Press, 1981), 21.
3. Ibid., 20–22. See also Campbell Craig, *Destroying the Village: Eisenhower and Thermonuclear War* (New York: Columbia University Press, 1998).
4. Ronald E. Powaski, *March to Armageddon: The United States and the Nuclear Arms Race, 1939 to the Present* (New York: Oxford University Press, 1987), 60–61; Melvyn P. Leffler, *For the Soul of Mankind: The United States, the Soviet Union and the Cold War* (New York: Hill and Wang, 2007), 98–100. For a range of views on the controversial John Foster Dulles, see Richard H. Immerman, ed., *John Foster Dulles and the Diplomacy of the Cold War* (Princeton, NJ: Princeton University Press, 1990).
5. Quoted in Leffler, *For the Soul of Mankind*, 99.
6. On McCarthy's fall, see David M. Oshinsky, *A Conspiracy So Immense: The World of Joe McCarthy* (New York: Oxford University Press, 2005), 472–494; Richard M. Fried, *Nightmare in Red: The McCarthy Era in Perspective* (New York: Oxford University Press, 1990), 138–42; Albert Fried, ed., *McCarthyism: The Great American Red Scare* (New York: Oxford University Press, 1997), 178–81.
7. The term "peaceful coexistence" appeared in the statements of Soviet leaders early in 1954 during the "campaign" for the Supreme Soviet. Nikita Khrushchev often used the term, but not all leaders did so. One senior Soviet official regarded it as a "slippery expression" and one that made it appear as if his country was weak by asking for peace. See David Holloway, *Stalin and the Bomb* (New Haven, CT: Yale University Press, 1994), 335–37. See also Vladislav M. Zubok, *A Failed Empire: The Soviet Union in the Cold War from Stalin to Gorbachev* (Chapel Hill, NC: University of North Carolina Press, 2007), 94–95.
8. H.W. Brands, *The Devil We Knew: Americans and the Cold War* (New York: Oxford University Press, 1993), 41–43; 656–58; Leffler, *For the Soul of Mankind*, 84–150.
9. Herring, *From Colony to Superpower*, 652.
10. David Alan Rosenberg, "The Origins of Overkill: Nuclear Weapons and American Strategy, 1945–1960," *International Security* 7 (Spring 1983): 27–28.
11. The argument is not persuasive, but there are some who believed that Eisenhower wanted to continue Truman's containment policy, only changing it rhetorically. Ike distanced himself from the Democratic president, therefore, only to be identified more clearly as a Republican. See John Newhouse, *War and Peace in the Nuclear Age* (New York: Alfred A. Knopf, 1989), 87–93; Gerard J. DeGroot, *The Bomb: A Life* (Cambridge, MA: Harvard University Press, 2005), 185.
12. Rosenberg, "The Origins of Overkill," 27.
13. Ibid., 28.
14. Quoted in Ira Chernus, "The Real Eisenhower," *George Mason University's History News Network*, online, http://hnn.us/articles/47326.html.
15. Powaski, *March to Armageddon*, 60; Saki Dockrill, *Eisenhower's New-Look National Security Policy, 1953–61* (London: Macmillan Press Ltd., 1996), 2–3.

16. Quoted in William S. Borgiasz, *The Strategic Air Command: Evolution and Consolidation of Nuclear Forces, 1945–1955* (Westport, CT: Praeger, 1996), 37.

17. According to Nina Tannenwald, the United States decision to develop tactical nuclear weapons in the 1950s was an effort "to counteract an emerging taboo against first use of nuclear weapons by creating an alternative norm that tactical nuclear weapons should be treated as ordinary weapons." In other words, they wanted to "conventionalize" nuclear weapons. See Nina Tannenwald, *The Nuclear Taboo: The United States and the Non-Use of Nuclear Weapons Since 1945* (New York: Cambridge University Press, 2007), 166. See also 165, 167–70, 177–84, 185–89.

18. Quoted in Allan M. Winkler, *Life Under a Cloud: American Anxiety About the Atom* (Urban, IL: University of Illinois Press, 1999), 81.

19. There is substantial debate as to whether the New Look was really a defensive military policy utilizing an offensive strategy in non-military areas, or a more aggressive offensive strategy overall. For the former position, see Dockrill, *Eisenhower's New-Look National Security Policy, 1953–61*, 4–5.

20. Rosenberg, "The Origins of Overkill," 29–32. In the next several years, priority was given to radar detection, interceptor aircraft, early warning alert systems, sound surveillance, missile batteries, and so forth. In December 1952 Truman had authorized construction of an early warning radar system.

21. Ibid., 33–35.

22. Holloway, *Stalin and the Bomb*, 325.

23. Dockrill, *Eisenhower's New-Look National Security Policy, 1953–61*, 5; Herring, *From Colony to Superpower*, 653; Gerald Clarfield, *Security with Solvency: Dwight D. Eisenhower and the Shaping of the American Military Establishment* (Westport, CT: Praeger, 1999), xi–xiv.

24. Quoted in Robert C. Williams and Philip L. Cantelon, eds., *The American Atom: A Documentary History of Nuclear Policies from the Discovery of Fission to the Present, 1939–1984* (Philadelphia, PA: University of Pennsylvania Press, 1984), 219.

25. Ibid.

26. Borgiasz, *The Strategic Air Command*, 36–49; Brands, *The Devil We Knew*, 43–46; Julian E. Zelizer, *Arsenal of Democracy: The Politics of National Security—From World War II to the War on Terrorism* (New York: Basic Books, 2010), 126–28; Divine, *Eisenhower and the Cold War*, 38.

27. Brands, *The Devil We Knew*, 46; Rosenberg, "The Origins of Overkill," 41.

28. Powaski, *March to Armageddon*, 63–64; Timothy J. Botti, *Ace in the Hole: Why the United States Did Not Use Nuclear Weapons in the Cold War, 1945 to 1965* (Westport, CT: Greenwood Press, 1996), 55.

29. Rosenberg, "The Origins of Overkill," 37. The number of targets in the Soviet Union grew from a few dozen in 1948 to more than 2,500 in 1960. See 66.

30. The air force also got a third missile, the solid-fueled *Minuteman*, but it did not become operational until 1962. Powaski, *March to Armageddon*, 63–64; Botti, *Ace in the Hole*, 55.

31. Botti, *Ace in the Hole*, 55; Powaski, *March to Armageddon*, 63–64.

32. Brands, *The Devil We Knew*, 47–50; Borgiasz, *The Strategic Air Command*, 50; Powaski, *March to Armageddon*, 63–65; Clarfield, *Security with Solvency*, xiv–xv.

33. Rodney P. Carlisle, ed., *Encyclopedia of the Atomic Age* (New York: Facts on File, 2001), 279. See also Francis Duncan, *Rickover and the Nuclear Navy: The Discipline of Technology* (Annapolis, MD: Naval Institute Press, 1990).

34. Herring, *From Colony to Superpower*, 660–61. A sticking point in the armistice negotiations was repatriation of North Korean prisoners of war.

35. Ibid., 661–62; Divine, *Eisenhower and the Cold War*, 39–55; Botti, *Ace in the Hole*, 58.

36. Herring, *From Colony to Superpower*, 661–62; Botti, *Ace in the Hole*, 58; Divine, *Eisenhower and the Cold War*, 39–55.

37. Chen Jian, *Mao's China and the Cold War* (Chapel Hill: University of North Carolina Press, 2001), 11, 202–04.

38. Herring, *From Colony to Superpower*, 662–63; Divine, *Eisenhower and the Cold War*, 55–66; Botti, *Ace in the Hole*, 73, 77.

39. Quoted in Herring, *From Colony to Superpower*, 663.

40. Jian, *Mao's China and the Cold War*, 161–71, 202–04; Herring, *From Colony to Superpower*, 662–63.

41. Jian, *Mao's China and the Cold War*, 203.

42. Divine, *Eisenhower and the Cold War*, 66–70.

43. Ibid., 55–66; Herring, *From Colony to Superpower*, 662–63; Botti, *Ace in the Hole,* 73, 77. See also Jian, *Mao's China and the Cold War*, 71, 189–90; Matthew Brzezinski, *Red Moon Rising: Sputnik and the Hidden Rivalries That Ignited the Space Race* (New York: Henry Holt, 2007), 208–09.

44. See Kenneth Osgood, *Total Cold War: Eisenhower's Secret Propaganda Battle at Home and Abroad* (Lawrence, KS: University Press of Kansas, 2006).

45. Ibid., 1–11.

46. Quoted in William Taubman, *Khrushchev: The Man and His Era* (New York: W.W. Norton, 2003), 43–44.

47. Ibid., xix–xx.

48. Ibid., xix–xx; Aleksandr Fursenko and Timothy Naftali, *"One Hell of a Gamble": Khrushchev, Castro, and Kennedy, 1958–1964: The Secret History of the Cuban Missile Crisis* (New York: W.W. Norton, 1997), 11.

49. Herring, *From Colony to Superpower*, 664–69; Zelizer, *Arsenal of Democracy*, 136.

50. Powaski, *March to Armageddon*, 66–67, 78–79; Herring, *From Colony to Superpower*, 669–70.

51. Quoted in Taubman, *Khrushchev*, 352.

52. Herring, *From Colony to Superpower*, 671–80; Brands, *The Devil We Knew*, 54–55; Zelizer, *Arsenal of Democracy*, 135–36; Divine, *Eisenhower and the Cold War*, 71–104.

53. Zelizer, *Arsenal of Democracy*, 135–36; Herring, *From Colony to Superpower*, 671–80; Brands, *The Devil We Knew*, 54–55; Divine, *Eisenhower and the Cold War*, 71–104.

54. Brands, *The Devil We Knew*, 54–55; Herring, *From Colony to Superpower*, 671–80; Zelizer, *Arsenal of Democracy*, 135–36; Divine, *Eisenhower and the Cold War*, 71–104.

55. Jenny Barker Devine, "'Mightier Than Missiles:' The Rhetoric of Civil Defense for Rural American Families, 1950–1970," in Rosemary B. Mariner and G. Kurt Piehler, eds., *The Atomic Bomb and American Society: New Perspectives* (Knoxville, TN: University of Tennessee Press, 2009), 186.

56. Laura McEnaney, *Civil Defense Begins at Home: Militarization Meets Everyday Life in the Fifties* (Princeton, NJ: Princeton University Press, 2000), 5.

57. Quoted in ibid., 23.

58. Quote from Jon Hunner, "Reinventing Los Alamos: Code Switching and Suburbia at America's Atomic City," in Scott C. Zeman and Michael A. Amundson, eds., *Atomic Culture: How We Learned to Stop Worrying and Love the Bomb* (Boulder, CO: University Press of Colorado, 2004), 39.

59. Andrew D. Grossman in *Neither Dead Nor Red: Civilian Defense and American Political Development During the Early Cold War* (New York: Routledge, 2001), 104.

60. JoAnne Brown, "'A Is for Atom, B Is for Bomb': Civil Defense in American Public Education, 1948–1963," *Journal of American History* 75 (June 1988): 69–70; Elaine Tyler May, *Homeward Bound: American Families in the Cold War Era* (New York: Basic Books, 2008), 99; Paul Boyer, *By Bomb's Early Light: American Thought and Culture at the Dawn of the Atomic Age* (Chapel Hill, NC: University of North Carolina Press, 1994), 175–77, 319–33; Margot A. Henriksen, *Dr. Strangelove's America: Society and Culture in the Atomic Age* (Berkeley, CA: University of California Press, 1997), 92–101.

61. McEnaney, *Civil Defense Begins at Home*, 7; Grossman, *Neither Dead Nor Red*, 128.

62. Stephen E. Atkins, ed., *Historical Encyclopedia of Atomic Energy* (Westport, CT: Greenwood Press), 88–89; Snead, *The Gaither Committee, Eisenhower, and the Cold War*, 43–46, 168–69; Hunner, "Reinventing Los Alamos," 41.

63. Quoted in Kenneth D. Rose, *One Nation Underground: The Fallout Shelter in American Culture* (New York: New York University Press, 2001), 27.

64. On a personal note, during the early 1960s my high school in California hired an actor to play a communist official who ranted at us about the failure of capitalism and the glories of communism. Not until "the official" completed his tirade and had thoroughly agitated the assembled student body did our principal tell us that the presentation was staged, and then warned us about the dangers of Soviet propaganda.

65. Executive Offices of the President, National Security Resources Board, Civil Defense Office, *Survival Under Atomic Attack: The Official U.S. Government Booklet*, NSRB Doc. 130 (Washington, D.C.: GPO, 1950). See also Robert A. Jacobs, *The Dragon's Tail: Americans Face the Atomic Age* (Amherst, MA: University of Massachusetts Press, 2010), 63–64.

66. Brown, "'A Is for Atom, B Is for Bomb,'" 69–80, 84; Hunner, "Reinventing Los Alamos," 43; Dee Garrison, *Bracing for Armageddon: Why Civil Defense Never Worked* (New York: Oxford University Press, 2006), 46.

67. Brown, "'A Is for Atom, B Is for Bomb,'" 81–89.

68. Ibid., 87–90.

69. Ibid., 90. See also Henriksen, *Dr. Strangelove's America*, 108–11. For information on civil defense from an architectural perspective, see Tom Vanderbilt, *Survival City: Adventures among the Ruins of Atomic America* (New York: Princeton Architectural Press, 2002).

70. Henriksen, *Dr. Strangelove's America*, 106–108; Garrison, *Bracing for Armageddon*, 31–103.

71. Quoted in A. Costandina Titus, *Bombs in the Backyard: Atomic Testing and American Politics* (Reno, NV: University of Nevada Press, 1986; 2nd ed.), 55.

72. Ibid., 57–58.

73. Ibid., 55–58. See also Howard Ball, *Justice Downwind: America's Atomic Testing Program in the 1950s* (New York: Oxford University Press, 1986), 29.

74. Titus, *Bombs in the Backyard*, 58, 65; Allan M. Winkler, *Life Under a Cloud: American Anxiety About the Atom* (Urban, IL: University of Illinois Press, 1999), 91; Atkins, ed., *Historical Encyclopedia of Atomic Energy*, 247–49.

75. Atkins, ed., *Historical Encyclopedia of Atomic Energy*, 159–60, 287–89, 360–61.

76. Quoted in Eileen Welsome, *The Plutonium Files: America's Secret Medical Experiments in the Cold War* (New York: Dell Publishing, 1999), 252.

77. Titus, *Bombs in the Backyard*, 59–64; Ball, *Justice Downwind*, 31.

78. Titus, *Bombs in the Backyard*, 63.

79. Ibid., 91; David Wingrove, ed., *Science Fiction Film Source Book* (New York: Longman, 1985), 30.

80. A megaton is a unit of explosive force that is equal to one million metric tons of TNT; a kiloton is equal to 1,000 metric tons of TNT.

81. Atkins, ed., *Historical Encyclopedia of Atomic Energy*, 252; Carlisle, ed., *Encyclopedia of the Atomic Age*, 217–18. For a list of Soviet nuclear explosions between 1949 and 1955, see Holloway, *Stalin and the Bomb*, 323.

82. Atkins, ed., *Historical Encyclopedia of Atomic Energy*, 86–87; Carlisle, ed., *Encyclopedia of the Atomic Age*, 41.

83. Titus, *Bombs in the Backyard*, 70.

84. Winkler, *Life Under a Cloud*, 108; Robert A. Divine, *Blowing on the Wind: The Nuclear Test Ban Debate, 1954–1960* (New York: Oxford University Press, 1978), 262–80.

85. Quoted in "Fallout," Merriam-Webster, online, http://www.merriam-webster.com/dictionary/fallout.

86. Carlisle, ed., *Encyclopedia of the Atomic Age*, 98–99.

87. Barton C. Hacker, "'Hotter Than a $2 Pistol:' Fallout, Sheep, and the Atomic Energy Commission, 1953–1986," in Bruce Hevly and John M. Finlay, eds., *The Atomic West* (Seattle: University of Washington Press, 1998), 158.

88. Ibid.

89. Mary Dickson, "Downwinders All," in John Bradley, ed., *Learning to Glow: A Nuclear Reader* (Tucson, AR: University of Arizona Press, 2000), 130.

90. Veterinarians identified iodine-131, a radioactive element produced in the blast, as the primary culprit. A report sent to the AEC about the findings was ignored or at least papered over. Atkins, ed., *Historical Encyclopedia of Atomic Energy*, 124.

91. Winkler, *Life Under a Cloud*, 93.

92. Ibid. Not until 25 years later did a federal judge rule that the government deliberately had suppressed critical evidence and misrepresented what transpired.

93. Titus, *Bombs in the Backyard*, 65, 86–88, 96–97; Winkler, *Life Under a Cloud*, 91–93; Hacker, "'Hotter Than a $2 Pistol,'" 157–60, 169.

94. Atkins, ed., *Historical Encyclopedia of Atomic Energy*, 58–59.

95. Ibid.; Winkler, *Life Under a Cloud*, 93–94.

96. A free lance writer, Giff Johnson, in "Micronesia: America's 'Strategic' Trust," *Bulletin of the Atomic Scientists* 35 (February 1979): 11, stated that the Rongelap people were exposed to 175 rems of gamma radiation, which was considered a high dose. See also Atkins, ed., *Historical Encyclopedia of Atomic Energy*, 58–59; Winkler, *Life Under a Cloud*, 93–94.

97. Quoted in Titus, *Bombs in the Backyard*, 47.

98. Jean E. Howard, Ashok Vaswani, and Peter Heotis, "Thyroid Disease Among the Rongelap and Utirik Population—An Update," *Journal of Health Physics* 73 (July 1997): 190. See also Johnson, in "Micronesia: America's 'Strategic' Trust," 11–15; Jacob Robbins and William H. Adams, "Radiation Effects in the Marshall Islands," Medical Research Center, Brookhaven National Laboratory, n.d., online, http://www.yokwe.net/ydownloads/RadiationEffectsintheMarshallIslands. pdf, "The Evacuation of Rongelap," Greenpeace International, online, http://www.greenpeace.org/ international/rainbow-warrior-bombing/the-evacuation-of-rongelap?mode=send.

99. Divine, *Blowing on the Wind*, 3–35; Winkler, *Life Under a Cloud*, 93–94; Carlisle, ed., *Encyclopedia of the Atomic Age*, 179–80; Shane J. Maddock, "Ideology and U.S. Nuclear Nonproliferation Policy Since 1945," in Mariner and Piehler, eds., *The Atomic Bomb and American Society*, 131–32; Powaski, *March to Armageddon*, 75.

100. A roentgen is the international unit of an emitted dose of x-radiation or gamma radiation. One roentgen equals one rem. See J. Samuel Walker, *Permissible Dose: A History of Radiation Protection in the Twentieth Century* (Berkeley, CA: University of California Press, 2000), 23. According to the Nuclear Regulatory Commission, the average annual radiation exposure from natural sources is about 310 millirem, US NRC, online, http://www.nrc.gov/reading-rm/doc-collections/fact-sheets/ bio-effects-radiation.html.

101. Quoted in Winkler, *Life Under a Cloud*, 94.

102. Maddock, "Ideology and U.S. Nuclear Nonproliferation Policy Since 1945," 131–32; Carlisle, ed., *Encyclopedia of the Atomic Age*, 179–80; Lawrence S. Wittner, *Resisting the Bomb: A History of the World Nuclear Disarmament Movement, 1954-1970*, vol. 2 of *The Struggle Against the Bomb* (Stanford, CA: Stanford University Press, 1997), 1–2; Atkins, ed., *Historical Encyclopedia of Atomic Energy*, 222; Divine, *Blowing on the Wind*, 3–35; DeGroot, *The Bomb*, 197.

103. Quoted in Walker, *Permissible Dose*, 20. See also 18–19; Winkler, *Life Under a Cloud*, 95; Tannenwald, *The Nuclear Taboo*, 157.

104. See Walker, *Permissible Dose*, for the protracted discussion concerning exactly what was a "permissible dose" of radiation.

105. Paul Boyer, "Nuclear Themes in American Culture, 1945 to the Present," in Mariner and Piehler, eds., *The Atomic Bomb and American Society*, 7.

106. Winkler, *Life Under a Cloud*, 102–03.

107. Atkins, ed., *Historical Encyclopedia of Atomic Energy*, 353; Boyer, "Nuclear Themes in American Culture, 1945 to the Present," 6–7; Titus, *Bombs in the Backyard*, 87; Wittner, *Resisting the Bomb*, 3; Winkler, *Life Under a Cloud*, 95–96, 103–05; Walker, *Permissible Dose*, 21.

108. Walker, *Permissible Dose*, 21–28.

109. Quoted in Joyce A. Evans, *Celluloid Mushroom Clouds: Hollywood and the Atomic Bomb* (Boulder, CO: Westview Press, 1998), 91. See also Jerome Shapiro, *Atomic Bomb Cinema* (New York: Routledge, 2002), 102–11; Toni A. Perrine, *Film and the Nuclear Age: Representing Cultural Anxiety* (New York: Garland Pub., 1998), 84–89.

110. Shapiro, *Atomic Bomb Cinema*, 272–90; Perrine, *Film and the Nuclear Age*, 77, 90–92.

111. Perrine, *Film and the Nuclear Age*, 92–94; Shapiro, *Atomic Bomb Cinema*, 113–25; Evans, *Celluloid Mushroom Clouds*, 112–13; Errol Vieth, *Screening Science: Contexts, Texts, and Science in Fifties Science Fiction Film* (Lanham, MD: The Scarecrow Press, 2001), 201–05.

112. For a good overview of the test ban debate between 1954 and 1960, see Divine, *Blowing on the Wind*, 31, 58–83.

113. Wittner, *Resisting the Bomb*, ix, 1, 29; Tannenwald, *The Nuclear Taboo*, 156–57.

114. Quoted in Wittner, *Resisting the Bomb*, 4–5.

115. Quoted in ibid., 7.

116. Ibid., 8–10, 14–22.

117. Ibid., 10–13; Tannenwald, *The Nuclear Taboo*, 158–62.

118. Powaski, *March to Armageddon*, 75; Wittner, *Resisting the Bomb*, 13–14; Winkler, *Life Under a Cloud*, 107–08; Divine, *Blowing on the Wind*, 84–112.

119. Wittner, *Resisting the Bomb*, 30–33.

120. Atkins, ed., *Historical Encyclopedia of Atomic Energy*, 295; Wittner, *Resisting the Bomb*, 34–37.

121. Tannenwald, *The Nuclear Taboo*, 159; Wittner, *Resisting the Bomb*, 37–39; Carlisle, ed., *Encyclopedia of the Atomic Age*, 243–44.

122. Wittner, *Resisting the Bomb*, 44–51; Atkins, ed., *Historical Encyclopedia of Atomic Energy*, 41, 70–71.

123. Quoted in Carlisle, ed., *Encyclopedia of the Atomic Age*, 297.

124. Wittner, *Resisting the Bomb*, 56–60; Tannenwald, *The Nuclear Taboo*, 158–60; Carlisle, ed., *Encyclopedia of the Atomic Age*, 295–98.

125. Winkler, *Life Under a Cloud*, 105–06; Titus, *Bombs in the Backyard*, 98–99; Wittner, *Resisting the Bomb*, 51–82; Maddock, "Ideology and U.S. Nuclear Nonproliferation Policy since 1945," 131–32; Tannenwald, *The Nuclear Taboo*, 158–60; Atkins, ed., *Historical Encyclopedia of Atomic Energy*, 321.

126. Wittner, *Resisting the Bomb*, 83–97, 336–82; Tannenwald, *The Nuclear Taboo*, 156.

127. James Robert Parish and Michael R. Pitts, *The Great Science Fiction Pictures* (Metuchen, NJ: Scarecrow Press, inc., 1977), 250–52; Stanley Kramer, "On the Beach: A Renewed Interest," in Danny Peary, ed., *Omni's Screen Flight/Screen Fantasies: the Future According to Science Fiction Cinema* (Garden City, NY: Doubleday & Co., 1984), 117–19; Evans, *Celluloid Mushroom Clouds*, 146–49; Perrine, *Film and the Nuclear Age*, 154–57; Shapiro, *Atomic Bomb Cinema*, 89–94.

128. Tannenwald, *The Nuclear Taboo*, 162.

129. Ibid., 162–65.

130. Powaski, *March to Armageddon*, 76–78, 80.

131. Tannenwald, *The Nuclear Taboo*, 163–64; Powaski, *March to Armageddon*, 76–78, 80.

132. Tannenwald, *The Nuclear Taboo*, 164–65; Powaski, *March to Armageddon*, 81–84.

133. Powaski, *March to Armageddon*, 81–84, 87–92; Tannenwald, *The Nuclear Taboo*, 184–85; Atkins, ed., *Historical Encyclopedia of Atomic Energy*, 15–16.

134. Carlisle, ed., *Encyclopedia of the Atomic Age*, 341–42; Herring, *From Colony to Superpower*, 692.

135. Zuoyue Wang, *In Sputnik's Shadow: The President's Science Advisory Committee and Cold War America* (New Brunswick, NJ: Rutgers University Press, 2008), 120–41; Powaski, *March to Armageddon*, 81–84, 87–92; Botti, *Ace in the Hole*, 87.

136. Quoted in Paul Dickson, *Sputnik: The Shock of the Century* (New York: Walker & Co., 2001), 1.

137. Martin Walker, *The Cold War: A History* (New York: Henry Holt and Company, 1993), 115.

138. Wang, *In Sputnik's Shadow*, 71–73.

139. Holloway, *Stalin and the Bomb*, 324–25; Brands, *The Devil We Knew*, 65.

140. Rosenberg, "The Origins of Overkill," 44.

141. Dickson, *Sputnik*, 2; Wang, *In Sputnik's Shadow*, 73; Newhouse, *War and Peace in the Nuclear Age*, 118; Carlisle, ed., *Encyclopedia of the Atomic Age*, 316; Rosenberg, "The Origins of Overkill," 44–48; Carlisle, ed., *Encyclopedia of the Atomic Age*, 316.

142. Zelizer, *Arsenal of Democracy*, 139; Herring, *From Colony to Superpower*, 691–92.

143. Robert A. Divine, *The Sputnik Challenge: Eisenhower's Response to the Soviet Satellite* (New York: Oxford Press, 1993), xiii–xviii, 205.

144. Quoted in Brzezinski, *Red Moon Rising*, 269.

145. Ibid., 268–69.

146. David L. Snead, *The Gaither Committee, Eisenhower, and the Cold War* (Columbus, OH: Ohio State University Press, 1999).

147. The full report was not declassified until the 1970s.

148. Snead, *The Gaither Committee, Eisenhower, and the Cold War*, 1–3; Brands, *The Devil We Knew*, 66–67; Wang, *In Sputnik's Shadow*, 80–81; Richard Rhodes, *Arsenal of Folly: the Making of the Nuclear Arms Race* (New York: Alfred A. Knopf, 2007), 108–10; Newhouse, *War and Peace in the Nuclear Age*,118–20; Zelizer, *Arsenal of Democracy*, 140.

CHAPTER 6

Too Cheap to Meter, Too Tempting to Ignore

Peaceful Uses of the Atom

INTRODUCTION: PEACEFUL ATOMS IN OUR TIME

Peaceful uses of the atom were a much lower priority than bomb development. Some saw them as little more than a rationale for bringing such destruction into the world. Over time, expectations for atomic energy/nuclear power resulted in some important scientific breakthroughs as well as some hair-brained schemes. Clichés about "letting the genie out of the bottle" proved to be more apt with respect to atomic science than anyone could have imagined. Whether the genie was beneficent, malicious, or unpredictable, it attracted promoters and dissenters. On a pragmatic level, encouraging peaceful uses in the Cold War clearly was in the national interest. Such efforts would keep the United States in the leadership role in atomic energy, bolster its reputation in science, and possibly limit weapons proliferation. As an adjunct to the arms race, space race, and any other race, American leaders did not want to lose ground to Soviet attempts to jump ahead in developing civilian nuclear power or any other new atomic technology.[1] On another level, just what it meant to harness the energy of the atom met only with general speculation during and after the war. Was this a case of discovery before need? Such was the uncertainty of the peaceful atom in the postwar world.

REIMAGINING ATOMIC POWER

Writing in the *Nation* in 1945, J.D. Bernal, the controversial Irish-born scientist, former communist, and author noted, "What the effect of the use of atomic energy is likely to be on society, we can now only dimly see." The atomic bomb, he added, was "only the first impressive practical utilization of knowledge that appeared almost as startlingly as the bomb itself in the scientific world of fifty years ago." While harnessing this new concentration of energy would not be easy, Bernal speculated that it would be possible to use atomic power economically "to provide directly or indirectly for immediate human wants." He saw the possibility of uses from the mundane to the sublime, from pumping water and making fertilizer to social transformation. But he counseled, "A mere increase in scientific activity, however, is not enough; it must be coordinated and directed to really worthy tasks."[2] For some, the risks of fallout and radiation were little understood or too quickly dismissed. But Bernal's cautionary note was rarely found among those who envisioned the endless possibilities of a new era of atomic technology.

Popular science writers regaled their audience with a variety of atomic dreams. Some of these made their way into the commercial world. In 1958, Ford Motor Company unveiled its design for the Ford *Nucleon,* a futuristic car run by a small reactor with a rechargeable core. More in line with "peaceful" and "unpeaceful" uses of atomic energy, various toymakers in the 1950s and early 1960s released a play reactor, a ballistic missile with a tiny toy warhead, and a replica of Flash Gordon's "atomic disintegrator."[3] Although not as numerous as the gloomy atomic films that began to multiply in the 1950s, a few attempted a lighter touch, such as *Dig That Uranium* (1957), in which Leo Gorcey, Huntz Hall, and the rest of the Bowery Boys spoofed bomb testing and uranium mining. Like Blake Edward's *The Atomic Kid* (1954), *Dig That Uranium* utilized a serious subject as backdrop for a classic matinee comedy.[4] Expectations for and benefits of the peaceful atom also infused a variety of popular publications and other media in, what one book called, the "Early Atomic Culture" of the late 1940s and the "High Atomic Culture" of the 1950s and early 1960s.[5] Children in particular were a prime audience for the promising atomic future. Schools added the benefits of the atom to their lesson plans. In one celebrated episode, King Features allowed its comic book couple, Dagwood and Blondie Bumstead, to be used to educate children about nuclear fission in *Dagwod Splits the Atom* (1949). General Leslie Groves wrote the foreword in which he encouraged the country's youth to seek a career in atomic energy to help produce, among other things, atomic-powered trains, planes, and automobiles. Other comic book publishers soon jumped into the market with their own "atomic superheroes," such as the short-lived *Atomic Man* and *Atoman,* but also the more enduring *Captain Marvel.*[6] Walt Disney was the most formidable promoter of the peaceful atom for children and even many adults. His film, *Our Friend the Atom,* explored the scientific background of atomic theory in elemental and very positive terms.[7]

In the first few years after World War II, surveys revealed that a sizable majority of college-educated Americans believed that atomic energy in the long run would "do more good than harm." Among the less-educated there was more skepticism, possibly because atomic energy was too easily equated with "the bomb."[8] Historian Paul Boyer's 11th chapter heading in the celebrated treatment of atomic culture, *By Bomb's Early Light,* says it best: "Bright Dreams and Disturbing Realities." For all the expectations about peaceful uses of atomic energy, the undercurrents of wariness still existed, be it chronic anxiety about the bomb or a desire to soften or assuage its destructiveness.[9] But most Americans seemed to share both anxiety and hope. Whether one or the other rose or fell at any given time, it was difficult to be absolutely certain about the place of atomic power in the postwar years.[10]

Battle lines were drawn over commercial development of nuclear power beginning in the 1960s, but its promise seemed more tangible than atomic cars or atomic weather control in the two previous decades. Atomic Energy Commission (AEC) chairman Lewis Strauss commented that energy produced by nuclear reactors could ultimately be "too cheap to meter." For Strauss such talk was pretty much tongue in cheek and no one else of any prominence uttered such words with any conviction.[11] Indeed, many have used the idea of nuclear power being "too cheap to meter" to mock the allegedly false optimism of pro-nuclear forces. Yet the phrase, although often unfairly attributed, echoes the excitement of new discovery and the hyperbole often associated with expectation. The prospect of abundant and cheap energy was too delicious to ignore. The "whys" and "hows" of the promotion and development of commercial nuclear power in the United States and elsewhere through the end of the 1950s is the heart of this chapter. Along with that story, other important applications of the peaceful atom also deserve attention, including nuclear submarines and airplanes, attempts at canal building and other construction through *Project Plowshare*, and the rise of nuclear medicine. Taken as a whole, the variety of applications of nuclear power as the Cold War evolved give a more complete rendering of the alternative to the arms race imbedded in the generic term, "peaceful uses."

In the late 1940s and 1950s expanding affluence and opportunity for many middle-class Americans was on the rise. The need to use the atom to augment existing energy supplies or to jumpstart the economy was just unnecessary then. Peaceful uses often balanced military applications for propaganda purposes. To be sure, the advocacy of peaceful uses did attract supporters who refused to believe that the bomb was the only suitable and tangible outcome of breakthroughs in atomic science. The earliest nuclear reactors, however, were designed expressly to make atomic weapons possible, not to produce never-ending supplies of energy.[12] That civilian nuclear power and other peaceful uses developed in the immediate postwar years speaks to the almost erratic momentum of a transforming technology.

FIRST TENTATIVE STEPS TOWARD COMMERCIAL NUCLEAR POWER

The AEC formally took control of the American atomic energy program and related infrastructure on January 1, 1947. Until that time, they had been under the authority of the US Army's Manhattan Engineer District. The AEC's responsibilities were fourfold: rehabilitate the existing production facilities built hastily during the war; expand the capacity for producing fissionable material; organize the production process on a long-term basis; and plan permanent facilities for the development of peacetime uses of atomic energy.[13] The immediate tasks of the AEC were not linked to civilian power, but to research, development, and production of more nuclear weapons. In fact, the Commission was required by law to give its highest priority to producing fissionable material and bombs.[14] Much to the surprise of many, the United States did not have a stockpile of weapons available for use in 1946. The desire to replenish the American nuclear stockpile and the USSR's detonation of its own A-bomb in 1949 left the AEC little time to think about power plants. Work on the hydrogen bomb also pushed back peaceful uses.[15]

The Division of Military Application (DMA) carried out weapons development and stockpiling. While the DMA "interpreted" military requirements, it did not have a role in formulating war planning and essentially limited its responsibilities to managing the technical operations under AEC control. The weapons production system itself remained somewhat decentralized. The Los Alamos Scientific Laboratory (operated by the University of California) conducted weapon research and development. The Sandia branch of Los Alamos carried out some weapons ordnance assignments. The Rock Island Arsenal produced bomb components. The AEC, however, had no permanent weapons testing area. It took two years for the picture to change, resulting in a more substantial industrial organization for developing, testing, and producing nuclear weapons.[16]

Who would maintain custody of the weapons (the AEC or the military) was an important, and sometimes bitter, point of debate during the last days of the Truman administration. Harry Truman decided in favor of civilian control, but Secretary of Defense James Forrestal continued to hunt for ways to expedite transfers in case of a military emergency.[17] Among the AEC's necessary tasks were major overhauls of the original reactors, construction of two new plutonium reactors at Hanford, scheduling of expansion of the existing K-25 plant at Oak Ridge, and construction of a third gaseous diffusion plant also at Oak Ridge.[18] Here too, turf guarding created problems. In late December 1947 the government centralized reactor development at the Argonne site, much to the shock and frustration of Oak Ridge.[19] Developing strategy and policy guidelines for a civilian nuclear power program proved more difficult for the AEC. It required building an effective organization and setting realistic goals. As the official historians of the AEC asserted, "[I]f the Commission were to achieve any success in giving atomic energy a peaceful, civilian image, there would have to be a clearly defined, forceful plan for research and development, not only in the Commission's laboratories, but also in industry and the universities."[20]

Establishing a control mechanism for dealing with nuclear power and the continuation of the weapons program preoccupied the AEC and the Joint Committee on Atomic Energy (JCAE) in the immediate postwar years. But during the war (and maybe because of it) interest in generating power through the use of atomic energy ran high. In fact, in meetings with President Franklin Roosevelt, Winston Churchill pressed for the rights to civilian use of atomic energy. The French also showed interest, but did not have the close working relationship with the United States that Britain enjoyed, and thus did not pursue civilian use at the time.[21] Among many US officials and others there was hope that the awe-inspiring but terrible images of mushroom clouds could be replaced with the promise of bountiful, cheap energy and other peaceful uses of the atom. But new laws granting statutory authority to the AEC (and the JCAE) rendered such prospects limited to all but a small group. The AEC, the JCAE, and its various confederates controlled program information and access to decision making in an effort to construct a positive image for atomic energy. This image stressed the atom's tremendous potential as a cheap and plentiful source of power and minimized fears of bombs and radiation.[22] Before the passage of the 1946 Atomic Energy Act, some members of Congress queried whether military and nonmilitary nuclear programs should be separated. President Truman rejected that idea, believing that the entire area of atomic energy development belonged within the proposed AEC.[23]

For two decades after World War II, questions related to atomic energy reflected "subgovernment dominance." The AEC, the JCAE, a burgeoning nuclear power industry, and scientists and engineers working in national laboratories and universities formed, what political scientist Robert Duffy rightly called "a classic policy subsystem" that "exercised a virtual monopoly over the nuclear program."[24] For the first few years after the war the JCAE's influence over the AEC was limited, partly because of the tension between the Truman-appointed commission and the Republican Congress, and partly because of the committee's lack of expertise. By 1949 the JCAE became more influential as the advisory role of the atomic scientists weakened. This was due in part to scientists' inaccurate predictions about when the Soviets would get the bomb. It also was due to the USSR's more aggressive actions in Germany and Czechoslovakia, and Mao's victory in China.[25] Subgovernments (also known as iron triangles or policy monopolies) are small groups of public and private actors who dominate a policy in a given area. In this case, government participants were not "captured" by the business interests they were supposed to monitor, since there was no industry until the government created it. The subgovernment phenomenon was due instead to an outgrowth of the World War II era's almost permanent link between scientific and technological expertise and federal institutions.[26]

Between 1946 and about 1954 there was little development in the field of commercial nuclear power.[27] In Washington, the military and civilian claimants to the control and use of atomic energy and its by-products needed to arrive at some workable accommodation. The role of the AEC and the JCAE required clarification and better focus. Federal bureaucrats had to woo and nurture partners from the private sector to literally buy into commercial nuclear power production. Beginning in the 1940s, therefore, the path to commercial power started with questions of clearer policy formation, and the mundane, but necessary, demands of organization. The most vigorous supporters of civilian nuclear power in the 1940s were nuclear scientists.[28] Testifying before the Senate in December 1945, nuclear physicist Alvin M. Weinberg (who patented the first design for the water-cooled reactor) said that "Atomic power can cure as well as kill. It can fertilize and enrich a region as well as devastate it. It can widen man's horizons as well as force him back into the cave."[29] As the Cold War deepened and international controls became less likely, the general scientific community was even more willing to support peaceful development. Nevertheless, military demand for reactor development continued to dominate the subgovernment's work in the 1950s.[30]

Some electric utility companies were cautious about participating in the emerging nuclear power industry, but others saw potential long-term financial benefits. In building the bomb, the army had turned not only to scientists at a variety of universities, but also to the private sector including DuPont at Hanford and Monsanto and Union Carbide at Oak Ridge. Equipment manufacturers often saw more immediate potential in nuclear power than utility companies. Coal interests, of course, opposed developing the new technology altogether.[31] The involvement of Westinghouse in the navy's nuclear submarine project raised the company's interest in commercial nuclear power. In 1948, CEO Gwilym Price called for the formation of its Atomic Power Division. General Electric's (GE) military work also led to an interest in civilian nuclear power. General Groves invited GE to take over the operation of the plutonium production plant in Hanford run by DuPont. Groves also allocated funds to help establish GE's own atomic research and development laboratory (Knolls Atomic Power Laboratory) in Schenectady, New York.[32]

Within the Commission, interest in potential new uses for atomic energy evolved through basic nuclear physics research, materials testing, engineering reactor components, experimental reactor construction, and other reactor development programs.[33] The idea of a breeder reactor, that is, a reactor that produces fissile material (primarily U-233, U-235, or PU-239) that could be made into fissionable material, also was very attractive.[34] Prior to 1948 there were at least six different plutonium reactors. Some were used for production, others simply for testing purposes. The wartime reactors at Hanford produced plutonium, and while they provided valuable experience about reactor design and use, the designs were not prototypes for the energy-generating reactors in subsequent years.[35] In the early 1950s the reactor development program remained committed to military projects especially propulsion reactors for the navy and air force, with only the slightest nod to the civilian power program.[36] For the AEC it was imperative to build public-private relationships for commercial-grade reactors. Consequently, Chair David Lilienthal set up an Industrial Advisory Group in 1947, which requested that the AEC declassify and publish more technical information as well as establish a reactor development program. In late 1948, the AEC created its first Reactor Development Program (RDP). Although modest, the RDP and the establishment of the Division of Reactor Development in 1949 made a beginning in the direction of generating commercial electricity with nuclear power.[37]

During the first phase of the Reactor Development Program, the Commission established the National Reactor Testing Station at Arco, Idaho, to study various reactor concepts. One of four initial projects was operated at the Idaho site. The Experimental Breeder Reactor-1 (EBR-1) was the first to generate electrical power in 1951, and the first to demonstrate the feasibility of breeder reactors themselves. Another early venture called for experimenting with breeders. The final project was a land-based prototype for a submarine reactor. The AEC added three additional projects between 1950 and 1952, but at this juncture only a limited number of industrial firms were involved.[38] The achievements of the initial reactor program were modest and the

Hanford Engineer Works, atomic energy plant, Richland, Washington. A major center of plutonium production for the Manhattan Project, it faced overhauls and new construction after World War II. *Source:* Library of Congress, Prints & Photographs Division, [LC-USZ62-101676]

costs high. In order to attract more interest from the private sector, the AEC began to ease its control over technical data. In 1949, it granted more liberal access to selected industrial representatives, and, in 1951, it initiated an Industrial Participation Program, which granted security clearances to industry personnel and offered funding for promising reactor designs. At least eight industrial study teams, including as many as 20 companies, sought security clearances in order to examine the secret data. The Industrial Participation Program stimulated some proposals, but overall nonmilitary use of nuclear power made little headway. Industrial demand for electricity from nuclear power was negligible in the short run because of the availability of cheap fossil fuels. Planning for the long run required great expenditures without immediate rewards, and the AEC's financial support for reactor research was insufficient to attract many private firms. Most important, civilian projects remained subordinate to weapons development in the postwar years. In 1948, more than 80 percent of the commission's budget went into military projects.[39]

ADMIRAL RICKOVER TO THE RESCUE

The development of a practical commercial reactor in the United States came by sea rather than by land. The efforts of the controversial Admiral Hyman G. Rickover in the area of naval propulsion set the stage for accelerating the sluggish start in harnessing nuclear power to generate electricity in the 1950s. Born in Poland, Rickover entered the US Naval Academy in 1918. Because he was Jewish he was shut off from much of the social life there, and after graduation did not have the opportunities of fellow cadets. He earned an MS degree in engineering from Columbia University and trained for submarine command. With the outbreak of World War II, he got a chance to hone his aggressive, meticulous, and single-minded (some would say ruthless and wildly independent) management style with the Bureau of Engineering. Several years later (1982) in a speech he delivered at Columbia University he clearly articulated his management philosophy, "Human experience shows that people, not organizations or management systems, get things done. For this reason, subordinates must be given authority and responsibility early in their careers. In this way they develop quickly and can help the manager do his work. The manager, of course, remains ultimately responsible and must accept the blame if subordinates make mistakes."[40] During the war he became acquainted with some key people at GE and Westinghouse while directing the production of electrical equipment for American ships. These relationships would be of great use later in his career, especially as he moved toward promoting reactor development and nuclear propulsion for the navy.[41]

After the war, Captain Rickover accepted an assignment at Oak Ridge, where he explored the prospects of nuclear propulsion of naval vessels. Some in the navy had expressed an interest in developing an atomic-powered submarine since 1939, believing that such a craft could stay submerged for long periods.[42] Rickover was not impressed with the alleged promise of commercial nuclear power at the time, arguing that naval propulsion held out the only immediate opportunity. He believed that the electric power industry saw no immediate economic incentive, and that AEC was preoccupied elsewhere. "[I]f we are going to have atomic power plants in naval vessels," he stated, "the inspiration, the program and the drive must come from the navy itself."[43] Rickover became a one-man campaigner for the building of a nuclear navy. The General Advisory Committee of the AEC, which had hoped for grander possibilities for commercial nuclear power, somewhat reluctantly sought to focus on the narrower goal of naval propulsion in late 1947. Hope for stimulating a public-private partnership in reactor development did not disappear, but it never reached significant levels in the late 1940s and early 1950s. On the other hand, here was a project that was concrete and incorporated practical engineering skills, essentially taking reactor development out of a purely experimental phase.[44]

In 1948, Rickover assumed both the headship of the navy's nuclear propulsion program and a similar post with the AEC's Naval Reactors Branch.[45] He not only developed leverage within navy circles, but also with the leadership of the AEC and the JCAE. Despite the fact that Rickover ran his programs as if they were his personal dominion, his most important asset, according to one writer, "was his ability to make complex technologies work."[46] He also was a great believer in "the discipline of technology," which emphasized that machines and products did not respond to leadership nor orders, but to testing and analysis. This view suited him well, but not always his superiors.[47]

Admiral Hyman G. Rickover, the father of the Nuclear Navy. *Source:* Library of Congress, Prints & Photographs Division, [LC-DIG-ds-01461]

Rickover took interest in a high-pressure, thermal, water-cooled reactor that a submarine could carry. Rather than turning to the theoretical scientists central to the Manhattan Project, he followed up on his connections at GE and Westinghouse. The admiral nevertheless kept direct control over the project, reviewing every technical decision himself and with his own staff. He planned to have a nuclear-propelled submarine ready in five years. In 1954 the navy launched the *Nautilus* with a Westinghouse-designed pressurized-light-water reactor. The vessel was named after Captain Nemo's wondrous (but fictional) vessel in Jules Verne's *Twenty Thousand Leagues Under the Sea* (1870). This was a fitting name for a twentieth century wonder of its own.[48] Rickover once said, "Good ideas are not adopted automatically. They must be driven into practice with courageous impatience. Once implemented they can be easily overturned or subverted through apathy or lack of follow-up, so a continuous effort is required."[49] Because of the success of the *USS Nautilus*, the pressurized-water reactor (PWR) also became the model for a civilian nuclear power plant. The standard design for most American nuclear reactors, especially in the early years, was the light-water reactor (LWR) originally introduced in the late 1940s. It used enriched uranium as fuel and water as the coolant. Westinghouse's version was marketed as the pressurized-water reactor (PWR) and GE's as the boiling-water reactor (BWR). A by-product of the LWR was the production of about 150 kilograms of unburned plutonium annually in its core. It was possible to use it for nuclear weapons. (However, the LWR proved not to be terribly efficient and developed safety problems as the equipment aged.)[50]

AEC took up the project before the launch of the submarine and implemented a five-year program in research and development. By September 1954 construction began (with government funds) on the first nuclear power plant in the United States at Shippingport, Pennsylvania. Duquesne Light Company provided the site and built the generating facility, while Westinghouse (who had played a major role in the production of the *Nautilus*) designed and built the reactor with PWR technology. The AEC asked Rickover himself to head up the building of the plant.[51] The development of commercial nuclear power was moving forward, but hardly at a frantic pace. With federal budget problems looming in 1953, the Eisenhower administration looked for cuts. The AEC temporarily canceled its four reactor projects and stopped development on a nuclear airplane. There also was a proposal to cancel the navy's aircraft carrier reactor project to which Rickover objected. A compromise was reached when the AEC agreed to continue the construction of the light water reactor, but to modify it for power production instead of for the carrier. Shippingport became the location. Thus a political choice, not a technical on, was a crucial element in this decision.[52] There is little doubt, however, that Rickover's role in promoting of atomic energy within the navy and its civilian offshoots were important steps forward in reactor development crucial to propulsion and the generation of electricity by employing nuclear power.

ATOMS FOR WAR/ATOMS FOR PEACE

As stated in Chapter 3, the issue of international control of nuclear power was preconditioned by US wartime experiences and shaped by the onset of the Cold War. Whatever earnest interest in peaceful uses for the atom may have surfaced in the 1940s and beyond was offset by the looming arms race. Speaking before the UN General Assembly on December 8, 1953, President Dwight D. Eisenhower presented an alternative approach to international control first included in the Baruch Plan during Truman years. It was essentially an "Atomic Marshall Plan."[53] He declared that "peaceful power from atomic energy is no dream of the future… that capability, already proved, is here today."[54] What followed was an intense public relations campaign in the United States and elsewhere, meant to proclaim a new era of atomic energy broadly applied.[55] Could the new slogan, "Atoms for Peace," counteract the fear and anxiety over nuclear weapons?

Some observers regarded Eisenhower's Atoms for Peace speech as the cornerstone of America's aspirations for the international management of atomic energy.[56] The president proposed that the nations capable of producing fissionable material contribute to a pool (or bank) from which other nations, who forswore production of nuclear weapons, could draw for nonmilitary purposes. This entailed sharing civilian nuclear information, and potentially stimulating an international market for commercial nuclear power. The speech called for an International Atomic Energy Agency (IAEA) through the United Nations to oversee the application of atomic energy not only for commercial power but also for programs in agriculture and medicine. IAEA would be involved in the on-site inspection to make sure that nuclear material was not being diverted to weapons use. Preceding IAEA, the United Nations created an Atomic Energy Commission (UNAEC) in January 1946, which had studied international control of atomic energy soon after the war ended. The United States had introduced the Baruch Plan before this body, and the Soviets highly criticized it. In 1948 UNAEC was disbanded.[57]

Eisenhower seemed truly committed to peaceful applications of the atom, and supporters regarded the proposal as the beginnings of a large-scale transfer of peaceful nuclear technology worldwide.[58] However, the sincerity and vigor with which the United States pursued a real program of international controls with respect to atomic bombs and in terms of the management of nonmilitary uses of nuclear power came into question.[59] Some presumed that Eisenhower, while pointing out the horrors of nuclear war, was simply trying to offer hope through expansion of the civilian uses of atomic energy. More cynical observers argued that Atoms for Peace principally advanced American Cold War aims, and was a way to create a large enough demand for fissionable material to limit what otherwise might be used to build weapons. On a more pragmatic level, Atoms for Peace allowed the United States to maintain its commanding lead over the Soviets, neutralize potential adversaries, provide a potential economic boost to industries especially in Western Europe, and afford a great propaganda outlet. Broadcast worldwide by Voice of America, the ideas about the peaceful use of the atom were nevertheless well-received. Aware of its value to the United States in several ways, the Soviets tried to ignore the plan. In turn, the United States proceeded with a series of bilateral agreements with cooperating nations.[60]

By taking the lead in developing peaceful uses for the atom, the United States could retain its overall dominance of atomic energy. This was particularly important because the British were moving rapidly toward large-scale nuclear power generation. The United States, therefore, supported ventures like Euratom, a Western European nuclear cooperative designed to establish an integrated program for development of an atomic energy industry. Euratom was significant because it offered energy-poor Europe an important new power source. In America, the public generally applauded Atoms for Peace. While the plan got favorable reaction in the Senate (possibly more for its propaganda value than for its feasibility), it stood little chance of full implementation in the 1950s. Suspicions and tensions stimulated by the Cold War were too great. Democrats criticized Atoms for Peace for not going far enough. Liberals did not sufficiently trust AEC leadership to back the plan wholeheartedly. Right-wing Republicans saw it as a "give-away" program through which Eastern European countries could acquire enriched uranium. In practice, the United States followed through with only mild support for the IAEA. When cheap oil became available from the Middle East, support for nonmilitary nuclear power development on a world scale began to decline.[61]

SERIOUS COMMITMENT TO NUCLEAR POWER:
THE ATOMIC ENERGY ACT OF 1954

Given the world events of the 1950s, it is amazing that commercialization of nuclear power occurred at all. With the passage of the Atomic Energy Act of 1954, nonmilitary applications received a substantial boost with an emphasis on the role of private companies in reactor and plant development. The AEC's initial five-year-plan for reactor development, launched in 1954 prior to the new law, called for five experimental reactors of which the Shippingport PWR was the most significant. Yet the plan did little to advance the role of private industry in the new field.[62] The energy market also had little to do with the passage of the law, since there was no pressing need for a new source of energy in the United States at the time. There was a strong interest in enhancing American prestige, however. Several government leaders believed that it was important that the United States lead the way in developing nuclear power, especially beyond weapons-making. Since nuclear power was a government-dominated and government-sponsored power source, the mechanism for carrying out that goal (the AEC and the JCAE) was already in place, but had yet to acquire the necessary tools to exploit the peaceful atom.[63]

Atoms for Peace set the tone for the commercialization of nuclear power. The election of a business-oriented Republican president in 1952 also gave a boost to civilian development. While American firms expressed interest in atomic power, and several large companies aggressively lobbied to enter of the nuclear power business, the call to revise laws for greater industrial participation primarily came from within the government. In late 1952, the JCAE issued a report outlining the history and problems involved in developing atomic power for peaceful uses; however, the lame duck period was inappropriate for bold new legislation. The launching of the navy's *USS Nautilus*, to the contrary, clearly brought attention to the possibilities of uses of nuclear energy.[64] Eisenhower's State of the Union address also in January 1954 stressed the need for greater development of commercial nuclear power, and in February the administration called for Congress to amend the 1946 Atomic Energy Act, a process in which the Joint Committee took an active role. After lengthy hearings and much partisan wrangling (Democrats thought the idea of commercial nuclear power would severely weaken the government monopoly; within AEC chair Lewis Strauss argued against government ownership of all fissionable materials), the Republican-controlled 82nd Congress passed the Atomic Energy Act of 1954, also known as the Cole-Hickenlooper Act.[65]

The 1954 act combined the spirit of Atoms for Peace with a call to develop civilian nuclear power in the private sector. It authorized greater international cooperation for the AEC, by providing for more latitude in the dissemination of scientific data and loosened restrictions on access to classified information.[66] It also sought to increase participation by private enterprise in the development and construction of reactors, and through economic incentives and revisions in patent law. The act allowed the private use of fissionable material through licensing, but the federal government retained ownership. All of the changes effectively demonstrated a significant adjustment in US atomic diplomacy, but also raised questions about potential security and safety risks and public versus private power.[67] The AEC was not retreating from commercial nuclear power. Instead it was partnering with private companies to underscore its mission as a developer, promoter, and regulator. It was the proverbial mouse guarding the cheese with an inherent conflict of interest.[68]

With the new law the AEC retained most of its regulatory powers, including issuing permits for entering the industry and constructing facilities; purchasing the industry's nuclear products or licensing such sales; and setting security and safety standards. It also opened the door to JCAE decision making in the commercial reactor program, especially since the legislative body was active in drafting key sections of the bill. Congress, however, gave the AEC little guidance in setting regulations or safety standards. For their part, the JCAE and industry wanted regulations as simple as possible. During the 1950s and 1960s, the AEC demonstrated an unswerving promotional role even though nuclear reactors did not prove to be economically competitive with other power plants.[69]

PROMOTION, INDEMNIFICATION, AND SAFETY—IN THAT ORDER

The 1954 act removed legal impediments to commercial development, but development itself was not automatic. While some large companies hoped to tap into the potential of the new energy source, the response of numerous privately owned utilities was lukewarm in the short term. Supplies of conventional fossil fuels for production of electricity were plentiful and cheap. Many scientific, technical, and economic questions about reactors and other components remained unanswered. In addition, the government and the fledgling industry had only begun to explore the potential hazards of nuclear power generation.[70] To move in a tangible direction, the AEC announced its Power Reactor Demonstration Program in January 1955. The program intended to go beyond the Commission's initial five-year-plan, which was inadequately funded and limited industrial participation. While the program provided more money for research and development, utility companies were to design, build, and operate the power reactors and plants. Since the emphasis was on large units, small utilities could not afford to take part. By April 1 the AEC received only four proposals, with three projects involved in the demonstration program—Yankee Rowe (Massachusetts), Hallam (Nebraska), and Fermi (Michigan). Little new research and development occurred during this first round, and some questioned the appropriate relationship between the government and the emerging nuclear power marketplace. What was the appropriate model? Should these plants be government projects with participation from industry like Shippingport or vice versa? What would be the limits of government investment in and responsibility for commercial technology?[71]

In September 1955, AEC announced a second round of demonstrations focused on smaller reactors, which were more attractive to foreign markets and public power groups. The total number of projects now stood at seven, and the general approach favored government-industry partnership like Shippingport.[72] Despite the move toward public-private partnerships, AEC chairman Lewis Strauss interpreted the policy behind the 1954 act to rely primarily on private companies to develop civilian reactor technology. This point of view deviated from those wanting to keep the federal government significantly involved in all phases of nuclear power development.[73] The reactor program, however, did not make the headway that some had expected. Dissatisfied with the pace of rounds one and two, and distrustful of both the AEC and the private sector's commitment to build new nuclear plants, Senator Albert Gore (Dem., TN) and Representative Chet Holifield (Dem., CA) of the JCAE introduced a bill to accelerate reactor construction in spring 1956. It called for a significant expansion in the reactor development program, namely to construct six nuclear power facilities of different designs that would be financed completely by the government (totaling $400 million). These facilities were in addition to projects already in operation. The Gore-Holifield bill set off a political storm which carried over into the 1956 presidential election campaign. Here was an opportunity for Democrats to challenge the Eisenhower administration's plan for private development of nuclear power, and, in essence, to intensify the public-versus-private debate raging over electrical utilities. It also brought into question the overall approach to reactor construction that the Democrats believed was too sluggish. They argued that such lack of commitment could result in the British and the Soviets surging ahead of the United States. The bill came up for a vote that year. It passed the Senate with ease but was narrowly defeated in the House when Republicans, some of whom saw the plan as "atomic socialism," closed ranks, bringing with them Democrats from the coal states. Chairman Strauss reveled in the outcome, since the AEC opposed the bill from the start.[74]

Possibly in spite of, rather than because of, the actions of the Democratic JCAE, the AEC soon moved more quickly toward commercialization. On December 18, 1957, the Shippingport plant delivered the first electricity to the city of Pittsburgh marking an actual beginning point for nuclear power generation in the United States. (Completed in October, Shippingport cost $100 million instead of the projected $70 million.)[75] The AEC announced a third round of the PDRP in January 1957, which encouraged both large and small companies to construct reactors of newer designs, especially moderated heavy water reactors and homogenous reactors. Technical problems similar to those in the previous rounds recurred. Trying to solve technical problems in a demonstration project was self-defeating. Thus, the AEC turned to a "modified" third round, announced in 1962, concentrating on proven designs to demonstrate that light-water reactors were reliable for commercial development.[76] In some respects, the meandering path to nuclear power production in the 1950s and early 1960s reflected the mixed objectives of the program. American prestige was on the line to demonstrate its leadership in the peaceful uses of the atom. But how to go about that task was not altogether settled. As nuclear power languidly moved toward commercialization in the 1950s, a concern for safety was one of several issues still needing to be addressed effectively. The AEC was aware of the potential dangers of a nuclear accident, but it often gave priority to its role as promoter of nuclear power. While encouraging construction of plants, it nonetheless attempted to anticipate safety issues, which it seemed to regard as manageable technical problems, and to develop some safety standards at a time when research data was sparse.[77]

The AEC established a mechanism for dealing with safety issues within the Commission as early as 1947. It was called the Reactor Safeguard Committee, chaired until 1953 by weapons expert Edward Teller. The group's primary responsibility was to evaluate and audit technical health and safety issues related to AEC reactors. Teller, a very curious choice for such a position, later commented that the committee was not very popular and had been referred to as the "Committee for Reactor Prevention." Colleagues even kidded him about the appointment. In 1950, a second advisory group, the Industrial Committee on Reactor Location Problems was established to deal with nontechnical considerations of siting plants. To avoid duplication and to refocus the safety program, the two bodies were merged into the Advisory Committee on Reactor Safeguards (ACRS) in 1953. It was an important independent committee, composed of scientists and engineers from industry, universities, and national laboratories, to serve as a sort of overseer group to keep tabs on AEC safety issues.[78] Shortly after the passage of the 1954 act, the AEC established a full-time reactor hazards review group called the Hazards Evaluation Staff and added a Division of Civilian Application to monitor safety aspects of construction applications. The act included at least 25 references to health and safety and greatly increased the scope of the licensing program. The AEC, however, did not apply universal safety standards to all reactor projects and had no independent safety research program (although safety research was being carried out in some areas in the 1950s). It judged reactor applications on a case-by-case basis, with some general regulations on radiation, operator training, special nuclear material, and on siting (in the early 1960s). The exact nature of how health and safety were to be protected was not spelled out clearly.[79]

Controversy over safety was first dramatized in the case of the Fermi Unit 1, built near Monroe, Michigan. In January 1956, the Power Reactor Development Company (PRDC), a consortium of 25 equipment manufacturers and utility companies headed by Detroit Edison, applied for a license to construct the first full-scale breeder system (a fast breeder). Fermi 1 generated steam for electricity, using uranium as its fuel and sodium as its coolant, while also producing plutonium. It became operational at low power in 1963. The AEC's interest was high because this would fulfill its goal of having the industry move beyond the light-water reactor, and seriously begin to develop breeders. ACRS raised some technical questions about the plant (sodium coolant could explode) including the proposed site 18 miles south of Detroit. On October 5, 1966, a zirconium plate at the bottom of the reactor broke away and blocked the flow of sodium to some fuel subassemblies which began to melt. The accident led to shutting down the reactor, which did not reopen until three years and nine months later.[80]

The case became political when the JCAE chairman Clinton Anderson (Senator, Dem., NM) learned of the existence of the Safeguards Committee report on the incident. The AEC refused to give him the document, arguing that it was an internal memorandum not a formal report. This set off a major jurisdictional battle. While the debate raged, the AEC issued a construction permit subject to later review (all were "conditional"), and stated that the safety problems would have to be solved prior to the operation of the plant. The idea of a "conditional" permit, although not supported unanimously within the Commission, became an accepted means for dealing with reactor safety problems at the construction permit stage.[81] Senator Anderson had another point of view. Incensed at AEC's secrecy, he threatened to seek separation of AEC's regulatory functions from its promotional functions. Short of accomplishing that, he secured passage of the Price-Anderson Act in 1957. The act established the ACRS as a statutory body to review every license application and to issue a public report in each case. In addition, it required public hearings for all application proceedings, contested or not. Anderson won a partial victory by forcing the AEC to conduct some of its business in the open.[82]

The best-known portion of the Price-Anderson Act, the indemnification measures, actually increased AEC's influence as a promoter of nuclear power. Many private companies were hesitant to invest heavily in nuclear power plants because of a concern over the potential liability should a major accident occur. A 1956 AEC report (WASH-740) prepared by scientists at the Brookhaven National Laboratory was delivered to the AEC in March 1957. It concluded that the likelihood of an accident resulting in a major release of radioactive material outside the reactor ranged from 1 in 100,000 to 1 in 1 billion per reactor year. In the worst case, such an accident might result in 34,000 fatalities, 43,000 injuries, $7 billion in property damage, and contamination of 150,000 square miles.[83] The AEC regarded these estimates as evidence of the need for insurance in case of a worse-case accident. As early as 1955, the AEC encouraged the insurance industry to explore its own ability to provide liability policies, leading to the formation of three insurance syndicates willing to extend such coverage. Utility companies remained hesitant until the passage of Price-Anderson. It committed the government to pay an indemnity of at least $560 million if that was ever necessary, and the industry paid premiums to cover potential costs of an accident. Price-Anderson, therefore, stimulated commercial development by offering coverage for those most directly engaged in building and operating nuclear power plants. It also provided some protection for people or communities who might suffer injury or damage from an accident. In the 1970s, the indemnification measure was singled out by opponents of nuclear power as a public admission of its dangers, and as a way to limit the liability of contractors and utilities.[84]

Despite the questions raised about reactor safety, there was little in the way of widespread public protest during the 1950s. Other issues, such as health hazards to uranium miners, were likewise ignored or overlooked. The activities that most closely resembled an organized protest were the efforts of the AFL-CIO and other unions to force a public hearing on the construction permit for the Fermi 1 plant over questions of safety and other issues. In 1957, the AEC held a public hearing on the PRDC permit, the first of its kind on licensing of nuclear plants. But the commission reconfirmed the permit in 1958, and the Supreme Court upheld it in a landmark decision in 1961.[85] Reactor orders nevertheless remained low for several years despite the Price-Anderson Act and increased confidence over the future potential of nuclear power. By 1958 the AEC helped develop 11 different types of reactors in association with private industry, with LWRs advancing more quickly than others. In 1959 the Reactor Development Division compared the various reactors and tried to determine which were most promising. As of yet there was no attempt at standardization.[86] At the time, nuclear power was not yet economically viable, and safety concerns remained unsettled. Still a few private utilities decided to purchase units in order to gain a competitive advantage years down the road; others simply wanted to keep the federal government out of the power business.[87] The introduction of nuclear power into the energy equation was evolving in ways that defied past practices. This is not to say that political decisions eluded the marketplace, but only that with commercial nuclear power everything was turned on its head, with demand for more electricity low on the list of priorities.

ATOMS FOR PEACE AND THE WORLD

Looking back on the 1950s, especially given the dramatic events surrounding the advent of the hydrogen bomb, it is easy to overlook the budding internationalization of nuclear power generation. Atoms for Peace, especially in helping to advance a new energy source, may have been more immediately significant outside the United States than within it. The tug-of-war between the AEC and the JCAE over controlling the atom, the administration's focus on the New Look and weapons strategy, and the readily available alternative energy sources kept civilian nuclear power in the United States from becoming little more than a sideshow for the moment. Yet, Atoms for Peace provided the United States with a variety of political and diplomatic opportunities beyond the prospects for energy generation.[88] Europe was still digging out from World War II in the 1950s. Both Western Europe and Eastern Europe were increasingly reliant on the United States and the USSR respectively to supply basic defense needs and other financial support. In such a setting, finding a new local source of energy was tempting, especially in Western Europe where constraints were economic and political. The Soviet Union also sought alternative energy sources, and viewed nuclear power generation as a powerful propaganda tool.[89]

American sponsorship of the 1955 Geneva Conference on Peaceful Uses of Atomic Energy (attracting scientists from 73 countries), and the 1957 creation of the International Atomic Energy Agency, were important assurances that nuclear materials and technology would not be diverted to military projects. The United States also supported the creation of Euratom. Announced in 1956, Euratom was intended to devise an integrated program for developing an atomic energy industry along the same lines as the European Coal and Steel Community. It planned to finance and coordinate research and development activities in the nuclear power area, but most especially it sought to generate industrial electrical power by offsetting the lack of new hydroelectric sites and the decline in coal production. The closure of the Suez Canal in 1956 brought attention to Europe's vulnerability to importing oil.[90] Aside from providing uranium and offering a market for American energy technology, Euratom played a role in keeping Western Europe cohesive and perpetually bound to the United States and its interests. However, the treaty that created Euratom never clearly drew the line between peaceful and military uses insofar as declaring that fuel controls prohibited or permitted military uses of nuclear energy.[91]

Beyond support for Western European cooperation, the United States was interested in forging bilateral agreements with foreign countries, which provided fuel, technology, and research reactors, and also offered isotopes for nuclear medicine, agriculture, and other research. Such agreements gave the United States more leverage than programs run through an international agency. The search for the peaceful atom, however, had Western European countries and others thinking about developing civilian nuclear power in their own right beyond the foreign policy implications of the Cold War.[92] Questions about safety and security were serious points of negotiation for all of the arrangements. Because the IAEA was large and politically diverse (the Soviets entered in May 1957), the United States was cautious about participation. The Senate moved slowly toward ratification, raising a familiar array of issues bound to isolationism, anti-communism, concerns about protecting military secrets, and even skepticism about the domestic reactor program. The Senate ratified the treaty, but included a provision that congressional approval would be necessary to transfer nuclear materials to the IAEA.[93]

WESTERN EUROPE AND SCANDINAVIA

Great Britain faced problems in securing energy independence after World War II. By the 1950s it became the first industrial country to devise a plan for utilizing nuclear power to generate electricity. In August 1945 the British government formed the Advisory Committee on Atomic Energy (ACAE) for the purpose of making recommendations on general atomic energy policy. The ACAE was very active through the middle of 1947 in formulating British atomic energy policy, but soon ceased to function. In 1946 the Atomic Energy Bill passed Parliament, giving legal authority to the government for a monopoly over atomic energy. The British made a decision to move forward with a civil program in 1952, after the testing of their first atomic bomb. They viewed a civilian power program as having practical value in generating electricity, but also important in increasing British prestige.[94] In 1954 the Atomic Energy Authority (UKAEA) was formed to take responsibility for all nuclear activity in Great Britain and to coordinate all initiatives. UKAEA controlled all expert knowledge on the subject and thus led government policy. In the 1950s the British were the leaders in nuclear power plant technology, emphasizing gas-graphite systems. They opened the world's first prototype nuclear power station at Calder Hall in 1956 for electricity and plutonium production. Some British leaders hoped that their country would export technology around the world and give Great Britain an economic and political boost. The gas-graphite reactor proved to be too expensive and unreliable, however. While British nuclear technology made international inroads over time, the burgeoning British program found itself under attack from a variety of protesters, presaging what would be more common around the world in the 1960s.[95]

The Windscale plutonium pile accident in 1957 (the same year as Price-Anderson) focused attention on safety problems and had a harmful impact on the whole field of nuclear power generation, not just in the United Kingdom. Windscale was a uranium reprocessing plant close to the Irish Sea, which produced much of Britain's plutonium for weapons use. A uranium fire started inside the reactor on October 8, resulting in serious radioactive iodine fallout released into the atmosphere. Heat kept rising in the matrix of graphite blocks surrounding the nuclear fuel and would not cool down. Worse yet, the fire destroyed the safety equipment needed to contain the pile. By October 10 the temperature reached 1,300 degrees centigrade. After three days of fighting the fire, the plant management made the risky decision to pump water into the reactor, increasing the possibility of a hydrogen explosion. Luckily the pile began to cool. The accident contaminated some 100 farms within a radius of 500 kilometers, and all of the milk produced had to be dumped (unfortunately into the Irish Sea). Fallout spread not only over England, but also France, the Low Countries, Germany, and Denmark. (The accident produced about one-tenth of the radioactivity experienced in Hiroshima.) Without a doubt Windscale was the worst environmental disaster in the British Isle's history. The plant was closed down permanently, and the news of the accident rekindled interest in safety measures everywhere. British authorities, however, withheld the details of the accident and its serious implications from the public, playing down its full significance.[96]

Some of the groundwork for reactor development and nuclear power generation in Europe and elsewhere predated the formal ratification of IAEA or roughly coincided with it. The French became the leading nuclear power generator in Europe and one of the most important in the world. France's goals were tied to a broad interest in science and technology and their political, economic, and military implications. Several leaders believed that unless France integrated science into its governmental institutions making science part of national policy goals, France ran the risk of falling seriously behind other world powers. The French also were sensitive to the fact that they had been excluded from Anglo-American collaboration, as weak at it was becoming, in the exchange of information about nuclear matters. On a cultural note, President DeGaulle wanted to demonstrate that the French technical skill reflected the nation's own abilities and expertise, not those of others.[97]

The French built their initial nuclear pile in December 1948. Called Zoè (*zero/oxide/eau*: meaning zero power, natural uranium oxide, and heavy water), it was the first of a series of heavy water reactors constructed in France and produced small amounts of plutonium 11 months after its startup. It was not surprising that the French applauded the accomplishment with such vigor. President Vincent Auriol who visited Zoè declared, "This achievement will add to the radiance of France."[98] As historian Gabrielle Hecht emphatically concluded, "The nuclear program epitomized the link between French radiance and technical prowess."[99] Whether the nuclear program was really "French" was beside the point. Many French came to embrace their expertise and their glory (radiance) in the promising new field.[100] France was the first country to patent a nuclear power station design (1939), which was reintroduced in 1952 as part of its new national energy strategy. In 1945, DeGaulle announced the Commissariat à l'Energie Atomique (CEA), which was the first civilian organization devoted to peaceful applications of nuclear power. It also was a vehicle through which the French president hoped to build an atomic bomb.[101] Although the French path to nuclear power (like others) began with nuclear weapons, France did not detonate a nuclear device until 1960. Unlike the United States, the French lacked local energy sources and had to look to nuclear power for domestic uses beyond its military interests. Nonetheless, the instability of the French government and the expense of fashioning a nuclear weapons program, slowed commercial development in France. The French built the first large nuclear facilities for military purposes, although they were the prototype for commercial plants. In 1955 the government secretly funded the CEA to work directly on an atomic bomb, which was only announced to the public three years later. The "G" series reactors (standing for gaz or gas-cooled) at Marcoule began operation in January 1956, producing plutonium for nuclear weapons. They also were adapted for electrical power generation. French recognition of its energy vulnerability after the Suez Crisis and the onset of the Fifth Republic under De Gaulle, spurred more interest in developing an independent nuclear policy.[102]

In the early 1950s, the CEA constructed several reactors capable of producing electricity. The French accomplished this goal following the turmoil caused by replacing (or purging) leftist scientists with engineer/bureaucrats. There was a long-standing tension between De Gaulle's wish for France to enter the "atomic club" and the reticence of scientists seriously concerned about the seeming bellicosity of the government. Rifts intensified because the French nuclear program did not exhibit clear objectives or priorities beyond building bombs, including what a civil program might look like. The expulsion of the scientists from CEA answered some of those questions forcefully.[103] In 1955 the French government established La Production d'Electricitè d'Origine Nuclèare Commission (PEON) to coordinate the program in civil nuclear power. The main participants were the CEA, France's public utility, Electricitè de France (EDF), and representatives from industry. In 1955 PEON prepared plans for a series of electrical-power producing reactors to be operated by EDF. In the late 1950s and early 1960s, EDF showed much of the same trepidation toward nuclear energy as experienced in the United States and Sweden before France made a major commitment to nuclear power some years later. France also moved warily to join Euratom, finally entering in 1955. The relationship with Euratom began to deteriorate after the French decided to enter an agreement with the United States in 1958 over adapting LWR technology to their needs.[104]

Civilian nuclear power made selective strides elsewhere in Western Europe and Scandinavia. Sweden made one of the most serious commitments to nuclear power, with reactors generating almost 50 percent of its electricity by the early 1990s. The Swedes began working on nuclear power production after World War II, but the Atombolaget Company did not complete a trial reactor until 1954. This accomplishment prompted the development of a full-scale program in Sweden through a joint venture with OKG, which was one of the major private utility companies in the country. OKG built the first commercial reactor in Sweden. Through the 1950s and 1960s the commitment to nuclear power drew strong political support.[105] Belgium played an important role in civilian nuclear power, but did not go as far as the Swedes. Belgian engineers participated in the start-up of the Shippingport plant in 1957. Five years later, Belgium commissioned its first reactor, and the first light-water reactor built outside the United States. In the 1950s, however, Belgium (and its neighbor the Netherlands) did not have sufficient resources to establish much of a nuclear program. Belgium's strength rested on the discovery of vast amounts of uranium in the Belgian Congo before World War II. During the war, with their government in exile, Belgium granted exclusive rights to this valuable resource to the United States and Great Britain, and in return received access to allied nuclear information. The arrangement ended in 1949, but the United States provided grants in exchange for uranium supplies. These grants stimulated Belgium's own nuclear program. Ultimately Belgium became second only to France among industrialized nations in relying on nuclear power.[106]

West Germany and Italy hinted at interest in nuclear power, but made few strides. The treaty that ended World War II in Europe restricted West Germany's use of nuclear power until 1956. After the war, nevertheless, German scientists again took up research on heavy-water moderated reactors primarily for industrial purposes and initiated other studies. But since the Germans did not face the problem of domestic energy shortages, as the French or British did (92 percent of Germany's energy consumption was linked to coal), interest in nuclear power was not as strong as elsewhere.[107] Increasing dependence on non-local sources of power led to the nation's early focus on nuclear research. Yet not until the late 1960s did Italy make any progress in nuclear power generation.[108]

EASTERN EUROPE

The USSR had a long history with atomic energy going back to the 1940s. The first controlled chain reaction took place on December 24, 1946, with the initial work on thermal reactor design taking place at the I.V. Kurchatov Institute of Atomic Energy. The Soviets explored several reactor types, including gas-cooled units and LWRs. The Radium Institute in St. Petersburg developed the nuclear fuel cycle for reactors. (The fuel cycle starts with the mining of uranium and ends with the disposal of nuclear waste.)[109] In 1952 the Soviets constructed a materials-testing reactor. The first Soviet power plant became operational on July 27, 1954, in Obninsk near

Moscow. It was the first reactor worldwide to produce usable electricity for a national grid, and proved to be a propaganda coup for the Soviets. All indications suggest that the USSR's reactor program to this point was more formalized and better focused than the US program, where a similar achievement in civilian nuclear power was not met until four years later.[110]

Like the French, the Soviets viewed the development of nuclear power as a strong indication of technical prowess and accomplishments, especially beginning with the leadership of Nikita Khrushchev in the 1950s. But as historian Paul Josephson claimed, "The same engineering culture that existed in the peaceful nuclear industry also held sway in the military industry. Technological hubris, unnecessary risk, and technological momentum combined in a cold war world to create an ethos where issues of health, safety, and environment received inadequate attention."[111] Something similar could be said about the United States, France, and other countries as well, but not to the same degree. The closed society of the USSR made total secrecy possible, but much less ironclad in the West.

In March 1956, the Soviets founded the Joint Institute for Nuclear Research with the expressed purpose of coordinating and controlling Soviet nuclear initiatives outside the country. Located north of Moscow in Dubna, the institute trained thousands of scientists and engineers from the Soviet Bloc. A large number of Soviet scientists got their start at Dubna, many of whom were sent out to build nuclear power plants in allied countries and also nonaligned countries like India.[112] In the late 1950s the Soviets experimented with sodium-cooled fast-breeder reactors at Dubna. They soon built a military nuclear station at Novotroitsk to produce weapons-grade plutonium. The USSR built the first commercial plants using graphite-moderated BWRs in the 1960s in Beloyarsk and Novovoronezh. To a greater degree than the United States, the Soviet's efforts in nuclear power intertwined its military and civilian interests throughout the Cold War.[113]

BEYOND EUROPE

Civilian nuclear power showed modest signs of development outside of the United States and Europe in the 1950s and early 1960s. The Canadian Parliament established the Atomic Energy Control Board (AECB) in May 1946. Unlike the American AEC, the AECB delegated many of its functions to a variety of agencies and governmental bodies. In April 1952 the AECB was integrated into the Atomic Energy of Canada Limited (AECL), a federal Crown Corporation. The Canadians built its first experimental reactor at Chalk River, Ontario in 1945, as a joint project with Great Britain. Admiral Rickover tested his design for the nuclear-powered submarine there. In 1952 the AECL took control of the Chalk River Nuclear Laboratories from the National Research Council. The laboratories were charged with developing peaceful uses of nuclear power.[114] In 1955 the AECL proposed that a civilian natural uranium, heavy-water reactor be built by Canadian General Electric, but replaced it with a larger reactor design in 1957. By 1958, the AECL established two special design teams which launched the first power plants in Canada. AECL is credited in the late 1950s and early 1960s with designing CANDU (Canada Deuterium Uranium) pressurized water reactors (fueled with natural uranium and moderated with heavy water) known worldwide for being fuel efficient and for being among the safest available.[115] The Canadians did not construct enrichment plants in these years, and thus turned to building reactors that utilized natural, nonenriched uranium requiring a moderator of heavy water. Canada sold CANDU reactors in Argentina, Romania, South Korea, Pakistan, and India.[116] Ironically, the older reactor at Chalk River suffered the first nuclear accident in history (and Canada's worst) on December 12, 1952, when a technician's error led to contamination of the reactor building and a partial meltdown. This accident shut down the reactor for 14 months. A second accident occurred at Chalk River on

May 25, 1958, when a fuel rod caught fire and contaminated the building. The facility, however, continued to be the site for reactor design tests including the CANDU.[117]

Before the formation of IAEA in 1957, countries in the western hemisphere began meetings culminating in the Inter-American Nuclear Energy Commission (IANEC). Chiefs of state from several American republics met in Panama City in July 1956 to celebrate the 130th anniversary of the Congress of Panama. At the meeting President Eisenhower proposed an Inter-American Committee of Presidential Representatives (CPR) to explore peaceful uses of atomic energy. The CPR held its first meeting in Washington in September and twice more in 1957. It prepared a report advocating peaceful applications of nuclear energy as "one of the most important fields for inter-American collaboration." A significant impact of IANEC was providing a forum to discuss common interests and objectives. But like other such bodies, it lacked the financial support and the leverage to accomplish much more. The United States soon turned its attention to a larger stage when IAEA was formed. Leadership changed often in Latin America and countries like Cuba were excluded. In addition, the all-too-common concerns about US domination of hemispheric issues worked against IANEC's influence.[118]

Among the developing countries throughout the world, India was the first to operate a nuclear power station. In 1945 interest in nuclear power began with the Tata Institute of Fundamental Research in Bombay. Three years later the government established the India Atomic Energy Commission. As a consequence of the Atoms for Peace program India was one of thirteen Third World countries to receive fissionable material and training support from the West. India's first experimental reactor (the *Apsara* reactor) went on line in 1956. The Canadians also had provided India with an experimental *Cirus 40* megawatt-thermal heavy water reactor in 1955. Built by Indian scientists, material supplied by the UK Atomic Energy Authority fueled it. As in many countries, India experienced a considerable gestation time before nuclear reactors could provide electrical power.[119] Pakistan set up an Atomic Energy Commission in 1955, receiving its first research reactor through an Atoms for Peace grant in 1960.[120]

In Japan the nuclear power program began in 1952 at the end of the US occupation. The ban on nuclear research was lifted at that time. Ultimately (and somewhat ironically) Japan eventually developed a far-reaching nuclear industry. In 1955 the Japanese government passed the Atomic Energy Basic Law requiring that the development of nuclear power be undertaken in the private sector. It established the Japanese Atomic Energy Commission (JAEC) to supervise the promotion of nuclear power much like its American counterpart. Given that Japan had few domestic power resources, and thus imported much of its energy, the JAEC set as its goal energy self-reliance through breeder reactors and a complete domestic fuel cycle, which it pursued for many years. Japan's first power reactor was a small unit imported from Great Britain in the 1950s and installed at Tokai Mura northeast of Tokyo. Early efforts to develop its own nuclear production took place between 1957 and 1963 with the construction of six experimental reactors. The first commercial reactor (a British gas-cooled unit with a plant built by a French contractor) was completed in 1966 at Tokai Mura.[121]

Whether spurred by Atoms for Peace, searching for an alternative energy source, or transforming destructive weapons into some positive ends, civilian nuclear power took a leap forward worldwide in the 1950s. The actions of several governments did not immediately lead to cheap and abundant electricity, but governments put agencies in place, and, in some cases, proponents in the private sector moved to create a burgeoning new industry. Various types of conflict accompanied these tentative steps: scientists versus bureaucrats, government versus private utilities, local versus national, or national versus transnational. Windscale and a few other disturbing accidents also set the stage for a much more serious and intense debate over safety in the 1960s.

ATOMIC HUBRIS: *PROJECT PLOWSHARE*

Geographer Scott Kirsch referred to *Project Plowshare* as "a study of hubris and failure."[122] Hubris ("wanton insolence or arrogance resulting from excessive pride or from passion")[123] is a term sometimes used to describe human efforts to tame the atom. In the case of *Plowshare,* it proved to be particularly apt. The AEC advanced *Plowshare* as an experimental program in massive earthmoving. The program had formerly been established in 1957 at its Livermore Laboratory at the request of three scientists working there—Harold Brown, Gerald Johnson, and Edward Teller. Teller became director of the lab in 1958, and the leading proponent of the *Plowshare* program.[124] He favored a new field, geographical engineering, stating, "We will change the earth's surface to suit us."[125] He later reflected somewhat disingenuously it seems, "In retrospect, given the steadily increasing public hysteria over radiation, we should have given up without beginning."[126] Teller strongly advocated developing a "clean bomb" in the 1950s and 1960s (not available at that time) to meet *Plowshare*'s objectives and as a way to rationalize the environmental impacts of the projects. Teller used the idea of developing a clean bomb to deter test ban agreements. Keeping a nuclear weapons program alive also extended the work of his lab.[127]

One explanation for *Plowshare*'s initiation is that the Suez Crisis (1956) raised concerns about the need to build a new canal through Israel. Physicists at Livermore advanced the idea of using a string of nuclear bombs to excavate a gigantic ditch extending from the gulf to the sea. (As early as 1949 the Soviets may have considered using nuclear explosions for developing similar projects.)[128] There were a variety of proposals associated with the *Plowshare* program, including *Project Chariot* (producing a harbor in Alaska) and *Project Carryall* (creating a railroad pass near Amboy, California). The idea was to use atomic and thermonuclear devices for underground mining and natural gas stimulation, energy production from explosions, production of heavy elements for scientific and medical research, and especially for projects requiring massive earthmoving. Colossal civil engineering was the most exciting venture for the promoters of *Plowshare* and potentially the most advantageous. *Plowshare* also was linked in some respects to national security and defense issues, insofar as some of the underground nuclear tests (such as *Ranier* in 1957) were designed to produce little or no radioactive fallout.[129]

The first *Plowshare* blast took place at Gnome, New Mexico (near Carlsbad), where a 160-foot hole was made in a salt bed. Workers buried the explosive more than 1,000 feet to test a power-production idea of igniting the device in an underground stream to produce steam to run an electric turbine on the surface. Resulting from the steam production, radioactive fallout drifted as far as Omaha, Nebraska.[130] In the early 1960s a most ambitious project targeted creating a canal in Central America with nuclear blasts. In 1964 Congress approved funds for a feasibility study, and President Lyndon Johnson got caught up in the enthusiasm for the idea. He authorized a five-year study costing more than $22 million. Ultimately, the high cost of the project and the uncertainty of the technology doomed the nuclear canal. Nevertheless, in 1967 *Operation Gasbuggy* led to an atomic explosion of 29 kilotons to release natural gas in New Mexico, and in 1969 *Project Rulison* (using a 40-ton device) attempted to liberate natural gas in western Colorado. The last of the big projects was *Operation Rio Blanco*, using three 33-kiloton devices, near Rifle, Colorado, in 1973. In all, approximately 30 blasts were attributed to *Plowshare* between 1957 and 1973, with a price tag of approximately $770 million (in 1996 dollars) before it was terminated in the Richard Nixon years.[131] Teller was particularly fond of *Project Chariot*, a plan to create a deepwater harbor at Cape Thompson in northwest Alaska in the late 1950s and early 1960s. This would require exploding several thermonuclear devices simultaneously. "We looked at the whole world—almost the whole world," Teller stated, "and tried to pick a spot where we could most effectively demonstrate the peaceful uses of [nuclear] energy."[132] This

Crater from the 1962 *Sedan* nuclear test as part of *Project Plowshare*. The 104 kiloton blast displaced 12 million tons of earth and created a crater 320 feet deep and 1,280 feet wide. (Look to the size of the roads in the bottom-right of the picture, and the observation deck at the lower-right edge of the crater, for a sense of scale.) *Source:* Photo courtesy of National Nuclear Security Administration/Nevada Site Office

project, if completed, would have been most dramatic, but at what risk? The Alaskans called the new breed of geographic engineers the "firecracker boys." Environmental activists, scientists, and others turned away the grandiose plan in Alaska—a signal defeat for *Plowshare*.[133]

Project Plowshare was not accepted blindly by everyone. Almost from its start people questioned the hazards related to radioactivity and the actual ability to implement geographic engineering itself. As Kirsch argued, putting science and technology to work into such uncharted areas "results in an impoverished sense of agency for all manner of social and political relations, including those of scientific work."[134] But *Plowshare*, surely misguided, was no simple pipe dream. The Soviets (in a different political and economic system, with different opportunities and constraints) continued their geographical engineering program until 1989. (Teller himself promoted the ideas into the 1980s.) They called it "Nuclear Explosions for the National Economy." However, the Soviets never accomplished their goal of using nuclear explosions to free Siberian water to irrigate the deserts of Central Asia. Yet they took much longer to give up on their vision than most American advocates of *Plowshare*.[135] *Plowshare* and the Soviet equivalent made it difficult to untangle what was folly from what was possible with atomic energy.

NUCLEAR MEDICINE: THE BEST OF THE PEACEFUL ATOM

On the other end of the peaceful-atom spectrum from *Plowshare* were advances made in nuclear medicine. Moving from the grandness of geographical engineering to the intimacy of internal medicine requires a historical reorientation. The formal nuclear medicine field is more than 50 years old, and has become a central tool in diagnosing and treating serious diseases. Essentially it is a specialty that internally administers radioactive materials (radioisotopes) for those purposes. The origins of nuclear medicine preceded the Manhattan Project and the bomb, actually beginning with the discovery of radioactivity by French physicist Antoine-Henri Becquerel (1896) and the discovery of radium and polonium by French chemists Marie and Pierre Curie. In the 1920s and 1930s, scientists administered radioactive phosphorus to animals. For the first time researchers determined that they could study the metabolic process in a living creature. Soon radioactive material was used for the first time to treat a human patient with leukemia. In the late 1930s, physicians could study thyroid physiology with radioactive iodine. Strontium-89, which was a compound localized in the bones, and now used to treat pain in cancer patients, was under study in 1939.[136]

In 1946 treatment of a thyroid cancer patient with radioactive iodine resulted in the complete disappearance of the cancer. World War II-era reactors ultimately led to the production of cheaper medical radioisotopes (radio nuclides) after the war, making their use more practical. The production of radio nuclides for medicine began at Oak Ridge National Laboratory in 1946. John Lawrence, the brother of Ernest Lawrence, may very well have been the father of nuclear medicine, and Berkeley Lab, the birthplace of the field. During World War II, John Lawrence and his colleagues used radioisotopes to help American pilots deal with the consequences of high-altitude flying. The widespread use of nuclear medicine, however, started in the 1950s as its use increased to measure thyroid function and for diagnosis and treatment of thyroid disease. Hal Anger revolutionized the field with his invention of the scintillation scanning camera in 1958, which evolved into modern imaging systems. The Anger Camera enabled physicians to detect tumors and also to conduct other diagnoses by imaging the gamma rays emitted by radioactive isotopes. In 1958 the Brookhaven National Laboratory opened the Medical Research Center to conduct basic research in nuclear medicine. Into the 1960s and 1970s the field expanded markedly, and in 1971 it was formally recognized as a medical specialty. Currently, the majority of radiopharmaceuticals available are used for diagnostic purposes. Among the leaders in the field, Dr. Marshall Brucer (1913-1994) was the first president of the Society of Nuclear Medicine.[137]

CONCLUSION

"By 1948," Paul Boyer observed, "radioactive isotopes had absorbed the full magical aura that earlier surrounded the atomic automobile, the artificial atomic sun, and the household atomic-power pack."[138] The popular press echoed the optimism of many who viewed these advances in medicine as indicators of other economic and social boons to come: increased crop yields through radioactive fertilizer, new plant species via radiation, generation of new energy sources by human-made photosynthesis, and new cures for a range of crippling diseases.[139] The mind raced at the prospects of further harnessing the atom. Thus even the tangible accomplishments of nuclear medicine, like the false hopes of *Plowshare* and the tepid start of nuclear power generation, were leading down the same path of hope, realistically or not. Such was the power of imagination linked to atomic energy. Such was the effort to think beyond its one massive success so far—nuclear weapons. Yet, nuclear power generation proved not to be a fantasy or unrealistic, growing in its development on a worldwide basis in fits and starts. In the 1960s expectations for generating plentiful electricity with nuclear reactors rose markedly, to be quickly dashed by the late 1970s. "Peaceful Uses" of the atom was a carnival ride. For some it was exhilarating. For others the starts and stops were difficult to manage. And for others still, the ticket price was just too steep.

MySearchLab Connections: Sources Online

READ AND REVIEW

Review this chapter by using the study aids and these related documents available on MySearchLab.

✓•⌐**Study** and **Review** on **mysearchlab.com**

Chapter Test

Essay Test

📖•⌐**Read** the **Document** on **mysearchlab.com**

Dwight D. Eisenhower, Press Release Announcing Atoms for Peace Policy (1953)

Nikita Khrushchev Challenges the West to Disarm and Advance World Prosperity (1960)

Dwight D. Eisenhower, Farewell Address (1961)

Price-Anderson Act of 1957

RESEARCH AND EXPLORE

Use the databases available within MySearchLab to find additional primary and secondary sources on the topics within this chapter.

Endnotes

1. Robert J. Duffy, *Nuclear Politics in America: A History and Theory of Government Regulation* (Lawrence, KS: University Press of Kansas, 1997), 31–32.
2. J.D. Bernal, "Everybody's Atom" (1945) in Robert Engler, ed., *America's Energy: Reports from the Nation on 100 Years of Struggles for the Democratic Control of Our Resources* (New York: Pantheon Books, 1980), 308–09.
3. Joseph J. Corn and Brian Horrigan, *Yesterday's Tomorrows: Past Visions of the American Future* (New York: Summit Books, 1984), 29, 102.
4. Michael A. Amundson, "Uranium on the Cranium: Uranium Mining and Popular Culture," in Scott C. Zeman and Michael A. Amundson, eds., *Atomic Culture: How We Learned to Stop Worrying and Love the Bomb* (Boulder, CO: University Press of Colorado, 2004), 52–54.
5. Zeman and Amundson, eds., *Atomic Culture*, 2–3.
6. Ferenc M. Szasz, "Atomic Comics: The Comic Book Industry Confronts the Nuclear Age," in Zeman and Aundson, eds., *Atomic Culture*, 12–15.
7. Allan M. Winkler, *Life Under a Cloud: American Anxiety About the Bomb* (Urbana, IL: University of Illinois Press, 1999), 140.
8. Paul Boyer, *By Bomb's Early Light: American Thought and Culture at the Dawn of the Atomic Age* (Chapel Hill, NC: University of North Carolina Press, 1994), 120–21.
9. Ibid., 122.
10. Winkler, *Life Under a Cloud*, 136.
11. Correspondence from Samuel Walker to Martin Melosi, December, 2010.
12. Jonathan Scurlock, "A Concise History of the Nuclear Industry Worldwide," in David Elliott, ed., *Nuclear or Not? Does Nuclear Power Have a Place in a Sustainable Energy Future?* (New York: Palgrave MacMillan, 2007), 24.

13. Corbin Allardice and Edward R. Trapnell, *The Atomic Energy Commission* (New York: Praeger Pub., 1974), 35; Richard G. Hewlett and Oscar E. Anderson, Jr., *The New World, 1939–1946*, Volume I of *A History of the United States Atomic Energy Commission* (University Park, PA: Pennsylvania State University Press, 1962), 651–55.

14. Richard G. Hewlett and Francis Duncan, *Atomic Shield, 1947–1952*, Volume II of *A History of the United States Atomic Energy Commission* (University Park, PA: Pennsylvania State University Press, 1969), 96.

15. Peter Stoler, *Decline and Fail: The Ailing Nuclear Power Industry* (New York: Dodd, Mead & Company, 1985), 25–27; Daniel Ford, *Meltdown: The Secret Papers of the Atomic Energy Commission* (New York: Simon & Schuster, 1986), 32.

16. Roger M. Anders, "Institutional Origins of the Department of Energy: The Office of Military Application," *Energy History Series* 1, Historian's Office, United States Department of Energy (November 1978; revised August 1980), 2–3.

17. Hewlett and Duncan, *Atomic Shield, 1947–1952*, 154–72.

18. Alice L. Buck, *A History of the Atomic Energy Commission* (Washington, D.C.: US Department of Energy, August 1982), 1; Anders, "Institutional Origins of the Department of Energy: The Office of Military Application," 3–4.

19. Hewlett and Duncan, *Atomic Shield, 1947–1952*, 185.

20. Ibid., 96.

21. Martin V. Melosi, *Coping with Abundance: Energy and Environment in Industrial America* (New York: Alfred A. Knopf, 1985), 226.

22. See Duffy, *Politics in America*, 21.

23. Melosi, *Coping with Abundance*, 226.

24. Duffy, *Nuclear Politics in America*, 1–2.

25. David Howard Davis, *Energy Politics* (New York: St. Martin's Press, 1982; 3d ed.), 208–09; Brian Balogh, *Chain Reaction: Expert Debate and Public Participation in American Commercial Nuclear Power, 1945–1975* (Cambridge, MA: Cambridge University Press, 1991), 66–68.

26. Duffy, *Nuclear Politics in America*, 3, 20. Historian Brian Balogh's version of the subgovernment system (the prominitrative state) was based on "a symbiotic relationship between professional experts and the nation's public bureaucracies." See Balogh, *Chain Reaction*, 12.

27. Joseph P. Tomain, *Nuclear Power Transformation* (Bloomington, IN: Indiana University Press, 1987), 7.

28. Ibid.; Balogh, *Chain Reaction*, 12–68.

29. Quoted in Nuclear Regulatory Commission, *A Short History of Nuclear Regulation, 1946–1999*, 3 (January 29, 2001), online, http://www.nrc.gov/SECY/smj/shorthis.htm. See also Ford, *Meltdown*, 25.

30. Balogh, *Chain Reaction*, 95.

31. Duffy, *Nuclear Politics in America*, 22; Balogh, *Chain Reaction*, 64–65.

32. Mark Hertsgaard, *Nuclear Inc.: The Men and Money Behind Nuclear Energy* (New York: Pantheon Books, 1983), 21–23.

33. They studied a variety of reactor designs including pressurized water and boiling water, fast and intermediate breeders, and gas-cooled reactors. Fritz F. Heimann, "How Can We Get the Nuclear Job Done?" in Arthur W. Murphy, ed., *The Nuclear Power Controversy* (Englewood Cliffs, NJ: Prentice-Hall, Inc., 1976), 91–92; George T. Mazuzan and J. Samuel Walker, *Controlling the Atom: The Beginnings of Nuclear Regulation, 1946–1962* (Berkeley, CA: University of California Press, 1984), 16.

34. J. Samuel Walker, *Containing the Atom: Nuclear Regulation in a Changing Environment, 1963–1971* (Berkeley, CA: University of California, 1992), 185; Wendy Allen, *Nuclear Reactors for Generating Electricity: U.S. Development from 1946 to 1963* (Santa Monica, CA: Rand Corporation, June 1977), v. According to the Nuclear Regulatory Commission, fissile material is "A nuclide that is capable of undergoing fission after capturing low-energy thermal (slow) neutrons. Although sometimes used as a synonym for fissionable material, this term has acquired its more-restrictive interpretation with

the limitation that the nuclide must be fissionable by *thermal neutrons*." "Fissile Material," August 1, 2011, *U.S. NRC*, online, http://www.nrc.gov/reading-rm/basic-ref/glossary/fissile-material.html.

35. Steven L. Del Sesto, *Science, Politics, and Controversy: Civilian Nuclear Power in the United States, 1946–1974* (Boulder, CO: Westview Press, 1979), 41.

36. Jack M. Holl, Roger M. Anders, and Alice L. Buck, *United States Civilian Nuclear Power Policy, 1954–1984: A Summary History* (Washington, D.C.: US Department of Energy, February 1986), 2.

37. Ibid., 39–40; Melosi, *Coping with Abundance*, 226–27.

38. Stephen E. Atkins, ed., *Historical Encyclopedia of Atomic Energy* (Westport, CT: Greenwood Press), 242; Scurlock, "A Concise History of the Nuclear Industry Worldwide," 25; Mazuzan and Walker, *Controlling the Atom*, 126; Hertsgaard, *Nuclear Inc.,* 24–25.

39. Hertsgaard, *Nuclear Inc.*, 24–25; Del Sesto, *Science, Politics, and Controversy,* 43–49; Melosi, *Coping with Abundance*, 227; Mazuzan and Walker, *Controlling the Atom*, 19; Buck, *A History of the Atomic Energy Commission*, 2–3.

40. Quoted in "Doing a Job," *GovLeader.org*, online, http://govleaders.org/rickover.htm.

41. See Francis Duncan, *Rickover and the Nuclear Navy: The Discipline of Technology* (Annapolis, MD: Naval Institute Press, 1990).

42. William Beaver, *Nuclear Power Goes On-Line: A History of Shippingport* (New York: Greenwood Press, 1990), 6.

43. Quoted in Balogh, *Chain Reaction*, 89.

44. Rodney P. Carlisle, ed., *Encyclopedia of the Atomic Age* (New York: Facts on File, 2001), 279–82; Duncan, *Rickover and the Nuclear Navy.*

45. Balogh, *Chain Reaction,* 89–90; Winkler, *Life Under a Cloud*, 147–48; Beaver, *Nuclear Power Goes On-Line*, 6; Richard G. Hewlett and Jack M. Holl, *Atoms for Peace and War, 1953–1961: Eisenhower and the Atomic Energy Commission* (Berkeley, CA: University of California Press, 1989), 186–89; Hewlett and Duncan, *Atomic Shield, 1947–1952*, 75–76; Atkins, ed., *Historical Encyclopedia of Atomic Energy*, 92, 310–11.

46. Beaver, *Nuclear Power Goes On-Line*, 7. See also Allardice and Trapnell, *The Atomic Energy Commission*, 92–93.

47. Carlisle, ed., *Encyclopedia of the Atomic Age*, 279–82; Duncan, *Rickover and the Nuclear Navy.*

48. Winkler, *Life Under a Cloud*, 148; Peter Pringle and James Spigelman, *The Nuclear Barons* (New York: Holt, Rinehart and Winston, 1981), 149–57; Buck, *A History of the Atomic Energy Commission*, 3; Atkins, ed., *Historical Encyclopedia of Atomic Energy*, 243–44. According to Samuel Walker, "In this model, water that circulated through the fuel was kept under pressure to prevent it from boiling. The heat from the water was transferred to a secondary loop through a steam generator, and the steam in the secondary loop drove turbines to create electric power." *Containing the Atom: Nuclear Regulation in a Changing Environment, 1963–1971* (Berkeley, CA: University of California, 1992), 20. For more on the nuclear navy, see Hewlett and Holl, *Atoms for Peace and War, 1953–1961*, 422–26, 506–08, 520–22.

49. Quoted from "Leadership Quotes," *GovLeaders.org*, online, http://govleaders.org/quotes7.htm.

50. Atkins, ed., *Historical Encyclopedia of Atomic Energy*, 210.

51. Winkler, *Life Under a Cloud,* 148–49; Ford, *Meltdown*, 47; Walker, *Containing the Atom*, 20; Balogh, *Chain Reaction*, 102.

52. Joseph G. Morone and Edward J. Woodhouse, *The Demise of Nuclear Energy? Lessons for Democratic Control of Technology* (New Haven, CT: Yale University Press, 1989), 42–46.

53. Winkler, *Life Under a Cloud*, 146.

54. Quoted in Buck, *A History of the Atomic Energy Commission*, 3.

55. Ira Chernus, *Eisenhower's Atoms for Peace* (College Station, TX; Texas A&M University Press, 2002); Joyce A. Evans, *Celluloid Mushroom Clouds: Hollywood and the Atomic Bomb* (Boulder, CO: Westview Press, 1998), 121.

56. Buck, *A History of the Atomic Energy Commission*, 3; Allardice and Trapnell, *The Atomic Energy Commission*, 103, 200–01.

57. Melosi, *Coping with Abundance*, 229–30; Davis, *Energy Politics*, 211; Hewlett and Holl, *Atoms for Peace and War, 1953–1961*, 66–67; Atkins, ed., *Historical Encyclopedia of Atomic Energy*, 31, 33; Carlisle, ed., *Encyclopedia of the Atomic Age*, 148–49.

58. Munir Ahmad Khan, "Nuclear Energy and International Cooperation: A Third World Perception of the Erosion of Confidence," in Ian Smart, ed., *World Nuclear Energy: Toward a Bargain of Confidence* (Baltimore: Johns Hopkins University Press, 1982), 53.

59. See Craig Doyle Nelson, "Nuclear Bonds: Atoms for Peace in the Cold War and in the Non-Western World" (MA Thesis, Ohio State University, 2009).

60. Ronald E. Powaski, *March to Armageddon: The United States and the Nuclear Arms Race, 1939 to the Present* (New York: Oxford University Press, 1987), 74; Duffy, *Nuclear Politics in America*, 32–33; Hewlett and Holl, *Atoms for Peace and War, 1953–1961*, 209; Joseph F. Pilat, ed., *Atoms for Peace: A Future after Fifty Years?* (Baltimore: Johns Hopkins University Press, 2007), 2–3, 15–20; Bertrand Goldschmidt and Myron B. Kratzer, "Peaceful Nuclear Relations: A Study of the Creation and the Erosion of Confidence," in Smart, ed., *World Nuclear Energy*, 21–22.

61. Melosi, *Coping with Abundance*, 230; Hewlett and Holl, *Atoms for Peace and War, 1953–1961*, 209–11.

62. Holl, Anders, and Buck, *United States Civilian Nuclear Power Policy, 1954–1984: A Summary History*, 3; Buck, *A History of the Atomic Energy Commission*, 3.

63. David E. Nye, *Consuming Power: A Social History of American Energies* (Cambridge, MA: MIT Press, 1998), 201.

64. Allardice and Trapnell, *The Atomic Energy Commission*, 45.

65. Duffy, *Nuclear Politics in America*, 34; Melosi, *Coping with Abundance*, 230–31; Winkler, *Life Under a Cloud*, 146–47; Davis, *Energy Politics*, 213; Allardice and Trapnell, *The Atomic Energy Commission*, 44–51; Carlisle, ed., *Encyclopedia of the Atomic Age*, 22–23; Buck, *A History of the Atomic Energy Commission*, 3; Holl, Anders, and Buck, *United States Civilian Nuclear Power Policy, 1954–1984: A Summary History*, 3.

66. In 1956, 40,000 kilograms of U-235 for nonmilitary uses became available to "friendly" governments. Powaski, *March to Armageddon*, 75.

67. Duffy, *Nuclear Politics in America*, 33–35.

68. Ibid., 36–37; Nuclear Regulatory Commission, *A Short History of Nuclear Regulation, 1946–1999*, 3.

69. Melosi, *Coping with Abundance*, 231; Duffy, *Nuclear Politics in America*, 36–37.

70. Walker, J. Samuel. *Three Mile Island: A Nuclear Crisis in Historical Perspective* (Berkeley, CA: University of California Press, 2004), 3–4.

71. Holl, Anders, and Buck, *United States Civilian Nuclear Power Policy, 1954–1984: A Summary History*, 4; Melosi, *Coping with Abundance*, 231; Duffy, *Nuclear Politics in America*, 39; Winkler, *Life Under a Cloud*, 149.

72. Del Sesto, *Science, Politics, and Controversy*, 53–61; Holl, Anders, and Buck, *United States Civilian Nuclear Power Policy, 1954–1984: A Summary History*, 5; Winkler, *Life Under a Cloud*, 150; Ford, *Meltdown*, 46–47.

73. Tomain, *Nuclear Power Transformation*, 8. See also Horace Herring, "Opposition to Nuclear Power: A Brief History," in Elliott, ed., *Nuclear or Not?* 35; Boyd Norton, "A Brief History," in Michio Kaku and Jennifer Trainer, eds., *Nuclear Power: Both Sides: The Best Arguments For and Against the Most Controversial Technology* New York: W.W. Norton & Co., 1982), 18–19.

74. Melosi, *Coping with Abundance*, 232.

75. Atkins, ed., *Historical Encyclopedia of Atomic Energy*, 329–30; Stoler, *Decline and Fail*, 36–37; Beaver, *Nuclear Power Goes On-Line*; Holl, Anders, and Buck, *United States Civilian Nuclear Power Policy, 1954–1984: A Summary History*, 3.

76. John W. Johnson, *Insuring Against Disaster: The Nuclear Industry on Trial* (Macon, GA: Mercer University Press, 1986), 51–52; Nuclear Regulatory Commission, *A Short History of Nuclear Regulation, 1946–1999*, 5–6; Hewlett and Holl, *Atoms for Peace and War, 1953–1961*, 344–45, 351, 411; Mazuzan and Walker, *Controlling the Atom*, 79–82; Morone and Woodhouse, *The Demise of Nuclear Energy?*, 51–52.

77. See Balogh, *Chain Reaction*, 131.

78. Winkler, *Life Under a Cloud*, 153; Mazuzan and Walker, *Controlling the Atom*, 60–65; Atkins, ed., *Historical Encyclopedia of Atomic Energy*, 3; Ford, *Meltdown*, 42–43, 51; Balogh, *Chain Reaction*, 131–33; Melosi, *Coping with Abundance*, 235.

79. Correspondence from Samuel Walker to Martin Melosi, December, 2010; Melosi, *Coping with Abundance*, 235.

80. "Fermi, Unit 1," US NRC, April 13, 2011, online, http://www.nrc.gov/info-finder/decommissioning/power-reactor/enrico-fermi-atomic-power-plant-unit-1.html; Allardice and Trapnell, *The Atomic Energy Commission*, 110-11; correspondence from Samuel Walker to Martin Melosi, December, 2010.

81. Melosi, *Coping with Abundance*, 235–36; Atkins, ed., *Historical Encyclopedia of Atomic Energy*, 3; Carlisle, ed., *Encyclopedia of the Atomic Age*, 103–04.

82. Johnson, *Insuring Against Disaster*; Nuclear Regulatory Commission, *A Short History of Nuclear Regulation, 1946–1999*, 9–10; Atkins, ed., *Historical Encyclopedia of Atomic Energy*, 293; Morone and Woodhouse, *The Demise of Nuclear Energy?* 54–57.

83. Duffy, *Nuclear Politics in America*, 40; Winkler, *Life Under a Cloud*, 154; Ford, *Meltdown*, 44–46, 51–54; Atkins, ed., *Historical Encyclopedia of Atomic Energy*, 398–99. Melosi, *Coping with Abundance*, 236.

84. Tomain, *Nuclear Power Transformation*, 9l; Melosi, *Coping with Abundance*, 235–36; Correspondence from Samuel Walker to Martin Melosi, December, 2010. See also Johnson, *Insuring Against Disaster*.

85. Melosi, *Coping with Abundance*, 235–37; Winkler, *Life Under a Cloud*, 155; Ford, *Meltdown*, 54–57.

86. Morone and Woodhouse, *The Demise of Nuclear Energy?* 62–65.

87. Duffy, *Nuclear Politics in America*, 36.

88. Irvin C. Bupp and Jean-Claude Derian, *Light Water: How the Nuclear Dream Dissolved* (New York: Basic Books, Inc., 1978), 21.

89. Powaski, *March to Armageddon*, 75; Holl, Anders, and Buck, *United States Civilian Nuclear Power Policy, 1954–1984: A Summary History*, 6–7.

90. Holl, Anders, and Buck, *United States Civilian Nuclear Power Policy, 1954–1984: A Summary History*, 6–7; Buck, *A History of the Atomic Energy Commission*, 4; Hewlett and Holl, *Atoms for Peace and War, 1953–1961*, 305–25, 403–05, 430–48, 489–92, 565–66; Atkins, ed., *Historical Encyclopedia of Atomic Energy*, 123; Carlisle, ed., *Encyclopedia of the Atomic Age*, 96–97;

91. Henry R. Nau, *National Politics and International Technology: Nuclear Reactor Development in Western Europe* (Baltimore: Johns Hopkins University Press, 1974), 99–100.

92. Holl, Anders, and Buck, *United States Civilian Nuclear Power Policy, 1954–1984: A Summary History*, 6–7; Powaski, *March to Armageddon*, 75; Goldschmidt and Kratzer, "Peaceful Nuclear Relations," 21–22.

93. Melosi, *Coping with Abundance*, 235–36; Carlisle, ed., *Encyclopedia of the Atomic Age*, 148.

94. Ian Welsh, *Mobilising Modernity: The Nuclear Moment* (London: Routledge, 2000), 45–56; Achille Albonetti, *Europe and Nuclear Energy*, (Paris: Atlantic Institute for International Affairs, 1972), 21.

95. Peter D. Dresser, ed., *Nuclear Power Plants Worldwide*, (Washington, D.C.: Gale Research Inc., 1993), 277–78; Atkins, ed., *Historical Encyclopedia of Atomic Energy*, 2–3, 28–29; Scurlock, "A Concise History of the Nuclear Industry Worldwide," 25, 36–37; Roger Williams, *The Nuclear Power Decisions: British Policies, 1953–78* (London: Croom Helm, Ltd, 1980), 18–48, 60–77, 80–103; Welsh, *Mobilising Modernity*, 56–94.

96. Lorna Arnold, *Windscale 1957: Anatomy of a Nuclear Accident* (New York: St. Martin's Press, 1992); Atkins, ed., *Historical Encyclopedia of Atomic Energy*, 404–05; Davis, *Energy Politics*, 217; Carlisle, ed., *Encyclopedia of the Atomic Age*, 364–65.

97. Nau, *National Politics and International Technology*, 68, 72.

98. Quoted in Gabrielle Hecht, *The Radiance of France: Nuclear Power and National Identity after World War II* (Cambridge, MA: MIT Press, 1998), 2.

99. Ibid., 2.

100. Radiance indeed referred to French "glory" but also played off nuclear power's own "radiance."

101. Atkins, ed., *Historical Encyclopedia of Atomic Energy*, 95–96.

102. Dresser, ed., *Nuclear Power Plants Worldwide*, 59–60; Atkins, ed., *Historical Encyclopedia of Atomic Energy*, 95–96, 136; Carlisle, ed., *Encyclopedia of the Atomic Age*, 109.

103. Bupp and Derian, *Light Water,* 22–29; Hecht, *The Radiance of France,* 58–59.

104. Dresser, ed., *Nuclear Power Plants Worldwide,* 59–60; Richard L. Garwin and Georges Charpak, *Megawatts and Megatons: The Future of Nuclear Power and Nuclear Weapons* (Chicago: University of Chicago Press, 2001), 127; Mary D. Davis, *The Military-Civilian Nuclear Link: A Guide to the French Nuclear Industry* (Boulder, CO: Westview Press, 1988), 7, 21; James M. Jasper, *Nuclear Politics: Energy and the State in the United States, Sweden, and France* (Princeton, NJ: Princeton University Press, 1990), 74–75; Atkins, ed., *Historical Encyclopedia of Atomic Energy,* 136–37, 283; Nau, *National Politics and International Technology,* 68–72, 81–82, 90–91.

105. Nau, *National Politics and International Technology,* 239–40.

106. Dresser, ed., *Nuclear Power Plants Worldwide,* 13; Nau, *National Politics and International Technology,* 79–81, 90.

107. Dresser, ed., *Nuclear Power Plants Worldwide,* 79; Nau, *National Politics and International Technology,* 72–74, 85–86; Albonetti, *Europe and Nuclear Energy,* 24.

108. Nau, *National Politics and International Technology,* 77–79; Albonetti, *Europe and Nuclear Energy,* 26.

109. The nuclear fuel cycle is "the series of industrial processes which involve the production of electricity from uranium in nuclear power reactors," that is, "the various activities associated with the production of electricity from nuclear reactions" are referred to collectively as "the nuclear fuel cycle." See "The Nuclear Fuel Cycle," World Nuclear Association, updated August 2010, online, http://www.world-nuclear.org/info/inf03.html.

110. Paul R. Josephson, *Red Atom: Russia's Nuclear Power Program from Stalin to Today* (Pittsburgh: University of Pittsburgh Press, 2000), 2. Interestingly, the Obninsk reactor was the forerunner of the Chernobyl model which exploded in 1986.

111. Ibid., 272.

112. Beaver, *Nuclear Power Goes On-Line,* 12–13; Atkins, ed., *Historical Encyclopedia of Atomic Energy,* 188–89.

113. Dresser, ed., *Nuclear Power Plants Worldwide,* 269.

114. Ibid., 27–28; Atkins, ed., *Historical Encyclopedia of Atomic Energy,* 33–34.

115. Unfortunately, the CANDU reactor had extensive down time in the 1960s because of design problems.

116. Dresser, ed., *Nuclear Power Plants Worldwide,* 27–28; Garwin and Charpak, *Megawatts and Megatons,* 115; Atkins, ed., *Historical Encyclopedia of Atomic Energy,* 73.

117. Dresser, ed., *Nuclear Power Plants Worldwide,* 27–28; Garwin and Charpak, *Megawatts and Megatons,* 115; Atkins, ed., *Historical Encyclopedia of Atomic Energy,* 77–78.

118. Donald Oscar Eichner, *The Inter-American Nuclear Energy Commission: Its Goals and Achievements* (New York: Arno Press, 1979), 1–9, 139–45.

119. Carlisle, ed., *Encyclopedia of the Atomic Age,* 143; Dresser, ed., *Nuclear Power Plants Worldwide,* 111. See also David Hart, *Nuclear Power in India: A Comparative Analysis* (London: George Allen & Unwin, 1983), 32–33, 39–40, 49–50, 60, 93.

120. Dresser, ed., *Nuclear Power Plants Worldwide,* 165.

121. In 1999 while preparing fuel for an experimental reactor, three workers received high doses of radiation in the Tokai Mura plant. The accident was caused by human error as a result of breaching safety rules. A total of 119 people were exposed to radiation at the plant, but only the three operators received doses above permissible limits; two of which, however, proved fatal. See "Tokaimura Criticality Accident," July 2007, World Nuclear Association, online, http://www.world-nuclear.org/info/inf37.html. See also Dresser, ed., *Nuclear Power Plants Worldwide,* 123; Atkins, ed., *Historical Encyclopedia of Atomic Energy,* 184–85; Carlisle, ed., *Encyclopedia of the Atomic Age,* 155.

122. Scott Kirsch, *Proving Grounds: Project Plowshare and the Unrealized Dream of Nuclear Earthmoving* (New Brunswick, NJ: Rutgers University Press, 2005), 4.

123. Michael Agnes, ed., *Webster's New World College Dictionary* (New York: Macmillan, 1999; 4th ed.), 694.

124. In his memoirs Teller stated, "In 1945, those of us working on nuclear energy expected that it would be applied in three broad fields. The first, of course, was its application to military weapons. Another…was to supply energy through the generation of electivity. The third logical application

was in civil engineering, the use of nuclear explosives to dig harbors and canals or to stimulate gas or oil production through explosives." See Edward Teller, *Memoirs: A Twentieth-Century Journey in Science and Politics* (Cambridge, MA: Perseus Publishing, 2001), 448.

125. Quoted in Dan O'Neill, *The Firecracker Boys: H-Bombs, Inupiat Eskimos, and the Roots of the Environmental Movement* (New York: Basic Books, 1994), 28. See also Paul Boyer, *Fallout: A Historian Reflects on America's Half-Century Encounter with Nuclear Weapons* (Columbus, OH: Ohio State University Press, 1998), 87–94; Kirsch, *Proving Grounds*, 3.

126. Teller, *Memoirs*, 449.

127. O'Neill, *The Firecracker Boys*, 296–97. See also Peter Goodchild, *Edward Teller: The Real Dr. Strangelove* (Cambridge, MA: Harvard University Press, 2004), 284–95.

128. O'Neill, *The Firecracker Boys*, 24–25.

129. Kirsch, *Proving Grounds*, 3; Atkins, ed., *Historical Encyclopedia of Atomic Energy*, 286; Richard T. Sylves, "U.S. Nuclear Exotica: Peaceful Use of Nuclear Explosives," in John Byrne and Daniel Rich, eds., *The Politics of Energy Research and Development*, Energy Policy Studies, v. 3 (New Brunswick, NJ: Transaction Books, 1986), 35–59; Nye, *Consuming Power*, 201. See also Frank Kreith and Catherine B. Wrenn, *The Nuclear Impact: A Case Study of the Plowshare Program to Produce Gas by Underground Nuclear Stimulation in the Rocky Mountains* (Boulder, CO: Westview Press, 1976).

130. Atkins, ed., *Historical Encyclopedia of Atomic Energy*, 286–87; Carlisle, ed., *Encyclopedia of the Atomic Age*, 247–48.

131. Kirsch, *Proving Grounds*, 6; Atkins, ed., *Historical Encyclopedia of Atomic Energy*, 286–87; James Mahaffey, *Atomic Awakening: A New Look at the History and Future of Nuclear Power* (New York: Pegasus Books, 2009), 236–37; Carlisle, ed., *Encyclopedia of the Atomic Age*, 247–48.

132. Quoted in O'Neill, *The Firecracker Boys*, 35. See also Kirsch, *Proving Grounds*, 53–69, 71.

133. See O'Neill, *The Firecracker Boys*, passim.

134. Kirsch, *Proving Grounds*, 5.

135. Ibid.; Josephson, *Red Atom*, 3; Mahaffey, *Atomic Awakening*, 232.

136. "Nuclear Medicine," *Chemistry Explained*, online, http://www.chemistryexplained.com/Ne-Nu/Nuclear-Medicine.html.

137. Ibid.; Jeffrey Kahn, "From Radioisotopes to Medical Imaging, History of Nuclear Medicine Written at Berkeley," September 9, 1996, Berkeley Lab, online, http://www.lbl.gov/Science-Articles/Archive/nuclear-med-history.html; "Nuclear Medicine," *Chemistry Encyclopedia*, online, http://www.chem-istryexplained.com/Ne-Nu/Nuclear-Medicine.html; "A History of Nuclear Medicine," *MedHunters*, online, http://www.medhunters.com/articles/historyOfNuclearMedicine.html; "History of Nuclear Medicine," *Imaginis,* online, http://www.imaginis.com/nuclear-medicine/history-of-nuclear-medicine.html; "Nuclear Medicine," Nuclear Museum, online, http://www.nuclearmuseum.org/online-museum/article/nuclear-medicine.html. For a thorough chronology of nuclear medicine, see "Important Moments in the History of Nuclear Medicine," *SNM*, online, http://interactive.snm.org/index.cfm?PageID=1107.

138. Boyer, *By Bomb's Early Light*, 119.

139. Ibid., 119–20.

CHAPTER 7

To the Brink in Berlin and Cuba

The Military-Industrial Complex and the Arms Race

INTRODUCTION

The Cuban Missile Crisis in October 1962 was a momentous event in the Cold War, maybe the closest the United States and the USSR came to a nuclear exchange. The tensions were enormous, but the good news was that missiles were never fired. The confrontation was unique because it represented a shift from peripheral bomb-rattling in parts of Europe, Korea, the China Sea, Indochina, and elsewhere, to direct squaring off between the two superpowers. Whether it was posturing or deadly serious, for the first time, as B-52 commander Major "King" Kong bluntly proclaimed in the apocalyptic *Dr. Strangelove* (1964), the United States faced nuclear combat "toe to toe with the Ruskies." Possible confrontation between the Americans and Soviets in the early 1960s not only centered in the Caribbean, but also in Berlin. A new crisis there intensified risk in the mounting nuclear arms race. Cooler heads prevailing in both crises, however, led to the test ban talks in 1963. Yet the Cold War was far from over. Neither John F. Kennedy nor Nikita Khrushchev completely toned down their rhetoric or altered their security objectives. Had anything changed? Living in the moment, it was difficult to tell. There were other problems and other distractions in the 1960s that eventually stole the limelight from US-Soviet tensions. The potential for a nuclear exchange, although diminishing for a time, did not altogether disappear. Yet the world was spared from nuclear holocaust for one more day.

THE MILITARY-INDUSTRIAL COMPLEX

In his farewell address on January 17, 1961, President Dwight Eisenhower gave what has become his most memorable speech. He referred to the military establishment as "a vital element in keeping the peace." He repeated a long-held sentiment of his administration: "Our arms must be mighty, ready for instant action, so that no potential aggressor may be tempted to risk his own destruction." He spoke about the size of the defense establishment and the expense, and how the United States in the postwar world had been "compelled to create a permanent armaments industry of vast proportions." This "conjunction of an immense military establishment and a large arms industry," he added, "is new in the American experience" and its influence total. While the president uttered these words with more than a hint of resignation, he quickly changed tone and emphatically stated:

> In the councils of government, we must guard against the acquisition of unwarranted influence, whether sought or unsought, by the military-industrial complex. The potential for the disastrous rise of misplaced power exists and will persist. We must never let the weight of this combination endanger our liberties or democratic processes.[1]

Although receiving little attention at the time, this was a powerful message coming from the former Supreme Allied Commander of World War II and a major architect of American Cold War policy in the 1950s. It spoke to an institutional change in American society, where the permanent state of military preparedness had become the hallmark of the postwar world and the Cold War itself. The "iron triangle" of Congress, defense contractors, and the Department of Defense (DOD) were central components in perpetuating the American military establishment and the arms race. (An earlier draft referred to the "military-industrial-congressional complex.") Nuclear weapons were deeply entrenched in the very structure of America's national government, economy, and security policy that produced this military-industrial complex. It should not be forgotten, however, that the Eisenhower administration helped entrench them.[2]

The process of creating such a system did not grow logically out of World War II, but was rationalized by the Cold War consensus that followed. The USSR, even with its differences in government, ideology, and history, developed an iron triangle of its own. In such a world the building, stockpiling, and deploying of nuclear weapons may not have been inevitable, but their extraction from the new institutional order was virtually impossible. One writer noted that the military-industrial complex "was a new term invented to identify a new threat. Never before, in peacetime, had the military and the arms industry wielded so much power in the everyday life of nations and in the determination of government policy."[3] Of course, elements of a domestic and international military-industrial complex predated Eisenhower's declaration, going back at least to World War I.[4] Yet World War II produced an historic high by employing 20 percent of the US civilian workforce in military service; still in recent years that number was 2 percent, representing 2.2 million people. The Department of Defense (DOD) budget for contractors reached $118 billion at the turn of the twenty-first century.[5]

To speak of the military-industrial complex as being restricted to an iron triangle is to lose sight of the large number of interested parties in this enduring institutional shift, including the Executive Branch, the Pentagon, the intelligence community, the Foreign Service, scientists, and the nation's universities. In his speech, Eisenhower recognized that "[i]n this revolution (changes in our industrial-military posture), research has become central; it also becomes more formalized, complex, and costly" and that an increasing amount was under the direction of the federal government. He was gravely concerned that "the prospect of domination of the nation's scholars by federal employment, project allocations, and the power of money is ever present…and is gravely to be regarded." On the opposite end of the scale, he also cautioned about policy likewise being dominated by "a scientific technological elite."[6]

In essence, the Manhattan Project had already determined the dependence of scientists on government employ. Despite the fact that several of the atomic scientists were critical of US arms policy, many did not abandon the prospect of work in the public sector. They not only were involved in future weapons development, but also in evaluating what other nations were doing (analyzing surveillance data, for example) and even helping to negotiate treaties including the Limited Nuclear Test Ban Treaty (on such issues as verification). Although Eisenhower was concerned that scientific expertise could leverage foreign policy, that observation was dashed with events such as J. Robert Oppenheimer's fall from grace. The publication of physicist Alvin Weinberg's essay, "Impact of Large-Scale Science on the United States," in 1961 coincided with Eisenhower's address and forcefully seconded the idea that much scientific research now focused on national security, and was making scientists and engineers increasingly linked to Cold War goals. This was true not only for American scientists, but others as well. Japan founded *Tsukuba Academic City* and the Soviets established *Akademgorodok*; both were whole communities to support scientific research.[7] The USSR also embarked on a vast expansion of its industrial capacity to feed its own growing militarization. With the German army advancing in 1941–42, the Soviets dismantled about 1,500 factories, shipping them to new areas beyond the Ural Mountains. They rebuilt about 700 of the plants in an industrial zone near Sverdlovsk and Chelyabinsk, which became a nuclear weapons development complex in the 1950s. They shipped another 550 plants to Kazakhstan and the Soviet Far East, several serving as arms and machine-tool facilities.[8]

"For better and for worse," one writer persuasively argued, "the Cold War redefined American science." The DOD became the single biggest patron, especially in the physical sciences and engineering, but also important in the natural and social sciences. Through the 1950s the DOD accounted for approximately 80 percent of the federal research and development budget. Funds for defense research in 1960 were $5.5 billion.[9] Research universities, such as MIT and Stanford, became partners in this process and central elements in the military-industrial complex. While the total dollars for research and development at the university level could not come close to matching R&D appropriations of contractors like Lockheed, General Electric, General Dynamics, and AT&T, the role of higher education was important, especially in providing the scientific expertise that made its way to the big contractors themselves. Another "triangle" ("golden triangle") seemed to build around the high technology industry, military agencies, and research universities. The Korean War completed the mobilization of American scientists and major universities for national security research that began with the Manhattan Project. In 1950 Congress created the National Science Foundation (NSF) which promised to diversify research opportunities, but fell short of having sufficient funding to become a major supporter of pure science. The military and the Atomic Energy Commission (AEC) remained the major sources of research dollars for the physical sciences for many years.[10]

Militarization of science was clearly a feature of the Cold War, as it had become in World War II. For scientists themselves, the question was whether federal funding came with the price of tethering the academy to the goals of the nuclear arms race. Could civilian science ever be civilian again? And were the research universities inextricably beholden to federal largesse? The consensus that linked the political parties to permanent waging of Cold War, a commitment to the arms race, and to military approaches to international affairs encompassed several institutions inside and outside government.[11] The military-industrial complex in Eisenhower's speech was but an expression of a postwar phenomenon that enveloped the United States, its allies, and its enemies alike. The arms race had acquired a logic and a momentum all its own.

AN AMERICAN CAMELOT

The Cold War consensus remained alive and well during the brief presidency of John F. Kennedy. Indeed, the nuclear arms race framed those early years of the 1960s, punctuated by two major confrontations, one in Berlin and the other in Cuba. While Truman, Eisenhower, and Kennedy shared common views about the relationship with the Soviet Union, the position and objectives of the United States in the world, and the important role of atomic and thermonuclear weapons, their differences were more than a matter of style. With the young and vibrant new president (at 43 the youngest ever to be elected) many found it difficult to look beyond style. He and his beautiful, chic wife Jackie came to be viewed as American royalty to reign over a "New Frontier."[12] Soon after JFK's assassination, Jackie referred to the brief presidential years as "Camelot," a historical reference to the days of King Arthur.[13] But just like the Arthurian tale, there was a dark side, especially Jack Kennedy's major health problems, his womanizing, and ultimately, his tragic death.[14] All of these images clouded understanding of his foreign policy, and the momentous turn that the world was taking in the early 1960s.[15]

Kennedy beat Vice President Richard M. Nixon for the presidency by a thin margin in 1960. The Democratic Senator from Massachusetts had done little in the halls of Congress, but he was bright, charming, ambitious, and a World War II hero—the second child of nine in a wealthy Boston Irish-Catholic family. During the campaign, Kennedy took the stance of a hard-line Cold Warrior. He harshly criticized the Eisenhower administration (and Nixon by association) for its tepid military spending and poor handling of Quemoy and Matsu, the rise of Fidel Castro in Cuba, and expansion of communism into the Third World. He charged that military spending under Eisenhower was required to fit into the nation's budget rather than adjusting the budget to support security needs. He railed about the so-called "missile gap" between the two powers, in which the United States was allegedly falling behind.[16] Sometimes Kennedy sounded more like John Foster Dulles than Dulles himself. Truth be told, there was not much difference between the foreign policy stances of Kennedy and Nixon, but the young Democrat tried to make the election a mandate on America's falling world prestige and the need to aggressively strengthen its military presence. Nixon, a seasoned politician with an extensive foreign affairs resume, painted his opponent as inexperienced and out of his league. But Nixon could not compete with Kennedy's charisma, nor escape his own subordinate role to the war hero Eisenhower. The vice-president's history as a notorious anticommunist was neither a strength nor a liability in the face of Kennedy's own Cold War attitudes.[17]

The new administration, however, was unwilling to heed the warnings of Eisenhower's farewell address. Military and diplomatic policy played directly to the heart of the iron triangle, although the executive more than the legislative branch had a central role in this case. Publicly, talking tough and taking a hard line in the 1960 election, no matter how politically viable, had a tendency to limit options. Privately, the talk was not always so tough for the new president, and thus his choices potentially less predictable. The national security system under JFK made stark adjustments to Eisenhower's. Kennedy did not agree with the former president's practice of sharing power with his cabinet, his chief of staff, or the National Security Council (NSC). He opposed delegating authority beyond his control and resisted expanding the executive bureaucracy to the extent that his predecessor had done. Kennedy turned to Dean Rusk, president of the Rockefeller Foundation, to be his Secretary of State. Rusk was a candidate chosen "by process of elimination," insuring that the president maintained firm control over foreign affairs. Rusk was viewed as "a sort of faceless, faithful bureaucrat" who would not challenge presidential decisions.[18]

NSC rather than the state department was the key group that Kennedy relied upon in foreign policy matters. But neither the NSC nor any other body shared authority with the president. Unlike the more ritualized committee system employed under Eisenhower, the national security apparatus under Kennedy was less proactive. In addition, civil-military relations deteriorated now that Eisenhower had retired from office. Among officers like General LeMay there was not only wariness but contempt for the inexperienced Commander in Chief.[19] JFK had few close confidants, with the exception of his brother Robert F. Kennedy, who served as attorney general, and Theodore "Ted" Sorenson, his "alter ego" who was his primary speech writer and special counsel. A student of history, the president listened to historians and political scientists, such as Arthur Schlesinger, Jr. and Richard Neustadt. He also brought into his administration civilian strategists from think tanks such as RAND. (RAND and other think tanks dealt with nuclear strategy, among other things.) McGeorge Bundy, special assistant for national security affairs, was crucial in the foreign policy area. Not many within the cabinet regularly got the president's attention, with the exception of brother Bobby, Secretary of Defense Robert McNamara (the former president of Ford Motor Company and registered Republican), and a few others. JFK often was suspicious of other experts and many in the military establishment.[20]

Kennedy never exhibited much ideological purity. Liberals did not hold down very many key slots in the administration. If Republicans exhibited the necessary credentials, the president was not opposed to appointing them to important positions. Like his predecessor, JFK wanted to distance himself from the previous administration in setting national security policy. The ever-present dangers in Berlin and West Germany and expansion of the Cold War into the Third World, also provoked the new president to change the emphasis of America's national security program. JFK accepted the basic assumptions of containment and framed his public rhetoric on national security around the abiding destiny of the American way. Rather than changing the goals, his administration, however, devoted considerable energy to changing the manner in which the Cold War was waged. Out went "massive retaliation," and in came "flexible response."[21] The idea of flexible response (attributed to or reinforced by General Maxwell Taylor) was predicated on the notion that the United States had to prepare for various kinds of threats, not only a nuclear exchange with the Soviet Union, but also small-scale and conventional warfare. Part of the justification was the president's growing awareness that change was underfoot in developing countries beyond the areas of traditional Cold War engagement. The United States needed to be prepared for such changes, affirmed by Nikita Khrushchev's promotion of wars of national liberation and revolutionary "struggles against imperialism" in the Third World.[22]

Flexible response required beefing up and modernizing conventional forces to strengthen defenses in Europe as well as to combat guerilla warfare in other parts of the world. Counterinsurgency measures and the establishment of elite units were primary tools for the latter. At the time there were 14 American combat divisions, 11 of which were combat ready and only three deployed in the United States. There also was a shortage of rifles, ammunition, and other equipment, as well as poor airlift capacity and few modern tactical airplanes. Economic and technical assistance programs needed to augment the military upgrade. Flexible response, therefore, reinforced a global approach to the Cold War.[23] Far from diminishing the dependence on nuclear weapons, flexible response built upon it. Harangues about a missile gap demanded an immediate response after the inauguration. There were, however, indications early in JFK's term that the Soviets were prepared to curtail, or at least scale down, their nuclear arms program. Kennedy at first refused to believe it, although in fall 1961 the administration admitted that Khrushchev was not pursuing any massive increase in ICBMs. The premier himself faced pressure from his military to increase defense spending when Kennedy was elected, but was hopeful that reducing tensions with the United States would stave off such pressure.[24]

The president was somewhat trapped by his own campaign rhetoric in supporting what amounted to the most rapid and large-scale military buildup in peacetime. He was convinced that the United States needed a clear advantage in missile deployment before he could effectively negotiate with the USSR. Whether he really believed that the missile gap ever existed, and when he changed his mind, is almost beside the point. Flexible response required more options and greater control of nuclear as well as conventional weapons, and would be guaranteed, some experts argued, with superior strength. The air force in particular lobbied for more nuclear weapons, insisting that they needed at least 3,000 ICBMs. McNamara's own calculations set the requirement at less than 1,000. He nonetheless requested 1,000 from Congress, believing that was the smallest number it would accept. By 1964 American nuclear warheads increased by 150 percent and the total in megatons more than doubled during Kennedy's time in office. In March 1961, the president asked Congress for an additional $3 billion in defense appropriations to expand conventional and nuclear military programs. The immediate impact was an increase in the defense budget by about 15 percent, doubling the number of combat-ready divisions in the army's strategic reserves, expanding the navy and marines' combat units, and deepening the nuclear arsenal. The United States devoted a much larger portion of its economic wealth than Europe to its defense budget. To say that the US economy was "militarized" by the Cold War was hardly an exaggeration.[25]

The missile gap turned out to be illusory. In November 1957, an intelligence estimate forecasted that the USSR would have 500 ICBMs (some said 1,000) operational by the end of 1962, while the US total would be approximately 65. After the 1960 election, new estimates placed Soviet ICBMs in 1962 at 150, with the number later falling to 50. In fall 1961, satellite reconnaissance showed that the Soviet Union actually had 44 ICBMs and 155 heavy bombers, compared to 156 ICBMs, 144 *Polaris* submarine launched ballistic missiles, and 1,300 strategic bombers for the United States. SAC contested the figures concerning the Soviet Union for obvious reasons.[26]

More weapons found their way to the western allies in the Kennedy years, but it is questionable that increased stability followed them. Secretary McNamara championed a "damage limitation" strategy to implement a more flexible nuclear response. Under the approach circumstances might lead to targeting the enemy's military installations rather than strikes against cities. While this appeared more measured than Eisenhower's Single Integrated Operational Plan (SIOP) that targeted both cities and military installations, it only raised the anxiety of NATO members who coveted the American nuclear shield.[27] To make amends, McNamara increased tactical nuclear weapons by 60 percent (many going to Europe), agreed to sell *Polaris* misses to the United Kingdom, and attempted to create a multilateral nuclear force with the British and French. The latter proved unworkable because the two European powers did not want to have their independent nuclear programs absorbed by the Americans. In essence, McNamara's attempts at creating a more flexible nuclear response did little to build Western European confidence in America's new strategic goals.[28] The objective of America's massive buildup in conventional and nuclear weapons was meant to intimidate the Soviets more than to catch up with them. Unfortunately, this approach sent an inadvertent message, prompting the USSR to respond in kind. Not intended to do so, the US buildup accelerated the arms race rather than slowing it down or ending it, and complicated the relationship between the superpowers.

CASTRO'S REVOLUTION IN CUBA

It is difficult to imagine the threat of nuclear war being any more acute than during the Kennedy years. Two "hot spots" were bound together: one in the New World, the other in the Old. What became the Cuban Missile Crisis had roots going back to Fidel Castro's revolution in Cuba. The Berlin Crisis built upon tensions from World War II. Both emerging crises entailed dangerous direct face-offs between the United States and the USSR, which was novel in the Cold War years and made the threat of nuclear war possible. The rise of Fidel Castro in Cuba set in motion a series of events that led to the Cuban Missile Crisis. Along the path, Cuba emerged as a successful example of indigenous revolution, and with it the stock of the United States in Latin America plummeted. Cuba long harbored a love-hate relationship with the United States, and in turn the United States viewed Cuba as an economic resource, a playground, and a security buffer. By the time Eisenhower became president Cuba had suffered under more than two decades of the oppressive rule of Fulgencio Batista, who courted organized crime to run Havana's casinos, made numerous concessions to US corporations such as IT&T, and did little to regain control of the sugar industry from American hands.[29]

The son of an affluent planter, holder of a law degree, and a professional-quality baseball pitcher, Fidel Castro was an unswerving nationalist and a social democrat, who viewed Batista and the United States as threats to Cuban sovereignty. While in his twenties he led two uprisings in his country in 1953 and 1956, but they ended badly. (His movement, *the July 26 movement,* took its name from the first attack in 1953.) Retreating to the Sierra Maestra Mountains in southeastern Cuba, Castro raised a guerilla army that drove Batista from power, and entered Havana victorious on January 1, 1959. The United States, preoccupied elsewhere, thought Batista capable of thwarting the rebels (Washington earlier had cut off aid and pressed him to resign). If Batista did not succeed, the Eisenhower government believed that Castro was incapable of surviving politically without American support.[30] US leaders were intrigued by Castro, but also wary of him. Was he a new Cuban strongman or a revolutionary? Vice President Nixon's view that Castro might be turned "in the right direction" failed to take into account the long history of American domination of Cuba and Castro's fervent desire to end it. The new Cuban leader's move toward the Soviet Union, while not inevitable, was hardly an unthinkable strategy. Fidel's brother Raul and their revolutionary compatriot, Che Guevara, were more doctrinaire than Fidel, and embraced communist ideology earlier. Neither was ready to blindly jump in bed with the USSR. They were part of a revolutionary generation inspired by Mao's successes in China, and very distrustful of US ambitions in Latin America, especially after the CIA's overthrow of Guatemala's Jacobo Arbenz in 1954 and its embrace of Batista.[31]

After the revolution in Cuba, and after it became clear to Fidel that good relations with the United States were a pipedream, he legalized the Communist Party and brought leftists into his government. Castro broke with moderates and tried and publicly executed Batista supporters. He expropriated land, nationalized basic industries, and began purchasing weapons from Eastern bloc nations. In a visit to the United States in late 1959, Castro condemned American imperialism in a speech before the United Nations. He also publicly called for revolution throughout Latin America. The leftists in Cuba began making contact with Moscow after the fall of Batista, but the Kremlin was cautious in 1959. They were uncertain what political direction the Castro government might take. Fearing further spread of revolution in Latin America, the Eisenhower administration extended an arms embargo to Cuba initiated while Batista was in power, and protested Castro's nationalization and expropriation of American holdings in sugar, mining, oil, and utilities. On March 4, 1960, the French freighter *La Coubre*, delivering Belgian arms to Cuba, exploded in Havana harbor. Castro blamed the United States, and used the incident as pretext to nationalize all US property. After Castro expelled most officials working at the American embassy in 1961, the United States severed diplomatic ties with Cuba.[32]

Earlier in 1960 in the wake of the American suspension of the sugar quota, Castro decided to seek a trade deal with the USSR, which brought Cuba into the Soviet camp. The new economic exchange and the entry of Soviet arms and military advisers into Cuba was subtext for much of the tension between the two neighboring countries. Once the Soviets sensed the strategic advantage of developing a relationship with Castro, its protection of the new ally was robust. Khrushchev was perpetually concerned about a possible US invasion in Cuba, and he wanted to assert Soviet leadership in the communist world (especially with respect to the prime challenger, China). In a June 9, 1960, he declared that the USSR would protect Cuba (read, with nuclear weapons): "…if need be, Soviet artillerymen can support the Cuban people with their rocket fire, should the aggressive forces in the Pentagon dare to start intervention against Cuba."[33] There was little doubt in American eyes that Castro must go.[34]

The approach of the Eisenhower administration was to eliminate Castro and improve relations with other Latin American countries. The question was whether these methods were mutually exclusive. Treating Cuba as if it were a wartime enemy was not going to win many friends in the hemisphere. An American-imposed trade embargo, the break in diplomatic ties, and opposing Castro's government would not inspire good public relations in other Latin American countries. Behind the scenes, the United States started a propaganda campaign to incite rebellion within Cuba itself, organized political opposition among anti-Castro exiles, and armed and trained an exile force for an eventual invasion. The CIA forged ahead with plots to assassinate the Cuban leader. In the Dominican Republic Rafael Trujillo faced many of the same harassments as Castro, although his rule more closely resembled Batista's.[35] Efforts to improve relations in Latin America, while worth trying, were a hard sell given American enmity for the Cuban leader. The Eisenhower administration toned down the backing of local dictators and began to support some moderate reformers. It turned substantial resources away from military aid to economic, medical, and education programs (as JFK also did).[36]

THE BAY OF PIGS

For the Kennedy administration, there was nothing inevitable about following through with the CIA invasion plan for Cuba. JFK's criticism of Cuban policy during the election put him squarely in the camp with those who saw no redemption in dealing directly with Castro. The bearded revolutionary, they believed, threw his lot with the Soviets, embraced communism, and posed a threat to stability in Latin America. With these predilections, Kennedy did not seriously reevaluate this dubious plan despite the reservations that some military men and others decided to keep private at the plan's inception. Furthermore, the president chose not to listen to skeptics in his own administration, and did not want to incur Republican criticism for indecisive action. Vice President Lyndon Johnson, Senate Foreign Relations Chair J. William Fulbright, and others viewed the plan as nonsensical. Yet CIA director Allen Dulles assured JFK that the invasion would work. Choosing to approve the plan, but refusing to provide air support for the land invasion (as a way of masking the US role) proved to be a serious blunder for Kennedy. Castro, himself, was convinced that an invasion was coming and made sure the Soviets heard about it often.[37]

The invasion in 1961 deepened the chasm in Cuba-American relations for years. With Kennedy's election, Castro had made gestures to lessen tensions before the ill-conceived attack. He hoped that the new American administration would abandon the CIA's plan for an assault he knew was likely to occur. After JFK's inauguration, the Cuban leader demobilized a large portion of his rural militia that he had called up in November 1960. This action was less a peace gesture, than a response to the economic strain of supporting the troops. His trusted adviser, Che Guevara, convinced Castro that without utilizing some of the militia to help with the sugar harvest, Cuba would be unable to deliver on its promised quotas to the Soviets. At this moment of opportunity, negotiations with Washington seemed propitious.[38] But negotiations never came, and the United States did not seek diplomatic options. Beyond the possibility of invasion, the Kennedy administration only considered economic embargo and a variety of covert actions.

The perception of Cuba entrenched in the communist sphere was much too fresh. The code name chosen for the invasion plan was ironically apt, *Bumpy Road*. An air strike would take place on April 15, 1961, and the land assault two days later. Many observers came to realize that the plan was conceptually flawed and feebly executed. The CIA assumed that the invasion by a modest force of Cuban exiles (trained in Guatemala) would trigger internal revolt that would bring down Castro. They had no substantiation for such an assertion and were proved unequivocally wrong. The amphibious landing force was poorly trained and badly organized. To make matters worse Kennedy placed constraints on the original invasion strategy. Rather than landing on a beach that provided an option to retreat if necessary into nearby mountains, he wanted a landing at night in an area where there would be "a minimum likelihood of opposition." The hastily revised decision put the troops on an isolated beach in the Bay of Pigs, where there was no easy with-drawal. JFK's reluctance to authorize air support also was a major drawback.[39]

The surprise air strike by eight obsolete CIA-supplied B-26 bombers flown in from Nicaragua (disguised to look like Cuban planes) destroyed more than 50 percent of the Cuban fighters at three main airfields, but with no follow-up strike. Castro therefore had time to disperse the remaining planes and sound an alert.[40] JFK called off a second air strike scheduled for April 17 now that the surprise element was gone (although he had authorized a minor air operation on April 19 that failed). Four of the CIA planes were themselves shot down by Cuban trainer jets, and a *Sea Fury* fighter plane sank a ship which was supplying ammunition and communication equipment to the invaders. Without effective air cover, the 1,500 assault troops were pinned down on the beach. The advantage quickly shifted to the Cuban military, which had superior ground forces supplemented by Soviet-built tanks and air support. They routed the invaders in three days, killing 140 and capturing another 1,189 of the American-trained troops. Even with air-cover support, the landing would not likely have been sufficient to route Castro's much larger force and defensive advantage.[41]

The inexperienced, but now tested president took full responsibility for the humiliating defeat (in doing so his approval ratings actually went up!). He was embarrassed by the fiasco, distressed by the loss of life, and angry with the military and the CIA. Both sides of the aisle in Washington reproached JFK after the Bay of Pigs. Conservatives believed that Kennedy could not finish what he started and was gutless. Liberals could not believe that he had intervened in the internal affairs of a sovereign nation, and that he also seriously bruised relations in Latin America. Newspaper reactions in the United States and among Western European allies, however, were supportive of the American action, given that they had long been bought into the era's Cold War rhetoric and believed that the invasion was understandable if not effective. The story was very different in Latin America, which experienced anti-American demonstrations and violent acts against US embassies. The Soviets were livid and incredulous, condemning the United States for violating Cuban territory, rebuking Kennedy, and promising future assistance to the island nation. The president went on the defensive, writing to Khrushchev that the action was a case of Cubans against Cubans, and publicly stated that if the Soviets intervened in Cuba the United States would respond. Although JFK had a weak case, tensions over the failed assault reduced the chances of improving relations between the superpowers.[42]

Kennedy's decision not to embark immediately on a prolonged military intervention in Cuba seemed prudent, and may have avoided more dire consequences. He developed serious doubts about professional experts, even as he rethought his own performance. In the wake of the "perfect failure," the president fired CIA Director Dulles, replacing him with California business-man, former government bureaucrat, and hard-line Republican, John McCone.[43] But rather than tempering the administration's long-term goals with respect to Cuba, the Bay of Pigs intensified

Kennedy's resolve to eliminate Castro.[44] What became a personal vendetta led to efforts to assassinate the Cuban leader by a variety of means, including Mafia hit men. (The degree to which the president knew about the details is unclear.) Kennedy authorized the CIA-led *Operation Mongoose*, a large-scale covert action to remove Castro, including intelligence gathering, sabotage, paramilitary activity, and assassination.[45] The CIA and the Pentagon developed contingency plans for military exercises in the South Atlantic and Caribbean, and—as if they had not learned a lesson from the Bay of Pigs—considered schemes for future military intervention. The United States also banned all Cuban imports, and sought to further isolate Castro diplomatically by forcing Cuba's expulsion from the Organization of American States.[46]

The Bay of Pigs fiasco exposed a failed foreign policy decision encompassing two presidential administrations. In the larger Cold War, it also played an important role in driving Cuba closer to the Soviet Union, deepening tension between the two superpowers, and doing little to quell the arms race. The Kremlin had to feel good about the outcome of the Bay of Pigs, at least in the short run. It was a propaganda bonanza. April 1961 proved to be auspicious in another way for the Cold War of words. On the April 12, the Soviets put the first man into space (Yuri Gagarin) harkening back to its earlier technical success with *Sputnik* in 1957.[47]

BERLIN-AGAIN

The Kennedy administration's fixation on Cuba had no immediate impact on the nuclear arms race. But the Bay of Pigs fiasco quickly intertwined with tensions in Europe, especially Berlin, which complicated and ultimately intensified frictions between the United States and the USSR. Events in the early 1960s, like no other, brought the two powers into uncomfortable proximity. Berlin remained a tinderbox in what was becoming the most dangerous time in the Cold War. Unlike Cuba (for the moment at least), Guatemala, the Congo, Southeast Asia, Iran, and elsewhere, Berlin directly involved the national security of the United States.[48] To Khrushchev's way of thinking, Berlin was a "bone in the throat."[49] Situated in the heart of East Germany, the Prussian city was a constant reminder of military pressure placed on his country by the United States and its NATO allies. It also was a painful exemplar of the economic success of West Germany. One has to wonder if the propaganda value of the contrast between West and East, as well as the regular flight of people across the border, was as much of an aggravation as the security issues implicit in a divided Berlin, a divided Germany, and a divided Europe.[50]

Khrushchev also knew that Western powers were most vulnerable in Berlin. He was a risk-taker by nature, and also was emboldened by what he believed to be Eisenhower's passive stance in other parts of the world. Therefore, he demanded in November 1958 that Berlin be made a free city with an autonomous government. If his rivals did not respond within six months, the premier asserted, he would break the World War II pact concerning the Third Reich and sign a separate peace with the East Germans. The new ally would determine access to West Berlin at that point. Eisenhower did not back down, but was intent on avoiding a provocative move. He ordered a military buildup without fanfare, and as the ultimatum deadline of May 27, 1959, approached tensions eased somewhat.[51]

During the mid- to late-1950s discussions about controlling nuclear weapons (including testing) were on again/off again. On the surface at least, both sides temporarily curtailing atmospheric testing was a positive step, but was more important for propaganda purposes than encouraging real dialogue. (See Chapter 5.) In fall 1959 Khrushchev visited the United States in what appeared to be an attempt to ease tensions. Eisenhower was reluctant to host the mercurial Soviet leader, who lived up to billing with several moments of political theater. At the president's

Camp David retreat, some progress seemed possible when Khrushchev vaguely withdrew his Berlin ultimatum and Eisenhower likewise obliquely intimated that the status of the city needed reconsideration. The four principle powers involved in the German partition agreed to meeting in Paris (scheduled for May 1960), with the American president making a follow-up visit to Moscow. The U-2 incident which Khrushchev made public on May 7 derailed this plan a week before the proposed summit. Tensions in Berlin mounted.[52]

The aftermath of the Bay of Pigs was another opportunity for the superpower leaders to seek some resolution of outstanding issues, particularly a nuclear test ban. The more practiced Soviet premier and the recently chastened and less experienced American president agreed to meet in Vienna on June 3, 1961. After his conference with Eisenhower at Camp David, Khrushchev came away believing that the ex-president was someone with whom he could work; Kennedy was another matter. JFK was pushing to accelerate the arms race again and the Cuban fiasco sent ominous signals about US policy. Khrushchev also believed that the young president was "too much of a light-weight," and rather than trying to feel him out, decided to bully him.[53] Although Kennedy came to Vienna prepared, at least on paper, he was in pain because of his various ailments, weakened by the medications he was taking, and not ready to confront a wily, seasoned adversary. He tried to stand his ground on Berlin, bucked up by Soviet experts such as George Kennan and hardliner Dean Acheson, who consulted with him before the trip. Yet Kennedy had to keep in mind that reducing tensions was a major objective of the meeting.[54]

President John F. Kennedy (right) meets Premier Nikita Khrushchev with Soviet ambassador Andrei Gromyko in Vienna, Austria, 1962. *Source:* Library of Congress, Prints & Photographs Division, NYWT&S Collection, [LC-USZ62-121341]

The president also had to take into account the interests of his anxious European allies. The meeting in Vienna exposed potential problems within NATO and with its relationship to the United States. President Charles DeGaulle wanted France and other Western European countries to play a more central role in defending their territory (and for France to obtain its own nuclear force). Kennedy already had made clear a desire to expand conventional forces in Europe, but with the United States wielding the nuclear shield independently. The shift to more conventional weapons, DeGaulle believed, meant that the Americans would not be willing to engage in nuclear retaliation against the USSR in case of an invasion of the Red Army. Kennedy's burden going into the Vienna talks, therefore, carried weight not only between the superpowers, but within a larger global community.[55] The talks themselves were a see-saw affair with many tense moments and little agreement, interspersed with a few lighter civilities. Early on the two leaders engaged in an ideological debate, which advisers told Kennedy to avoid. He criticized the Soviets for denying self-determination in places like Laos and the Congo. Khrushchev in return trumpeted the inevitability of communist victories in the Third World. Kennedy raised the issue of Laos, strongly desirous of seeing the Southeast Asian country not succumb to communist takeover; Khrushchev appeared disinterested. Kennedy tried to get a reaction over the current Sino-Soviet falling-out that the United States hoped to exploit, but Khrushchev did not take the bait.[56] More central were discussions over a nuclear test ban, an issue that went right to the heart of tensions between the two nations. But the gap was too wide. The Soviet premier would not budge on inspections, instead wanting to make the ban contingent upon pursuit of general disarmament. Equally frustrating, and more dangerous at the time, was the question of Germany. Khrushchev discussed the same set of issues he had raised with Eisenhower prior to his Berlin ultimatum. He trotted out another ultimatum over Berlin as a free city, asserting that the USSR would sign a separate treaty with East Germany at the end of 1961 if the Western powers refused to abandon the city. The United States was risking nuclear war, he threatened, if it did not honor the new arrangement. Bluster or no, Kennedy bluntly responded, "It will be a cold winter."[57]

What had the Vienna meeting accomplished? In the short run, the results seemed inconclusive, if not bleak. Kennedy was clearly shaken by the encounter and his self-assessment not favorable. He privately told journalist James Reston that Khrushchev "just beat [the] hell out of me" over his actions in the Bay of Pigs.[58] Fortunately, his performance remained outside the public gaze. He returned home and began to construct a response to the new ultimatum on Berlin, pressing his advisers for solutions. Acheson probably took the hardest line, calling for the United States to initiate a major military buildup, to declare a national emergency, to order an airlift into West Berlin if necessary, and to prepare for war. The president wanted to pursue a middle course, hinting in a nationally televised speech on July 25 that he was willing to negotiate. Not going as far as Acheson suggested, he nonetheless announced a further military buildup, increased draft calls, brought up reserves, pushed for more enlistments, and urged a federal program to build fallout shelters.[59]

The Soviet premier regarded the speech as "warlike." He was under pressure at home to do more than threaten action, and wanted some kind of victory in Berlin to weaken US influence globally. He also wanted the western powers to negotiate in earnest, and not hide behind its containment strategy. Khrushchev decided that the Soviet Union would resume atmospheric testing in September (the Americans would soon follow), although this would not play well in most countries. The premier was faced with what he regarded as provocation and the unresolved issues over Berlin, and thus called for an increase in the Soviets' military spending. On August 13, Khrushchev ordered the building of the famous Berlin Wall. The wall was not only a symbol, but a way to stop the flow of East Germans to the West (at the time about 100,000 per year). It also was an indicator of a hardened military policy for the Soviet Union. In another respect, erecting the wall conceded what the West had been saying all along, that the "Peoples' Paradise" on the east side was not reality. East Germans were chancing escape because of what they believed was a much more promising life on the western side.[60]

US officials were caught off guard by the act. Kennedy, who may have sensed that something like this might occur, believed that he could not back down from the explicit challenge that the wall represented. He chose to respond by making the military presence in West Berlin more visible. Although he did not think a nuclear exchange was likely, he thought it was possible. In a certain respect, the Berlin Wall presented a temporary reduction of tensions between the superpowers, and it may have offered a solution to a possible confrontation with the Soviets if things destabilized in East Germany. For better or for worse, the modest response from the United States and NATO indicated a willingness to accept the idea of two Germanys. By fall 1961 the intensity of the Berlin Crisis subsided. Both Kennedy and Khrushchev (if not the hardliners in their own government) wanted to back off from this risky game of chicken. The two parties thus agreed to quiet negotiations in September. Kennedy wanted a "neutral" Laos prior to another summit and publicly supported the idea of disarmament talks. In October Khrushchev agreed to terminate the deadline for the proposed German peace treaty. Talks were one thing, but diplomacy did not end the arms race at this juncture. Bilateral diplomacy tended to leave Kennedy's NATO allies on the outside looking in, having to adjust to a shifting American policy without much direct participation. Khrushchev did not believe that he had to take into account the views of Warsaw Pact countries in his negotiations with the US president. The USSR would discover, however, that such neglect ultimately had consequences. Neither side went to the brink, but the temporary lull in Berlin did not remove the ever-present risk of nuclear war along a hardened border in Europe.[61]

THE OCTOBER CRISIS

In the fall of 1962 the threat of nuclear war became real—in the Caribbean, not in Europe. The events surrounding the Cuban Missile Crisis have been debated, studied, and second-guessed ever since. Robert F. Kennedy's *Thirteen Days*, although clearly biased in favor of his own and his brother's performance, made clear the tensions created by the crisis within the president's inner circle.[62] Although short on context and several details, the 1974 docudrama, *The Missiles of October*, and the 2000 film, *Thirteen Days*, clearly demonstrated that this riveting episode was well-suited for dramatic rendering. In recent years newly opened records in both the United States and Russia have clarified the actions of participants, especially in the USSR, while some issues remain murky. It is no exaggeration that in October 1962, the world came perilously close to heading toward nuclear war, even though the major participants did not want that to happen. The setting in which the crisis occurred clearly grew out of the distrust that all three parties had toward each other. Castro believed that it was just a matter of time before the United States would mount another invasion of his country. Kennedy wanted Castro out of Cuba, wanted to demonstrate that he could stand firm in the wake of other potential communist revolutions, and steadfastly believed that the Soviet Union did not favor peaceful coexistence. Khrushchev was happy to see Cuba fall into his country's lap, and did not trust the motives of the United States in Cuba or elsewhere for that matter. He later recalled in his memoirs, "Everyone agreed that America would not leave Cuba alone unless we did something…It was clear to me that we might lose Cuba if we didn't take some decisive steps in her defense."[63] In addition, leaders in the Soviet Union believed that the Americans were capable, and possibly willing, to launch a first strike against the USSR because of its nuclear superiority. Khrushchev told the Presidium, "In addition to protecting Cuba, our missiles would equalize what the West likes to call 'the balance of power.'"[64]

Because of these feelings and beliefs, Khrushchev gambled. He was a survivor; he was clever; he was impulsive. And his audacious plan was his alone.[65] The Soviet Union had never placed ballistic missiles outside of its borders, but now was resolved to do so. The premier's decision to deploy medium- and intermediate-range ballistic missiles in Cuba rested both on the need to protect Cuba from a US invasion, and to offset the imbalance in the size and potency of the superpowers' nuclear arsenals.[66] (Although Castro did not encourage the placement of offensive missiles in Cuba, he requested surface-to-air missiles from the Soviets even before the Bay of Pigs fiasco.) The Soviets now could deliver strategic nuclear weapons to several American cities, especially on the East Coast, in less than a minute. Khrushchev, however, was more interested in intimidation than causing a nuclear exchange. Brinksmanship became a tool of the lesser superpower, but it was a dangerous ploy to be sure.[67]

The larger Cold War tensions between the two powers also brought into play (although how much is merely conjecture) the volatile situation in Berlin and the existence of US *Jupiter* missiles in Turkey aimed at the USSR.[68] Berlin as an immediate cause was unlikely since the building of the wall in its own strange way temporarily diffused conflict there. In the eyes of the Kremlin, Kennedy more than Khrushchev believed that Cuba and Berlin were intertwined. The *Jupiter* missiles in Turkey were obsolete, and they primarily served as an irritant and a wound to Soviet prestige rather than posing a real threat. Taken as a whole, Khrushchev's actions were based on perception not delusion, borne of the momentum of the Cold War in the 1960s. However, he had not thought through his plan very carefully (nor developed alternatives) and misjudged the response of the Americans.[69] Kennedy and the United States were not blameless in leading Khrushchev to his decision. The Bay of Pigs, *Operation Mongoose*, the Berlin crisis, rebuke of Soviet actions in the Third World beyond Cuba, the words and actions of the president's political opponents, and the American arms buildup all contributed to the belief among Soviet leaders that the Americans challenged (or even thwarted) the USSR's interests on multiple continents.[70]

Both sides had only recently deployed ICBMs and other ballistic missiles to deliver nuclear warheads. Whether pipedream or no, the talk of a missile gap was constantly in the air.[71] How such weapons would influence the military strategies of both countries was unclear. Any new missile installation, or any new missile-delivery device such as a submarine, was a cause for concern. But unlike some of the dramatic retellings of the crisis, the placement of missiles in Cuba *per se* was not news to the United States in October 1962. What was shocking was that offensive weapons had been added to the defensive arsenal already there. *Operation Anadyr* was the code name for the Soviets plan to deploy offensive missiles (mid-range, intermediate-range, and cruise missiles armed with nuclear warheads), bombers, and additional troops in Cuba in fulfillment of Khrushchev's dramatic scheme.[72] As early as summer 1962, the CIA and others reported that a heavy volume of Soviet ships in the Baltic and Black Seas were heading for Cuba, at least 10 of them carrying military equipment. In early August, CIA director McCone raised the possibility that the Soviets were transporting not just antiaircraft missiles, but nuclear weapons. Despite the president's distrust of CIA reports growing out of his Bay of Pigs experience, he instructed the NSC to study the impact of Cuba installing defensive as opposed to offensive missiles. On August 29 a U-2 surveillance airplane confirmed the construction of a surface-to-air missile network with anti-aircraft weapons having a 30-mile range. Also, the Soviets provided *Komar* patrol boats with short range surface-to-surface missiles to guard Cuban waters, as well as coastal cruise missile sites (missiles ultimately discovered to be tipped with nuclear warheads) to protect against amphibious landings. What the Americans did not know was that Khrushchev had sent the antiaircraft equipment to discourage US surveillance and to distract attention from the offensive missiles to come.[73]

Various Soviet statements tried to assure the American leadership that the military buildup in Cuba was strictly defensive in nature. Since May or June, however, Khrushchev and other Soviet leaders had agreed to send to Cuba (despite the wariness of Castro) 24 medium-range *R-12* missiles, with a range of 1,050 miles, and 16 intermediate *R-14* missiles, with a range of 2,100 miles. *ILB-28* bombers, 44,000 troops, and 1,300 civilian workers supported the deployment of the missiles. This plan doubled the number of Soviet missiles capable of directly attacking the US mainland.[74] The strategy called for having the missiles operational by mid-October. Khrushchev gamble was predicated on the notion that the Kennedy administration would not suspect him of sending offensive nuclear weapons into the Caribbean. He planned to make public his actions after the American mid-term elections in November to trumpet Soviet superiority. He ultimately realized that the initial operational goal might take too long, and authorized his military leaders to consider sending tactical (or battlefield) nuclear weapons as a temporary solution, which they decided to do. Such tactical weapons would be placed in the hands of local commanders, who could be given authority to use them as needed against approaching ships and invasion forces. Khrushchev withheld that authorization for the time being. The first shipment of nuclear warheads from the USSR reached Cuba on October 4. The freighter *Indigirka* carried 45 one-megaton warheads for the medium-range ballistic missiles (MRBMs), 12 2-kiloton warheads for *Luna* short-range missiles, six 12-kiloton bombs for the *Il-28* bombers, and 36 12-kiloton warheads for the cruise missiles. This was amazingly large firepower. Cuban leaders were thrilled with the Soviet commitment to them.[75]

Kennedy and his advisers were suspicious of Soviet actions, and also were under stout pressure from Republicans, some Democratic leaders, and elements in the press to take a stronger stance on Cuba. In addition, Castro claimed that an invasion was looming. On September 4, the president released a statement noting the presence of antiaircraft missiles, torpedo boats, and 3,500 Soviet support technicians in Cuba. He warned the Soviets about deploying offensive weapons. On September 6 the president learned that 9 *SA-2* SAM (surface-to-air missile) sites were under construction. On September 7 he requested stand-by authority from Congress to call up 150,000 army reserves (which got bogged down in the House), and announced amphibious operations near Cuba in mid-October. The following day, the CIA reported three more SAM sites were added to the network. Kennedy's actions were followed by more assurances from Khrushchev in late September. Some in Congress called for a much tougher stance against the prospect of offensive weapons in Cuba.[76]

JFK continued to show restraint, aware of the political ramifications (especially during an election year) of publicizing unsubstantiated rumors about the presence of offensive weapons. He pressed Bobby to increase the activity of *Mongoose* in trying to upend Castro, but he was not going to incite problems over Cuba just to silence political opponents.[77] The Senate approved a resolution by a vote of 86 to 1 on September 20 stating that the United States by whatever means necessary, "including the use of arms" would keep Cuba from becoming a threat to American security. Republicans kept hounding Kennedy, while others called for a naval blockade. JFK accepted the Senate resolution, but he continued to believe that without offensive nuclear weapons in Cuba, the threat to American security was not grave. Moscow, therefore, believed that the American president would not respond aggressively to their actions.[78] Reports about possible MRBM sites at San Cristobal in western Cuba led administration leaders to call for additional U-2 flights over the island.[79] An October 14 U-2 flights provided incontrovertible evidence of offensive weapons. The camera recorded two MRBM sites in the San Cristobal area. (Subsequent flights discovered four additional MRBM sites and three intermediate-range ballistic missile sites under construction.) NSC Adviser McGeorge Bundy received the information on the two initial sites on the evening of October 15, and he presented it to a stunned president the next morning.[80]

The only issue on the table was to get the missiles out of Cuba, if possible before they were operational. This latter concern remained a sticking point, since none of Kennedy's advisers were in any position to state with confidence when the missiles would be ready to fire. The administration did not know for a fact that nuclear warheads themselves had reached Cuba by that point, but they worked under the assumption that the weapons were on the island but not yet usable. Soviet sources later revealed that 20 nuclear warheads had arrived in Cuba in September or early October, and another 20 were on the way by the time of the crisis. The warheads *in situ* were for MRBMs.[81] The same sources claimed that during the 13-day crisis it would have taken only about four hours to target and launch the missiles in place.[82] The IRBMs (Soviet *R-14s*) were en route to Cuba when the U-2 spy planes discovered the intermediate missiles, but they did not reach their destination.[83]

Believing there was time to act, but precious little time, the president considered his options. Kennedy chose not to convene his cabinet, but brought together a select group of 14 advisers called the Executive Committee of the National Security Council (Ex Comm) who met regularly during the crisis. They came primarily from State, Defense, the Joint Chiefs, and the CIA, along with the Attorney General. The old Cold Warrior, Dean Acheson, was invited to some sessions.[84] Conducting foreign affairs by committee was not a typical practice for JFK, but ExComm gave him access to necessary opinions immediately, especially since time clearly appeared to be of the essence. JFK wanted the advisers to explain why Khrushchev decided to take the action he did, to help him frame his own response. "What is the advantage?" the president queried. "Must be some major reason for the Soviets to set this up."[85] ExComm held several formal meetings to take up policy options, some with him and some without him. For security sake, Kennedy appeared to follow his normal routine. In some instances, it was better for him to be out of the room in order to solicit frank comments from the group. The first order of business was to decide on how, not if, the United States would eliminate the missiles from Cuba. Military considerations immediately trumped politics or diplomacy.[86]

JFK posed four options: a surgical air strike against the missile installations, a more general air strike, invasion, and a naval blockade. (Later someone floated the option of mining of Havana harbor, but that was quickly dismissed.) The president had to decide upon a course of actions that would remove the threat (linked to when the missiles might become operational), but not start a war. Each plan had potential strengths, but also some clear liabilities. A surgical strike would not guarantee 100 percent success, especially if there were unknown missile sites. A general air strike and/or an invasion would court further conflict or could trigger a nuclear exchange. By international law, a blockade was an act of war and also did not guarantee that the Soviet Union would honor it. Opinions varied widely and several ExComm members vacillated from one approach to another. McNamara raised the question as to whether offensive weapons in Cuba actually threatened the United States. The group quickly dismissed this query because even if the military risk was low, the political implications were quite high. With evidence of additional missile sites in place on October 18, the Joint Chiefs were even more adamant that a full-scale invasion was necessary.[87] Characteristically General LeMay, interpreting Kennedy's caution as weakness, took the most extreme view, pushing hard for an all-out effort in Cuba. He believed the Soviets would back down, not try to use American action as an excuse to take Berlin, and would avoid full-scale nuclear conflict. The general told the president in a rather condescending way, "You're in a pretty bad fix" to which Kennedy responded, "You're in there with me." [88]

On October 18 Kennedy was leaning toward a blockade and considering ways to open talks with Khrushchev. He, however, had not given up on the possibility that military action ultimately would be required. On October 20, JFK ordered the navy to place a defensive "quarantine" (a term that seemed more tempered than "blockade") around the island nation as soon as possible. The interdiction line included 16 destroyers, 3 cruisers, an aircraft carrier, and more than 150 ships in reserve. The president planned to address the American people on Monday, October 22, announcing that the US Navy would turn back Soviet vessels carrying offensive weapons and material related to deploying such weapons. Kennedy received immediate backing from his international allies and got a unanimous vote from the Organization of American States. Support was not forthcoming from congressional leaders, who viewed the president's response as flimsy. He nevertheless went forward with his address as planned.[89] He concluded his speech by saying, "Our goal is not the victory of might, but the vindication of right—not peace at the expense of freedom, but both peace *and* freedom, here in this hemisphere, and, we hope, around the world. God willing, that goal will be achieved."[90]

What US officials did not know at the time (although the CIA had speculated about it) was that MRBMs with warheads were in Cuba and almost operational. Khrushchev hoped that at this point Kennedy would back down. But Tuesday morning came with no reply from the premier. JFK and his brother wanted to get word to the Soviet leader through back channels about solutions short of direct confrontation, including the possibility of dismantling the *Jupiter* missiles in Turkey in exchange for the removal of offensive missiles in Cuba. Time was running short. On October 24 American forces, which were currently on Defense Condition 3 (DEFCON 3), were put on DEFCON 2, which was only one stage lower than a state of readiness for general war. The United States had never reached this level of alert during Cold War. SAC fitted bombers with almost 3,000 nuclear weapons, and the navy deployed 112 *Polaris SLBMs* (almost the entire US strategic force). The Soviet vessels had yet to show any signs of turning back and were being screened by Soviet submarines.[91]

Posturing between the superpowers was making it appear that the quarantine as a buffer against more serious actions had become a thin veneer. On October 24, however, some ships began to reverse course. Two days later Kennedy received a long, somewhat disjointed, emotive letter from Khrushchev. It did not admit that the missiles in Cuba were offensive in nature, nor did it make specific promises or commitments for the USSR and Castro. But in essence it maintained that if the United States agreed not to invade Cuba, Khrushchev saw no need for missiles there. "I think that one could rapidly eliminate the conflict and normalize the situation," he stated. "The people would heave a sigh of relief, considering that the statesmen who bear the responsibility have sober minds, an awareness of their responsibility, and an ability to solve complicated problems and not allow matters to slide to the disaster of war."[92] US officials believed that the letter was probably impulsive and written without input from other Soviet leaders. Apparently, Khrushchev had not developed any contingency plan for use if the Americans reacted to his gambit with aggression of its own. Beyond the letter, the premier provided more concrete assurances to Kennedy through an intermediary, KGB official Aleksandr Fomin, who passed along information to American journalist John Scali. Just when it looked as if the ice was breaking, a second tougher letter arrived from the USSR the following morning (also broadcast over Radio Moscow). It was clearly a formal letter, one produced with more care, which called for the removal of missiles in Turkey in exchange for those in Cuba. Was Khrushchev being pressured to harden his line? Was he being deposed and thus forced to make changes in his stance? Or was he now less panicky, less willing to believe that an invasion of Cuba was imminent? His American counterparts were in no position to know his motives, but the new letter clearly turned up the heat. A U-2 aircraft shot down over Cuba exacerbated tensions, which particularly incensed US military advisers now wanting retribution. Castro, himself, was communicating with Moscow. He later stated that he feared an invasion was coming and wanted to remind Khrushchev to remain assertive and not let the Americans strike first. (It is unclear, however, if Castro was calling for a nuclear attack on the United States or simply trying to buck up Khrushchev's resolve.) [93]

As a back channel approach, Robert Kennedy had a private meeting (known only to the president and a few others) with Soviet Ambassador Anatoly Dobrynin. RFK told Dobrynin that at some future date the United States would remove its missiles from Turkey, but would make no formal or public pledge to do so, in exchange for removal of the offensive missiles from Cuba. At the ExComm meeting on "Black Saturday," October 27, the trade was the central topic. President Kennedy appeared to be the strongest proponent, with many of the others unaware of Bobby's meeting. At this late date building consensus in ExComm still was difficult. Bobby later recalled that "There were sharp disagreements. Everyone was tense; some were already near exhaustion; all were weighted down with concern and worry."[94] But JFK then decided to publicly accept the conditions of the first letter, trading removal of the missiles from Cuba for a pledge of no invasion. This decision, along with the private assurance to Dobrynin, gave the president political clout at home for appearing decisive, but also relied on careful, nuanced diplomacy.[95] Tensions remained high as American leaders awaited Khrushchev's response. They pressed forward with plans for more direct action in Cuba, with US forces poised for such an eventuality. Accompanying private correspondence to Kennedy was a broadcast on Radio Moscow

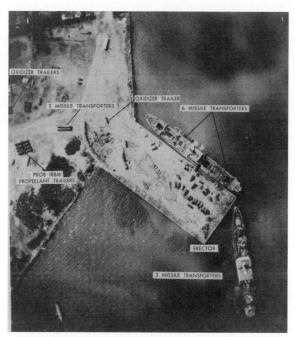

This 1962 aerial photograph of Mariel naval port in Cuba shows Soviet missile equipment being loaded onto the Soviet freighter *Anesov*. *Source:* Library of Congress, Prints & Photographs Division, NYWT&S Collection, [LC-USZ62-128473]

Soviet freighter *Anesov* being escorted by a Navy plane and the destroyer *USS Barry*, while it leaves Cuba, probably loaded with missiles under the canvas cover seen on deck during the Cuban Missile Crisis. *Source:* Library of Congress, Prints & Photographs Division, NYWT&S Collection, [LC-USZ62-128471]

at 9:00 a.m. Sunday, October 28, in which Khrushchev accepted the American conditions. The immediate danger of war was over.[96]

A period of tedious negotiations followed, as the sides tried to determine how to implement the terms of Soviet withdrawal of its offensive weapons from Cuba. Castro resisted UN supervision of the missile withdrawals, argued against removal of the bombers, and contested American air reconnaissance to verify packing and shipping the missiles. Khrushchev worked to get concessions from the Americans on the bombers and other weapons such as the torpedo boats. The naval quarantine continued until US officials were satisfied that the agreement between the countries had been completed. Kennedy backed off a little on verification, but he insisted that the bombers and Soviet forces had to go. On November 20 the United States terminated the quarantine, and by early 1963 the USSR evacuated most of their troops (some remained until 1979). Kennedy, however, modified his no-invasion pledge by tightening the conditions. Beyond the agreement that no more offensive weapons would be based in Cuba and that verification was assured, there could be no attempt by Cuba to "export revolution." This left the door ajar for possible future American action.[97]

In the aftermath, Kennedy received high praise from a grateful nation for his handling of the crisis, even among Republican critics. The president's political position at home strengthened, reinforced by an atypical increase in Democratic seats in the Senate during the November mid-term elections. Was this a real victory for Kennedy and a possible first step toward détente with the Soviet Union? Since Khrushchev had backed down, the immediate results of the crisis seemed obvious. Hard-liners like Acheson and several military leaders felt that the United States should have pressed its advantage. LeMay expressed the most extreme view with his dire assessment that the United States had "lost," since the president failed to finish the job in Cuba and did little to weaken the Soviets. In retrospect, both sides came very close to losing control of events that could have plunged them into nuclear war. Little decisions mattered, and constraint where there could have been action probably saved the day.[98]

Although sobered by the experience, both Kennedy and Khrushchev boasted of success. For the latter, especially, this claim rang hollow. The USSR's relations with China and Cuba were damaged. Khrushchev was angered that the Chinese especially had contributed little to the stand-off in Cuba, and that Castro was not easily consoled. Hanging on for a time, Khrushchev was ousted from power in 1964 by Soviet rivals. That he survived for so long after the crisis was impressive. Khrushchev had secured two objectives with his gamble: the promise that Americans would not invade Cuba and the removal of the *Jupiters* from Turkey. Whether the humiliation of backing down was worth the extraordinary risk is debatable. In reality Khrushchev could not be certain that the United States would honor its pledge about Cuba or that the *Jupiters* would have remained if there was not a back channel deal.[99] The Cuban Missile Crisis also had implications beyond the superpower face-off. The possibility of the Cold War turning hot was a frightening prospect. One estimate suggested that as many as 10 million Americans fled their homes after the president's public address. Many listened to the radio or watched television worrying and wondering if and when a nuclear attack might actually occur. The civil defense program, such as it was, seemed to be inadequate to meet the pending conflagration. A sense of vulnerability never felt before became commonplace among American citizens.[100]

For the policy makers on both sides, time had to be the greatest enemy. In an era of ballistic missiles careful deliberation took a backseat to quick judgment, which was a frightening revelation. Beyond the superpowers, the impact of the October Crisis was a mixed legacy. Fidel Castro was enraged with his Soviet allies for exposing his country to such a grave risk and then backing away. He was angry that Khrushchev had not consulted with him before agreeing to withdraw the missiles, having heard about it on the radio like everyone else. Both superpowers treated Cuba like a pawn and not a sovereign nation. Ultimately, the Castro regime became more deeply entrenched, despite efforts by the United States to continue covert *Mongoose* operations.[101] NATO countries sensed that the United States was willing to act alone in security issues, which did not speak well for the alliance. The Chinese, on the other hand, came to believe that the crisis provided opportunities for them in the Third World, since the Soviet Union could not be trusted to stand by its commitments.[102] And the nuclear arms race went on. There was movement toward a limited test ban treaty that was the first promising sign that the two powers would take up the issue of arms control more seriously. The Soviets understood that the US nuclear arsenal outmatched theirs. The missile gap existed, but was exactly opposite what Kennedy believed at the start of his term. By 1965 the USSR undertook a massive arms buildup, and despite the fact that attention turned away from Cuba and Berlin (to places like Indochina for the United States and Afghanistan for the USSR) the American arms program proceeded apace. Weapons technology also changed the complexion of nuclear war and made the installation of missiles in Cuba an obsolete way to close the nuclear gap. A new era of nuclear submarines in particular added deadly mobility to positioning weapons of mass destruction. For the United States, the *Minuteman* and *Polaris* programs were extremely important. In the end, there was no one lesson coming out of the October Crisis. Leaders changed and circumstances fluctuated, but the possibility of mutual annihilation was temporarily put to rest.[103]

GOING UNDERGROUND: THE BOMB SHELTER CRAZE IN THE 1960S

The intensity of the Cold War in the early 1960s reinvigorated the dialogue about civil defense. At the center was a controversy over fallout (or bomb) shelters. The "flashpoint" for the national debate was Kennedy's speech of July 25, 1961, concerning Khrushchev's threat to make a separate peace with East Germany and to declare Berlin a neutral city. In lieu of the rising tensions over Berlin, and to accentuate the seriousness with which he took the warnings, the president requested an increase of $3.24 billion in the military budget and another $207 million for civil defense. The latter amount was to be used especially to identify public and private space to be used for fallout shelters "in case of attack."[104] Today fallout shelters are quaint artifacts of the bygone Cold War era. Kennedy's desire to commit major resources to them highlighted the authentic anxieties of the time. There is little doubt that Americans struggled to determine how and if it was feasible to protect themselves from the unthinkable. The possibility of total war could be coming to the United States, just as it did in Japan less than two decades before.[105]

The argument over fallout shelters in the United States can be traced to the Truman years. There was no clear-cut civil defense policy in place after World War II. Some pushed for evacuation (see Chapter 5), while others argued for shelters. Since initial concerns focused on the explosion, several planners sought ways to make buildings blast-proof. This proved too expensive for the government to implement. Public shelters also were costly, but study continued in various governmental agencies on windowless buildings and on the conversion of basements and garages to shelters. Government-sponsored technicians drew plans for a White House shelter. At Los Alamos, workers retrofitted 40 buildings with underground facilities to accommodate 17,000 people in case of attack. The law creating the Federal Civil Defense Administration (FCDA) included a recommendation for a modestly funded shelter program, but no money from Congress was forthcoming. Some municipalities saw the development of shelters as a means to spur urban redevelopment, which blended civil defense with community improvements.[106]

The hydrogen bomb changed the thinking about civil defense. Not only were such weapons of monumentally grander scale and potency, but now many viewed the issue of fallout as the overriding threat to surviving a thermonuclear assault. If you could live through the devastation of such "city killers" (a colossal "if") the fallout not the blast would prove to be the greater risk. When the Eisenhower administration took up the issue of defense against the H-bomb, it turned to evacuation as the most realistic approach. But Eisenhower was not persuaded by the Gaither Report's call for a major fallout-shelter program, especially the idea of spending public funds on the shelters. Other measures of civil defense undertaken by the government were intended to pacify Americans, not spare them from nuclear attack. Characteristically Eisenhower favored a private approach. Citizens had primary responsibility for their own protection, he argued. In May 1958 the president issued the National Shelter Policy, emphasizing homeowner-built shelters. A new Office of Civil and Defense Mobilization (OCDM) was in charge of stockpiling materials and coordinating state and local planning, and an Office of Civil Defense (OCD, under the Department of Defense) directed the fallout shelter program. The federal government provided modest funding support for state projects and offered leadership in civil defense, but did not underwrite the cost of private shelters. Instead government employees distributed do-it-yourself publications such as *Family Fallout Shelter* and *How It Was Done*. These activities piqued public interest in shelters, and raised interest about the best type to build. Popular magazines such as *Good Housekeeping* and *Life* publicized on-going projects and sometimes promoted the need for shelters. Not everyone was quick to extol their value, but neither the government nor the press publicly aired the concerns and skepticism. People could surmise that there might be little escape if all-out nuclear war occurred.[107]

Because of the widespread public discussion of civil defense and shelters in the 1950s, American citizens were not wholly unprepared for Kennedy's pronouncements in the early 1960s. The congressional allocation of more than $200 million for fallout shelters in 1961 far exceeded the financial commitment of the previous administration. In January 1962 JFK proposed a five-year shelter program with the ultimate goal of protecting all citizens. Funding such a project was prohibitive, and the new Democratic leaders began to back off from this unrealistic objective. The Kennedy civil defense program also faced public opposition. Scientists, intellectuals, peace activists, and anti-nuclear protesters placed little stock in the ability of fallout shelters

View of the "Sleeping Quarters" inside a 100-man Fallout Shelter at the Naval Air Station in Churchill County, NV. *Source:* Library of Congress, Prints & Photographs Division, [HABS NEV,1-FALL,2A—13]

or other civil defense measures to protect the public from thermonuclear weapons, and lobbied hard against the bombs. Reducing or eliminating those weapons, they argued, would be a sounder course.[108] The Shelter Incentive Plan (meant to build community shelters and provide funding to non-profit organizations) gained few supporters in Congress, with only modest financial support. The decision of Senator Henry Jackson (Dem., Washington) not to hold Senate hearings on the plan effectively killed it in 1964. In several respects public-sponsored civil defense (particularly fallout shelters) was linked to debates over the anti-ballistic missile (ABM) program. Those supporting the defense system clearly saw the need for complementary shelters. Those with little faith in the ABM program did not support them.[109]

Like its predecessor, the Kennedy administration tried to encourage private initiative through such Office of Civil Defense booklets as *Fallout Protection: What to Know and Do about Nuclear Attack*. Public reaction to the call for fallout shelters in the early 1960s remained high for some, but not all. The shelters themselves took many forms, including new construction in backyards and basements, or stocking food and other supplies in existing storm cellars and windowless buildings. The OCD also designated public spaces as shelters. Smelling a profit, new companies took advantage of the situation, such as Acme Bomb and Fallout Shelters Company and Peace-O-Mind Shelter Co. in Texas and the Atlas Bomb Shelter Company and Nuclear Survival Corp. in California. Banks saw an opportunity to make money by offering loans to homeowners for building shelters.[110] Private fallout shelters built by homeowners also had deep cultural significance. Since the 1950s civil defense approaches focused markedly on the security of the home. In the post-World War II world, as historian Elaine Tyler May argued, "The home seemed to offer a secure, private nest removed from the dangers of the outside world." She added, "The self-contained home held out the promise of security in an insecure world."[111] Such a sense of security was restricted largely to the American property owner, not to renters or tenants. The private shelter corresponded well with an American free market perspective as opposed to

Soviet collectivism. Even though the idea of a privately built and privately controlled shelter had appeal, many still believed that civil defense *per se* was a government responsibility. Home shelters were never a complete answer.[112] The FCDA's Jean Wood Fuller developed a campaign that was intended to appeal to middle-class Americans. With the help of the National Grocer's Association, the American National Dietetic Association, and pharmaceutical companies, she promoted "Grandma's Pantry." Relying on nostalgia and traditional (albeit overly simplified) ideals, the resulting brochure stressed the need to be prepared through designing a well-stocked kitchen for the family shelter. To enhance self-sufficiency, the brochure also called for adding a first-aid kit, a flashlight, and a portable radio to the necessary supplies. Implicitly such an appeal was meant to help insure survival of a family, but also to perpetuate traditional family values and gender roles.[113]

Not everyone shared the view that fallout shelters were the answer to personal security in the atomic age. Much of the public actually rejected the media hype about them, despite stories to the contrary.[114] Celebrated anthropologist Margaret Mead argued that the embrace of bomb shelters was morally disturbing and terribly self-centered. In practice, no real nationwide construction boom in fallout shelters ever occurred. Some very uncomfortable issues, however, arose as a result of the debate. Private investment in shelters clearly favored the middle and upper classes. Many of those with shelters or planning to build them were caught in arguments over who should be allowed to benefit and who might be excluded. In some cases, community shelters were racially segregated, mirroring the contemporary social norms, especially in the Jim Crow South. A number of people came to believe that fire arms, as much as food and water, were necessary supplies for keeping unwanted guests out of shelters and for dealing with the post-apocalyptic world. Some stretched ethical and moral considerations to the limit: kill or be killed, scare off intruders so your family can survive. The lessening of international tensions after the end of the Berlin and Cuban crises and more realistic appraisal of fallout shelters led to diminishing interest in them by the mid-1960s. During the short period of the bomb shelter craze, people built only about 200,000 shelters in the United States. In many respects, the shelters were a desperate attempt to find relief from the horrors of possible nuclear annihilation, rather than a practical means of self-preservation.[115]

AFTERMATH: THE LIMITED TEST BAN TREATY

Since the early 1950s both the United States and the Soviet Union discussed a test ban (and even arms control), but pronouncements often were little more than good propaganda. Internationally protest after protest called for a ban, and some groups in the anti-nuclear movement received a positive hearing in several nations. Celebrated anti-nuclear activist Norman Cousins played an important role in exchanging information on a proposed ban between Khrushchev and Kennedy in 1962–1963.[116] In the wake of the Berlin and Cuba crises, the arms race was spinning out of control, or at the least very soon could be. There was growing concern about the cost, in dollars and rubles, as well as stress; disquiet over nuclear proliferation; and alarm over the return to the brink. (Few gave attention at the time to what the tests were doing to poison the environment—an issue that awaited future public discussion.) There had been some hope for the voluntary cessation of testing in 1958, but its resumption by the USSR in September 1961 and the United States in March 1962 defeated that hope.

Sustained opposition to a test ban from within the governments of both countries made matters even more difficult. On the US side, the Joint Chiefs railed about national security. Democrats were unenthusiastic about a comprehensive treaty for fear that the United States would get nothing in return. The Republican right and a variety of conservative activists criticized Secretary McNamara for wanting to use the ban to reduce production of nuclear weapons and to divert funds to conventional military forces. They questioned McNamara's idea of Mutual Assured Destruction (MAD), which called for nuclear balance between the superpowers (rather than nuclear primacy) as the best way to avoid confrontation.[117] Such an idea, critics argued, came at a time when the Soviets were bent on accelerating their arms program. In the USSR, Khrushchev faced the pressure of his hard liners who wanted to move toward parity with the Americans.[118]

Troubled by the Cuban Missile Crisis and pushed to move their countries beyond tensions of October 1962, both Kennedy and Khrushchev earnestly wanted some form of test ban. Both hoped to decrease the chance for war. Both wanted to curb spiraling military costs. And both saw China on the horizon to join the nuclear club. In early 1963 negotiations were underway over a test ban treaty, but a wide gap existed between the parties regarding a verification program and other issues. Khrushchev was probably less concerned about the espionage that might result from on-site inspections than about exposing the limitations of the Soviet nuclear buildup. Kennedy's stirring commencement address at American University on June 10 helped to break the stalemate. He called for a test ban by the United States, the Soviet Union, and Great Britain, declaring that

> While we proceed to safeguard our national interests, let us also safeguard human interests. And the elimination of war and arms is clearly in the interest of both. No treaty, however much it may be to the advantage of all, however tightly it may be worded, can provide absolute security against the risks of deception and evasion. But it can—if it is sufficiently effective in its enforcement and if it is sufficiently in the interests of its signers—offer far more security and far fewer risks than an unabated, uncontrolled, unpredictable arms race.[119]

The speech had a profound effect on Khrushchev, who viewed it as conciliatory and a genuine indicator of the president's commitment to a treaty.[120]

Some days after the address, the two powers struck an agreement to install a "hot line" to provide quicker and more direct communication during a crisis. Their continual clashes over the years had created a somewhat begrudging bond between the two leaders that was at once political and personal. The realization that little things mattered when pushed to the brink made the hot line most welcomed. After the American University speech, Kennedy fulfilled a promise to visit Europe in hopes of building support for negotiations over the test ban. The British had been active in the earlier negotiations with the Soviets, and obviously had a stake in any new treaty. Kennedy avoided DeGaulle and Paris because of the streak of independence that the French had shown in NATO affairs, and made only a brief stop in England. JFK also visited Ireland, Italy, and Germany. In Berlin he made his famous announcement, "Ich bin ein Berliner!" in support of America's West German allies. The Kennedy of this triumphal tour appeared to come of age on the international scene, with more advantage to pursue an accommodation with the USSR.[121]

Negotiators on a proposed treaty met in Moscow beginning on July 15. On October 7, Kennedy signed the Partial Limited Test Ban Treaty (PLBT), which was agreed upon by the United States, the USSR, and Great Britain. It prohibited atmospheric, outer space, and underwater testing of nuclear weapons. Eventually over 100 countries signed it. The treaty was a compromise. Kennedy would have preferred a complete ban on testing and help from the Soviets in curbing China's nuclear ambitions. Khrushchev wanted a non-aggression treaty between NATO and the Warsaw Pact. The sides settled for what they could get. Opposition from the Joint Chiefs and Republicans in the Senate was weak since the treaty did not end testing per se, and did not address the use of the weapons themselves. Kennedy aggressively lobbied the Senate for passage, effectively confronted air force criticism, sought public support, and held off Edward Teller, who believed the treaty would interfere with the development of an anti-ballistic missile system he envisioned.[122] Aside from its symbolic importance and modest impact on testing, the treaty did little to influence weapons technology, end the arms race, or curb proliferation. Nevertheless, given the intensity of their rivalry, the United States and the Soviet Union were at the least reflexive point in their mutual histories since the earlier days of the Cold War. Some would argue that the Soviet Union was at a greater state of accommodation with the United States at the moment than it was with China. Because of Kennedy's tragic death just a few months after the signing of the treaty, and Khrushchev's ouster in 1964, there is no way to know if the new relationship between the two leaders would have blossomed into anything more than it did in the immediate post-Cuban Missile Crisis period.[123]

DID WE LEARN TO STOP WORRYING AND LOVE THE BOMB?

The early- to mid-1960s saw the release of several films that picked up on central themes in the Cold War and the nuclear arms race. In some respects, the narrative of the unnerving events in this period is clearly enhanced by images, albeit fictional, that get to the heart of the anxiety (and in some cases the absurdity) of the relentless competition between the superpowers. Many films of the time or those attempting to depict the period in more recent movies, used key events in the early 1960s simply as backdrop or broad parody. The era continued the trend of irradiated-monster movies, possibly none as bad as *The Beast of Yucca Flats* (1961). In the film a defecting Soviet scientist received serious radiation exposure because of a nuclear explosion at the Nevada Test Site and roams the area as a fiend. Movies with apocalyptic themes became new stock and trade. *Panic in Year Zero!* (1962) played on fears of lawlessness and chaos after a nuclear attack on Los Angeles and elsewhere. Harry Baldwin (Ray Milland) is bent on protecting his wife and two teenage children at gun point from looting, rape, and hooliganism in an environment where survival is everything. Some of the talk surrounding the necessity of keeping guns in the family fallout shelter seems to connect fiction and reality here.[124] George Pal's *The Time Machine* (1960) turns H.G. Wells's 1895 book into a cautionary tale about a future transformed by perpetual wars and nuclear holocaust. Although the film lapses into a love story between the time machine's inventor and a girl from the distant future, it raises questions about the caprice of technology (be it an atomic bomb or a time machine) and the uncertain consequences of actions too closely associated with mistakes of the past.[125] *The Day the Earth Caught Fire* (1962) is more ambivalent about the end of the world than portrayed in other apocalyptic films, but strives to answer "What if?" (although rather implausibly) about the excesses of the arms race. In the movie, London faces wild fluctuations in unseasonable weather. Newspapermen discover that simultaneous atomic tests at the North and South Poles have knocked the Earth off its axis, thus accounting for the furious winds, rain, and hurricanes. Unlike *On the Beach* (1959), in which the hapless Australians resign themselves to their fate, the Londoners panic, try to escape, or throw wild orgies. The only possible solution to the pending end of the world is to use nuclear weapons to bring the earth back into alignment. The movie ends before we know if it worked. There is no little irony in using the same bombs to save the earth as those that came close to destroying it.[126]

Some more recent films put up to derision experiences and social behaviors with respect to the bomb and the arms race of the early 1960s. Nothing is more iconic and easy to lampoon than the fallout shelter craze. The superb satirical documentary, *The Atomic Café* (1982) takes on government propaganda about the bomb and civil defense, the immersion of the culture in all things atomic, and (in a more serious vein) the human consequences of atomic warfare in Hiroshima and Nagasaki. The film is quick to ridicule and trivialize the rush to build the shelters and to stock them with necessary provisions to survive a missile attack. People are portrayed as naïve, susceptible to hucksterism or salesmanship, and too easily duped into believing that fallout shelters actually could save their lives. The reality of the fear and anxiety that the arms race engendered is lost here, and some necessary balance is missing. *Blast from the Past* (1999) is not as bent on lampooning the early 1960s as it is using the fallout shelter as a device to establish its plot line. Adam (Brendan Fraser) has spent his whole 35 years growing up in an elaborate fallout shelter, and is innocent and unsophisticated. Once let loose in the world, he is unprepared for the changes. Not surprisingly he meets Eve (Alicia Silverstone) and together they start a new life for themselves, and for Adam's long-sheltered parents. The film plays not only on Adam's awakening to the new world, but on nostalgia that in many respects (conversely) is truly cynical about the corruptibility of the present. In this instance, the fallout shelter is a cocoon, a safe haven, and thus taken totally out of context.[127]

Probably the most original contributions to atomic films in the early- to mid-1960s were dramas rich in the plausibility of something going terribly wrong in the Cold War, especially an event that could trigger nuclear holocaust. There was much historical fodder in the period, and some filmmakers made the most of it. *The War Game* (1965) is frighteningly realistic. The BBC funded the documentary, but never aired it. Executives at the communications giant regarded the film as "too horrifying" for television. As a theatrical release, it won an Oscar for Best Documentary Feature. County Kent was the setting, and the director shot *The War Game* in newsreel style. Graphic by the standards of the day, it depicts the immediate aftermath of a nuclear attack replete with the chaos that such an event would cause. The imposition of ferocious martial law makes for a strong political statement, probably as abhorrent to the BBC as the violence.[128] *Seven Days in May* (1964) is more "Hollywood," with big-name stars like Burt Lancaster, Kirk Douglas, and Frederic March, but unnerving in its own way. Director John Frankenheimer portrays a military coup that almost succeeds. Heroic, but tilted, General James Mattoon Scott (Lancaster) and his cohorts are adamantly opposed to a disarmament agreement that President Jordan Lyman (March) has signed. They regard such dovish behavior in the midst of the Cold War as unacceptable and threatening to American security. Through the efforts of Scott's former friend, Colonel Martin "Jiggs" Casey (Douglas), the plot is exposed and the coup thwarted. The central debate over disarmament is a core issue of the era, making the film quite timely.[129] In the same year, *Fail-Safe* (1964) also opened in theaters. It was based on the successful book of the same name published in 1962 by Eugene Burdick and Harvey Wheeler. The story line is straight-forward and chilling: An electronic-device malfunction leads to SAC's Group Six of bombers passing their fail-safe points and heading for Moscow. The real hawk in the story is the scientist, Professor Groeteschele (Walter Matthau), who advises the president to press his advantage. Unable to stop the attack, even with Soviet help, the president (Henry Fonda) decides to trade the destruction of New York City for Moscow as a way to avoid World War III. The movie was remade for live television (2000), but without the same eerie and unsettling power.[130]

The consummate atomic age film of the 1960s (and well beyond it) is *Dr. Strangelove or: How I Learned to Stopped Worrying and Love the Bomb* (1964). Much has been written about Stanley Kubrick's superb black comedy. We know that Kubrick initially wanted to adapt to the screen Peter George's 1958 novel about nuclear war, *Red Alert*. Instead, with George and Terry Southern, he produced a screenplay that spares no one and no group from derision and satire.

Dr. Strangelove is an indictment of governmental and military institutions complicit in the military-industrial complex and the arms race. It also is a denunciation of the mentality of the era which by accident or by intention could set off nuclear destruction of the world. The context in which thermonuclear weapons were built, stored, and readied for deployment could only be dealt with, Kubrick surmised, within the theater of the absurd. Few if any could get their minds around the use of such weapons in a rational world. The plot line runs parallel with *Fail-Safe*. In this case, a deranged SAC commander, Brigadier General Jack Ripper (Sterling Hayden), makes sure a squadron of his B-52s goes past the fail-safe point to bomb the USSR. He is convinced that a first strike against the Soviets is the only way for the United States to escape the impasse of the Cold War. The remainder of the film focuses on a lone B-52 that is not shot down nor successfully called back, and is heading directly to its target. The efforts of the inept, utterly bureaucratic President Merkin Muffley (Peter Sellers) and his advisers in the war room to capture the deranged Ripper and stop the attack prove futile. Most of the other main characters also represent the flawed elements in the Cold War era, which have led to this crisis. The inept, utterly bureaucratic president; General "Buck" Turgidson (George C. Scott) is a LeMay-like adviser, although Ripper also is channeling some LeMay characteristics. The Cold War-hardened Soviet ambassador Alexi de Sadesky is played by Peter Bull, and a drunk Premier Kisoff is not shown on screen. Major "King" Kong and his racially and religiously balanced crew dutifully fly their B-52 into certain peril. Dr. Strangelove himself (Peter Sellers again) is the maniacal ex-Nazi scientist central to the American bomb program.[131]

Vital to Strangelove's portrayal is a boiled-down version of nuclear-war dogma under debate at the time by a cohort of nuclear planners. Nuclear planning itself developed in the 1950s. Individuals such as Bernard Brodie of the RAND Corporation (the BLAND Corporation in the film) raised issues about restrained responses to an enemy attack as opposed to all-out war, especially what was understood to be the inflexibility of MAD. In 1957 Professor Henry Kissinger of Harvard was among the first to argue that limited nuclear war was a legitimate alternative to total commitment of nuclear forces. But it was physicist Herman Kahn, with his wildly popular book, *On Thermonuclear War* (1960), who made nuclear war seem "logical, palatable and winnable." As several observers noted, Kahn also made nuclear war "thinkable." Frightening at its core, Kahn's analysis was taken seriously by a number of planners and policy makers. Individuals like Paul Nitze, who helped shape American defense policy and was involved in Ex Comm during the Cuban Missile Crisis, debated the limits of deterrence and other approaches to developing a viable nuclear strategy, if viability was actually possible.[132] The best known model of nuclear deterrence in the mid-1960s that challenged MAD was "nuclear utilization target selection" (derisively called by opponents, NUTS). This theory built upon the work of experts like Kahn, which emphasized the need to hit enemy military targets first, rather than cities, and thus make it possible to fight a limited nuclear war.[133]

Kahn changed his views occasionally, but had been deeply tied to first strike deterrence coupled with effective civil defense. He delighted in stirring the pot and even portraying himself as some kind of mad scientist, which was in stark contrast with his methodical use of systems analysis to present his ideas. Many found it difficult to separate what was rational or irrational from the arguments and what was indeed moral or immoral. Kahn's discussion of a "doomsday machine" that could be triggered automatically in case of an attack, put faith in technology rather than policy makers to make tough decisions about nuclear war. Such a device becomes a centerpiece in the fictional *Dr. Strangelove*.[134] In the film, the Soviets possessed such a doomsday machine, but had not yet revealed it publicly when Ripper executes his plan. With a B-52 poised to attack the Soviet Union, more is at stake than one Russian city as we see in *Fail-Safe*. Instead of two mega-cities being destroyed, the world will end as becomes obvious when Major Kong rides his H-bomb to its target. The strains of Vera Lynn's "We'll Meet Again" indicate that we will not. The apparent absurdity of the film's premise and the cartoonish portrayals of the principle characters detract little from the underlying theme. Using bombers to deliver nuclear warheads instead of missiles certainly belies the likely result (and limited time interval) of nuclear exchange in 1964. Yet the extension of time and tension that the much slower aircraft achieves makes the final scenes all the more excruciating. There is much that is all too real in *Dr. Strangelove*. It is a nightmare borne of warped logic to be sure, but not so far removed from the irrationality of the arms race itself.

CAN WE DISARM? THE STRATEGIC ARMS LIMITATIONS TALKS

In many respects, policy makers in the early 1960s made less time for contemplation and more time for crisis management. Such a disparity was part of the problem with this friction-driven era. "Risk" was the word of the day, but some hope emerged with the Limited Test Ban Treaty (LTBT), if but the faintest kind. By comparison, the remainder of the decade seemed to favor less direct tensions between the superpowers. Kennedy's assassination and Khrushchev's departure from government did not signal an abrupt break that the death of Stalin and the rise of Eisenhower appeared to promise in 1953. The strange dual track of weapons buildup and arms talks was center stage for the United States and the USSR. In June 1967 President Lyndon B. Johnson met with Soviet Premier Aleksey Kosygin in Glassboro, New Jersey, to discuss nuclear arms limitations, including restrictions on ABMs, but came away without any clear resolution to their differences. The United States still maintained a commanding lead in nuclear weapons in that year (2,400 delivery systems compared to the Soviet's 900), and an almost 5 to 1 lead in nuclear warheads (4,500 to 1,000). But the Soviets remained committed to closing the gap, and began doing so as early as 1964.[135]

More than JFK, Johnson relied on Secretary of Defense McNamara on questions related to nuclear weapons. With the Soviets catching up on ICBMs by the late 1960s, McNamara began to understand that relying on mutual assured destruction and damage limitation would be less feasible in the near future. It would be futile to assume that staying ahead of the USSR (at least in the number of missile launchers or warheads) was a realistic approach. Nor was it realistic to build up a counterforce capable of responding to a Soviet first attack, since it would be ineffective and would increase superpower tensions unnecessarily. Allowing the Soviets to have a rough parity in missile launchers was economically and strategically practical. Making the American nuclear inventory more potent was not out of the question either. In addition, technology had a way of complicating the arms race. In 1964 the United States began to equip *Polaris* A-3 SLBMs with multiple warheads (three for each missile) or MRVs (multiple reentry vehicles). It also began introducing additional nuclear weapons, which included multiple "independently targetable" reentry vehicles (MIRVs) that added tremendous flexibility to the use of ICBMs. Part of the explanation for the new technology was a "keeping up with the Jones's" mentality. The rationale was that since the Soviets were deploying anti-ballistic missile systems for defense (built around Moscow and briefly around Leningrad), the MIRVs were necessary. Cost was an additional reason, especially when defense dollars were being diverted to the escalating Vietnam War. Not only the Soviets, but domestic opponents began to weigh in on the new multiple warhead technology. Advocates of an American ABM program (especially in the Pentagon and Congress) believed that expenditures on new warheads threatened the development of a missile defense system. Ultimately the Johnson administration proceeded with both the MIRVs and an ABM program. This did little to reduce skepticism about the sincerity of the United States pursuing arms control, or its willingness to control its military budget.[136]

Despite the desires of both superpowers, nuclear proliferation accelerated in the 1960s. On October 15, 1964, China exploded its first atomic bomb at Lop Nur Testing Site, and in June 1967 it detonated an airborne hydrogen bomb. France tested its first H-bomb in September 1966. A positive event was the Treaty of Tlatelolco, which 14 Latin American countries signed (Cuba and Guyana did not) on February 14, 1967. It banned nuclear weapons in their respective countries. The major powers also signed the Treaty on the Nonproliferation of Nuclear Weapons (NPT) on July 1, 1968, with little opposition. The US Senate ratified it in March 1969 quite easily, largely because it did not address a reduction in American nuclear arms. In 1975 there were 111 signatories. Refusing to sign were France, China, and India. While this did not portend well for the future of non-proliferation, the NPT provided impetus for the Strategic Arms Limitations Talks (SALT), negotiated between 1968 and 1972. SALT would begin the next phase in arms control.[137] The question lingered: Was all of this for show or did it reflect a new sincerity over arms control?

There is good reason to believe that Johnson was sincere about nuclear arms limitation. While the Vietnam War frustrated him in his last months in office, LBJ turned more assertively to détente (the easing of tensions or strained relations) with the Soviet Union. Most promising were the NPT and a meeting with Soviet leaders scheduled for September 1968 that would begin strategic arms control negotiations. But the Soviet invasion of Czechoslovakia in August derailed the plans. Fearing a breach in the control of Eastern bloc countries, Soviet leaders acted quickly and aggressively to quell democratization among the Czechs. They proclaimed the "Brezhnev Doctrine" (named for Premier Leonid Brezhnev) which stated the right to intercede anywhere that socialism was threatened. That LBJ had pushed to negotiate with the Soviets on arms control, and avoided confrontation by failing to act on the invasion of Czechoslovakia, set the stage for extending détente in the 1970s. His apparent appeasement outraged critics, but also laid the foundation for meaningful negotiations leading to SALT I.[138]

The new president, Richard M. Nixon, and his National Security Advisor, Henry A. Kissinger, looked back on Cold War diplomacy and sought to devise something different, even radically different. Nixon wanted a "new approach to foreign policy to match a new era of international relations."[139] The "new approach" was in many ways old-style *realpolitik* (pragmatism). Concerned that the United States faced a relative decline in power in recent years, Nixon and Kissinger hoped to take advantage of the rivalry between the USSR and China. They mapped out plans to reduce American commitments abroad, to use regional powers to stabilize order, and to seek nuclear arms control. All of this was to be done within the White House often in secrecy, virtually shutting out the State Department and Congress from participating in foreign policy. In order to move toward détente with the USSR and to get closer to the People's Republic, the war in Vietnam had to be resolved. Yet Nixon's approach (Vietnamization and his offensive in Cambodia) only deepened the quagmire there. Interestingly, Nixon had mentioned to Kissinger that the use of "the nuclear bomb" might be necessary in Vietnam, and it came up again in 1973 with respect to the Israeli war with Egypt. But they took no action in that direction.[140]

Despite the serious distraction (and tragedy) of Vietnam, the president and Kissinger moved to secure improved relations with Moscow. Like McNamara, they accepted nuclear parity and built deterrence on MAD, but remained wary that the Soviets would seek nuclear superiority if they could. Leaders in the USSR also had their anxieties, but were very interested in reducing the risk of nuclear war and seemed emboldened to do so as a balance in arms became possible. The need to strengthen their economy also demanded a reduction in the cost of nuclear weaponry and its infrastructure. But progress was slow, particularly because of ABMs and MIRVs. The Soviets had a fairly elaborate ABM system surrounding Moscow, but for the United States to upgrade its very modest system would require vast sums. Nixon, while somewhat dubious about a major upgrade, publicly supported a new ABM system primarily as a bargaining tool. Throughout the 1970s, any discussion of ABMs automatically raised new debates over civil defense. How to control offensive weapons was another matter.[141]

Arms control agreements were the centerpiece of the first détente summit that took place in Moscow in May, 1972. This came on the heels of Nixon's more famous trip to Beijing in February publicly proclaimed as "ping-pong diplomacy." The United States had been withdrawing forces from Vietnam, while the Soviet Union increased its troop strength along its border with China. Nixon wanted to find a way to reconcile with China to isolate North Vietnam, and also to drive a wedge between the two major communist countries. Nixon became the first US president to use the term "People's Republic of China" and also removed special passport restrictions for travel to China, both regarded as important symbolic gestures. In its turn, Beijing leaders invited the US table tennis team to China, which was another gesture indicating a thaw in the relations between the two countries.[142] Little things mattered in diplomacy.

After substantial wrangling (essentially 30 months), the United States and the Soviet Union made progress in arms control in 1972. This occurred partly because of the Nixon administration's efforts to work both sides of the street in dealing with Beijing and Moscow. Preliminary talks in Helsinki and Vienna, and back-channel negotiations between Kissinger and Ambassador Dobrynin, preceded the final negotiations on arms control and ceremonial events at the summit. A treaty limited each country to two ABM systems: one for the respective capitals, another for a major missile installation. This first Strategic Arms Limitation Agreement (SALT I) did not limit MIRVs, but a second agreement to run for five years set an upper limit on offensive missiles at 1,600 for the USSR and 1,054 for the United States. It also allowed improvements in existing missile systems and limited the number of submarines capable of launching SLBMs. That Nixon and Kissinger wanted a bargain at almost any cost, led to some disadvantages for the United States. But the apparent disparity in missiles, which critics harped upon, did not take into account the substantial advantage that the Americans retained in the accuracy of its weapons and the sheer number advantage in MIRVed warheads. SALT I was an important first step in arms control, but one that neither the Left (viewed as too limited) nor the Right (viewed as too extensive) at home completely embraced, and one that did not end the peril of nuclear war.[143]

For all its historic significance and controversy, SALT I was not disarmament, and not really a step toward it either. The treaties fit within narrow tolerances of what was acceptable to the United States and the USSR at the time without threatening their national security. The treaties represented a change in the Cold War and the signals that the respective parties sent about backing away from years of confrontation. While the anti-nuclear movement heated up over civilian nuclear power in the early 1970s, the swirl of debate concerning nuclear weapons was causing nary a ripple among peace and disarmament groups (outside of the South Pacific). The Vietnam War preoccupied many protesters. Nevertheless, détente and current arms control negotiations in the early 1970s signaled positive steps in the arms race even if those steps were modest.[144]

When Vice President Gerald R. Ford assumed the presidency from a disgraced Nixon after the Watergate scandal, Soviet-American relations already were strained. Long-standing differences between the two countries could not be cured by a generalized notion of cooling off the Cold War. Watergate undermined the efforts to buck up détente, as did challenges to détente by strong congressional Cold Warriors, supporters of Israel, and advocates for human rights. Senator Henry Jackson (Dem., Washington) led the charge. A failed trade agreement, in particular, stymied further arms control talks. Jackson complicated the process by demanding that future SALT agreements be built on a principle of equal numbers of missiles on both sides. This was a simplistic notion, especially because the weapons systems of the two countries were so different. The Soviet arsenal primarily was land-based, with larger and slower missiles, while the US weapons were faster, smaller, and generally more mobile. Under this constraint Kissinger (retained by Ford) attempted to hammer out a new agreement over a period of two years. In late 1974, Ford met with Brezhnev near Vladivostok, and remarkably the Soviets agreed to a policy of equality to be set at 2,400 strategic delivery vehicles and 1,300 MIRVs. The deal ran afoul at home, and despite incorporating Jackson's equality principle, congressional hawks opposed it or at least tried to slow it down. Such action, Kissinger's vacillation on private discussions with the Soviets, and failed trade talks undermined détente. Further efforts to expand on the forward motion of SALT I and even the Vladivostok Accord went no further for the rest of Ford's term in office.[145]

Ford and Kissinger had tried to sustain the advances in détente and arms control growing out of SALT I, but the immediate results were marginal at best. Talks on SALT II began in fall 1972, but moved slowly. Conditions worsened on détente, and ultimately on arms control, under the new president, Democrat Jimmy Carter. He had campaigned as a Washington outsider to win the election, but what worked in the campaign failed in the White House. His political instincts, not his social and moral standards, weakened his performance. Carter often was contradictory, that is, wanting to decrease tensions with the Soviets, but downplaying the importance of Soviet-American relations. In arms control, he spoke about moving beyond the Cold War and SALT I. He promoted a goal of limits as opposed to reductions, however, and did so publicly without consulting with his counterparts in the USSR. In his inaugural address he called for "elimination of all nuclear weapons." Carter was well-meaning, but he was often impractical. For example, he called for deeply cutting land-based missiles, where the Soviets had superior numbers.[146]

Despite the reservations of many over his foreign policy approach and perspective, Carter made progress on SALT II. Like presidents before him, the toughest crowd he faced seemed to be in his own backyard. In June 1979, Carter and Brezhnev came to agreement on a new (and more complex) nuclear arms treaty which capped the number of warheads on missiles, limited MIRVed missiles, froze the number of delivery systems, and discussed some new systems and their deployment. Liberal critics believed it did not go far enough to reduce armaments, and were incensed with the president's sop to conservatives of adding a very costly new missile system. Conservatives saw the new treaty as weakening the United States. Rumors of a brigade of Soviet troops newly based in Cuba did not help matters (it was discovered that they had been there since 1962). More importantly, problems in Iran and elsewhere, and the Soviet invasion of Afghanistan in December, made passage of SALT II untenable.[147] Afghanistan, in particular, pushed Carter closer to the hawks, as the Cold War heated up again. Détente crumbled in the face of Carter's own policymaking shortcomings, intense conservative opposition, and events external to the arms-control debate itself.[148] For the world anti-nuclear movement, things changed by the mid- to late-1970s. With the end of the Vietnam War in April 1975 and the sour turn of events as the SALT II talks withered, anti-bomb protests began to revive. In 1978 the United Nations held a Special Session on Disarmament in New York City, which became somewhat of a rallying point. The revelation that the US government planned to build a "neutron bomb" only added to the renewed disquiet. The neutron bomb was meant to be a tactical weapon that maximized human casualties, but minimized destruction of buildings and equipment. A specialized thermonuclear weapon, it produced a relatively small blast but released large amounts of lethal radiation which could penetrate armor or yards of earth. The focus on obliterating people, but saving buildings and other infrastructure, was distasteful to many.[149]

CONCLUSION: DIVERGING PATHS OF NUCLEAR WEAPONS AND CIVILIAN POWER

Rise and fall in expectations about nuclear arms control in the 1960s and 1970s mirrored the rise and fall in expectations about civilian nuclear power in the same period. As we will see in Chapter 8, optimism about the range of peaceful uses of the atom rose in the 1960s and early 1970s, only to stumble in 1979 with the accident at Three Mile Island. Did splitting the atom lead to transforming technologies that revised the map of history? Such undulations in the hopes and fears associated with nuclear weapons and nuclear power led Americans and others to consider the intensity of the changes they wrought, but not their importance. At the very center of the development of nuclear weapons and nuclear power was the need for some sort of control, control in their use and control in their limits. In the little more than three decades since Hiroshima such a goal remained elusive. Events in the early 1960s at the heart of the Cold War had made clear, for nuclear weapons at least, that no one wanted a repeat of the Cuban Missile Crisis or the Berlin Crisis, even though the fragile détente of the 1970s gave a glimmer of hope that such calamities could not occur. As long as thermonuclear weapons existed in such abundance, hope had to trump promise, but risk remained constant. What the future held for harnessing nuclear power was unknown.

MySearchLab Connections: Sources Online

READ AND REVIEW

Review this chapter by using the study aids and these related documents available on MySearchLab.

✓•─[**Study** and **Review** on **mysearchlab.com**

Chapter Test

Essay Test

▢•─[**Read** the **Document** on **mysearchlab.com**

John F. Kennedy, Inaugural Address (1961)

John F. Kennedy, Address Before the General Assembly of the United Nations (1961)

Executive Discussions on the Cuban Missile Crisis (1962)

President Kennedy's Address to the People of Berlin (1963)

RESEARCH AND EXPLORE

Use the databases available within MySearchLab to find additional primary and secondary sources on the topics within this chapter.

👁─[**Watch** the **Video** on **mysearchlab.com**

The Cuban Missile Crisis

Cold War Connections: Russia, America, Berlin and Cuba

Endnotes

1. "Military-Industrial Complex Speech, Dwight D. Eisenhower, 1961," online, http://www.h-net. org/~hst306/documents/indust.html.

2. "Military-Industrial Complex" from *SourceWatch*, online, http://www.sourcewatch.org/index. php?title=Military-industrial_complex; James Alden Barber, Jr., "The Military-Industrial Complex," in Stephen E. Ambrose and James Alden Barber, Jr., eds., *The Military and American Society* (New York: Free Press, 1972), 43–44.

3. Gerard J. DeGroot, *The Bomb: A Life* (Cambridge, MA: Harvard University Press, 2005), 214.

4. Britannica Online Encyclopedia, online, http://www.britannica.com/EBchecked/topic/382349/ military-industrial-complex.

5. "Military Industrial Complex," Center for Defense Information, online, http://www.cdi.org/issues/ usmi/complex; Barber, Jr., "The Military-Industrial Complex," 46–60.

6. "Military-Industrial Complex Speech, Dwight D. Eisenhower, 1961."

7. "Big Science," online, http://www.britannica.com/EBchecked/topic/64995/Big-Science?view=print; Alvin Weinberg, "Impact of Large-Scale Science on the United States," *Science* 134 (July 21, 1961): 161–64; Lawrence Badash, *Scientists and the Development of Nuclear Weapons: From Fission to the Limited Test Ban Treaty, 1939–1963* (Amherst, NY: Humanity Books, 1995), 91.

8. Martin Walker, *The Cold War: A History* (New York: Henry Holt and Co., 1993), 140.

9. Stuart W. Leslie, *The Cold War and American Science: The Military-Industrial-Academic Complex at MIT and Stanford* (New York: Columbia University Press, 1993), 1; Daniel J. Kevles, "K_1S_2: Korea, Science, and the State," in Peter Galison and Bruce Hevly, eds., *Big Science: The Growth of Large-Scale Research* (Stanford, CA: Stanford University Press, 1992), 314–15.

10. Leslie, *The Cold War and American Science*, 1–12; Jessica Wang, *American Science in an Age of Anxiety: Scientists, Anticommunism, and the Cold War* (Chapel Hill, NC: University of North Carolina Press, 1999), 254, 260–61; Kevles, "K_1S_2: Korea, Science, and the State," 329, 332.

11. John Kenneth Galbraith, "Characteristics of the Military-Industrial Complex," in Ambrose and Barber, Jr., eds., *The Military and American Society*, 64–67; Leslie, *The Cold War and American Science*, 3; Kevles, "K_1S_2: Korea, Science, and the State," 316; David C. Cassidy, *J. Robert Oppenheimer and the American Century* (New York: Pi Press, 2005), 253–54, 257, 259, 306.

12. Kennedy used "New Frontier" in his acceptance speech at the 1960 Democratic National Convention in Los Angeles. It became the label for his administration's foreign and domestic policies and programs. See Richard Dean Burns and Joseph M. Siracusa, *Historical Dictionary of the Kennedy-Johnson Era* (Lanham, MD: The Scarecrow Press, 2007), 230.

13. Nick Bryant, "The Jackie Kennedy Tapes: Sharp, Tender and Gossipy," *BBC*, online, http://www.bbc. co.uk/news/world-us-canada-14913829.

14. Robert Dallek made a persuasive case that JFK was fearful in the 1960 presidential campaign that his Addison's disease, colitis, back troubles, and prostatitus would make him appear unfit for office. He thus placed these issues under wraps, as he did with his "reckless womanizing." Dallek also argued that beginning in fall 1961 Kennedy's exercise regimen and new medications helped keep his health issues from hampering his performance as president. See Robert Dallek, *An Unfinished Life: John F. Kennedy, 1917–1963* (New York: Little Brown and Company, 2003), 704–07.

15. One observer noted that JFK had "a fixation with image," which may explain in part his reluctance to make public some of his vulnerabilities. However, to some extent all politicians have had or were compelled to have such a fixation. See Mark J. White, *The Cuban Missile Crisis* (London: Macmillan Press, 1996), 1–18. Among the most recent expansive accounts of Kennedy and his times—but clearly with different emphases—are Dallek, *An Unfinished Life*; Michael O'Brien, *John F. Kennedy: A Biography* (New York: St. Martin's Press, 2005); Michael R. Beschloss, *The Crisis Years: Kennedy and Khrushchev, 1960–1963* (New York: Harper-Collins, 1991); Richard Reeves, *President Kennedy: Profile of Power* (New York: Simon & Schuster, 1993).

16. The idea of a missile gap came from studies during the Eisenhower years. Then Secretary of Defense Neil McElroy predicted in 1959 that the Soviets would have a 3 to 1 edge in ICBMs by the early 1960s, which proved to be very wrong. James N. Giglio, *The Presidency of John F. Kennedy* (Lawrence, KS: University Press of Kansas, 1991), 45.

17. George C. Herring, *From Colony to Superpower: U.S. Foreign Relations Since 1776* (New York: Oxford University Press, 2008), 699, 702–703; Ronald E. Powaski, *March to Armageddon: The United States and the Nuclear Arms Race, 1939 to the Present* (New York: Oxford University Press, 1987), 95; Julian E. Zelizer, *Arsenal of Democracy: The Politics of National Security—From World War II to the War on Terrorism* (New York: Basic Books, 2010), 145–46; W.J. Rorabaugh, *Kennedy and the Promise of the Sixties* (Cambridge, MA: Cambridge University Press, 2002), 1, 9, 19; Giglio, *The Presidency of John F. Kennedy*, 1–13, 17, 45.

18. Dallek, *An Unfinished Life*, 314–15.

19. Herring, *From Colony to Superpower*, 703–04; Rorabaugh, *Kennedy and the Promise of the Sixties*, 20; Giglio, *The Presidency of John F. Kennedy*, 30, 45.

20. Jonathan Stevenson, *Thinking Beyond the Unthinkable: Harnessing Doom from the Cold War to the Age of Terror* (New York: Viking, 2008), 101–04; Giglio, *The Presidency of John F. Kennedy*, 30–36; Burns and Siracusa, *Historical Dictionary of the Kennedy-Johnson Era*, 99.

21. Giglio, *The Presidency of John F. Kennedy*, 45–46; Herring, *From Colony to Superpower*, 704; Georg Schild, "The Berlin Crisis," in Mark J. White, ed., *Kennedy: The New Frontier Revisited* (New York: New York University Press, 1998), 92–93.

22. Herring, *From Colony to Superpower*, 704–05.

23. Ibid. See also Philip Nash, "Bear Any Burden? John F. Kennedy and Nuclear Weapons," in John Lewis Gaddis, et al., eds., *Cold War Statesmen Confront the Bomb: Nuclear Diplomacy Since 1945* (New York: Oxford University Press, 1999), 124–34.

24. White, *The Cuban Missile Crisis*, 23.

25. Powaski, *March to Armageddon*, 93–95, 98; Walker, *The Cold War*, 141–45; Giglio, *The Presidency of John F. Kennedy*, 46–47; Schild, "The Berlin Crisis," 93. By July 1964 the arsenal stood at 5,007 (although the development of some weapons had been underway during the prior administration). See Nash, "Bear Any Burden? John F. Kennedy and Nuclear Weapons," 124–25.

26. Richard Rhodes, *Dark Sun: The Making of the Hydrogen Bomb* (New York: Simon and Schuster, 1995), 570. For slightly different figures in that time period, but grossly in favor of the US, see DeGroot, *The Bomb*, 256; Richard Rhodes, *Arsenals of Folly: The Making of the Nuclear Arms Race* (New York: Alfred A. Knopf, 2007), 93.

27. On the origins of the first SIOP, see David Alan Rosenberg, "The Origins of Overkill: Nuclear Weapons and American Strategy, 1945–1960," *International Security* 7 (Spring 1983), 64–71.

28. Powaski, *March to Armageddon*, 95–97.

29. Zelizer, *Arsenal of Democracy*, 143; Herring, *From Colony to Superpower*, 686–87.

30. Aleksandr Fursenko and Timothy Naftali, *"One Hell of a Gamble": Khrushchev, Castro, and Kennedy, 1958–1964: The Secret History of the Cuban Missile Crisis* (New York: W.W. Norton, 1997), 5–9; Louis A. Perez, Jr., *Cuba and the United States: Ties of Singular Intimacy* (Athens, GA: University of Georgia Press, 2003; 3rd ed.), 223–38. See also Tad Szulc, *Fidel: A Critical Portrait* (New York: William Morrow and Co., 1986).

31. Herring, *From Colony to Superpower*, 687–88.

32. Ibid.; Giglio, *The Presidency of John F. Kennedy*, 48; Fursenko and Timothy Naftali, *"One Hell of a Gamble,"* 9–16, 24, 35–55.

33. Quoted in William Taubman, *Khrushchev:The Man and His Era* (New York: W.W. Norton, 2003), 533.

34. Giglio, *The Presidency of John F. Kennedy*, 48; Fursenko and Timothy Naftali, *"One Hell of a Gamble,"* 9–16, 24, 35–55.

35. Herring, *From Colony to Superpower*, 687–89.

36. Ibid., 688–89; Walker, *The Cold War*, 23–24.

37. Fursenko and Timothy Naftali, *"One Hell of a Gamble,"* 65–67, 82–83; Walker, *The Cold War*, 25–36.

38. Walker, *The Cold War,* 23–24.

39. Fursenko and Naftali, *"One Hell of a Gamble,"* 84–85.

40. Garry Wills argued that Castro knew about preparations for the invasion in advance. See Garry Wills, *Bomb Power: The Modern Presidency and the National Security State* (New York: Penguin Press, 2010), 166.

41. Ibid., 148–49; Herring, *From Colony to Superpower*, 705–06; Zelizer, *Arsenal of Democracy*, 150; Powaski, *March to Armageddon*, 100; Giglio, *The Presidency of John F. Kennedy*, 48–60.

42. Walker, *The Cold War,* 37–38; Zelizer, *Arsenal of Democracy*, 151.

43. Trumbull Higgins, *The Perfect Failure: Kennedy, Eisenhower, and the CIA at the Bay of Pigs* (New York: W.W. Norton, 1987), 173.

44. Herring, *From Colony to Superpower*, 707.

45. The CIA operated the plan primarily out of Miami. About four hundred Americans and two thousand Cubans participated. Along with the assassination plots, the CIA engaged in or promoted numerous sabotage activities in Cuban meant to disrupt the economy, including contaminating a shipment of sugar, encouraging importers to sell defective materials to the Cubans, and smuggling in arms. See Giglio, *The Presidency of John F. Kennedy*, 190; Fursenko and Naftali, *"One Hell of a Gamble,"* 132–48; White, *The Cuban Missile Crisis*, 52–57.

46. White, *The Cuban Missile Crisis*, 20–23; Herring, *From Colony to Superpower*, 706–08; Walker, *The Cold War,* 149–50; Higgins, *The Perfect Failure*, 173–76.

47. Fursenko and Naftali, *"One Hell of a Gamble,"*102.

48. Giglio, *The Presidency of John F. Kennedy*, 69; W.R. Smyser, *Kennedy and the Berlin Wall* (Lanham, MD: Rowman & Littlefield Pubs., 2009), xiii–xiv; Dallek, *An Unfinished Life*, 418.

49. Quoted in Herring, *From Colony to Superpower*, 695.

50. Ibid., 695–96.

51. Ibid., 695–96.

52. Ibid., 696–99.

53. Quoted in Rorabaugh, *Kennedy and the Promise of the Sixties*, 34.

54. Giglio, *The Presidency of John F. Kennedy*, 71–74; Dallek, *An Unfinished Life,* 401–02.

55. Giglio, *The Presidency of John F. Kennedy*, 71–74; Dallek, *An Unfinished Life,* 401–02; Smyser, *Kennedy and the Berlin Wall,* 43.

56. Smyser, *Kennedy and the Berlin Wall,* 64–68, 75–77; Herring, *From Colony to Superpower*, 708–09; Dallek, *An Unfinished Life,* 403–14.

57. Quoted in Herring, *From Colony to Superpower*, 709. See also Smyser, *Kennedy and the Berlin Wall*, 64–68, 75–77; Dallek, *An Unfinished Life,* 403–14.

58. Quoted in Dallek, *An Unfinished Life,* 413.

59. Herring, *From Colony to Superpower*, 708–09.

60. Schild, "The Berlin Crisis," x, 109–111, 116–19, 123; Vladislav M. Zubok, *A Failed Empire: The Soviet Union in the Cold War from Stalin to Gorbachev* (Chapel Hill, NC: University of North Carolina Press, 2007), 133, 140; Powaski, *March to Armageddon*, 100–101; Rorabaugh, *Kennedy and the Promise of the Sixties*, 34–6; Giglio, *The Presidency of John F. Kennedy*, 78–88.

61. Powaski, *March to Armageddon*, 100–101; Rorabaugh, *Kennedy and the Promise of the Sixties*, 34–6; Giglio, *The Presidency of John F. Kennedy*, 78–88; Herring, *From Colony to Superpower*, 708–10; Schild, "The Berlin Crisis," x, 109–11, 116–19, 123; Dallek, *An Unfinished Life*, 426–35.

62. Robert F. Kennedy, *Thirteen Days: A Memoir of the Cuban Missile Crisis* (New York: W.W. Norton, 1969).

63. Quoted in Jeffrey Porro, ed., *The Nuclear Age Reader* (New York: Alfred A. Knopf, 1989), 149.

64. Quoted in Zubok, *A Failed Empire*, 144.

65. Taubman, *Khrushchev*, 546.

66. Ibid., 535–546; Vladislav M. Zubok and Hope M. Harrison, "The Nuclear Education of Nikita Khrushchev," in Gaddis, et al., eds., *Cold War Statesmen Confront the Bomb*, 141–68. See also Aleksandr Fursenko and Timothy Naftali, *Khrushchev's Cold War: The Inside Story of an American Adversary* (New York: W.W. Norton, 2006).

67. Taubman, *Khrushchev*, 535; Zubok and Harrison, "The Nuclear Education of Nikita Khrushchev," 141–68; Fursenko and Naftali, *Khrushchev's Cold War*, 6; Fursenko and Naftali, *"One Hell of a Gamble,"* 139, 171, 178–83.

68. For a detailed discussion of the *Jupiters* in Turkey, see Philip Nash, *The Other Missiles of October: Eisenhower, Kennedy, and the Jupiters, 1957–1963* (Chapel Hill: University of North Carolina Press, 1997).

69. James G. Blight and David A. Welch, *On the Brink: Americans and Soviets Reexamine the Cuban Missile Crisis* (New York: Farrar, Strauss and Giroux, 1989), 327–34; Taubman, *Khrushchev*, 530–32, 541–47; Walker, *The Cold War*, 166–70. W.R. Smyser, who served at the US Mission in Berlin from 1960 to 1964, put more stock than most in the importance of the Soviet interest in Berlin in relation to the Cuban Missile Crisis. He argued that Khrushchev "left a number of clues that he saw Berlin as a key objective for putting missiles in Cuba, and perhaps even as his main objective." See Smyser, *Kennedy and the Berlin Wall*, 192–204.

70. Giglio, *The Presidency of John F. Kennedy*, 190; White, ed., *Kennedy*, 8; White, *The Cuban Missile Crisis*, 232–37; Taubman, *Khrushchev*, 532, 553–55.

71. Len Scott, *The Cuban Missile Crisis and the Threat of Nuclear War Lessons from History* (London: Continuum, 2007), 12.

72. Fursenko and Naftali, *"One Hell of a Gamble,"* 184–97.

73. Dino A. Brugioni, "The Invasion of Cuba," in Robert Cowley, ed., *The Cold War: A Military History* (New York: Random House, 2006), 211; Dallek, *An Unfinished Life*, 537; Zelizer, *Arsenal of Democracy*, 156–58; Fursenko and Naftali, *"One Hell of a Gamble,"* 198–203.

74. Fursenko and Naftali, *"One Hell of a Gamble,"* 206–19; Dallek, *An Unfinished Life*, 537; Herring, *From Colony to Superpower*, 719–20; Zelizer, *Arsenal of Democracy*, 155.

75. Dallek, *An Unfinished Life*, 537; Herring, *From Colony to Superpower*, 719–21; Zelizer, *Arsenal of Democracy*, 155; Fursenko and Naftali, *"One Hell of a Gamble,"* 206–12, 217, 219.

76. Dallek, *An Unfinished Life*, 538–42; Zelizer, *Arsenal of Democracy*, 155–60.

77. Thomas G. Patterson and William J. Brophy, ""October Missiles and November Elections: The Cuban Missile Crisis and American Politics, 1962," *Journal of American History* 73 (June 1986), 87–119.

78. Dallek, *An Unfinished Life*, 538–42; Zelizer, *Arsenal of Democracy*, 155–60; Giglio, *The Presidency of John F. Kennedy*, 191.

79. They had been suspended since September 10 because a Taiwanese U-2 was shot down over mainland China the day before. The president finally bent to his advisers' wishes and resumed reconnaissance flights over Cuba on October 9.

80. Dallek, *An Unfinished Life*, 542–45; Zelizer, *Arsenal of Democracy*, 160–63; Brugioni, "The Invasion of Cuba," 211–12; White, *The Cuban Missile Crisis*, 171; Burns and Siracusa, *Historical Dictionary of the Kennedy-Johnson Era*, 66.

81. Even the Cubans themselves were not well informed about this crucial issue.

82. Blight and Welch, *On the Brink*, 335; Fursenko and Naftali, *"One Hell of a Gamble,"* 224–25. The ship carrying IRBMs turned back when the blockade was mounted, while IRBM warheads were cargoed in a Soviet ship in a Cuban port. By October 14 eighty cruise missile warheads, six atomic bombs for the bombers, and twelve *Luna* warheads were in Cuba along with the warheads for the MRBMs. See Taubman, *Khrushchev*, 551.

83. Norman Polmar and John D. Gresham, *DEFCON-2: Standing on the Brink of Nuclear War During the Cuban Missile Crisis* (New York: John Wiley & Sons, 2006), xxiii.

84. Herring, *From Colony to Superpower*, 720–21.

85. Quoted in Dallek, *An Unfinished Life*, 546. The Cabinet Room where Ex Comm met was equipped with a recording device, and Kennedy turned it on for the meetings, thus providing rare insight into the policy making process.

86. Fursenko and Naftali, *"One Hell of a Gamble,"* 224–34; Dallek, *An Unfinished Life*, 546– 55; Walker, *The Cold War*, 172–74.

87. Dallek, *An Unfinished Life*, 546– 55; Walker, *The Cold War*, 172–74; Fursenko and Naftali, *"One Hell of a Gamble,"* 224–34.

88. Quoted in Dallek, *An Unfinished Life,* 553.

89. Blight and Welch, *On the Brink,* 336–44, 411–17; Dallek, *An Unfinished Life,* 556–565; Zelizer, *Arsenal of Democracy,* 163–69; Giglio, *The Presidency of John F. Kennedy,* 193–213; Fursenko and Naftali, *"One Hell of a Gamble,"* 234–39.

90. Quoted in Porro, ed., *The Nuclear Age Reader,* 154.

91. Rhodes, *Dark Sun,* 572–73; Fursenko and Naftali, *"One Hell of a Gamble,"*256; Dallek, *An Unfinished Life,* 561–71. See also Dino A. Brugioni, *Eyeball to Eyeball: The Inside Story of the Cuban Missile Crisis* (New York: Random House, 1991; updated ed.).

92. Quoted in Porro, ed., *The Nuclear Age Reader,* 158.

93. Fursenko and Naftali, *"One Hell of a Gamble,"* 240–89; Walker, *The Cold War,* 174–77; Dallek, *An Unfinished Life,* 560–71.

94. Quoted in Michael Dobbs, *One Minute to Midnight: Kennedy, Khrushchev, and Castro on the Brink of Nuclear War* (New York: Vintage Books, 2008), 304.

95. Blight and Welch, *On the Brink,* 336–44; Taubman, *Khrushchev,* 569–77.

96. Dallek, *An Unfinished Life,* 535–571; Herring, *From Colony to Superpower,* 720–22; Blight and Welch, *On the Brink,* 336–44, 411–17; Zelizer, *Arsenal of Democracy,* 163–69; Giglio, *The Presidency of John F. Kennedy,* 193–213; Taubman, *Khrushchev,* 556–58, 561–65, 569–77; Walker, *The Cold War,* 177–79.

97. Giglio, *The Presidency of John F. Kennedy,* 214–15; O'Brien, *John F. Kennedy,* 671.

98. Powaski, *March to Armageddon,* 105–06; Walker, *The Cold War,* 178–79.

99. Walker, *The Cold War,* 178–79; Fursenko and Naftali, *"One Hell of a Gamble,"* 324–25; Zubok, *A Failed Empire,* 151–52.

100. Alice L. George, *Awaiting Armageddon: How Americans Faced the Cuban Missile Crisis* (Chapel Hill, NC: University of North Carolina Press, 2003), 1–2, 167–70; Tracy C. Davis, "Continuity of Government Measures for Civil Defense During the Cuban Missile Crisis," in Rosemary B. Mariner and G. Kurt Piehler, eds., *The Atomic Bomb and American Society: New Perspectives* (Knoxville, TN: University of Tennessee Press, 2009), 153–84.

101. Blight and Welch, *On the Brink,* 344–46; Philip Brenner, "Cuba and the Missile Crisis," *Journal of Latin American Studies* 22 (February 1990): 115–42.

102. Stephen E. Ambrose, *Rise to Globalism: American Foreign Policy Since 1939* (New York: Penguin Books, 1983; 3rd rev. ed.), 267.

103. Giglio, *The Presidency of John F. Kennedy,* 214; Walker, *The Cold War,* 178–79; White, *The Cuban Missile Crisis,* 238; Fursenko and Naftali, *"One Hell of a Gamble,"* 324–25; Polmar and Gresham, *DEFCON-2,* xiv–xv, xxiii; Blight and Welch, *On the Brink,* 344–50; Graham Allison and Philip Zelikov, *Essence of Decision: Explaining the Cuban Missile Crisis* (New York: Longman, 1999; second ed.), 379–407; Melvyn P. Leffler, *For the Soul of Mankind: The United States, the Soviet Union and the Cold War* (New York: Hill and Wang, 2007), 151–233.

104. Quoted in Kenneth D. Rose, *One Nation Underground: The Fallout Shelter in American Culture* (New York: New York University Press, 2001), 2.

105. Ibid., 1.

106. Allan M. Winkler, *Life Under a Cloud: American Anxiety About the Atom* (Urban, IL: University of Illinois Press, 1999), 113–16; Jon Hunner, "Reinventing Los Alamos: Code Switching and Suburbia at America's Atomic City," in Scott C. Zeman and Michael A. Amundson, eds., *Atomic Culture: How We Learned to Stop Worrying and Love the Bomb* (Boulder, CO: University Press of Colorado, 2004), 46; Laura McEnaney, *Civil Defense Begins at Home: Militarization Meets Everyday Life in the Fifties* (Princeton, NJ: Princeton University Press, 2000), 42–44.

107. Winkler, *Life Under a Cloud,* 117–23; McEnaney, *Civil Defense Begins at Home,* 47–52, 56–60; Rose, *One Nation Underground,* 37; David L. Snead, *The Gaither Committee, Eisenhower, and the Cold War* (Columbus, OH: Ohio State University Press, 1999), 45–46, 109–10, 118–19, 135, 168–70, 185–88.

108. Dee Garrison, *Bracing for Armageddon: Why Civil Defense Never Worked* (New York: Oxford University Press, 2006), 105.

109. Winkler, *Life Under a Cloud*, 125–29; Snead, *The Gaither Committee, Eisenhower, and the Cold War*, 179; Garrison, *Bracing for Armageddon*, 133–34.
110. Stephen E. Atkins, ed., *Historical Encyclopedia of Atomic Energy* (Westport, CT: Greenwood Press), 125; Garrison, *Bracing for Armageddon*, 105–06;Winkler, *Life Under a Cloud*, 125–29; Margot A. Henriksen, *Dr. Strangelove's America: Society and Culture in the Atomic Age* (Berkeley, CA: University of California Press, 1997), 206.
111. Elaine Tyler May, *Homeward Bound: American Families in the Cold War Era* (New York: Basic Books, 2008), 1.
112. Ibid., 100–03.
113. Ibid. See also Winkler, *Life Under a Cloud*, 122; McEnaney, *Civil Defense Begins at Home*, 40–41, 52–55, 60–67; Henriksen, *Dr. Strangelove's America*, 215–16.
114. Garrison, *Bracing for Armageddon*, 105–06.
115. Andrew D. Grossman in *Neither Dead Nor Red: Civilian Defense and American Political Development During the Early Cold War* (New York: Routledge, 2001), 92, 94–95; McEnaney, *Civil Defense Begins at Home*, 64–67; Henriksen, *Dr. Strangelove's America*, 216–18, 223; Rose, *One Nation Underground*, 10–11, 112, 224; Winkler, *Life Under a Cloud*, 129–32.
116. Lawrence S. Wittner, *Resisting the Bomb: A History of the World Nuclear Disarmament Movement, 1954–1970*, vol. 2 of *The Struggle Against the Bomb* (Stanford, CA: Stanford University Press, 1997), 415–28.
117. In the early 1960s McNamara also promoted PALs, permissive action links. These were electronic devices to prevent nuclear weapons from being fired without the authorization of the president. The Joint Chiefs objected, believing that such devices would slow down a missile launch. Nina Tannenwald, *The Nuclear Taboo: The United States and the Non-Use of Nuclear Weapons Since 1945* (New York: Cambridge University Press, 2007), 254.
118. Zelizer, *Arsenal of Democracy*, 172–73; Powaski, *March to Armageddon*, 106–09.
119. "Commencement Address at American University, June 10, 1963," *John F. Kennedy Presidential Library and Museum*, online, http://www.jfklibrary.org/Research/Ready-Reference/JFK-Speeches/Commencement-Address-at-American-University-June-10-1963.aspx.
120. Zelizer, *Arsenal of Democracy*, 173; John Mueller, *Atomic Obsession: Nuclear Alarmism from Hiroshima to Al-Qaeda* (New York: Oxford University Press,2010), 75–77.
121. Dallek, *An Unfinished Life,* 607–28; Taubman, *Khrushchev*, 602; Giglio, *The Presidency of John F. Kennedy*, 216–20; Mueller, *Atomic Obsession*, 75–77.
122. Giglio, *The Presidency of John F. Kennedy*, 216–20; Powaski, *March to Armageddon*, 109–12; Zelizer, *Arsenal of Democracy*, 173–74; Stevenson, *Thinking Beyond the Unthinkable*, 107–08; Dallek, *An Unfinished Life,* 628–30.
123. Taubman, *Khrushchev*, 582–85, 602–05; Fursenko and Naftali, *Khrushchev's Cold War*, 510–11, 517–23, 526–28; Giglio, *The Presidency of John F. Kennedy*, 216–20; Powaski, *March to Armageddon*, 109–12. See also Beschloss, *The Crisis Years*, 576–77, 586–88. 596–97, 618–38; Reeves, *President Kennedy*, 548–56, 593–94, 606, 618; Dallek, *An Unfinished Life*, 607–630; Wittner, *Resisting the Bomb*, 419–29; Mueller, *Atomic Obsession*, 75–77.
124. Toni A. Perrine, *Film and the Nuclear Age: Representing Cultural Anxiety* (New York: Garland Pub., 1998), 172–74; Wheeler Winston Dixon, *Visions of the Apocalypse: Spectacles of Destruction in American Cinema* (London: Wallflower Press, 2003), 42–43; Kim Newman, *Apocalypse Movies: End of the World Cinema* (New York: St. Martin's Griffin, 2000), 151–53; Joyce A. Evans, *Celluloid Mushroom Clouds: Hollywood and the Atomic Bomb* (Boulder, CO: Westview Press, 1998), 144.
125. Jerome Shapiro, *Atomic Bomb Cinema* (New York: Routledge, 2002), 124–39; Perrine, *Film and the Nuclear Age*, 207–08.
126. Henriksen, *Dr. Strangelove's America*, 235–36; Newman, *Apocalypse Movies*, 134–35; Mark Hall, "The Day the Earth Caught Fire," in Jack Shaheen, ed., *Nuclear War Films* (Carbondale, IL: Southern Illinois University Press, 1978), 39–45.

127. A. Costandina Titus, "The Mushroom Cloud as Kitsch," Scott C. Zeman and Michael A. Amundson, eds., *Atomic Culture: How We Learned to Stop Worrying and Love the Bomb* (Boulder, CO: University Press of Colorado, 2004), 110–12; Mick Broderick, "Is This the Sum of Our Fears? Nuclear Imagery in Post-Cold War Cinema," in Zeman and Amundson, eds., *Atomic Culture*, 136–41.

128. Newman, *Apocalypse Movies*, 154–55; Editorial review, online, http:www.amazon.com; David Wingrove, ed., *Science Fiction Film Source Book* (New York: Longman, 1985), 255.

129. Henriksen, *Dr. Strangelove's America*, 331, 336–39; Wingrove, ed., *Science Fiction Film Source Book*, 200–01.

130. Rose, *One Nation Underground,* 48; Winkler, *Life Under a Cloud, 176–77*; Henriksen, *Dr. Strangelove's America*, 331, 336–39; Evans, *Celluloid Mushroom Clouds*, 159–63.

131. Shapiro, *Atomic Bomb Cinema*, 143–53; Ronnie D. Lipschutz, *Cold War Fantasies: Film, Fiction, and Foreign Policy* (Lanham: Rowman & Littlefield, 2001), 88–89, 91–92; Rose, *One Nation Underground,*74–76; Winkler, *Life Under a Cloud*, 177–78; Evans, *Celluloid Mushroom Clouds*, 163–68; David Seed, *American Science Fiction and the Cold War: Literature and Film* (Edinburgh: Edinburgh University Press, 1999), 145–56; Newman, *Apocalypse Movies*, 156–62; Henriksen, *Dr. Strangelove's America*, 317–27; Anthony Macklin, "Sex And Dr. Strangelove," *Film Comment* 3 (Summer 1965): 55–57; Charles Maland, "Dr. Strangelove: Nightmare Comedy and the Ideology of Liberal Consensus," in Steven Mintz and Randy Roberts, eds., *Hollywood's America: United States History Through Its Films* (St. James, NY: Brandywine Press, 1993), 252–64; Lawrence Suid, "The Pentagon and Hollywood: Dr. Strangelove or: How I Learned to Stop Worrying and Love the Bomb (1964)," in John E. O'Connor and Martin A. Jackson, eds., *American History/American Film: Interpreting the Hollywood Image* (New York: Frederick Ungar Pub., 1988), 219–35.

132. DeGroot, *The Bomb*, 206–07. See also Walker, *The Cold War*, 166–67; Winkler, *Life Under a Cloud*, 82–83, 167; Paul Boyer, *Fallout: A Historian Reflects on America's Half-Century Encounter with Nuclear Weapons* (Columbus, OH: Ohio State University Press, 1998), 95–102.

133. Garrison, *Bracing for Armageddon*, 133–35; "Nuclear Utilization Target Selection," online, http://www.sccs.swarthmore.edu/users/08/ajb/tmve/wiki100k/docs/Nuclear_utilization_target_selection.html.

134. Rose, *One Nation Underground,* 67–68; Stevenson, *Thinking Beyond the Unthinkable*, 75–96. See Sharon Ghamari-Tabrizi, *The Worlds of Herman Kahn: The Intuitive Science of Thermonuclear War* (Cambridge, MA: Harvard University Press, 2005), which looks at the many faces of Kahn.

135. By 1969 the USSR had 1,060 ICBMs—slightly more than the 1,054 of the United States. See Powaski, *March to Armageddon*, 114.

136. Ibid., 113–21.

137. Ibid., 121–26; Rodney P. Carlisle, ed., *Encyclopedia of the Atomic Age* (New York: Facts on File, 2001), 292–93.

138. Herring, *From Colony to Superpower*, 755–56; Powaski, *March to Armageddon*, 113. See also Harland B. Moulton, *From Superiority to Parity: The United States and the Strategic Arms Race, 1961–1971* (Westport, CT: Greenwood Press, Inc., 1973).

139. Quoted in Herring, *From Colony to Superpower*, 760.

140. Lawrence S. Wittner, *Toward Nuclear Abolition: A History of the World Nuclear Disarmament Movement, 1971 to the Present*, vol. 3 of *The Struggle Against the Bomb* (Stanford, CA: Stanford University Press, 2003), 4. See also Robert Dallek, *Nixon and Kissinger: Partners in Power* (New York: HarperCollins Pub., 2007).

141. Garrison, *Bracing for Armageddon*, 133–52; Herring, *From Colony to Superpower*, 760–75.

142. Herring, *From Colony to Superpower*, 776–77.

143. Ibid., 794–95, 807, 818–19; Powaski, *March to Armageddon*, 127–45; Winkler, *Life Under a Cloud*, 183–85; Zelizer, *Arsenal of Democracy*, 245, 259; Andrew Rojecki, *Silencing the Opposition: Antinuclear Movements and the Media in the Cold War* (Urbana, IL: University of Illinois Press, 1999), 103–07.

144. Wittner, *Toward Nuclear Abolition,* 1–2.

145. Herring, *From Colony to Superpower*, 818–20, 826–27; Powaski, *March to Armageddon*, 146–61. Kissinger had not made things easier by arranging informal agreements with the Soviets, as he was want to do, to omit some contentious issues from the deal, like the Soviet *Backfire* bomber.

146. Herring, *From Colony to Superpower*, 835–36; Wittner, *Toward Nuclear Abolition,* 41–42.

147. Carter and Ronald Reagan both said that they would honor the limits of the treaty even if it was not ratified.

148. Rojecki, *Silencing the Opposition*, 107–08; Winkler, *Life Under a Cloud*, 185–86; Herring, *From Colony to Superpower*, 851–52, 854; Powaski, *March to Armageddon*, 162–83; Carlisle, ed., *Encyclopedia of the Atomic Age*, 293–94.

149. NuclearFiles.org, Project of the Nuclear Age Peace Foundation, online, http://www.nuclearfiles. org/menu/key-issues/nuclear-weapons/basics/neutron-bomb.htm; Wittner, *Toward Nuclear Abolition*, 7, 21.

CHAPTER 8

Nuclear Power versus The Environment

The Bandwagon Market and the Energy Crisis

INTRODUCTION

About the time that nuclear-generated electricity became viable in the United States in the 1960s, the modern environmental movement was on the rise. One did not offset the other, but their futures were intertwined. Civilian nuclear power emerged in the 1950s as the most promising aspect of Atoms for Peace. But, during the late 1950s and early 1960s using nuclear power to generate electricity was still novel and not well established.[1] In addition, civilian nuclear power was a technology that competed with the federal government's passionate interest in nuclear weapons. The intensification of the Cold War at that time was not about to change the equation very much. Yet the signing of the Limited Test Ban Treaty in 1963 and its diplomatic aftermath had positive impacts on the civilian power program.[2] In such a setting there was some promise for commercial nuclear power, although it proved inhospitable in all but a very few countries. Promotion of newly constructed reactors and accompanying power plants wriggled through the maze of constraints as the emerging energy source made strides by the late 1960s. The upward trajectory did not last long. The forces against nuclear power were much too strong, especially economic woes, environmental challenges, and an energy crisis. Dramatic accidents, especially Three Mile Island (1979), also derailed its momentum. From high expectations to utter dejection, the 1960s and 1970s were erratic years for civilian nuclear power.

"THE GREAT BANDWAGON MARKET"

After a series of fits and starts, the nuclear power industry enjoyed rapid growth by the mid-1960s. It faced tough competition from technologies that had many years' advantage over it, but some people were optimistic about its chances. Utility executive Philip Sporn called it the "great bandwagon market," with a boom in reactors peaking in 1966–1967. In 1966 utilities placed orders for 20 nuclear plants (36 percent of generating capacity purchased) and in 1967 they bought 31 more (49 percent of capacity purchased).[3] By the end of 1969, 97 nuclear plants were in operation, under construction, or had been built. For the moment, the future of the fledgling industry looked bright. Babcock and Wilcox, Combustion Engineering, and General Atomic Corporation joined GE and Westinghouse (the two leaders in the industry) in building reactors and selling them to large private and public utilities.[4] After a modest slowdown at the end of the decade, the boom picked up again in the early 1970s. By 1974, 37 nuclear plants were producing commercial power or had operating licenses, with anticipation of many more.[5]

The breakthrough occurred in December 1963. Jersey Central Power and Light Company agreed to purchase a 515,000-kilowatt nuclear plant with a boiling water reactor from General Electric (GE) to be built at Oyster Creek, New Jersey. This was the first of the "turnkey" projects, that is, GE would build the facility at a fixed price and simply turn it over to Jersey Central when it was ready. Jersey Central claimed that the new facility would produce electricity more cheaply over the long run than a coal-fired plant. This was the first time that anyone used that rationale to construct a nuclear power plant. The turnkey approach was risky and expensive for GE (Westinghouse also built turnkey facilities). It used the technique as a "loss leader" to develop a market for nuclear power plants, hoping that costs would soon fall. GE and Westinghouse signed 11 more turnkey contracts by 1967. Cost overruns and flawed projections of energy demand eventually forced the companies to withdraw from the turnkey market, or to increase their bid prices. They lost $1 billion from such ventures.[6]

Despite the failure of the turnkey strategy, the intense competition between GE and Westinghouse and the optimistic claims about the future of nuclear power prompted the bandwagon market. In the 1960s demand for electricity grew at an annual rate of 7 percent, doubling every 10 years. This spurred utilities to order new plants. It did not hurt that the Northeast blackout (the largest in history) occurred in 1965, when a system failure in the electrical grid literally darkened the East Coast and parts of Canada.[7] The highly dramatic event caused the Federal Power Commission to push for more and better interconnections and power-pooling. Power pools would allow for several electrical utilities to invest collectively in new plant construction and to share the electricity. In turn this encouraged the building of large generating stations and eased concern about excess capacity or overexpansion. In this setting nuclear power seemed more financially viable than ever before.[8]

Both the vendors and the Atomic Energy Commission (AEC) were predicting unrealistically optimistic cost projections for nuclear plants. "Economies of scale" for building large plants, they believed, would cut capital costs per unit of power and enhance efficiency. This presumption placed major emphasis on the minimum demand for power needs over a specific period (base load), while leaving high demand on the system (peak load) or fluctuating demand (intermediate load) to be filled by other providers. Nuclear power plants had to compete with fossil-fuel plants responsible for base loads that had doubled in capacity since the 1950s. Reactors ordered in the late 1960s were gigantic by the standards of the day, possibly as much as seven times as large as any in operation at the time. This brought to light the practice of "design by extrapolation" which meant scaling up the size of plants based on the experiences with smaller ones. Engineers and planners employed the approach for fossil-fuel units going back to the 1950s. In 1963 the largest commercial plant had a capacity of 200 megawatts, but four years later utilties ordered reactors of 1,100 to 1,200 megawatts. Design by extrapolation raised some serious problems because of the assumption that bigger was better and just as easy to manage. That often proved untrue.[9]

THE AEC IN TRANSITION

After John F. Kennedy's inauguration in 1961, the new president appointed Nobel Laureate in Chemistry and discoverer of plutoniun, Glenn T. Seaborg, to replace John McCone as chairman of the AEC. Seaborg favored a strong program of civilian nuclear power, but not with an all-or-nothing zeal for private control. Other national priorities, such as the invigorated space program, left the AEC following more than leading, even in the bandwagon years. Environmental challenges to nuclear power also created problems for the Commission. The Joint Committee on Atomic Energy (JCAE) urged Seaborg and the AEC to seek stronger support from the Kennedy administration for a robust civilian reactor program. Nuclear power advocates hoped that the president would help push power projects forward in a vigorous way, especially by reversing Eisenhower's cutbacks in the demonstration reactor programs. Kennedy agreed to a "new and hard look" at the role of nuclear power, but only within the context of total energy projections for the United States. In November 1962 Seaborg submitted a report to the White House, which focused on long-term goals. It suggested that with only moderate assistance the federal government could help make nuclear power economically competitive. It hailed the light water reactor (LWR) as close to fulfilling this goal for the utility industry.[10] In essence, the report claimed that nuclear power was now commercially feasible. This was expectation more than reality. The AEC's production reactor program in 1961, however, still focused on weapons and other military applications.[11]

A major concern of the Commission was that the scarcity of U-235 (which was less than 1 percent of natural uranium) would scuttle the growth of commercial nuclear power. Breeder reactors, on the other hand, could utilize U-238 (the other 99 percent of natural uranium) and thorium. Experts believed that reliance on breeders would make actual supplies of fissionable material "almost limitless."[12] Working with the Federal Power Commission (FPC), the AEC began developing long-range estimates for American power demand, concluding that nuclear energy could help supplement and even conserve supplies of fossil fuels. This conclusion was based on the assumption that energy demand in the United States would double by 1990 (much too modest), and that "low cost" fossils fuels would be depleted in 100 years (a little too dire). Such estimates, and the limited supply of U-235, led to the belief that breeder reactors were a major part of the solution in the long run. Until breeders could become practicable, the Commission intended to develop advanced converter reactors along with breeders. Converters used fuel more efficiently than LWRs, but produced less fissionable material than they used. Stories of possible shortages of uranium worldwide also spurred interest in breeders.[13]

The Kennedy administration did not believe that the AEC's goals as expressed in the report deserved priority status. The spark in civilian nuclear power would have to come from elsewhere. In 1962 six large central-station plants were operating. The largest of these was Indian Point Unit-1 (New York), and along with the Dresden Nuclear Power Station (Illinois), they had been built with private funds. None of the six were economically competitive. Given these modest beginnings and the limits of AEC stimulus, the daring turnkey approach was all the more precarious. The AEC's converter program never got going, and the breeder experiments were only modest successes. The AEC's future appeared no brighter when Lyndon Johnson assumed the presidency in 1963. Johnson soon called for a cutback of 25 percent in the production of enriched uranium and ordered four plutonium piles shut down. This was a strategic decision which he hoped would challenge the Soviets to do the same. Nikita Khrushchev obliged and announced cutbacks of his own. Momentum in the AEC shifted toward a further reduction in the federal government's role in nuclear power development. With the support of the AEC and the JCAE, Johnson signed the Private Ownership of Special Nuclear Materials Act (August 1964), which terminated the government's monopoly over enriched uranium. Private companies then were able to assume ownership of fissionable material, and after June 20, 1973, the government made mandatory the private ownership of power reactor fuels. The law also allowed the AEC to provide uranium enriching services to domestic and foreign customers beginning in 1969. These changes, along with the fact that the AEC had been declassifying literature on reactor technology since 1955, was meant to stimulate the quicker development of the private nuclear industry.[14]

As the bandwagon market gathered momentum, the AEC did not allow private companies to make all decisions about nuclear power's future. There remained much on the Commission's plate, including developing the breeder and challenges in the areas of safety, health, and the environment. Throughout the 1960s, however, the regulatory apparatus of the AEC was quite small. Of the approximately 7,000 employees, only 339 were regulators in 1964, and only 540 by 1971. Licensing policy, as unsexy as that appeared, eventually became a central debate point for the AEC's detractors. In the 1950s and much of the 1960s, the AEC licensed nuclear plants as if they were research facilities rather than housing commercial reactors. Its rationale was that the reactors could not be licensed until they proved "practical value" after a long period of operation. As historian Samuel Walker observed, "This placed the AEC in the awkward, and to some minds, ridiculous, position of proclaiming the arrival of nuclear power on the one hand while refusing to subject it to [commercial reactor licensing] requirements on the other."[15] It also allowed utilities to sidestep a prelicensing anti-trust review required for commercial reactors. The loophole was not closed until 1970.[16]

SAFETY FIRST OR SAFETY SECOND?

Economic viability was a high priority for the civilian nuclear program, as was the worldwide prestige in promoting the peaceful atom. However, nuclear power in the 1960s and beyond faced its greatest obstacle from unmet environmental challenges. Being able to produce electricity safely entailed layers of concerns that had been modestly addressed, rationalized away, or given low priority. These included everything from plant siting to reactor performance, from thermal pollution to a core meltdown, from public health to energy consumption. For years, the government underplayed radiation and fallout or hid it from the public when testing nuclear weapons, and defended the building of reactors. In both cases, advocates presented the promise of the new technology as outweighing the potential risks. In the 1960s, the biggest financial challenge to nuclear power was the capital costs of building plants. But the biggest environmental debate focused on reactor safety. For nuclear power generation, accidents could occur anywhere along the fuel cycle from uranium mining to waste disposal.[17]

More than other parts of the fuel cycle, reactor safety was a central problem related to licensing, siting, performance, and the potential for mishaps and even disasters. Proponents of nuclear power armed themselves with statistics attesting to the virtues of the technology they embraced, while opponents consistently tried to discredit those findings. Pro-nuclear forces were aware of the dangers of radioactivity, but they were confident that reactors were (or could be) built to avert or at least significantly reduce risk. Opponents eventually collected their own statistics to repudiate the other side, which was a proverbial half-empty instead of half-full glass. More importantly, they pointed to accidents past and present to demonstrate the unpredictability of reactor performance and the potential long-term dangers of radiation. The sides waged the battle on several fronts. The established nuclear community faced an array of upstarts from the burgeoning environmental movement, grassroots activists, and NIMBY (Not in My Backyard) groups (who may not have been opposed to nuclear power per se, but wanted nuclear reactors away from their neighborhoods). And they hoped to fend off as much as possible new government regulations or active citizen participation in decision making. The AEC, enmeshed in its dual role as promoter and regulator, sometimes found itself between the industry and the opposing forces as it attempted to develop siting guidelines, safety standards, and other regulations.[18]

Critics of nuclear power raised questions about health and safety as early as Hiroshima, but such trepidation in no way upstaged or restrained the momentum of nuclear technology. When commercial nuclear power was introduced in the 1950s polls suggested that there was widespread support for it. A 1956 survey stated that 69 percent of the respondents had "no fear" of nuclear power, and for many years after questioning its safety was a minority point of view.[19] Even fallout, which brought the dangers of radiation to America's own backyards, was disassociated from commercial nuclear power as quickly as possible. Reactors are not bombs, proponents would argue, but a means to deliver the blessings of electricity to an energy-hungry society. As the conflict over nuclear power accelerated and points of view polarized, the debate became a "dialogue of the deaf," a comment attributed to Claude Zangger, Switzerland's deputy energy director in the early 1970s.[20]

Beginning in the 1960s and then exploding (so to speak) in the 1970s, nuclear power safety took on greater importance as a public issue. Some nuclear accidents received attention in the late 1950s, no matter how hard governments tried to hide them. The nuclear fire at England's Windscale Pile Number One in 1957 was a good example. Other accidents were not so well documented. According to an exiled Russian scientist, radioactive waste buried in the Ural Mountains blew up in 1958, causing hundreds of deaths. In the United States, an experimental reactor exploded at the National Reactor Testing Station in Idaho Falls in 1961. The AEC maintained that the resulting three fatalities were due to an electrical power-surge blast, but union officials claimed the workers died from radiation. Whatever the cause, publicly acknowledged accidents took the safety question out of the realm of the abstract and into the real world where opponents linked them to flaws associated with nuclear technology.[21]

In 1962 the AEC issued new guidelines for siting nuclear reactors away from large urban populations. The question of siting had been a serious safety concern since the 1940s, but also involved economics. Transmission costs had to be factored into any construction and operating budget. The earliest government-built reactors were constructed at a distance from cities, presumably for security reasons. Newer reactors, constructed with public and private funds, tended to be near cities. Utilities, in particular, wanted their nuclear plants as close to their urban markets as possible. Informally, the AEC tried to compromise over "metropolitan siting" between reactors at extreme distances and those close to or in cities themselves. The implementation of a containment building was one technical fix employed as a hedge against accidents. Beginning with Shippingport in 1957, most nuclear facilities had containment structures of some kind.[22] The AEC guidelines for relating plant size to distance from dense populations formalized the concept of "remote location." The criteria did not include calculations on siting, but they were available.[23] The AEC assumed its procedure provided sufficient latitude for safety as well as allowing commercial reactors to be located in reasonable proximity to customers. Its faith in safety through improved technology and its practice of judging site criteria on a case-by-case basis was a balance between extremes, but not a concrete policy. In fact, such an approach encouraged utilities to request approval of plants as close to urban centers as possible.[24]

A celebrated case emerged when Consolidated Edison of New York sought a permit in late 1962 to build a reactor along the East River in Queens. Since Consolidated Edison announced the site for its Ravenswood plant before the new siting guidelines went into effect, the AEC based its decision to grant the permit on the technical merits of the plant's engineered safeguards. The company's rationale about the safety of the plant rested on its decision to use a double containment design to withstand even a worst-case accident. Complicating matters was the fact that the plant, if constructed, would be the largest power reactor in the world. The AEC's evaluation balanced the growing faith in the economic feasibility of commercial LWRs against the all-too-obvious siting problem. The gigantic reactor would be situated in the heart of the nation's largest city, a fact not lost on locals. Con Ed's desire to build the new plant thus ran headlong into a strong NIMBY response. The public reaction and informal reservations by regulators and Advisory Committee on Reactor Safeguards (ACRS) members led Con Ed to withdraw its application for Ravenswood in January 1964.[25] Such an outcome left the urban siting issue in limbo, although utilities backed off from reactors close to urban areas on their own by the end of the 1960s. The AEC granted some plants permits for falling within Commission guidelines for both safety and distance. Other plants remained in gray areas, meeting one but not both criteria. In those cases, the AEC made judgment calls based on a flexible approach to defining metropolitan siting. Con Ed's 1965 plan to construct a second reactor at Indian Point on the Hudson River (24 miles north of New York City) is a good example of how the AEC reached a decision by attempting to balance reactor size, improved safety features, and population density. Indian Point II, as it turned out, became the unofficial measure of size versus location for judging future reactors.[26]

Another kind of siting issue was encountered in California, but in this case the problem was earthquakes. Local opposition to siting of nuclear power plants was most prevalent in the Golden State. Pacific Gas and Electric Company (PG&E), short on new hydroelectric sources in the late 1950s, planned to build a nuclear power plant on the California coast north of San Francisco near the village of Bodega Bay. Its president had stated around this time that PG&E "will depend on the atom more and more as time passes."[27] Aside from fishing, the tiny Bodega Bay community was famous as the location for Alfred Hitchcock's movie, *The Birds*. In this case, population size was not the problem as it had been in New York. Locals objected because of the area's beauty, and the impact that such a plant would have on fishing and marine research.[28] The Sierra Club became involved in the protest. But it was the local Northern California Association to Preserve Bodega Head and Harbor, which lead the challenge largely over NIMBY concerns.[29] While not the immediate cause for concern, the group rallied around a potent weapon—seismic evidence—to discredit the project. PG&E had not overlooked the possibility of earthquakes affecting its plant, but that the site would be within several hundred feet of the notorious San Andreas fault had to give everyone pause. The utility company's experts dismissed the concern as not warranted by the available evidence. Opponents had a much different view.[30] The fallout controversy also got tangled in the debate. In one of her songs, communist folksinger Malvina Reynolds declared that PG&E was

> . . . *spreading atomic poison stuff*
> *Over all of the Golden West.*
> *They're starting a plant at Bodega,*
> *A place that was wild and pure,*
> *They call it an atomic park,*
> *But it's an atomic sewer.*[31]

The AEC found itself considering the conflicting claims, but with little expertise to do so. The plant's opponents chose not to wait for an AEC ruling and turned to the Department of Interior, especially its conservation-minded secretary Stuart Udall, for help. Without clear authority, Udall weighed in with strong reservations about the safety of the proposed plant. The

US Geological Survey also became involved in evaluating the practicability of the location. All of this was played out in the press. During the site investigation both the Geological Survey and PG&E's consultant discovered a recent fault line in the bedrock at the proposed site, with other information leading to the conclusion that the site might not be tenable. At best, the new findings and conflicting and inconclusive reports fueled the controversy.[32] Once again the regulatory staff at AEC and the ACRS could not agree on a decision, but in this case the regulators came down in opposition while the latter did not. With no clear-cut support for its plan, and the possibility of further negative publicity, PG&E withdrew its application in 1964 six years after its initial announcement of the project. As in the case of Ravenswood, the decision did not lead to clear policy formation for the AEC. At Malibu in Southern California some of the same conflicting issues plagued nuclear plant siting as well, and controversy over siting on or near fault lines persisted throughout the 1960s.[33]

The glut of licensing requests after 1966 strained AEC's review system, caused serious delays, and limited its efforts to order safety modifications in existing plants. The partial meltdown at the Fermi Unit 1 demonstration breeder reactor plant near Detroit in 1967 was a clear warning about the need for careful licensing and monitoring practices, and it raised concern about breeders specifically. In this instance, the safety controls worked, no large-scale release of radioactivity occurred, and no one was injured. But the plant had been plagued with problems from the start. While the AEC issued it a new operating permit and the plant functioned on a limited basis through 1972, Fermi Unit 1 was ultimately shut down and dismantled. By the time anti-nuclear protests heated up in the 1970s, the meltdown there became the centerpiece of suspicion.[34] The building of the Monticello reactor in Minnesota also publicized the nuclear safety issue. The Northern States Power Company (NSP) requested permission to construct the reactor in 1966. What made the project distinctive was the extent to which well-organized citizen involvement (maybe the first of its kind) played a role in the construction and operating license stages. The controversy revolved around a battle between the AEC and a state pollution control agency over regulating plant emissions. The Minnesota Pollution Control Agency (MPCA), formed in May 1967, attempted to gain jurisdiction over radioactive emissions from the plant. The AEC claimed sole responsibility, which was a position the Supreme Court concurred with in 1971. But public protest went beyond the jurisdictional battle, raising the question of citizens' rights in the licensing and siting of plants. The University of Minnesota held a Symposium on Nuclear Power and the Public in response to the dissent. The protesters did not win their battle, but the safety issue percolated as public awareness and concern over nuclear power intensified.[35]

Siting was the proverbial tip of the iceberg when it came to addressing the general question of nuclear power safety. Through the early 1970s, the AEC treated safety programs as in-house matters. The Commission appointed Milton Shaw, a former aide to Admiral Rickover, to head the new Division of Reactor Development and Technology. Shaw's critics complained that he was more interested in commercialization of the breeder reactor than in safety issues per se. Under Shaw the AEC focused primarily on reactor safety research. It gave little attention to other potential problems along the fuel cycle. It also embraced the general position that reactors had enough engineered safeguards built into the plants and containment structures to make them safe. By 1966 internal debates made clear that risk assessments based on the small experimental plants were not adequate for the larger facilities going on line. Yet major changes in the AEC safety program failed to materialize through the remaining years of the 1960s.[36]

Concern over core meltdowns caused by the loss of coolant in a reactor had led to focus on the core cooling system. This issue was moving the AEC closer to emphasizing accident prevention as a goal of its safety program, but still within the realm of engineered safety. Although the

AEC staff and the ACRS considered it unlikely, a severe loss of coolant in the reactor could lead to the buildup of heat in such a way that the core of radioactive fuel melted and possibly breached the containment structure. This was a particular worry as plants began to increase in size. In the worst case, the fuel could burn through the foundation of the building causing a so-called "China Syndrome," that is, metaphorically at least, searing through the earth all the way to China. The eventuality of a serious core meltdown became a major concern in the mid-1960s because of the design of much larger plants than before.[37] The AEC did not have good data available to determine the adequacy of the untried Emergency Core Cooling System (ECCS) that engineers were installing in the larger plants. The ECCS was based on a redundant technology designed to provide water rapidly to the core. Tests run in 1970 and early 1971 showed that ECCS in light-water reactors might not work as designed. There were sharp disagreements about ECCS among researchers at the national laboratories and at the AEC itself. But rather than conduct new research, the AEC impulsively issued interim criteria on cooling systems in 1973, hoping the system would "perform adequately." The newly formed Union of Concerned Scientists (1969) forcefully questioned the reliability of the ECCS technology, the adequacy of testing, and the criteria that AEC established. While some AEC scientists had reservations about the system, the Commission held firm that its critics exaggerated problems of ECCS and that its engineers could correct the flaws. The damage to AEC's credibility was done. Instead of demonstrating that it needed to rethink ECCS, the controversy reinforced the belief that the AEC was merely a promoter of nuclear power.[38]

Particularly disturbing on the larger issue of radiation was an internal debate over the dangers of radioactivity, which erupted after the publication of a highly controversial story, "The Death of All Children," in *Esquire* (1969) written by physicist Ernest J. Sternglass. According to Sternglass, radioactive fallout from atmospheric nuclear tests in Nevada in the 1950s could cause the deaths of 400,000 babies, especially in the South (which lay in the path of the fallout from the Nevada bomb site).[39] In response, two staff members of the Lawrence Radiation Laboratory, Drs. Arthur R. Tramplin and John W. Gofman, were assigned the task of evaluating (and discrediting) the Sternglass story and the effects of low-level radiation. Tamplin and Gofman only disagreed with the size of Sternglass's numbers, not his general conclusions. In the fall of 1969, they charged the AEC with poor radiation protection standards and repeated their claims before congressional investigating committees.[40]

The AEC took great exception to what they believed to be Tamplin and Gofman's exaggerated findings about cancer deaths from radiation and stood by the adequacy of the Commission's radiation regulations themselves. By this time, the controversy turned Tamplin and Gofman into public critics.[41] Whether the two scientists were overzealous whistle-blowers or martyred detractors, the controversy pushed the AEC to revise its radiation protection regulations. Tamplin and Gofman's resignations several years later (1976) gave the anti-nuclear movement a significant publicity tool.[42] However, the open airing of debates among experts, at least in the public mind, raised questions about expert authority in general. It is not that Tramplin and Gofman were publicly heralded or discredited, but that the fallibility of expertise was observable in clashes between professionals, and between employees and their employers.[43]

ENVIRONMENTAL CHALLENGE: LOCAL PROTEST AND THE NATIONAL MOVEMENT

The anti-nuclear movement began in the late 1940s in opposition to the bomb, and sometime later it focused on atomic testing. The movement soon became a worldwide phenomenon. Protest against civilian nuclear power arose as early as the late 1950s as an admixture of conflicting scientific expertise (Tamplin and Gofman), local protests (a la Bodega Bay), concerns by labor unions over worker safety, and criticism from environmental organizations. By the early 1960s anti-nuclear protests intensified particularly in industrial countries like the United States and Great Britain. What came to be known generically as the "anti-nuclear movement" was not a coherent body, but a collection of individuals and groups questioning the use of atomic energy in almost any form. Unraveling the various pieces of anti-nuclear protest is complex and somewhat inexact. The movement is intriguing because scientists and engineers conducted the development of the bomb and much of the work on nuclear power technology behind closed doors. Civilian nuclear power also is a commercial technology intentionally developed and promoted by governments in an essentially undemocratic (or anti-democratic) way.[44]

The binding issue in the United States had to be the fallout scare in the 1950s. Hiroshima and Nagasaki were events "over there," as had been the earliest bomb tests in the Pacific. Turning areas at or near Hanford and Oak Ridge into radioactive waste sites was not public knowledge at the time. But fallout threatened the air, water, and even the milk supply in America's backyard. While national alarm over fallout subsided in the 1960s, memories of radiation as risky business never totally disappeared. Within the AEC, scientists like Tomplin and Gofman questioned the seeming unified front of the nuclear iron triangle on the safety of nuclear power. As with Ravenswood and Bodega Bay, protests directed at proposed nuclear plant sites were a blend of wariness about potential accidents and a desire to see a location other than their community absorb the risk.[45] With a few exceptions, public protests were limited during the nuclear power boom between 1962 and 1966. During that period local citizen groups challenged only 12 percent of the license applications, compared with 32 percent from 1967 to 1971. Between 1970 and 1972, they challenged 73 percent of the applications.[46]

The late 1960s brought another issue for nuclear protesters to rally around—thermal pollution. This involved the effects of waste heat generated by nuclear power plants along rivers, lakes, and oceans. In the case of both nuclear and fossil fuel power plants the steam that drove electricity-producing turbines had to be cooled. Condensing the steam occurred by circulating cool water through the system from a nearby source. The process heated the cooling water by 10 to 20 degrees Fahrenheit before it was returned to its place of origin. Greatly heated water was not unique to nuclear power plants, but it was a serious problem because the plants used cooling water less efficiently than fossil-fuel facilities. Biologists and ecologists discovered that the major impact of thermal pollution was on aquatic life. The heat killed fish and caused some plants to flourish to the point of changing the biological balance of a river or stream. Cooling ponds or cooling towers eased many of the problems that thermal pollution caused, but utilities were reluctant to add them because of the cost.[47]

In the mid-1960s the Fish and Wildlife Service of the Department of Interior suggested that the AEC take up the issue of thermal pollution. The AEC was unwilling to do so, let alone mandate such technologies. They claimed that their regulatory power did not extend beyond radiation hazards (upheld in federal courts), and resisted expert and citizen complaints that came to them. Important public challenges to thermal pollution, nevertheless, occurred in 1968 and 1969. One of the best known protests was staged against the Bell Station on Cayuga Lake in upstate New York, where New York State Gas and Electric (NYSGE) filed an application for the station in 1967. As with Bodega Bay, a number of groups participated in the event including scientists from Cornell and the Cayuga Lake Association. Although NYSGE believed that the impact of the new reactor would be negligible on the adjacent water, and are convinced that opponents were using evidence selectively, they postponed their request. This incident was another early example of local site opposition, but turned on a different issue than in the California cases. Change in the general practice of cooling plant water soon came. Under public pressure and because of the frustration of utilities facing construction postponements, Congress passed legislation to extend the AEC's authority over thermal pollution. By 1970–1971 most new plants built (or planning to be built) on inland waterways added cooling ponds or towers. The positive response to public pressure and congressional action did little to improve the AEC's image. Debate over thermal pollution also brought into question claims that nuclear power was better for the environment than fossil-fuels.[48]

The rise of the modern environmental movement widened the challenging of nuclear power beyond local site protests. The environmental critique grew beyond NIMBYism and even beyond specific challenges to the AEC's radiation data. The bigger question was whether nuclear power under any circumstances should be part of the nation's energy mix. While the modern environmental movement emerged after World War II, not until the late 1960s did environmentalists have the tools to take on subgovernment dominance of nuclear power. The conservation movement of the pre-World War II era had focused on conserving, preserving, managing, and otherwise protecting the nation's resources. Thus it expressed a different mindset than what was to follow.[49] At the heart of modern environmentalism was the emergence of the "new ecology." The basic concept revolved around "the relationship between the environment and living organisms," particularly the reciprocal relationship between the two.[50] The emergence of ecology as a science coincided with the industrial era beginning in the early twentieth century. By the 1960s, ecology changed from a scientific concept to a popular one, questioning traditional notions of progress and economic growth.[51] Rachel Carson's *Silent Spring* (1962), a grim warning of the dangers of pesticides, seemed to best capture the new spirit. Career ecologists were beginning to make it clear that "respect for the biosphere, like respect for justice, must continuously have a place in law and government."[52]

Ecology was helpful in guiding reformers from the utilitarian conservationism of the past to an era emphasizing environmental quality and personal health and well-being. At the forefront of the new environmentalism were citizen and public-interest groups as well as a variety of experts. Between 1901 and 1960, an average of three new public-interest conservation groups appeared annually; from 1961, 18 per year. The 1970s witnessed the most dramatic rise in the modern environmental movement. Major victories like the Wilderness Act (1964) and several new pieces of environmental-friendly legislation during the Johnson administration were important. Also significant were the efforts of preservationist groups such as the Sierra Club (1892) and the National Audubon Society (1905), and newer organizations such as the Conservation Foundation (1948), Resources for the Future (1952), and the Environmental Defense Fund (1967).[53]

Although environmental groups had yet to identify a common agenda, the tone and spirit of environmentalism were changing. Quality-of-life issues, pollution control, critique of consumerism, growing interest in the preservation of natural places, and distrust of nuclear power indicated a giant step away from "wise use" of resources and a challenge to traditional faith in economic growth and progress. Despite the persistent battles fought in the public arena over preservation versus development, and economic growth versus environmental quality, the 1960s saw major strides in national environmental legislation.[54] Legislation alone did not guarantee improved conditions. What connected the older movement with the new one was the faith in science and technology to solve environmental problems. This was rather schizophrenic, since some had blamed science and technology for the excesses of the new consumer culture, while at the same time government and other leaders sought the advice of scientists and technical experts to help eradicate pollution and improve quality of life. Nuclear power fell into this category of environmental schizophrenia, that is, criticism of the technology and promotion of science and technology to help solve its problems. For a number of political activists, no technical fix could correct the potential harm of a nuclear accident, and they could settle for nothing less than its abandonment.

Girl Scout in canoe, picking trash out of the Potomac River during Earth Week, April 22, 1970. Beginning in the 1960s, the modern environmental movement was on the rise in the United States, and its future intertwined with the viability of nuclear-generated electricity. *Source:* Library of Congress, Prints & Photographs Division, [LC-DIG-ds-00750]

What made the modern environmental movement so remarkable was the speed with which it gained national attention beginning in the 1960s. Nothing epitomized that appeal better than Earth Day. The idea began as a "teach-in" on the model of an anti-Vietnam War event. The staff of Environmental Action declared, "On April 22, [1970,] a generation dedicated itself to reclaiming the planet. A new kind of movement was born—a bizarre alliance that spans the ideological spectrum from campus militants to middle Americans. Its aim: to reverse our rush toward extinction." Across the country, on 2,000 college campuses, in 10,000 high schools, and in parks and various open areas, as many as 20 million people celebrated purportedly "the largest, cleanest, most peaceful demonstration in America's history."[55] As a symbol of the new enthusiasm for environmental matters, and as a public recognition of a trend well underway, Earth Day served its purpose.[56] While Earth Day was good publicity for environmentalism, the tool that changed the anti-nuclear movement in the United States from simple public outrage to formidable opponent was the National Environmental Policy Act (NEPA, 1969). On January 1, 1970, four months before Earth Day, President Richard Nixon signed NEPA. While opposing the bill until it cleared the congressional conferees, the Nixon administration ultimately embraced it as its own. Going on record against "clean air, clean water, and open spaces" served no purpose. Many people trumpeted their approval of the president's gesture; others reserved judgment or remained cynical.[57]

NEPA, emerging in the wake of the Santa Barbara oil spill, was primarily the work of several congressional Democrats, especially Senators Edmund Muskie (Maine), Henry Jackson (Washington), and Gaylord Nelson (Wisconsin), Rep. John Dingell (Michigan), and Jackson's chief advisor, Indiana University professor Lynton K. Caldwell. While far from "the Magna Carta of environmental protection" that some people proclaimed, NEPA called for a new national responsibility for the environment. It was not simply a restatement of resource management, but promoted efforts to preserve and enhance the natural world. NEPA particularly emphasized the application of science and technology in the decision-making process and in the search for environmental solutions. The provision mandating action required federal agencies to prepare environmental impact statements (EISs) assessing the environmental effects of proposed projects and legislation. These were to be made public. NEPA provided substantial opportunity for citizen participation, especially through access to information in agency files. It established the Council on Environmental Quality (CEQ) to review government activities pertaining to the environment, to develop impact-statement guidelines, and to advise the president on environmental matters. While NEPA could be manipulated, it increased accountability for environmental actions. In June 1970 government officials announced that pollution-control programs and the evaluation of impact statements would be the responsibility of a new body, the Environmental Protection Agency (EPA).[58]

EPA began operations in December 1970 under the direction of William Ruckelshaus. Initially, it included divisions of water pollution, air pollution, pesticides, solid waste, and radiation. Other natural resource and environmental programs remained in a variety of agencies and departments, especially the Departments of Commerce and Interior. More significantly, EPA did not have single overall statutory authority for environmental protection; it simply administered a series of specific statutes directed at particular environmental problems. The agency was soon inundated by its regulatory responsibility and pressured by efforts of industry to skirt around those regulations.[59] It was NEPA, and especially the ability to utilize environmental impact statements, that gave the anti-nuclear movement leverage. In many respects, the anti-nuclear movement was a precedent-setter for the modern environmental movement as a whole. The anti-nukes of the 1960s and 1970s had their own precursors, including liberal and internationalist scientists like Leo Szilard and members of the Federation of American Scientists, who questioned the use of the bomb and/or did not want it under military control; ban-the-bomb groups and pacifists like Bertrand Russell; and expert critics of fallout and radiation who came to their positions through testing nuclear weapons, such as Linus Pauling and Herman Muller and then Sternglass, Gofman, and Tamplin. In the 1970s anti-nuclear groups may have been the first to use nonviolent protest in activist politics, and may have been among the largest protest groups of the era.[60]

As time passed, the debate over reactor safety especially drew together more coalitions of local citizen groups and national environmental organizations, some who were new to the anti-nuclear debate. For example, the Consolidated National Intervenors (CNI), which organized to participate in the hearings over ECCS, was composed of some 60 local and national bodies, including the Union of Concerned Scientists, the Sierra Club, Chicago's Businessmen in the Public Interest, and the New England Coalition on Nuclear Pollution. In this case, the primary concern was not banning nuclear-generated power, but making the plants safe.[61] Other groups, such as Friends of the Earth (FOE), became involved in opposition to nuclear power, and peace groups like Mothers for Peace added civilian nuclear power to its list of grievances. Greenpeace, an international environmentalist group formed in the early 1970s, initially organized around efforts to block nuclear testing in the Pacific, and became known for its direct-action tactics. Consumer advocate and corporate critic Ralph Nader became part of the anti-nuke movement, and in 1974 with FOE and UCS, sponsored the first "Critical Mass" conference of anti-nuclear groups to establish a new network.[62]

ENVIRONMENTAL LEVERAGE: CALVERT CLIFFS

While the protests grew more intense during the 1970s, environmentalists scored some important victories for the anti-nuclear cause. Pro-nuclear groups countered the growing tide of criticism. They strongly asserted that by obstructing the development of nuclear power, opponents were undermining an alternative to expensive foreign oil, and ignoring a clean source of energy that was "smog free."[63] The impact of the environmental movement, nevertheless, was felt in the changing role of the AEC and governmental participation in nuclear power development. The Commission became vulnerable to the charges of nuclear detractors that it could not be promoter and watchdog at the same time. Limited operating experience to make judgments at the very least challenged the regulatory staff.[64] Critics also assailed the AEC for not providing sufficient public input into policy decisions and limiting disclosure of information about nuclear power. The dynamic growth of the industry and the emergence of new technologies were probably placing more pressure on the AEC at the time than its critics. It could not keep up with licensing requests, which compounded the need to examine and possibly rethink the slew of regulatory issues.[65]

On the 25th anniversary of the AEC in 1971, a major reorganization and consolidation was finally deemed necessary. This was proposed partly to maintain the federal hold over nuclear development and partly in response to criticism. President Nixon appointed an economist and former assistant director of the Bureau of the Budget, James Schlesinger, as the new chairman of AEC in August. Nixon was exchanging an eminent scientist from within the nuclear establishment (Glenn Seaborg), for a proven administrator outside of that establishment. The goal was to oversee tightening internal operations and refining AEC's regulatory functions. Nixon did not intend to change the Commission's role entirely, but to meet the increasingly complex demands on the civilian side of its responsibilities. Settling on a fixed set of licensing measures was a goal.[66]

The biggest change of all, and one that gave the anti-nuclear movement one of its most powerful tools, came through a 1971 court case. The Federal Court of Appeals nudged the AEC toward paying greater attention to environmental costs. *Calvert Cliffs Coordinating Committee v. AEC* grew out of a citizen protest against Baltimore Gas and Electric Company's Calvert Cliffs Nuclear Generating Station near Lusby, Maryland. It focused on thermal pollution and other issues. The court found AEC regulations in violation of the NEPA mandate to make a detailed assessment of costs, benefits, and the environmental impact of nuclear power plants before licensing them.[67] Until that time, the AEC and the nuclear industry interpreted NEPA very narrowly and supported very limited citizen input in the licensing procedures. The Commission had been reluctant to judge environmental effects of applications beyond safety issues, or to increase citizen participation in public hearings. Hoping to avoid more litigation and to restore the AEC's faltering image, Schlesinger announced that he would not appeal the decision. Instead the Commission would make substantive changes in environmental review and licensing practices. Industry leaders were not pleased with his response or with his unwillingness to appeal. The *Calvert Cliffs* decision did not transform the nuclear industry or destroy the AEC's authority. It established an important precedent, ultimately reaching beyond nuclear power itself, with respect to environmental impact statements and the increasing role of citizen participation in environmental issues. To nuclear advocates, *Calvert Cliffs* made licensing more cumbersome and politically volatile, because it slowed down the process significantly.[68] Yet the AEC's problems went deeper than compliance with the court's decision. The potential for additional public participation related to nuclear power certainly affected the way the Commission did future business. In some ways the effort to promote commercial nuclear power heavily depended upon claims of its competitive advantages over other types of energy. Such a stance was walking a thin line if something did go wrong.[69]

THE ENERGY CRISIS: AN OPPORTUNITY FOR NUCLEAR POWER?

The early 1970s was a contradictory time for nuclear power. The market appeared strong for new and future reactor orders. But the AEC was suffering through a period of heightened criticism as licensing and siting issues ran up against local resistance and environmental challenges. Somewhat contrarily, signs of an energy shortfall in 1969 and 1970 (natural gas and heating oil shortages in the winter and chronic "brownouts" along the electrical grid in the summer) led President Nixon to advocate strongly for the AEC's breeder program. In his June 4, 1971, "Special Message to the Congress on Energy Resources" he spoke about a rising energy problem, the need to modernize and expand the nation's uranium enrichment program, and called for a successful liquid metal breeder reactor by 1980. "Our best hope today for meeting the Nation's growing demand for economical clean energy," he stated, "lies with the fast breeder reactor. Because of its highly efficient use of nuclear fuel, the breeder reactor could extend the life of our natural uranium fuel supply from decades to centuries, with far less impact on the environment than the power plants which are operating today." He also called for a new energy agency, a Department of Natural Resources (not established), and two new agencies to replace the AEC. The reorganization of AEC, for the moment at least, did not go farther than Schlesinger's changes within the Commission.[70]

When Schlesinger left AEC in January 1973 to become head of the Central Intelligence Agency, Nixon replaced him with Dr. Dixy Lee Ray, who had been appointed to the Commission in 1972. A marine biologist from Washington State, Ray became the first woman chairman of the AEC (the term *chairman* is used for all chairs of the Commission, male or female). The administration viewed her gender and her advocacy of ecological concerns as having important symbolic value. She faced the daunting task of reviewing the government's energy bureaucracy along the lines that the president had outlined in his 1971 speech. In April 1973, Nixon issued his second energy message, which was a strong plea to step up production through natural gas decontrol, leasing of the outer continental shelf, pushing ahead with the Alaska pipeline, a new oil import program, and easing environmental standards. In November, the president announced *Project Independence*, a plan for the United States to meet its own energy needs, with nuclear power as a major component. On December 1 Ray submitted a report to Congress calling for substantial expansion in energy research and development funding and the formation of an Energy Research and Development Administration. Energy self-sufficiency was the goal, and nuclear power was a big part of the mix.[71]

By this time, a world-wide convulsion which came to be known as the "energy crisis" was underway. The energy reorganization goals of Nixon and Ray were complicated by this most recent impasse. They were stymied as well by intense political debates over controlling inflation at home (as oil prices rose internationally), and as the Nixon administration itself imploded because of the Watergate scandal. Needless to say, framing a coherent energy policy became almost impossible. Furthermore, while the energy crisis seemed to have little to do with nuclear power, it simultaneously raised new hopes about future opportunities, but also aroused further suspicion.[72] The link between national and international energy issues was never clearer than in the 1970s. By 1973, the oil producing countries of the Persian Gulf and North Africa, through the Organization of Petroleum Exporting Countries (OPEC), virtually completed the process of controlling their own oil supplies and strongly influencing oil prices. These had been the province of multinational oil companies until now. The outbreak of the Arab-Israeli War (the "Yom Kippur war") in October 1973 provided the catalyst for OPEC's emergence as the world leader in crude oil pricing and production. A month earlier, King Faisal of Saudi Arabia warned multinational ARAMCO that the United States faced serious consequences if it continued to aid Israel. He assured the Egyptians that his country would restrict oil exports and cut production if American policy did not change.[73]

The war was, among other things, an expression of Arab unwillingness to accept the pro-Israeli stance of the United States, and their warnings were revealed through a new oil policy. The war did not cause the change, but came on the heels of global price inflation, rising nationalism in the Third World, and the shift in oil exploitation from the Western Hemisphere to the Middle East and North Africa. It was simply the last straw. The Arab-Israeli conflict prompted an oil embargo directed at the West. The Saudis threatened to use the embargo if the United States and others extended further aid to Israel. Ignoring the warnings, President Nixon authorized a $2.2 billion weapons airlift to its ally. The result was an Arab boycott of oil against the United States and other supporters of Israel. While the emotional impact was great, the embargo itself was never airtight. Iraq and Libya soon sent oil across the Atlantic, and even some Saudi oil (and non-Arab sources) found its way to the United States.[74]

The embargo produced tangible results for OPEC. Imports of oil in the United States dropped from 6 million barrels a day in September to 5 million in subsequent months. Japan and European nations dependent on Middle East crude yielded to Arab demands, putting distance between themselves and Israel. By December, the price per barrel rose 130 percent. After six-months the embargo ended on March 18, 1974. To the industrialized nations, the rapidly rising prices and the embargo were, at least, an overreaction by OPEC, and, at most, exploitation by a dangerous cartel. From OPEC's vantage point, the price revolution was a justifiable response to inflationary economic policies of industrialized nations and to the efforts of multinationals to control supplies and exploit developing countries. Into 1974, there was an uneasy lull for both producers and consumers. The future was more uncertain than ever, although through 1978 the world oil market remained relatively orderly. The stability was broken in the late 1970s by events beyond the control of either OPEC or the multinationals—events including the overthrow of the pro-American Shah of Iran and the rise of a conservative theocracy there under Ayatollah Khomeini.[75]

The United States was never more dependent on oil and natural gas than in the 1970s. The Iranian crisis and a whole new round of rising prices and scarce supplies reinforced the notion that cheap energy was coming to an end. On the eve of the embargo in 1973, oil furnished almost half of the total energy needs of the nation; by 1977 petroleum and natural gas provided 75 percent. The United States was particularly exposed to these changes because of massive consumption (which had doubled since 1950) and dwindling domestic supplies of oil. Excluding Alaska, oil production steadily declined. In 1972 and 1973, oil companies at home produced an average of 360,000 barrels less than in the previous year; in 1968, for the first time in US history, more natural gas was sold than was discovered. In 1970, foreign oil accounted for approximately 22 percent of American consumption; by 1973, 36 percent. For nuclear power, the energy crisis first appeared to offer utiltiy companies a real alternative to constricted oil markets and high fuel costs. Few options other than fossil fuels existed at the time, and a substantial energy conservation program was not politically viable in the United States. AEC's promotional support for nuclear power, even as the bandwagon years were waning, added to a belief among advocates that nuclear power was the answer to OPEC.[76]

Yet there were substantial constraints for nuclear power to take advantage of the energy crisis. Rising fuel prices encouraged coalitions of anti-nuclear groups, consumer activists, and large industrial users of electricity to oppose demands for rate increases. In some states, public utility commissions prohibited utilities from passing on the costs of construction work underway to ratepayers, which particularly hurt nuclear plant projects. Compared to an earlier period, states approved few rate increases for utilities in the 1970s at the point when nuclear power attempted to fill the gap in current American energy needs. Not only did the energy crisis drive up the price of oil and other fuels, but it aggravated the existing problem of inflation, which made money much more expensive to borrow. An economic slump, accompanied by increased unemployment, cut deeply into demand for electricity. Future projections for opportunity in the nuclear power field were dashed as capital costs rose, plans to build new plants were postponed or cancelled, and more than 100 ongoing projects deferred. Between 1975 and 1978 utilities in the United States ordered only 11 new nuclear facilities. In essence, the immediate reason for the nuclear power industry's decline was an energy crisis that appeared at first to promise a rebound in fortunes, but did not. The wheels on the bandwagon had fallen off.[77]

FALL FROM GRACE: NUCLEAR POWER SLUMPS IN THE LATE 1970s

If the economic news of the energy crisis had not been bad enough for the nuclear power industry, critics questioned its credibility on several fronts. The subgovernment supporting nuclear power was losing control of defining nuclear power issues, and its expert authority was being challenged. Anti-nukes dogged the AEC and the nuclear industry itself by using delaying tactics possible through demands for better environmental assessments and through calls for greater accountability and regulatory rigor.[78] On another level, the debate over nuclear power during the 1970s was an encounter between advocates of large, centralized systems and those suspicious of high technology and uncontrolled economic growth. The controversy over centralized power was rife with broad societal, political, environmental, and institutional import. Advocates proposed larger and more numerous nuclear power plants not only as a hedge against rising oil prices and OPEC control, but also as a way to divert petroleum use from stationary power production to other essential needs like transportation. The technology of light-water reactors had already proven itself, they claimed. And, turning the environmental debate on its head, they argued that fossil fuels were a much greater pollution risk than nuclear power.[79]

Opponents were suspicious of centralized power production as represented by large, powerful utilities (including those that employed nuclear systems). Embedded in the anti-growth sentiments of the time was the "Small is Beautiful" mantra of the counterculture. In this context, centralized power production kept energy development in the hands of government and big business, and left consumers vulnerable to their whims. A move toward decentralized (and passive) systems, especially solar and wind energy, would not only reduce the need for nuclear power, but weaken the trend toward corporate control of society.[80] The works of physicist and Friends of the Earth activist Amory Lovins was particularly popular. In a 1976 article in *Foreign Affairs*, he focused the anti-growth debate on energy issues, drawing a distinction between "soft" and "hard" energy paths. Soft paths were "low-energy, fission-free, decentralized, less electrified," while hard paths were "high-energy, nuclear, centralized, electric." Lovins was more concerned with the society that used the energy, than the source itself, but had grave concerns about a "plutonium economy" that failed to take into account questions of unlimited growth and uncontrolled energy use. In this instance, nuclear power was more than risky technology; it was a symbol of capitalist centralization and bloated consumption.[81]

THE RASMUSSEN REPORT: A QUESTION OF CREDIBILITY

The defense of nuclear power took another decided blow in 1974 with the release of the Rasmussen Report. Before he left the chair of the AEC, James Schlesinger tried to meet the criticism concerning reactor safety by initiating a new study with at least the appearance of independence. He selected MIT nuclear engineer Dr. Norman C. Rasmussen to direct the work. While not on the AEC payroll, Rasmussen had links to the nuclear industry, which was almost unavoidable in this field. He had substantial expertise in civilian nuclear power, but had little specialized training in reactor safety. The taskforce he headed included about 60 scientists and engineers, and the nine-volume study took two years to complete. Rasmussen was under pressure to produce the report as quickly as possible, indicating the extent to which the AEC wanted to use the results to defend its safety position. The AEC published the draft report in August 1974; the final version appeared in 1975. The *Reactor Safety Study* (WASH-1400) concluded that the risks from nuclear reactors were very small (1 in 5 billion). The report made the following statement about core meltdowns: "The value obtained was about one in 20,000 per reactor per year. With 100 reactors operating, as is anticipated for the U.S. by about 1980, this means that the chance for one such accident is one in 200 per year [or about 1 in 10 over a period of 20 years]."[82] The fail-safe systems like the ECCS, it added, made a serious accident highly improbable. The most vivid, and most quotable, claim was that the chance of one person dying from a nuclear accident was about the same as being struck by a meteor. To reinforce its conclusions, the report included data for comparative risks, including car fatalities and tornadoes. According to the report, if 100 nuclear reactors were in operation, the average number of deaths per year would be 2 (there were about 200 on line or in construction at the time). Using the same formula, 50,000 people per year would die in automobile accidents, 18,000 from falls, and 6,000 from drowning. These calculations applied to one person per one plant per year, which was misleading because it extrapolated risk across the nation, not taking into account proximity to a faulty reactor.[83]

The AEC and industry officials broadcast the report's findings widely and received favorable press attention. Criticism began almost immediately, however, challenging every aspect of the study from its methodology to its estimates. Its fault-tree method (developed by the aerospace industry) calculated the possibility of events that had never occurred, and focused on errors that could lead to malfunctions rather than to accidents themselves. The Rasmussen team had not made a general study of all US nuclear plants; it focused on a couple of "representative" plants instead. Critics claimed that the report was too theoretical, did not account sufficiently for human error, and did not look beyond reactor safety per se. The American Physics Society completed a "counter-analysis" in 1978, asserting that the probability of a plant breakdown was much higher than claimed in WASH-1400. In 1979, the Nuclear Regulatory Commission qualified its earlier endorsement of the executive summary of the report.[84] Through the mid-1970s at least, nuclear advocates regarded the rather hopeful conclusions of the Rasmussen Report (and the lack of any catastrophic accidents) as confirmation that criticism of the safety program was alarmist. By taking a defensive position on safety, the credibility of the nuclear power industry was vulnerable to any highly publicized accidents to come, which proved to be not very far into the future.

THE AEC BECOMES EXTINCT

The reorganization of the AEC under Schlesinger and Ray was just the starting point for eliminating what was the bureaucratic parent of the nuclear power industry. Nixon's departure after Watergate left the major change to his successor, Vice President Gerald Ford. In response to Ford's request, Congress passed the Energy Reorganization Act of 1974, with the broad purpose of reducing US dependence on foreign oil and other energy supplies. The act was not intended as a denunciation of the AEC, and initially had little to do with the issue of reactor safety. As a consequence, subsystem participants did not oppose what became the dismantling and replacement of the AEC. While not a controversial piece of legislation, the impact of the 1974 act on nuclear power regulation was important. The reorganization dealt with the long-standing conflict of interest within AEC between promotion and regulation by abolishing the Commission and dividing its functions between the Energy Research and Development Administration (ERDA) and the Nuclear Regulatory Commission (NRC). ERDA initially handled the non-regulatory issues (reactor development, physical research, and military applications),[85] while the NRC had responsibility for nuclear reactor regulations, materials safety, and regulatory research.[86]

Supporters of nuclear power viewed the reorganization as an expansion in research and development as well as reduction in bases for criticism that were perpetual with the AEC. Opponents supported the changes, hoping that cleaving the promotional role from the regulatory function would result in better oversight. In Congress the bill gained widespread support in its final form. President Ford signed it on October 11, 1974, and the AEC was formally abolished on January 19, 1975. The institutional changes, in and of themselves, did not end the battles over civilian nuclear power, and even may have made things more complicated. Although the NRC was prohibited from promotional activities that were a statutory obligation of AEC, it still maintained strong support for the nuclear industry.[87] The reorganization also affected the Joint Committee on Atomic Energy. The JCAE had never demonstrated much sympathy for environmental issues and sustained itself only as a promotional force. Yet those in Congress not connected with the JCAE increasingly resented its sweeping power. House Democrats began to whittle away at that power, ultimately splitting its functions among five standing committees. The Senate supported the House action, and a revision of the 1977 Atomic Energy Act abolished the JCAE.[88]

CONFIDENCE UNDERMINED: NUCLEAR ACCIDENTS FROM BROWN FERRY TO SEABROOK

As a near-future alternative to oil and gas during the energy woes of the 1970s, nuclear power first appeared to be in an opportunistic position to bolster its place in the civilian power world. It also weathered changes in the subgovernment with the Energy Reorganization Act of 1974. But slumping construction after 1974, pressure from anti-nuclear forces, and cracks in its credibility set up nuclear power for decline. This was especially true if anything else tipped the scales away from its delicate balancing act. The Three Mile Island disaster in 1979 was the most dramatic event in undermining further nuclear power development. But other accidents and public protests before then weakened the foundations of the industry well in advance of the bad news out of Pennsylvania. A peculiar, but prophetic accident occurred in 1975 at the Browns Ferry nuclear plant near Athens, Alabama. The TVA plant was the world's largest nuclear generating facility. On March 22, a fire broke out under the control room: a candle used by an employee to check for ventilation leaks ignited a sealant on electrical cables that controlled the emergency core-cooling system. The burning of the ECCS cable controls caused malfunctions in seven of the 12 safety systems. The fire lasted longer than it should because (through a misunderstanding) water was not used to douse it. Fortunately, no major water coolant pipe broke during the eight-hour fire, and no core meltdown occurred. Widely reported in the press, the NRC played down the accident while anti-nuclear forces played it up. Nuclear advocates pointed to the success of the redundancies built into the plant, but critics had legitimate cause for concern. Within its first year of operation (beginning in August 1974), 65 "abnormal occurrences" took place in the plant. Those who argued that the Rasmussen Report failed to take into consideration "Murphy's law" (whatever can go wrong, will go wrong) had a point, especially in a situation where inadequate safety regulations were in place.[89]

Tensions mounted between pro- and anti-nuclear groups, with public opinion still split over the advisability of civilian nuclear power. Between 1972 and 1976, efforts to stop nuclear expansion and the operation of plants in California failed at the polls. Nuclear critics blamed the

Stop Nuclear Power poster, circa 1977–78. *Source:* Clamshell Alliance/Library of Congress, Prints & Photographs Division, [LC-USZC4-2866]

well-financed pro-nuclear groups (especially the Atomic Industrial Forum) and stated that the legislature undermined the initiative by passing unfavorable laws. The vote on the latest measure was two-to-one against the proposition. In 12 other western states, only one nuclear initiative passed. Frustration over the defeats stimulated direct-action protests in 1976, although groups initiated some small ones as early as 1974.[90] In the East, dissent harkened back to the sixties in the form of mass demonstrations and acts of civil disobedience. Unlike the delaying tactics of environmental groups during licensing hearings, some anti-nuclear groups took to the streets. The best known event happened in Seabrook, New Hampshire. In July 1976, under the leadership of the Clamshell Alliance, 18 protesters staged a sit-in at the Seabrook plant site, followed by 180 in August. In April 1977 about 2,400 protesters temporarily occupied the site. (Prior to the initial demonstration, residents tried legal channels to voice their grievances. Failing that, the mass demonstration followed.) The sit-in was inspired by an immense occupation of a nuclear plant site, possibly as many as 28,000, in Whyl, West Germany.[91] Formed in July 1976 as a coalition opposed to the Seabrook plant, the Clamshell Alliance clearly was the most important anti-nuclear group in New England, serving as a model for future nonviolent protests of its kind. The name, symbolically at least, grew out of concern that the plant's pollution would destroy local clam beds. Ultimately factionalism tore the group apart, but its actions at Seabrook spread across the nation.[92]

The incident at Seabrook became a media event, with many of the demonstrators arrested and carted away. Although the EPA refused to approve a permit for the plant's cooling system, and the NRC suspended all but excavation at the site, another mass demonstration took place in April 1977. The tension was greater, with Governor Meldrim Thompson taking a hard line against what he called the "terrorists."[93] Seabrook and the Clamshell Alliance, if not successful in thwarting development of nuclear power in the East, inspired others to assume the same tactics elsewhere. One of the best known was the Abalone Alliance in California. The group formed in the late 1970s to protest the proposed nuclear power plant at Diablo Canyon in San Luis Obispo County in southern California.[94]

Typically, the initial reactions came from local residents some years before, not unlike Bodega Bay and Malibu. Mothers for Peace, which had been active in the anti-Vietnam War movement, also focused on the issue of earthquake fault lines and plant construction. The Abalone Alliance, like its eastern counterpart, grew out of a coalition of locals, anti-war activists, and environmentalists. It also received extensive publicity, especially after the accident at Three Mile Island. Between April and June of 1979 the Abalone Alliance staged several major protests. In one event, 25,000 people picketed numerous offices of PG&E in San Francisco. On June 30, the Abalones staged the largest anti-nuclear demonstration at the time in San Luis Obisbo with 40,000 protesters. They not only wanted to stop the Diablo plant but came out in favor of alternative energy sources such as solar power. Increasingly demands for a moratorium on new plants elevated to shutting down existing ones. (Direct-action did not appease labor groups who believed that the interests of working people, especially those dependent on jobs in the nuclear industry, were not served.) Within more conservative circles, utility rate reformers, who were hardly inclined to chant in the streets, also questioned the viability of nuclear plants.[95] Such a double hit suggested a growing lack of confidence in a technology once thought to be an answer to future energy needs. More demonstrations followed. On the 32nd anniversary of the bombing of Hiroshima (August 1977), Japanese held several events throughout their country. The dissent indicated the degree to which the anti-nuclear campaign increasingly linked commercial nuclear development with the proliferation of atomic weapons. Anti-nuclear activism continued throughout the decade and into the 1980s.[96]

THREE MILE ISLAND

The state of commercial nuclear power was unstable in the late 1970s, due to changes in the economic fortunes of the industry, a range of environmental and regulatory challenges, and a shift in American confidence in an apparent answer to energy needs. Polls show public support for nuclear power in decline by the mid-1970s, and by the end of the decade it tumbled to less than 50 percent.[97] The growing number of safety problems at a variety of installations (pipe cracks in reactors, flaws in electrical systems, and siting problems along earthquake faults) aggravated uncertainty about the construction of nuclear plants. The nuclear power industry and its regulators viewed their safety systems as having two purposes: to prevent accidents from occurring and to contain them if they did occur. To do this they relied on "defense-in-depth," that is, redundant safety systems and multiple barriers to protect against serious releases of radiation. The idea was rational and based upon solid, conservative engineering expertise. The building of increasingly larger plants beginning in the 1960s raised questions as to whether safety systems could protect against the worst case.[98] To complicate matters, the rational, pragmatic, technology-driven approach to safety did not give sufficient attention to the possibility of human error, which is not always governed by that rational, pragmatic, and technology-driven approach.

The fate of the industry, an industry that had gone on record touting the reliability and absolute safety of its product, was threatened in more ways than it could imagine by the accident at Three Mile Island. The plant may have escaped the "worst case" scenario, but the greatest loss

Three Mile Island, Dauphin County, Pennsylvania, April 1979. *Source:* AFP/GETTY IMAGES/Newscom

in that spring of 1979 was nuclear power's credibility. The accident at Three Mile Island was not the departure point for a loss of faith in nuclear power, but its climax. America's worst nuclear accident took place at a station owned by General Public Utilities (GPU) and located on Three Mile Island in the Susquehanna River, about 10 miles southeast of Harrisburg, Pennsylvania. The accident occurred on March 28, during the graveyard shift. The power station had two units. At 4:00 a.m., a pump on the main water-feed system malfunctioned in *TMI-2*. Within two seconds, the flow of water going to the secondary system (the steam generator) stopped, and the plant safety system automatically shut down the steam turbine and electric generator. A relief valve popped open to reduce the pressure in the reactor and stayed in that position. Unfortunately, an indicator on the control panel led the crew to believe that the valve was shut. Thus, it drained water from the reactor for more than two hours. Believing that the reactor was adequately supplied with cooling water, the operators shut off the emergency pump. The loss of cooling water caused the fuel to overheat. When the top of the core became uncovered, hydrogen was generated, and the core began to melt.[99] As Walker stated, "The uncovering of the core at TMI-2 produced a meltdown that was unprecedented and, at that point, undetected, although officials…gradually realized that they faced a serious challenge in finding a way to cool the heated core."[100]

By about 7:30 a.m., the station manager declared a general emergency. The auxiliary building had been evacuated, teams of technical personnel were sent into the surrounding neighborhoods to monitor radioactivity levels, and Route 441 near the plant was closed.[101] By Thursday night, operators still did not know exactly what was happening in the containment area. They detected a hydrogen bubble that had formed in the pressure vessel of the reactor, and were concerned that it could inhibit cooling of the core. On Friday, March 30, the situation at the site worsened when more radiation releases took place, although no major discharge occurred. A more serious worry arose that the hydrogen bubble could be flammable or potentially explosive. Although later proved unfounded, the chance for a major release of radioactivity through an explosion seemed possible at the time. The event created great uncertainty. Coordination between on-site, company, NRC, and state and local government officials proved difficult. Governor Richard Thornburgh and his aides considered several possible evacuation plans, but without notifying or consulting the plant staff. Rampant rumors of large-scale evacuation led the governor to recommend that pregnant women and preschool children leave the area within a five-mile radius of the plant. Independently of his announcement, many people had left their homes already. Chaos did not break out, but citizens were scared and confused. The site inspection by President Jimmy Carter and Mrs. Carter on Sunday, April 1, eased public fears, especially about the dangers of the hydrogen bubble. At a press conference after the plant tour, Governor Thornburgh thanked the president for "the courage and concern he had demonstrated at a time when such a personal gesture is most helpful."[102] However, the visit did not remedy all of the problems in communication among the various parties that were dealing with the accident.[103]

The aftermath was almost as dramatic as the episode itself. The event was "red meat" for the press; coverage was extensive and editorial responses were frequent and often clamorous. The press vilified plant officials, state authorities, and the NRC, but not always fairly. Public reaction ranged from relief to indignation. Anti-nuclear groups held a new round of protests. On April 7 and 8, 1979, they staged 10 demonstrations across the country railing against nuclear power. On May 6, some 65,000–75,000 protesters held an event at the Capitol in Washington, D.C., to strains of "Hell No, We Won't Glow!" In attendance was Jane Fonda, actress and anti-war protester. She had recently starred in *The China Syndrome*, a thriller about a mythical nuclear accident which had been released days before the real occurrence at Three Mile Island. A surge of black humor also surfaced, manifest in T-shirt slogans such as "I SURVIVED THREE MILE ISLAND—I THINK."[104]

Trying to sort out the nature of the accident, its causes, and the consequent responsibility, Dartmouth College President John Kemeny headed a presidential commission on Three Mile Island. Carter established the commission "to investigate the causes" of TMI and "to make recommendations on how we can improve the safety of nuclear power plants."[105] The final report placed blame on plant builders and managers, operators, and federal regulators but fell short of recommending a moratorium on construction of new plants. Several congressional investigations, as well as NRC and industry inquiries, followed. Both critics and supporters of nuclear power leveled charges of human error and poor training—the former, as a means to demonstrate that "it could happen again" and the latter, as a way of defending a sound technology. A Department of Energy (DOE)-sponsored study made an obvious but important point:

> Three Mile Island was not so much a technological event as a human and historical one… It was not only a mechanical breakdown but a series of human choices that crippled a nuclear reactor and threatened injury and death to the public. What escaped at Three Mile Island was not only radiation, but, more importantly for the nuclear power industry, public confidence in technology and technocracy.[106]

The long-term impacts of Three Mile Island would be felt for years in the on-going debate over the reliability of the technology and the safety system, and the necessity of commercial nuclear power itself. The event shocked and clearly humbled the industry and the government regulators at the NRC. Both attempted, in the coming years, to improve reactor safety and to rethink regulations. Internationally, the accident set off new debates about the future of nuclear power. But the accident was a setback in very different ways for both the industry and the anti-nuclear movement. Three Mile Island was the culmination of an era for nuclear civilian power dominated by decline in demand for electricity, a restrictive credit market, construction delays and cost overruns, and severe pressure from protesters. In 1973, the AEC predicted that more than 1,000 nuclear plants would be operating by 2000; the DOE scaled that back to 500, and as few as 200. In 1979 there were perhaps 70 operating nuclear plants, contributing only 13–14 percent of electricity production and only about 4 percent of total energy consumption. No one was willing to order a new plant after that year. By 1985 the construction of many plants that had been scheduled before TMI were cancelled. Times would have to change precipitously for this downhill slide to end.[107]

Validated by the TMI accident, the initial impact on the anti-nuclear movement was to energize it. But as historian Thomas Wellock persuasively argued, "After Three Mile Island, the nuclear industry collapsed and took the anti-nuclear movement with it. The industry's ruin took away the immediacy of the nuclear issue." Whether the anti-nuclear power movement was "reincarnated as an anti-nuclear weapons movement," as he further argued, is debatable, but it is nonetheless evident that the targets of criticism (for the time at least) were not available to rally the anti-nuke forces as they had in the past.[108] Nuclear power plants remained (at home and abroad) and the prospect of a revitalized nuclear power industry was inspiration for future protests. As of the late 1970s and the coming decade, the nuclear power landscape had mightily changed.[109]

NUCLEAR POWER ABROAD

In several respects civilian nuclear power in Europe and elsewhere mirrored the history of the US experience in the 1960s and 1970s largely because of worldwide economic realities. But there were differences as well. As in the United States, the 1960s and early 1970s brought high expectations for nuclear power. Many believed the technology had advanced sufficiently to make it reliable and economically feasible (especially LWRs) and electric utilities were poised to invest in new systems. Several developed countries, in particular, established a nuclear fuel cycle from uranium to waste disposal. The energy crisis of 1973–1974 contributed to "the expansionist mood" of the time as orders for plants boomed. By 1979, the future was not so bright with the exception of continued strong sponsorship of nuclear power in France, the Soviet Union, and some other Eastern European nations.[110]

Access to uranium, the availability of indigenous technical expertise, and military aims limited choices in the early stages of the nuclear power industry in Europe during the 1940s and 1950s. Those countries with little or no access to enriched uranium relied on reactor designs that could utilize natural uranium (if that was available) such as heavy water reactors in Canada and then Sweden, or graphite-moderated reactors in England and France. These reactors were meant to produce plutonium, with electrical power production as a secondary by-product. Reactors remained dual-purpose in France, England, and to some degree Sweden into the 1960s. The LWR became the foundation technology for the US civilian nuclear program, which it marketed especially in Western Europe and Japan, and ultimately spread throughout the world. The Atoms for Peace program provided the key justification for the United States to move from its strict monopoly over nuclear technology to promoting civilian nuclear power beyond its borders. Still, such a program demanded interdependency between the United States and other countries wanting to go nuclear. Developing an independent nuclear fuel cycle and home-grown technologies were complex and sometimes politically sensitive issues for most industrialized nations. Yet, such steps were essential in the long run for those seeking to move forward with civilian nuclear power. In the 1970s, the quest proved more difficult when the Americans became obsessed with fears of military nuclear proliferation and began to restrict civilian nuclear trade (especially in enriched uranium). Such intertwining of military and civilian issues consistently strained relations between the United States and its allies, and made nuclear programs in several countries vulnerable to breaks in trade.[111]

For most countries the state dominated the development of civilian nuclear power. The reasons for this were fairly obvious. A national government was best positioned to deal with the scale of the enterprise and (hopefully) the safety requirements, and needed to be in control of the military implications. Several groups in the international nuclear power community moved away from strict and exclusive government control in the 1960s and early 1970s. On the extremes were the United States, which came closest to private ownership, and the USSR, where state ownership was absolute. Somewhere in between, West Germany and Japan were closer to private ownership, and France, Great Britain, Sweden, and Canada closer to state ownership. Increasing demand for electricity especially bolstered nuclear power in the western world. The upsurge in demand began in the 1950s and continued through the mid-1970s, with future expectations also running high. But by 1974 the market for electricity in most developed countries flattened out or declined.[112] By 1978, the United States still was responsible for half of the nuclear generating capacity in non-communist countries worldwide. Western Europe as a whole was second (about 33 percent), Japan third (11 percent), and Canada fourth (4.7 percent). In the Soviet sphere, the USSR far and away had the largest nuclear generating capacity.[113]

The energy crisis had a variety of impacts on nuclear power generation in industrialized nations. The first OPEC oil price increase in 1973 led to a rush of new orders for reactors. Since uranium resources were more widely distributed than oil, and more easily traded than bulky resources like coal, nuclear power appeared to be an answer to comparative energy independence. This seemed particularly true to nations heavily dependent on foreign oil. The state-controlled Électricité de France (EDF) moved on a massive expansion of nuclear power in the 1970s in order to shuck French dependence on imported oil and gas. EDF relied on standardized construction using US-style pressurized water reactor (PWR) technology. Nuclear power generation in France went from 7 percent of total electricity production to 20 percent in 1980 (and 78 percent by 1994). Belgium, West Germany, and Sweden developed important nuclear programs, relying on LWRs. The commercial CANDU reactors came on line in Canada in 1971. These were the first reactors designed exclusively for power production with no military implications.[114] Nuclear power development outside of North America and Europe in the 1960s and 1970s was nascent or uneven by comparison.[115]

The initial upsurge in civilian nuclear power internationally did not last long, paralleling what had happened in the United States. A worldwide economic recession that followed on the heels of the oil crisis bloated the cost of nuclear construction and simultaneously reduced demand for electricity. The anticipation for vast expansion of gigawattage from nuclear power globally fell far short, to less than 10 percent of predictions.[116] The accident at Three Mile Island not only added to the economic woes for nuclear power produced during the energy crisis, but placed renewed and intense emphasis on risk and clearly eroded public confidence in nuclear power with a consequent disruption in future planning.[117]

THE WORLDWIDE ANTI-NUCLEAR MOVEMENT

Civilian nuclear power in Europe in particular, but also elsewhere, inspired a wave of anti-nuclear protest especially by the early 1970s. The timing was important since the anti-nuclear movement came into prominence internationally at the point at which energy policy was emerging as a central issue of concern spurred by the energy crisis. As in the United Sates, fears about oil short-ages actually brought hope to those who promoted nuclear power and trepidation to those who scorned it. Sharing many characteristics of its American counterpart, the anti-nuclear movement abroad also built upon local political, institutional, and cultural roots. One study, comparing nuclear power in the United States, Sweden, and France, suggested that all three countries faced "scattered but growing anti-nuclear forces" by 1973, and that the political and regulatory systems responded in similar fashion. Protests grew out of common roots: NIMBYism, environmental-ism, and occasionally from scientists and technicians or leftist critics of the capitalist state. Some groups developed into loose coalitions with a common cause. Protesters in France and Sweden, although embracing their own particular social and value base, used information gained from their American counterparts to fight their own battles. The French movement included some religious connections that were not apparent elsewhere.[118]

Tactics also evolved over time, beginning with grievances raised through conventional political channels and moving toward direct-action. In both France and the United States, the demonstration in Whyl, West Germany, inspired site occupations in 1976 and 1977. American protesters tended to devote more time to regulatory hearings, but activists in all three countries used the courts through a variety of lawsuits. Protesters in Sweden relied more heavily on elec-toral politics with special appeals to the Center Party. The particular structure of the state deter-mined anti-nuclear tactics. While Swedish and American activists explored legal routes. French demonstrators focused on illegal occupations of sites and even violence because they were most often blocked from access to state authorities. These are broad generalizations, but they raise the important point that differing political and social infrastructures influenced protest behavior. While government leaders assumed that the protests would die down (and often believed that the demonstrators were acting emotionally and were uninformed) they ultimately allowed for public participation in venting concerns. They did not allow for a formal role in the prioritizing issues, licensing, or siting. But there is little doubt that the public debates influenced nuclear poli-cymaking, especially in light of concerns raised during the energy crisis. The unease about the safety of nuclear power plants was supported by queries as to whether building those plants was a viable alternative to fossil fuels or a superior approach to a workable energy conservation plan. Environmental risk met "Small is Beautiful."[119]

By 1976 and 1977, efforts to silence criticism of nuclear power were most strident in France. Unlike in the United States, France was moving to make nuclear power the centerpiece in its energy future, and the public either resigned itself to that reality or supported it because of the untenable dependence on foreign oil and gas. The event that most directly ignited the strong government response to the protesters was the proposed breeder reactor, the *Super-Phenix*, to be built at Creys-Malville on the beautiful Rhone River. This would not be a commercial plant, but would also be used for research linked to weapons production. Two or three thousand people convened at the site in July 1976 for a peaceful "anti-nuclear fete." After a few days the Compagnies Républicaines de Sécurité (CRS, riot control forces and the general reserve of the French National Police) used tear gas and clubs to clear out the demonstrators. The vio-lence drew sympathy for the protesters and led to hearings on nuclear energy.[120] Things got worse in the summer of 1977. By that time the French anti-nuclear movement reached its great-est strength, and public opinion turned in its direction for the first time. A new demonstration near Malville at the end of July drew tens of thousands of protesters, and the CRS again moved in this time with brutal force. One person was killed, five were seriously injured, and hundreds suffered minor injuries. Some of the protesters were from the violent fringe of the movement, but most were not. For the government this repressive act had a beneficial outcome because the anti-nuclear groups retreated from site occupations. It also divided the country politically, between those who viewed the protesters as criminals ("les marginaux") and those who were sympathetic. As the French government's commitment to nuclear power became stronger, the anti-nuclear movement got weaker. Failing to gain much support among the political parties, and barred from participation in regulatory programs, anti-nuclear activism began to fade as did the recent anti-nuclear tone of French opinion polls.[121]

In West Germany direct-action protest started earlier than anywhere else. Some would argue it was because of "an unresponsive state," but also because the anti-nuclear movement there attracted elements who had espoused those tactics before, such as urban left-wing groups of the old student movement. "Anti-statism" was as much a part of the anti-nuclear movement in West Germany as criticism of nuclear power. Here too, the disruptive impact of the energy crisis played an important role in focusing debate over nuclear power, essentially politicizing the issue as it had done in the United States and elsewhere. Whereas nuclear opposition in the United States grew beyond the environmental movement before the mid-1970s, it moved to direct-action only after several protesting groups exhausted political remedies. The West German anti-nuclear movement actually began with direct-action, and was more radical in the wake of the energy crisis as the government became more entrenched. The energy crisis and the worldwide recession that it inspired led the Bonn government to embrace nuclear power as an avenue to renew economic growth. The government of the Social Democratic Party (SPD), which promoted nuclear power as equitable with economic growth, left no room at the table for the anti-nuclear movement. The possibility of "a nuclear security state" was troubling to critics.[122]

Such circumstances led to conflict, even violence, which became a hallmark of the protesters attempting to turn national policy away from nuclear power. Two immovable forces, the SPD on one side and the radical groups who opposed nuclear power and who asserted a strong anti-statist perspective on the other, headed relentlessly toward confrontation. The demonstration at Whyl had set intensive, direct-action public protests in motion, not only in West Germany but in the United States and elsewhere. Similar protests occurred in Brokdorf (70,000 to 80,000 people), at Grohnde (the first major defeat for the German anti-nukes), and at Kalkar (the last national demonstration for several years to come). Eventually, the rise of the anti-nuclear Green Party (originating in 1980) and the split in the SPD over nuclear power weakened pro-nuclear political resolve, but resulted in indecisiveness. Green Parties or "Greens" were political parties whose policies were based on concerns about the environment. They emerged out of the modern environmental movement in the 1970s and 1980s. They first achieved electoral success in Germany in the early 1980s, and could be found in other parts of Western Europe and beyond.[123] In this setting the administrative courts essentially produced a moratorium on new nuclear construction, and thus ended the intensity of the confrontation at least temporarily. By the early 1980s the anti-nuclear movement in West Germany was in decline.[124]

The anti-nuclear path was somewhat different in Great Britain. The year 1970 saw the first local protest against a nuclear power station in a decade, with increasing media attention similar to activities in the United States. Like elsewhere, protests started rather tepidly, were local and scattered, and included efforts to manipulate the political system. The first national campaign occurred in 1974. Led by the Conservation Society, it was a conventional lobbying effort that attracted little support from some of the more youthful environmentalists. Friends of the Earth, newly formed in the UK in 1970, used the leading environmental magazine, *Undercurrents*, to criticize nuclear power in an early 1975 issue and to launch its own anti-nuclear campaign. British Nuclear Fuels' decision in the early 1970s to construct the Thermal Oxide Reprocessing Plant (THORP) at Windscale became the movement's *cause célèbre*. The purpose of the plant was to reprocess spent reactor fuel, extracting plutonium that could be used for nuclear weapons and fast breeder reactors. This had real and symbolic significance, especially because this was the site of the previous, notorious Windscale accident. THORP had enthusiastic government support, but a series of accidents beginning in 1973 at Windscale raised serious concerns. In 1977 the government held an inquiry to determine the nature of those radioactive leaks, which not only brought the problem to public attention but incited the anti-nuclear movement. The *Windscale Report* (1978) rejected environmentalists' criticisms of leaks at the site and openly favored THORP. Parliament concurred. Construction started in 1985 and was completed in 1992. THORP remained a touchstone for anti-nuclear activity throughout the period.[125]

Context is everything in understanding the anti-nuclear movement. Public opposition to nuclear power in the United States, Great Britain, West Germany, France, and elsewhere was modestly successful because of the prevailing political, economic, and social environment. Protest in industrial democracies was one thing, but quite another in the Eastern bloc. While the USSR and some Eastern European countries adopted ambitious nuclear programs, public protest was non-existent through the early 1980s. Not until the governing system changed under Mikhail Gorbachev (1985), and the almost unthinkable accident occurred in Chernobyl in 1986, did opposition there get some sort of public face.[126] In other places, protests would have to await government decisions to go nuclear. In Mexico construction began on a nuclear power plant at Laguna Verde in the state of Veracruz in the early 1970s. Initial reactions by many locals were positive, since such construction brought jobs. The government's pro-nuclear policy, however was undermined by critical public reactions to nuclear power after Three Mile Island and Chernobyl.[127]

CONCLUSION: WHAT DID IT ALL MEAN FOR COMMERCIAL POWER?

The clash over reactor safety and related issues in the 1970s pitted the AEC/NRC and the nuclear power industry against a loosely organized, but increasingly committed anti-nuclear movement. Yet the polar positions do not give an accurate read on the state of nuclear power, at least in the United States. While showing declining support, polls indicated no clear majority for one side over the other, and suggested instead some ambivalence in thinking about the revolutionary power source. Even Three Mile Island itself did not lead to outright rejection. Two films released in the period, *Silkwood* (1983) based on a real incident, and *China Syndrome* (1979), a fictional scenario, subtly reflect the ambivalence toward nuclear power—an ambivalence that was sometimes lost in the commotion of the era.

Silkwood, directed by Mike Nichols and starring Meryl Streep, Cher, and Kurt Russell, was nominated for five Academy Awards. It told the story of Texas-born Karen Gay Silkwood, a metallographic laboratory technician who worked at Kerr-McGee's Cimarron River plutonium plant in Oklahoma in the early 1970s. The company made plutonium pellet fuel rods there. Silkwood had joined the Oil, Chemical and Atomic Worker Union (eventually the first female member of the bargaining committee) and participated in its protracted strike of the plant. As a member of the union she investigated and reported spills, leaks, and missing plutonium, and then testified before the AEC that she had been exposed to radiation on at least three occasions under unusual circumstances. On November 13, 1974, on her way to meet with an AEC official and a reporter for the *New York Times*, she was killed in an automobile crash. The highway troopers believed she had fallen asleep at the wheel, but a private investigator hired by the union was suspicious. He found

Karen Silkwood was a metallographic laboratory technician who worked at Kerr-McGee's Cimarron River plutonium plant in Oklahoma in the early 1970s. She was the center of controversy over spills, leaks, and plutonium and health risks at the plant. *Source:* AP Photo

curious the way in which the car crashed, and also discovered that the papers she was carrying to the meeting were missing from the wreckage. Union colleagues thought the worst.

Silkwood had been harassed by the company like other union activists at the plant, and her multiple exposures (and the contamination of her apartment) were not easily explained.[128] A story in *Time Magazine* on January 20, 1975, reported that AEC findings suggested that Silkwood's contamination "probably did not result from an accident or incident within the plant." It was possible, it stated, that she contaminated herself. Such a far-fetched conclusion only added to a sense of conspiracy.[129] Karen Silkwood's family settled an $11.5 million lawsuit against Kerr-McGee for one-tenth of the amount in 1986. The company did not admit liability in settling the case, and viewed Silkwood as nothing more than a union informer. But charges of manufacturing faulty fuel rods dogged them until the plant was closed.[130] To her union friends and to the National Organization of Women (NOW), Silkwood was a martyr and a key feminist. Her case brought both groups into the anti-nuclear movement, especially on questions of worker safety.[131]

Kerr-McGee (and possibly the AEC) involvement in the murder of Karen Silkwood is implied in the movie, but there is little evidence to support such serious charges. *Silkwood* plays strongly on the conspiratorial tone of the original events. As such, the first impression is that Nichols is taking a strong stand, much in line with union accusations. The realistic interaction between characters in the film and the concentration on people more than action reinforces that notion. But digging a little deeper into the film reveals some of the ambivalence over the nuclear power debate felt in society as a whole. The fact that the film was released in 1983, three years before the final court settlement of the case, may account for some of Nichols's hesitance. However, he (or more precisely the screenplay) devalues Silkwood's union ties and thus much of her motivation for questioning the practices of Kerr-McGee. The movie pits a courageous and tragic individual against a corporation and its minions. But even here, Nichols throttles back. As Karen is driving to her appointment, she sees headlights in her rearview mirror, but we do not know for certain if these are the headlights of an assassin or merely a coincidence that intensifies Karen's growing paranoia. There is an unseen, but not unfelt, danger lurking around the plant site, at Karen Silkwood's home, and on the highway, but it is not made tangible in the story line.[132]

The Oscar-nominated and critically touted *China Syndrome* was made famous for its release 12 days before the accident at Three Mile Island. It is a classic Hollywood feature film with a star-studded cast: Jane Fonda, Jack Lemmon, and Michael Douglas. The backdrop is a nuclear power plant in southern California, where a comely puff-piece reporter, Kimberly Wells (Fonda), is sent with her brash cameraman, Richard Adams (Douglas), to do an interview and to get a standard, public relations tour. While there, a shutter in the system (suggesting something much more serious) is caught on camera while the crew in the control room scrambles to correct the problem. A spokeman for the plant brushes off the incident. However, Jack Godell (Lemmon)—shift supervisor, chief technician, and long-time nuclear veteran—senses something is potentially more dangerous. It is. He fails to get his superiors to respond to his concerns, faces one road-block after another, and learns that he is being watched himself. Godell decides to take drastic action, and barricades himself in the control room with a gun demanding that he tell his side of the story on television. This is a big chance for Wells and she proves up to the task, but plant and company officials take a hardline, believe that Godell has gone off the deep end, and want him silenced. He is killed by a SWAT team that liberates the control room.

Director James Bridges turns *China Syndrome* into a fast-paced thriller for much of the second half of the film, and clearly seems to stir the pot about the dangers of nuclear power. This is reinforced by its coincidental release date. Critics of the film and supporters of the industry attacked *China Syndrome* as not being fair-minded by inaccurately portraying that a possible core meltdown would lead to a breach of containment. But like *Silkwood*, this film also suggests ambivalence not immediately apparent with a first view. Wells is a pretty face, turned crack reporter. Adams is a cynical, anti-establishment throwback to the 1960s. Utility officials and plant workers are clearly stereotyped, as are anti-nuclear protesters attending a licensing and safety meeting. Most importantly, Godell in particular is no nuclear critic, not even at death. He is true to his beliefs that nuclear power is safe and those running the reactor very competent. He trusts his long experience, and until these extraordinary events, has been a company man. The true villain in the story is a contractor who has faked x-rays of its construction work to save money and to increase profits. This is neither the failing of an ECCS system nor a problem in reactor design. It is the result of basic greed and dishonesty, which comes from outside the plant. Such avarice is hardly unique to nuclear power production. In making this point emphatic, *China Syndrome* is not the anti-nuclear film it has been made out to be. Those that view the film as essentially anti-nuclear have not looked at it closely enough.[133]

Neither film is the last word on civilian nuclear power, nor are a few others that took on the subject at the time.[134] But their tone and hesitancy are instructive. From the visceral perspective, at least, TMI and the events leading up to it in the 1960s and 1970s did not shut the door on nuclear power production for the future. At the very least, the French embrace of nuclear power made that clear. What could possibly transpire was that different times and different circumstances might reinvigorate the enthusiasm for nuclear power that seemed so lively in the bandwagon years, but so moribund in 1979. At the moment, ambivalence about nuclear power was a legitimate response to the hope and trepidation that it constantly stirred in people. Three Mile Island was not so much a defining event, as one inspiring introspection.

MySearchLab Connections: Sources Online

READ AND REVIEW

Review this chapter by using the study aids and these related documents available on MySearchLab.

✓•⌐**Study** and **Review** on **mysearchlab.com**

Chapter Test

Essay Test

📖•⌐**Read** the **Document** on **mysearchlab.com**

Calvert Cliffs' Coordinating Committee, Inc., et al., Petitioners, v. United States Atomic Energy Commission and United States of America (1971)

Excerpts from The Rasmussen Report (1974)

Thomas Tuohy: Windscale Manager Who Doused the Flames of the 1957 Fire (2008)

Nun Sentenced to Jail in Nuclear Plant Hassle (2012)

RESEARCH AND EXPLORE

Use the databases available within MySearchLab to find additional primary and secondary sources on the topics within this chapter.

👁•⌐**Watch** the **Video** on **mysearchlab.com**

Silkwood versus Kerr-McGee

Endnotes

1. J. Samuel Walker, *Containing the Atom: Nuclear Regulation in a Changing Environment, 1963–1971* (Berkeley, CA: University of California, 1992), 18.
2. Alice L. Buck, *A History of the Atomic Energy Commission* (Washington, D.C.: US Department of Energy, August 1982), 5.
3. J. Samuel Walker, *Permissable Dose: A History of Radiation Protection in the Twnetieth Centry* (Berekeley, CA: University of California Press, 2000), 30.
4. Babcock and Wilcox and Combustion Engineering had a small share of the power reactor market. The former's first major commercial project was the reactor for the Indian Point nuclear plant opened by Consolidated Edison of New York on the Hudson River in 1962. See Walker, *Containing the Atom*, 21–22.
5. Martin V. Melosi, *Coping with Abundance: Energy and Environment in Industrial America* (New York: Alfred A. Knopf, 1985), 233–34; James M. Jasper, *Nuclear Politics: Energy and the State in the United States, Sweden, and France* (Princeton, NJ: Princeton University Press, 1990), 47–48; J. Samuel Walker, *Three Mile Island: A Nuclear Crisis in Historical Perspective* (Berkeley, CA: University of California Press, 2004), 6–7; Steven L. Del Sesto, *Science, Politics, and Controversy: Civilian Nuclear Power in the United States, 1946–1974* (Boulder, CO: Westview Press, 1979), 79–80; Walker, *Containing the Atom*, 18–19, 21–22.
6. Jasper, *Nuclear Politics*, 45; Melosi, *Coping with Abundance*, 233.

7. David E. Nye, *When the Lights Went Out: A History of Blackouts in America* (Cambridge, MA: MIT Press, 2010), 69–72, 81–95, 101–07.

8. In the late 1960s there were more than 800 separate entities in the utility business, most private but some public like the Tennessee Valley Authority and the Bonnneville Power Administration. Approximately 200 investor-owned utilities controlled 75 percent of the generating capacity of the country. Those in the private sector always worried about increased competition from public bodies, and they generally tended to be conservative in managing their business to limit risk and to stifle competition. In this environment, nuclear power generators had to tread carefully. See Walker, *Containing the Atom*, 27–28.

9. Robert J. Duffy, *Nuclear Politics in America: A History and Theory of Government Regulation* (Lawrence, KS: University Press of Kansas, 1997), 57; Walker, *Three Mile Island*, 4–7; Jasper, *Nuclear Politics*, 46; Joseph G. Morone and Edward J. Woodhouse, *The Demise of Nuclear Energy? Lessons for Democratic Control of Technology* (New Haven, CT: Yale University Press, 1989), 78–79; Del Sesto, *Science, Politics, and Controversy*, 85–93; Walker, *Containing the Atom*, 28–34.

10. Jack M. Holl, Roger M. Anders, and Alice L. Buck, *United States Civilian Nuclear Power Policy, 1954–1984: A Summary History* (Washington, D.C.: US Department of Energy, February 1986), 9–10.

11. Melosi, *Coping with Abundance*, 233.

12. Quoted in Holl, Anders, and Buck, *United States Civilian Nuclear Power Policy, 1954–1984*, 10. See also Buck, *A History of the Atomic Energy Commission*, 5.

13. Holl, Anders, and Buck, *United States Civilian Nuclear Power Policy, 1954–1984*, 10; Stephen E. Atkins, ed., *Historical Encyclopedia of Atomic Energy* (Westport, CT: Greenwood Press), 60.

14. Buck, *A History of the Atomic Energy Commission*, 5; Holl, Anders, and Buck, *United States Civilian Nuclear Power Policy, 1954–1984*, 11–12.

15. Walker, *Containing the Atom*, 35.

16. Ibid., 35–37.

17. The waste disposal issue will be taken up more thoroughly in Chapter 9.

18. Duffy, *Nuclear Politics in America*, 49; Walker, *Three Mile Island*, 10–11.

19. Thomas R. Wellock, *Preserving the Nation: The Conservation and Environmental Movements, 1870–2000* (Wheeling, IL: Harlan Davidson, Inc., 2007), 199; Walker, *Three Mile Island*, 3–4.

20. Quoted in Duffy, *Nuclear Politics in America*, 79.

21. Melosi, *Coping with Abundance*, 237. See also William McKeown, *Idaho Falls: The Untold Story of America's First Nuclear Accident* (Toronto: ECW Press, 203).

22. Walker, *Containing the Atom*, 57–59.

23. Correspondence from Samuel Walker to Martin Melosi, February 4, 2011.

24. Walker, *Containing the Atom*, 58–59.

25. Ibid., 61–83.

26. Ibid.; David Okrent, *Nuclear Reactor Safety: On the History of the Regulatory Process* (Madison, WI: University of Wisconsin Press, 1981), 17, 70–84; Del Sesto, *Science, Politics, and Controversy*, 94–95.

27. Quoted in Thomas Raymond Wellock, *Critical Masses: Opposition to Nuclear Power in California, 1958–1978* (Madison, WI: University of Wisconsin Press, 1998), 29.

28. Walker, *Containing the Atom*, 84–85.

29. The Bodega issue ignited a major split within the Sierra Club between those who wanted to side with the locals and those who wanted to find a compromise position with PG&E. For some, taking a stand on aesthetic considerations only (preserving the beauty of the area) was a small part of a larger environmental dilemma. To conservatives in the group, maintaing a bargaining position through compromise was essential to the overall success of protecting California's coastline. The Northern California Association to Preserve Bodega Head and Harbor carried the fight to stop the building of the plant, not the Sierra Club. See Wellock, *Preserving the Nation*, 17–67.

30. Walker, *Containing the Atom*, 85–88.

31. Quoted in Wellock, *Critical Masses*, 47.

32. Walker, *Containing the Atom*, 88–90.

33. Richard L. Meehan, *The Atom and the Fault: Experts, Earthquaes, and Nuclear Power* (Cambridge, MA: MIT Press, 1984), 39–43; Walker, *Containing the Atom*, 90–112.

34. Melosi, *Coping with Abundance*, 237–38; Walker, *Three Mile Island*, 9; Nuclear Regulatory Commission, *A Short History of Nuclear Regulation, 1946–1999*, 15 (January 29, 2001), online http://www.nrc.gov/SECY/smj/shorthis.htm. See also Wellock, *Critical Masses*, 17–67; Brian Balogh, *Chain Reaction: Expert Debate and Public Participation in American Commercial Nuclear Power, 1945–1975* (Cambridge, MA: Cambridge University Press, 1991), 240–53.

35. Del Sesto, *Science, Politics, and Controversy,* 146–50; Melosi, *Coping with Abundance*, 238; Balogh, *Chain Reaction*, 265–66; Walker, *Containing the Atom*, 309–30.

36. Del Sesto, *Science, Politics, and Controversy,* 98–101.

37. Correspondence from Samuel Walker to Martin Melosi, February 4, 2011.

38. Del Sesto, *Science, Politics, and Controversy,* 98–101, 169–71. See also Walker, *Containing the Atom*, 139–232; Holl, Anders, and Buck, *United States Civilian Nuclear Power Policy, 1954–1984,* 12–13; Nuclear Regulatory Commission, *A Short History of Nuclear Regulation, 1946–1999*, 15–19; Atkins, ed., *Historical Encyclopedia of Atomic Energy*, 119; Duffy, *Nuclear Politics in America*, 63. For more information on the AEC in the 1960s and 1970s, see Glenn T. Seaborg with Benjamin S. Loeb, *The Atomic Energy Commission Under Nixon: Adjusting to Troubled Times* (New York: St. Martin's Press, 1993); Balogh, *Chain Reaction.*

39. Spencer R. Weart, *Nuclear Fear: A History of Images* (Cambridge, MA: Harvard University Press, 1988), 313.

40. Del Sesto, *Science, Politics, and Controversy,* 151–56; Duffy, *Nuclear Politics in America*, 60; Melosi, *Coping with Abundance*, 306–07. See also John W. Gofman, *An Irreverent Illustrated View of Nuclear Power* (San Francisco: Committee for Nuclear Responsibility, 1979); Arthur R. Tramplin and John W. Gofman, *'Population Control' Through Nuclear Pollution* (Chicago: Nelson-Hall Company, 1970).

41. For an extended interview with Gofman, see Leslie J. Freeman, *Nuclear Witnesses: Insiders Speak Out* (New York: W.W. Norton,1981), 96–114. For a critical view of Gofman and Tramplin's approach to the radiation issue, see Walker, *Containing the Atom*, 331–62.

42. David Howard Davis, *Energy Politics* (New York: St. Martin's Press, 1982; 3d ed.), 223–24; Walker, *Permissable Dose*, 36–47; Melosi, *Coping with Abundance*, 306. In 1972, the National Academy of Sciences issued the BEIR (Biological Effects of Ionizing Radiations) Report which urged that public exposure to radiation be minimized, but to do so without limiting the benefits of nuclear medicine, nuclear power, or other worthwhile applications. See Walker, *Permissable Dose*, 47–56.

43. Balogh, *Chain Reaction,*16; Christian Joppke, *Mobilizing Against Nuclear Energy: A Comparison of Germany and the United States* (Berkeley, CA: University of Clalifornia Press, 1993), 27–28.

44. Joppke, *Mobilizing Against Nuclear Energy,* 21–31; Robert Gottlieb, *Forcing the Spring: The Transformation of the American Environmental Movement* (Washington, D.C.: Island Press, 2005; rev.ed.), 135; Del Sesto, *Science, Politics, and Controversy,* 144–46.

45. Atkins, ed., *Historical Encyclopedia of Atomic Energy*, 17.

46. Joppke, *Mobilizing Against Nuclear Energy*, 31.

47. Walker, *Containing the Atom*, 267–73.

48. Ibid., 273–96; Balogh, *Chain Reaction*, 261–65, 284–87, 299; Duffy, *Nuclear Politics in America*, 55; Nuclear Regulatory Commission, *A Short History of Nuclear Regulation, 1946–1999*, 19–20; Wellock, *Preserving the Nation,* 198. J. Samuel Walker, "Nuclear Power and the Environment: The Atomic Energy Commission and Thermal Pollution, 1965–1971," *Technology and Culture* 30 (October 1989): 964–92.

49. Martin V. Melosi, "Battling Pollution in the Progressive Era," *Landscape* 26 (1982), 35–41.

50. Clifford B. Knight, *Basic Concepts of Ecology* (New York: Macmillan, 1965), 2.

51. Donald Worster, *Nature's Economy: A History of Ecological Ideas* (New York: Cambridge. University Press, 1977), 289, 339–40, 378.

52. Victor B. Scheffer, *The Shaping of Environmentalism in America* (Seattle, WA: University of Washington Press, 1991), 4.

53. Ibid., 113; Martin V. Melosi, "Lyndon Johnson and Environmental Policy," in Robert A. Divine, *The Johnson Years,* Volume Two. (Larence, KS: University Press of Kansas, 1987), 113–49; Michael

Egan, *Barry Commoner and the Science of Survival: The Remaking of American Environmentalism* (Cambridge, MA: MIT Press, 2007), 1–3.

54. Carolyn Merchant, *The Columbia Guide to American Environmental History* (New York: Columbia University Press, 2002), 177–79; Samuel P. and Barbara D. Hays, *Beauty, Health, and Permanence: Environmental Politics in the United States, 1955–1985* (Cambridge, MA: Cambridge University Press, 1987), 13–14, 21–39; Stephen R. Fox, *The American Conservation Movement: John Muir and His Legacy* (Madison, WI: University of Wisconsin. Press, 1981), 299, 302, 311; Hal K. Rothman, *The Greening of America? Environmentalism in the United States Since 1945* (New York: Harcourt Brace, 1998), xi, 29–31; Wellock, *Preserving the Nation,* 135–88; Gottlieb, *Forcing the Spring,* 34–36.

55. Quoted in Melosi, *Coping with Abundance,* 297.

56. Gottlieb, *Forcing the Spring,* 139–40, 148–58; Joppke, *Mobilizing Against Nuclear Energy,* 31–32.

57. Melosi, *Coping with Abundance,* 297.

58. Martin V. Melosi, *The Sanitary City: Urban Infrastructure in America from Colonial Times to the Present* (Baltimore, MD: Johns Hopkins University Press, 2000), 363–64.

59. Ibid., 364.

60. Jerome Price, *The Anti-nuclear Movement* (Boston: Twayne Publishers, 1990; rev. ed.), 2–6; Wellock, *Preserving the Nation,*190–91, 197; Rothman, *The Greening of America?* 143; Gottlieb, *Forcing the Spring,* 235.

61. Joppke, *Mobilizing Against Nuclear Energy,* 30–31.

62. Wellock, *Preserving the Nation,*198–99; Philip Shabecoff, *A Fierce Green Fire: The American Environmental Movement* (New York: Hill and Wang, 1993), 121–22; Gottlieb, *Forcing the Spring,* 236–37; Walker, *Three Mile Island,* 13–15; Duffy, *Nuclear Politics in America,* 61–71. See also Ralph Nader and John Abbotts, *The Menace of Atomic Energy* (New York: W.W. Norton & Co., 1977). Othe anti-nuclear critiques of the 1970s can be found in Friends of the Earth's John J. Berger in *Nuclear Power: the Unviable Option* (New York: Dell Pub. Co., 1977; rev. ed.); Richard Munson, ed., *Countdown to a Nuclear Moratorium* (Washington, D.C.: Environmental Action Foundation, 1976). Pro-nuclear material appeared as well. See Dr. Petr Beckmann, *The Health Hazards of Not Going Nuclear* (New York: Golem Press, 1976).

63. Joppke, *Mobilizing Against Nuclear Energy,* 33.

64. Correspondence from Samuel Walker to Martin Melosi, February 4, 2011.

65. Melosi, *Coping with Abundance,* 305.

66. Buck, *A History of the Atomic Energy Commission,* 7; Del Sesto, *Science, Politics, and Controversy,* 104–07.

67. Melosi, *Coping with Abundance,* 305–06.

68. To skirt around the letter of the law, the AEC got the power to issue temporary licenses through a bill signed by President Nixon in 1972 and bolstered by the conservative Burger Supreme Court in 1978.

69. Holl, Anders, and Buck, *United States Civilian Nuclear Power Policy, 1954–1984,* 15; Jasper, *Nuclear Politics,* 54–61; Buck, *A History of the Atomic Energy Commission,* 7; Davis, *Energy Politics,* 22–23; Del Sesto, *Science, Politics, and Controversy,* 156–59; Walker, *Containing the Atom,* 363–86; Atkins, ed., *Historical Encyclopedia of Atomic Energy,* 69–70.

70. Richard Nixon, "Special Message to the Congress on Energy Resources, *June 4, 1971*," The American Presidency Project, online, http://www.presidency.ucsb.edu/ws/?pid=3038#axzz1p1jjuXHD.

71. Holl, Anders, and Buck, *United States Civilian Nuclear Power Policy, 1954–1984,* 15, 17–18; Jasper, *Nuclear Politics,* 16–17; Buck, *A History of the Atomic Energy Commission,* 7–8; Walker, *Three Mile Island,* 7; Davis, *Energy Politics,* 229–30.

72. Melosi, *Coping with Abundance,* 282–84.

73. On October 6, 1973 (Yom Kippur), Egypt and Syria launched a coordinated attack on Israel. Egyptian troops entered the eastern bank of the Suez Canal, while Syrian forces claimed a foothold in the Golan Heights. The initial advantage went to the oil-financed and Soviet-armed Arabs, but by mid-October an Israeli counterattack reversed the fortunes of the war. Like the Soviets, the Americans gave up any pretense of neutrality, airlifting weapons, tanks, and planes into Israel. Melosi, *Coping with Abundance,* 277–79.

74. Ibid., 278–80.

75. Ibid., 280–82.
76. Ibid., 282; Del Sesto, *Science, Politics, and Controversy,* 103.
77. Walker, *Three Mile Island,* 7–8; Duffy, *Nuclear Politics in America,* 69–71, 79.
78. Duffy, *Nuclear Politics in America,* 71–75, 77; Balogh, *Chain Reaction,* 306–07.
79. Melosi, *Coping with Abundance,* 314–16.
80. Ibid. See also E.F. Schumacher, *Small is Beautiful: Economics as if People Mattered* (New York: Harper & Row, 1973).
81. Quotes in Amory B. Lovins, *Soft Energy Paths: Toward A Durable Peace* (Cambridge, MA: Harper Colophon Books, 1977), 3ff. See also Melosi, *Coping with Abundance,* 316–19; John Opie, *Nature's Nation: An Environmental History of the United States* (New York: Harcourt Brace, 1998), 472–74; Walker, *Three Mile Island,* 15–20.
82. "Excerpts from the Reactor Safety Study (WASH-1400) (commonly known as the Rasmussen Report) published by the US Nuclear Regulatory Commission 1974," online, http://www.ccnr.org/rasmussen.html#2.11.
83. Melosi, *Coping with Abundance,* 306–07.
84. Morone and Woodhouse, *The Demise of Nuclear Energy?* 91–92; Davis, *Energy Politics,* 219; Atkins, ed., *Historical Encyclopedia of Atomic Energy,* 306–07; Correspondence from Samuel Walker to Martin Melosi, February 4, 2011.
85. ERDA was transferred to the Department of Energy in 1977.
86. Atkins, ed., *Historical Encyclopedia of Atomic Energy,* 119–20.
87. Duffy, *Nuclear Politics in America,* 112–20; Nuclear Regulatory Commission, *A Short History of Nuclear Regulation, 1946-1999,* 24; Atkins, ed., *Historical Encyclopedia of Atomic Energy,* 119–20; Correspondence from Samuel Walker to Martin Melosi, February 4, 2011.
88. Melosi, *Coping with Abundance,* 308; Duncan Burn, *Nuclear Power and the Energy Crisis: Politics and the Atomic Industry* (New York: NYU Press, 1978), 82–83.
89. Burn, *Nuclear Power and the Energy Crisis,* 86–87; Melosi, *Coping with Abundance,* 309; Atkins, ed., *Historical Encyclopedia of Atomic Energy,* 64. See also, largely from the perspective of Friends of the Earth, David Dinsmore Comey, "The Incident at Browns Ferry," in Peter Faulkner, ed., *The Silent Bomb: A Guide to the Nuclear Energy Controversy* (New York: Random House, 1977), 3–22.
90. Wellock, *Preserving the Nation,* 199–200; Melosi, *Coping with Abundance,* 309; Gottlieb, *Forcing the Spring,* 237. See also Wellock, *Critical Masses,* 68–113, 147–72.
91. About 300 local inhabitants of the small village of Whyl in the Rhineland started what would be the first major protest against the construction of a nuclear power plant in Feberuary 1975. They sat down in front of heavy equipment, thus stopping the work. German police tried to force the protesters away, but without much luck. What started as a way to protect local vineyards from the humidity of the cooling towers turned into a major protest of reactor safety with more than 20,000 demonstrators. Protesters occupied the site itself for one year. The government tried legal action, but the reactor was never constructed. See Atkins, ed., *Historical Encyclopedia of Atomic Energy,* 401–02.
92. Atkins, ed., *Historical Encyclopedia of Atomic Energy,* 89–90; Wellock, *Preserving the Nation,* 200; Gottlieb, *Forcing the Spring,* 237–39; Henry F. Binford, *Seabrook Station: Citizen Politics and Nuclear Power* (Amherst, MA: University of Massachusetts Press, 1990).
93. Melosi, *Coping with Abundance,* 309–10; Atkins, ed., *Historical Encyclopedia of Atomic Energy,* 326–27.
94. Interestingly, some Sierra Club members (including photographer Ansel Adams) had acquiesced on the Diablo Canyon site in discussions with PG&E in the late 1960s. This created a rupture in the environmental organization's leadership and also among its rank and file members. See Meehan, *The Atom and the Fault,* 43. See also John Wills, *Conservation Fallout: Nuclear Protest at Diablo Canyon* (Reno, NV: University of Nevada Press, 2006).
95. Gottlieb, *Forcing the Spring,* 239–43; Wellock, *Preserving the Nation,* 200–01.
96. Melosi, *Coping with Abundance,* 308–10.
97. Morone and Woodhouse, *The Demise of Nuclear Energy?,* 96.
98. Walker, *Three Mile Island,* 51–52.

99. Melosi, *Coping with Abundance*, 310–11.

100. Walker, *Three Mile Island*, 78.

101. Melosi, *Coping with Abundance*, 311.

102. Quoted in Walker, *Three Mile Island*, 182.

103. Ibid., 190.

104. Melosi, *Coping with Abundance*, 311; Walker, *Three Mile Island*, 196–204.

105. Quoted in Walker, *Three Mile Island*, 210.

106. Quoted in Philip L. Cantelon and Robert C. Williams, *Crisis Contained: The Department of Energy at Three Mile Island* (Carbondale, IL: Southern Illinois University Press, 1982), xi. See also Ford, *Meltdown*, 230–34; Duffy, *Nuclear Politics in America*, 168–74; Union for Concerned Scientists, *Safety Second: The NRC and America's Nuclear Power Plants* (Bloomington, IN: Indiana University Press, 1987), 17–18, 43–45, 62, 139–42.

107. Nuclear Regulatory Commission, *A Short History of Nuclear Regulation, 1946–1999*, 25–27; Walker, *Three Mile Island*, 8–9, 222–24; Morone and Woodhouse, *The Demise of Nuclear Energy?* 96–98; Davis, *Energy Politics*, 220–21; Bonnie A. Osif, Anthony J. Baratta, and Thomas W. Conkling, *TMI 25 Years Later: The Three Mile Island Nuclear Power Plant Accident and Its Impact* (University Park, PA: Pennsylvania States University Press, 2004), 84–86; Jasper, *Nuclear Politics*, 213; Joppke, *Mobilizing Against Nuclear Energy*, 135–36.

108. Wellock, *Preserving the Nation*, 201. See also for critiques and journalists' reactions, Raymond Goldsteen and John K. Schorr, *Demanding Democracy After Three Mile Island* (Gainesville, FL: University of Florida Press, 1991); Mike Gray and Ira Rosen, *The Warning: Accident at Three Mile Island* (New York: W.W. Norton & Co., 1982); Wilborn Hampton, *Meltdown: A Race Against Nuclear Disaster at Three Mile Island* (Cambridge, MA: Candlewick Press, 2001).

109. Stanley M. Nealey, Barbara D. Melber, and Willian L. Rankin, *Public Opnion and Nuclear Energy* (Lexington, MA: Lexington Books, 1983), especially 169–82.

110. Mans Lonnroth and William Walker, "The Viability of the Civil Nuclear Industry," in Ian Smart, ed., *World Nuclear Energy: Toward a Bargain of Confidence* (Baltimore, MD: Johns Hopkins University Press, 1982), 148–49.

111. Ibid., 151–52. See also Joseph A. Camilleri, *The State and Nuclear Power: Conflict and Control in the Western World* (Seattle, WA: University of Washington Press, 1984), 186–226.

112. Lonnroth and Walker, "The Viability of the Civil Nuclear Industry," 155–56, 162–64.

113. Joseph A. Yager, *International Cooperation in Nuclear Energy* (Washington, D.C.: Brookings institution, 1981), 10, 13.

114. Jonathan Scurlock, "A Concise History of the Nuclear Industry Worldwide," in David Elliott, ed., *Nuclear or Not? Does Nuclear Power Have a Place in a Sustainable Energy Future?* (New York: Palgrave Macmillan, 2007), 28. For more information on the French, West German, Swedish, British, and Soviet nuclear power programs in the 1960s and 1970s, see Gabrielle Hecht, *The Radiance of France: Nuclear Power and National Identity after World War II* (Cambridge, MA: MIT Press, 1998); Camilleri, *The State and Nuclear Power*; Michael T. Hatch, *Politics and Nuclear Power: Energy Policy in Western Europe* (Lexington, KY: University Press of Kentucky, 1986); S.D. Thomas, *The Realities of Nuclear Power: International Economic and Regulatory Experience* (Cambridge: Cambridge University Press, 1988); Paul R. Josephson, *Red Atom: Russia's Nuclear Power Program from Stalin to Today* (Pittsburgh: University of Pittsburgh Press, 2000); Peter D. Dresser, ed., *Nuclear Power Plants Worldwide* (Washington, D.C.: Gale Research Inc., 1993).

115. Dresser, ed., *Nuclear Power Plants Worldwide*, 19, 11, 157, 165, 203, 251.

116. Scurlock, "A Concise History of the Nuclear Industry Worldwide," 28–29. See also Camilleri, *The State and Nuclear Power*, 133–35.

117. Lonnroth and Walker, "The Viability of the Civil Nuclear Industry," 167–68, 211–12. See also Jasper, *Nuclear Politics*, 107–28.

118. Jasper, *Nuclear Politics*, 3–4, 120–24, 127–28.

119. Ibid., 131–38, 148–54, 163–71, 177–81.

120. Ibid., 241.

121. Ibid., 237–41; Alaine Touraine, et al., *Anti-nuclear Protest: The Opposition to Nuclear Energy in France* (London: Cambrdige University Press, 1980).

122. Horst Mewes, "A Brief History of the German Green Party," in Margit Mayer and John Ely, ed., *The German Greens: Paradox Between Movement and Party* (Philadlephia, PA: Temple University Press, 1998), 31–32; Joppke, *Mobilizing Against Nuclear Energy*, 17, 49–50, 91.

123. "Green Party," *The Free Dictionary*, online, http://legal-dictionary.thefreedictionary.com/Green+Party.

124. Joppke, *Mobilizing Against Nuclear Energy*, 91–129; Roland Roth and Detlef Murphy, "From Competing Factions to the Rise of the Realos," in Mayer and Ely, eds., *The German Greens,* 66–67; Joachim Hirsch, "A Party Is Not a Movement and Vice Versa," in Mayer and Ely, eds., *The German Greens,* 182.

125. Horace Herring, "Opposition to Nuclear Power: A Brief History," in Elliott, ed., *Nuclear or Not?* 39–49.

126. Jane I. Dawson, *Eco-Nationalism: Anti-Nuclear Activisim and National Identity in Russia, Lithuania, and Ukraine* (Durham, NC: Duke University Press, 1996).

127. Velma Garcia-Gorena, *Mothers and the Mexican Anti-nuclear Power Movement* (Tucson, AR: University of Arizona Press, 1999), 2, 15–16.

128. Diana J. Kleiner, "Karen Gay Silkwood," Texas State Historical Association, *The Handbook of Texas Online*, online, http://www.tshaonline.org/handbook/online/article/fsi35; Price, *The Anti-nuclear Movement*, 95–97; Atkins, ed., *Historical Encyclopedia of Atomic Energy*, 331–32.

129. "Environment: The Silkwood Mystery," January 20, 1975, *Time*, online, http://www.time.com/time/printout/0,8816,912701,00.html.

130. Kleiner, "Karen Gay Silkwood."

131. Price, *The Anti-nuclear Movement*, 12, 15, 17, 31, 79, 98–99. For details on the medical aspects of the case, see "The Karen Silkwood Story," *Frontline* (first published in Los Alamos Science XXIII (November 23, 1995), online, http://www.pbs.org/wgbh/pages/frntline/shows/reaction/interact/silkwood.html. See also for more on the conspiratorial nature of the case, Howard Kohn, "Malignant Giant: The Nuclear Industry's Terrible Power and How It Silenced Karen Silkwood," in Faulkner, ed., *The Silent Bomb*, 23–42.

132. See Richard Schickel, "A Tissue of Implications," *Time*, December 19, 1983, 73; Dave Kehr, "Silkwood," *On Film: Brief Reviews*, Chicago Reader, online, http://onfilm.chireader.com/MovieCaps/S/SI/11109_Silkwood.html; "Remembering Karen Silkwood, Union Martyr," *UE News*, online, http://www.ranknfile-ue.org/uen_0100_slkwd.html.

133. Reviews of the film seem to miss the point about how nuclear power is portrayed in this film. See "The China Syndrome," *DVDLaser.com*, April 24, 2001, online, http://www.dvdlaser.com/cf/detail.cfm?ID=22895; Damian Cannon, "The China Syndrome (1979), April 24, 2001, *Movie Reviews UK 1997*, online, http://www.film.u-net.com/Movies/Reviews/China_Syndrome.html; "Critique of *The China Syndrome*," Nuclear Energy Institute, March 27, 2006, online, http://www.nei.org/doc.asp?catnum=&docid=565&format=print; Roger Ebert, "The China Syndrome," January 1, 1979, rogerebert.com, online, http://rogerebert.suntimes.com/apps/pbcs.dll/article?AID=/19790101/REVIEWS/9010103. Richard Schickel picks up on a little of the complexity of Lemmon's character in "An Atom-Powered Thriller," *Time*, March 26, 1979, 54.

134. See Kim Newman, *Apocalypse Movies: End of the World Cinema* (New York: St. Martin's Griffin, 2000), 210–15.

CHAPTER 9

The Post-TMI World, Chernobyl, and the Future of Nuclear Power

INTRODUCTION

In the late 1970s, Three Mile Island riveted attention on the hot debate over reactor safety and future energy needs. The 1980s did not begin with such an atomic drama. Civilian nuclear power was moribund by then; its ultimate fate uncertain. By mid-decade, however, nuclear power was again front-page news with the horror of Chernobyl, renewing the fear of fallout that rocked an earlier generation. As usual, hope, promise, and risk told the tale of atomic power in the 1980s and the decade which followed. Some of the issues debated were startlingly similar. Was a reactor accident a terrible aberration or something to anticipate again? Could commercial nuclear power live up to the dream of an abundant (and clean) new source of electricity? But other questions introduced new or seldom considered issues: Beyond reactor safety what do we do with other potentially risky elements of the fuel cycle such as waste? These concerns, the familiar and the unfamiliar, made it clear that atomic energy was woven tightly into the fabric of the coming decades as it had been since the 1940s. This chapter examines the post-TMI world of commercial nuclear power to the end of the twentieth century, pointing toward its uncertain future in the twenty-first.

CLEANING UP AFTER TMI

How and if the Three Mile Island facility returned to normal was a bellwether of things to come for the nuclear power industry. The initial prognosis was tentative. The cleanup of TMI-2 moved along slowly, but yielded significant new information. Researchers learned early on that the reactor suffered a major core meltdown, but without the release of a large amount of radiation. While this was a positive outcome in an otherwise unsettling event, there was no complete understanding of why the pressure vessel did not fail. That it did not was good news; that such findings were difficult to generalize for other plants under a variety of circumstances was not. A quick restart of the undamaged TMI-1 was not to be. Mired in controversy and facing practical and bureaucratic setbacks, the reopening did not take place until October, 1985. General Public Utilities Corporation (GPU) was desperate to generate revenue, especially because of the losses incurred over the accident. The long delay was due to several issues: court cases that anti-nuclear protesters brought against the Nuclear Regulatory Commission (NRC), opposition by a variety of forces in Pennsylvania (including Governor Richard Thornburgh), and the GPU's own management deficiencies, blunders, and the criminal conduct of reactor operators (who had been caught cheating on examinations in 1981).[1]

Beyond the controversies over reopening TMI-1 and the cleanup of TMI-2, concern remained about the public health impacts of the 1979 accident. This brought attention to a central point of debate which dogged nuclear power from its inception. Public health investigations took place during the cleanup period. Several of them corroborated the findings of the Kemeny Commission and other bodies suggesting that the accident had limited or no negative health impacts of residents in the vicinity of TMI. Yet allegations to the contrary continued to surface, and the state health department in Pennsylvania between 1974 and 1983 set out to evaluate cancer deaths within 20 miles of the plant. The conclusion was that cancer deaths were no higher than normal for the area. A 1990 study sponsored by the Three Mile Island Public Health Fund supported these results. Dissenting studies questioned these findings, but they were based on less reliable data.[2] More research followed. One study released in 1998 found a minor increase in cancer incidents, but most sustained earlier conclusions.[3] Despite the many studies, people were going to believe what they wanted to believe. It often came down to whether one trusted a particular expert analysis or not. The unseen impact of radiation made it very different from an automobile accident, for example, especially since it was difficult to determine results with the naked eye. Suspicions lingered.

The effort to seek answers to the causes and consequences of the accident at Three Mile Island was almost beside the point in determining the financial impact. GPU faced the most severe but not the only shortfall. While the cleanup took place and the wrangling over the restart of TMI-1 persisted, GPU had to buy power from other utilities to meet customer demand. The cost of the cleanup itself reached approximate $1 billion by the early 1990s. The price tag for GPU was $367 million, or about one-third of the total. Insurance payments ($306 million), other nuclear utilities ($171 million), the federal government ($76 million), and the states of Pennsylvania and New Jersey ($42 million) bore the remaining costs. At the time of the accident, demand for new nuclear plants already was in decline; utilities were guilty of overbuilding not only nuclear but coal-fired plants; and demand for electricity fell short of predictions. Between 1979 and 2001, 71 nuclear plant orders were cancelled, several were never completed, and no new plants were ordered since 1978.[4] (See Figure 9-1 for data on orders and cancellations between 1953 and 2001.) In the 1980s more than $30 billion in sunk costs in nuclear power plants were lost because of cancelled projects. The downturn had an impact on other institutions. For example, the number of students enrolled in nuclear engineering programs at universities dropped considerably despite the existence of 70 operating plants in 1979 generating 13-14 percent of the nation's electricity.[5]

There was plenty of blame to go around for the sad state of the industry. Critics were quick to point out poor utility management, "fast-tracking" plants, and impulsive commercialization of nuclear technology. Supporters looked askance at federal regulators and the anti-nukes. Everyone understood the corrosive influence of inflation, declining demand for electricity, and escalating capital costs of plant construction. Complicating the story was that

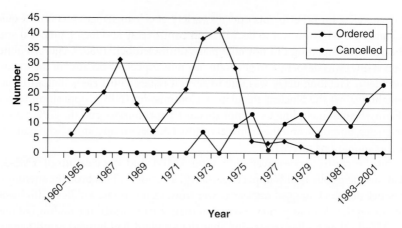

FIGURE 9-1 Reactors in the United States Ordered/Cancelled, 1953–2001. *Source:* Energy Information Administration, in Osif, Baratta, and Conkling, *TMI 25 Years Later*, 86

almost all existing plants remained geographically concentrated in the Northeast, Upper Midwest, and Southeast, but with related challenges such as future siting and waste disposal that were national in scope.[6]

NUCLEAR POWER POLITICS IN THE 1980s

A political bright spot for commercial nuclear power was the election of Ronald Reagan as president in 1981. Riding a wave of anti-big government sentiment and frustration with the idealistic, and seemingly ineffectual Jimmy Carter, the charismatic former actor and former governor of California offered leadership qualities that many voters craved. Campaign rhetoric strongly implied a turn toward fiscal restraint, but Republican administration's "New Federalism" essentially reshuffled priorities, strongly emphasizing federal support for national security and rebuilding the nation's defense system at the expense of many domestic programs. Reagan also pounded away on the issue that excessive regulation weakened the American economy, thus requiring a reduction in the role of federal government intervention, and moving those actions to the states or the private sector itself. His successor, George H.W. Bush, assumed a similar anti-regulation stance.[7]

Support for nuclear power (and the federal role in sustaining it) was one area where the new president drew the line about limited government involvement in non-military activities. In his acceptance speech at the Republican National Convention in July 1980, Reagan stated, "Coal offers great potential. So does nuclear energy produced under rigorous safety standards. It could supply electricity for thousands of industries and millions of jobs and homes. It must not be thwarted by a tiny minority opposed to economic growth which often finds friendly ears in regulatory agencies for its obstructionist campaigns."[8] Reagan's pro-nuclear rhetoric was matched by supporting action. The Carter program favoring energy conservation and research on renewable energy was all but abandoned. Department of Energy (DOE) funds for research on conservation dropped from $779 million in 1980 to $22 million in 1983. Allocations for solar and renewable energy fell from $751 million to $82 million. While funds for nuclear power dipped slightly during the same period, its percentage of DOE total research rose sharply from 39 percent in 1980 to 86 percent in 1983. The administration also supported efforts by pro-nuclear forces to limit state and local regulation of nuclear power, to reduce the number of public hearings, and to limit access to the NRC information and proceedings. The new administration acted to create a less restrictive regulatory environment and to streamline licensing practices. Reagan's appointments to the NRC also favored nuclear industry insiders.[9] As one critic of the Reagan/Bush nuclear policy argued, "Indeed, an inventory of White House actions in this period reads like a nuclear industry wish list."[10] Yet the new Republican administration could hardly control the NRC with an iron hand, and not all members of the Commission blindly fell in line with the industry.[11]

The Reagan and George H.W. Bush White Houses were unable to jumpstart the moribund nuclear power industry, but made progress by clearing away several obstacles to an eventual upturn in the 1990s. The Reagan administration slashed the budgets of regulatory agencies, thus reducing enforcement actions. It also employed the Office of Management and Budget to assess costs and benefits of proposed major regulations, stopping or slowing the implementation of new rules. Government action (or inaction) alone could not save the nuclear power industry.[12] Its problems were not only political, but structural, economic, and environmental. Also, much of the debate over nuclear power tended to focus on future promise rather than current realities. If commercial nuclear power was to move beyond the post-TMI doldrums in the United States, several issues had to be addressed, including management, regulations, competition with other energy sources, public opinion, and questions beyond reactor safety (particularly waste disposal). Change for the better or for the worse would likely be more incremental than revolutionary.

THE POST-TMI REGULATORY ENVIRONMENT

How regulatory authority would change after Three Mile Island and how the industry itself would respond to the accident were of immediate concern. The NRC faced the task of responding to the exposed weaknesses of plant operations, radiation protection, and emergency preparedness. It developed an Action Plan in response to the various investigations, producing dozens of proposals (and controversy) at every turn.[13] There was agreement that "human factors" in plant performance needed addressing. With industry cooperation, the NRC required the use of reactor simulators and improved operator training. It also upgraded control room design and related instrumentation, and reviewed information concerning the operation of plants. TMI had not been the only facility to experience the lapses under evaluation, but it did stimulate a general assessment of generic problems that the accident exposed. The Commission also sponsored surveys on radiation protection procedures, and expanded research programs on fuel damage, radiation releases, and hydrogen generation. Recognizing the confusion and indecision over evacuation procedures, it sought ways to improve emergency preparedness.[14]

A variety of licensing issues were among the most difficult to resolve, especially because licensing had been so controversial. Licensing was at once a point of convergence over safety, siting, public-private cooperation, and viable energy generation. However, there was no serious push to standardize reactor designs as the French had done, which worked against uniform safety standards. In November 1979, the NRC initiated a "licensing pause" to suspend granting of operating licenses for plants under construction. That policy ended in February 1980, and in August it issued the first full-power license to a plant in Virginia. In the remainder of the decade, the NRC granted more than 40 full-power licenses, most of which carried construction permits issued in the mid-1970s. During the early 1980s, licensing controversies became intense with Long Island's Shoreham plant and New Hampshire's Seabrook facility. Under pressure from Congress, the NRC adopted a requirement that all nuclear facilities develop an evacuation plan in case they experienced a reactor accident, which included an emergency planning zone (EPZ) with a 10-mile radius. A DOE study showed that more than 60 percent of the reactors in the United States lacked a formally approved evacuation plan.[15]

Much to NRC's consternation some state and local governments used the new rule as a means of preventing operation of nuclear plants. They simply refused to participate in emergency preparations. Both New York and Massachusetts used this tactic in the cases of Shoreham and Seabrook. State officials in New York argued that evacuation of Long Island was not possible, while proponents stated that the whole area would not require evacuation if an accident occurred. While the NRC granted Shoreham a low-power operating license, the state and the utility agreed on a settlement after a long and protracted political battle lasting until 1990. This ultimately led to Shoreham not being put into operation. At Seabrook the situation was different because the plant was located in New Hampshire, but parts of Massachusetts were in the evacuation radius. By balking, Massachusetts state officials essentially threatened the operation of a nuclear plant in its neighbor state. The NRC got around this dilemma (and the one occurring at Shoreham) by modifying its rule concerning emergency plans, and granted Seabrook an operating license. This smacked of selective enforcement. After the rule change in 1988 ("realism rule") utilities would be allowed to develop their own evacuation plans if state and local governments did not offer one. The argument went that states "realistically" would help in an emergency even if they had not drafted the plan. The rule was challenged in court, but the NRC prevailed. President Reagan ultimately intervened, issuing an executive order in late 1988 that gave the federal government extensive authority to draft emergency evacuation plans. The announcement was postponed until after the 1988 presidential election to avoid giving a hot issue to the Democratic nominee Michael Dukakis, governor of Massachusetts. Delays in opening Seabrook put the utility company into bankruptcy. It went on line in July 1990 after Northeast Utilities bought the plant[16]

Seabrook, Shoreham, and other cases reignited questions concerning the complexity of the licensing process and its impacts on moving nuclear power plants forward. For opponents such bureaucratic red tape and the regulatory morass were allies in stalling construction. To proponents the licensing process was a disincentive. From hindsight, the debate over licensing in the 1980s was somewhat academic given that new construction ground to a halt. At the time, no one knew for sure what the immediate or long-term future for nuclear power held. The NRC and proponents favored a "one-step" licensing system to make the application procedure simpler. Congress did not give the proposal serious attention, and the agency looked to administrative alternatives to streamline the process.[17]

CHERNOBYL

The worst nuclear plant accident in history occurred on April 25-26, 1986 at Chernobyl. In 1985, Lev Feoktistov, deputy director of the Kurchatov nuclear energy institute, confidently stated, "Thorough studies conducted in the Soviet Union have proved completely that nuclear power plants do not affect the health of the population."[18] Eyewitness accounts of the Chernobyl disaster seemed to point to the contrary. One observer noted,

> From above, from the helicopter, when I was flying near the reactor, I could see roes and wild boars. They were thin and sleepy, like they were moving in slow motion. They were eating the grass that grew there, and they didn't understand, they didn't understand that they should leave. That they should leave with the people.[19]

And reminiscent of Hiroshima and Nagasaki, one survivor stated,

> I'm not afraid of death anymore. Of death itself. But I don't know how I'm going to die. My friend died. He got huge, fat, like a barrel. And my neighbor—he was also there, he worked

a crane. He got black, like coal, and shrunk, so that he was wearing kids' clothes. I don't know how I'm going to die. I do know this: you don't last long with my diagnosis. But I'd like to feel it when it happens. Like if I got a bullet in the head. I was in Afghanistan, too. It was easier there. They just shoot you.[20]

The Chernobyl accident replaced Three Mile Island as the most publicized nuclear accident, and far exceeded its destructive impact on humans and the environment. The events were forever linked as warnings about the risks of nuclear power, yet neither was solely responsible for derailing it. Both also were grist for the anti-nuclear movement and stirred debate (and action) in some countries. Given the severity of Chernobyl in particular, it is amazing that the accident came to be regarded by many as an anomaly rather than a portent for the future of nuclear power. Yet on a purely visceral level, the name "Chernobyl" evokes something much more than a passing moment hopefully not repeated. "Why did that Chernobyl break down?" one contemporary inquired, "Some say it was the scientists' fault. They grabbed God by the beard, and now he's laughing. But we're the ones who pay for it."[21]

The Chernobyl Nuclear Power Station is located in Prypiat (or Pripiat), a town of 35,000, on the banks of the Prypiat River in the northern part of Ukraine close to Byelorussia and about 90 miles from Kiev. The facility construction began in the 1970s, and the four units came on line between 1977 and 1983. The plant included four graphite-moderated, water-cooled reactors, known as *RBMK-1000s*, and had two additional reactors under construction. (There were 14 such reactors in the former Soviet Union.) Although the reactors were equipped with emergency cooling systems, they had design weaknesses including susceptibility to electrical interruptions and the lack of protective containment structures.[22] A major contributor to the accident was related to "a prompt-temperature coefficient." In this case, the reactor had a positive temperature coefficient that the operators were not aware of. In other words, when the reactor got hotter, it became more reactive and power increased. The hotter the core, the more it wanted to shut down. A negative temperature coefficient meant the opposite, and all reactors today need to have an engineered negative temperature coefficient.[23]

Beyond the reactor's serious design flaws, the accident occurred primarily because of human error that was more egregious than at TMI. At the time of the accident, neither the plant director nor the chief engineer was on site. Instead, an electrical engineer with no previous experience with nuclear power plants was in charge. Also, some of the personnel in the control room had just met for the first time. Staff performed subsequent actions without adequate safety precautions, and deviated from standard procedures. When asked about what happened, one of the early shift operators, Igor Kazachkov explained, "We didn't have any foolproof safeguards against that particular thing happening…And we still don't have…"[24] The exact sequence of events varies slightly from source to source, but in general problems started with a decision to run a safety test on Unit 4 at the complex on Friday, April 25. For the sake of the experiment, the operating crew turned off the emergency core cooling system (in order to avoid flooding the core and shutting down the reactor). The exercise was started and stopped and started again with a new crew of operators failing to turn the water-cooling system back on or setting the automatic power. When the power level began dropping, the crew tried to drive it back up by withdrawing control rods and using more steam pressure. Despite computer warnings that the reactor needed to be shut down, they continued the test. A sudden power surge put the crew in panic mode as they attempted the shut down, but it was too late. The various actions resulted in losing control of the reactor causing explosions that tore off the roof of the building and spewed an array of radioactive material into the air. The explosions also ignited graphite in the core itself and

On April 26, 1986, a terrible catastrophe occurred at the Chernobyl nuclear plant from an atomic explosion in reactor unit 4. By one account, the surrounding area within a radius of 30 kilometers of the plant was rendered radioactive. Today only dead villages and neglected farmland remain, along with the dead city of Prypiat, where the engineers of the plant once lived. *Source:* ITAR TASS Photo Agency/Alamy

resulted in rampant fires in the reactor building, causing more radiation to be released into the atmosphere. Possibly as much as seven tons of radioactive fuel was blown out of the building.[25]

The immediate emergency responses were too limited, too late, and also had to be conducted at night. The fire in Unit 4 was now threatening Unit 3. Firefighters were brought in, but the intensity of the blaze at first made the water hoses useless. They ultimately contained the spread of the fire using 37 crews with 186 firemen and 81 machines. To also deal with the burning core the Russian air force bombed it with a variety of fire retardants, possibly as much as 5,000 tons of material by May 2. Not until May 10 did temperatures at the accident site begin to drop. An additional remedy was placing a sealed 12 story containment building ("the sarcophagus") around Unit 4 several weeks later. A structure that was meant to survive for years, was already failing in 1990.[26]

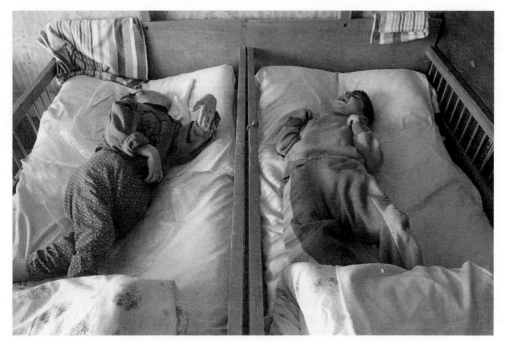

Novinki Asylum, located in Minsk Belarus, is a large asylum for children; most have handicaps apparently caused by the Chernobyl accident. *Source:* jon challicom/Alamy

In February 1986 Vitaly Sklyarov, Minister of Power for the Ukrainian SSR declared, "The odds of a meltdown are one in 10,000 years. The plants have safe and reliable controls that are protected from any breakdown with three safety systems."[27] But 10,000 years had arrived. For several hours after the accident, government officials had no accurate information about what occurred at the site. Thirty-one plant employees and emergency workers died quickly from radiation poisoning, and another 238 endured acute radiation syndrome. The main casualties were among the firefighters. Approximately 50,000 people living near the plant received radiation exposure. About 130,000 residents of the area (maybe many more) were resettled because of contamination to their homes, farms, adjacent forests, and water supplies. Long-term implications of the accident are still debated, particularly the potential threat of leukemia. While World Health Organization studies showed no increases in leukemia by 1993, it was clear that thyroid cases rose sharply, especially among children. Between 1990 and 1999, authorities detected approximately 1,800 cases of childhood thyroid illness in Byelorussia, Ukraine, and Russia. In the Gomel region of Byelorussia, slightly north of Chernobyl, they discovered 38 cases in 1991, where one or two cases per year were normal. In addition, thyroid cancer among teens more than doubled between 1996 and 2001, while childhood cases declined after 1995. In some instances, children of parents exposed to radiation at Chernobyl exhibited genetic mutations.[28]

In the USSR there was a mighty effort to keep the accident under wraps, but that was impossible. The radioactive cloud built up over Chernobyl drifted north to Eastern Europe and into Scandinavia. For obvious reasons detection over Eastern Europe led to no public announcements. But once the radioactivity reached Sweden the world knew the sad truth. On the morning of April 28 Swedish scientists and technicians at Forsmark Nuclear Power Plant learned about the irradiated cloud and Swedish diplomats in Moscow insisted on answers. The composition of the materials in the cloud gave them the response they were seeking—a meltdown of some significance had occurred. The radiation soon spread to Germany, Norway, Denmark, France, and Great Britain, and later to most of southeastern Europe, the Balkans, and Greece. Lesser amounts of radiation also appeared in Canada and the United States. In all, the residue of the explosions exposed people in over 20 countries. While most European nations instituted emergency responses to the fallout, lack of detailed information hampered them.[29]

The Soviets eventually took some action to quell the uproar, including the evacuations, beginning the process of reclamation at the site, and establishing a 25-mile exclusion zone around Chernobyl (May 3, 1986). Radiation exposure was high (sometimes 10 times the recommended dose) for the first responders, many of whom faced certain death. (From 30 to 50 first responders died.) Also, approximately 200,000 clean-up workers, or "liquidators," from throughout the Soviet Union were involved in recovery and cleanup action at Chernobyl in 1986–1987. The total number probably reached 600,000 over time. Many of them, including thousands of Armenian soldiers, had rushed into the "zone" to put out fires and tried to contain the radioactivity. Death rates were classified, but rough estimates suggest as high as 2,000 in total.[30]

Initially officials blamed the local operators, but they eventually loaded responsibility on top administrators in the Soviet atomic energy program. Several lost their jobs. All along the way, spokespersons denied that deficiencies in the design features of the station caused the disaster. They were very defensive about criticism from the West that the Soviets lagged behind in nuclear technology. In July 1987, six of the senior administrators who were declared directly responsible were tried and convicted of criminal negligence; the former director and two assistants received sentences of 10 years hard labor. The operators who conducted the experiment suffered a different fate, dying within a few days of the accident. The *RBMK* reactor was phased out.[31] The direct financial losses due to the disaster were great (at least $3 billion) and the long term loss of agricultural lands was grave. Some even wanted to link the Chernobyl accident to the tearing down of the Berlin Wall in 1989 and the downfall of the Soviet Union in 1991, but this may be stretching the point.[32]

The health effects of Chernobyl have been a serious topic of interest for several years. Between 2003 and 2005, the World Health Organization (WHO) within the UN Chernobyl Forum initiative conducted a series of expert meetings to review scientific evidence on health effects related to the accident. The WHO Expert Group began with the 2000 *Report of the United Nations Scientific Committee on the Effects of Atomic Radiation* (UNSCEAR), which they updated with critical research from peer-reviewed journals and other information provided by the governments of the three affected countries in the old Soviet Union. Some of their findings suggested that: (1) Effective doses of radiation for most of the residents of the contaminated areas were low, but for many people doses to the thyroid gland were large from ingesting milk contaminated with radioactive iodine; (2) Only liquidators who worked around the reactor in the first two years after the accident, the evacuees, and others in highly contaminated areas received doses "significantly above typical natural background levels;" (3) There was a large increase in thyroid cancer among people who had been young children and teens at the time of the accident and who lived in the most contaminated areas of Belarus, the Russian Federation, and Ukraine; (4) Recent studies suggest a doubling of the incidence of leukemia among "the most highly exposed Chernobyl liquidators," but "no such increase has been clearly demonstrated among children or adults resident in any of the contaminated areas;" and (5) "Given the low radiation doses received by most people exposed to the Chernobyl accident, no effects on fertility, numbers of stillbirths, adverse pregnancy outcomes, or delivery complications have been demonstrated nor are there expected to be any."[33] WHO's response to Chernobyl as of 2011 continued to revolve around the UN Action Plan "to overcome the negative legacy of the Chernobyl accident in the third decade of joint recovery efforts."[34] Yet debate over the health and environmental effects of the disaster have not resulted in any clear consensus about the safety of nuclear power.

In addition, 25 years after the disaster the Emergency Ministry in Ukraine was set to open the "dead zone" at Chernobyl to tourists, along with completing an ongoing project to build a safer shell around the reactor by 2015. The mixed signals are extraordinary. A story in a March 2011 issue of the *Telegraph* intoned, "As Ukraine prepares to mark the 25th anniversary of the Chernobyl disaster next month, its legacy remains as divisive as ever..." The first Soviet TV reporter on the scene of the accident recalled, "Chernobyl was a warning for the future. It was not just a banal disaster, it was a message that nuclear power is not safe. It is time to think, consider alternatives, and bring the industry under tight international control. Otherwise, humankind will destroy itself." On the other hand, a spokesperson for tourism in Ukraine stated, "The Chernobyl zone is not as scary as the whole world thinks. We want to work with big tour operators and attract Western tourists, from whom there is great demand." Promoters of tourism point to the widely blossoming plant life and the remarkable return of wildlife to the area. Studies indicate that in the "dead zone" there are 66 different species of mammals, including elk, wild boar, wolves, deer, beaver, foxes, and lynx.[35] Chernobyl remains a nuclear paradox.

A SECOND NUCLEAR POWER ERA?

In 1985, nuclear physicist and former administrator of Oak Ridge National Laboratory, Alvin M. Weinberg, and his colleagues published *The Second Nuclear Era: A New Start for Nuclear Power*.[36] Based on a previous study, the book argued that it was possible to build safe reactors where risks were manageable and even reduced. This could usher in a new day for civilian nuclear power. The timing of the book was not good, since the following April the greatest nuclear accident in history took place at Chernobyl. Nevertheless, Weinberg and others continued to argue that nuclear power had a real future, especially if new technological innovations were employed. In 1989, Weinberg called for worldwide construction of 2,000 to 6,000 additional reactors to be built in the following 40 to 60 years. MIT's Center for Energy Policy Research published a report also in that year stating that an industry-wide pattern of poor management obvious in the pre-TMI days was correctable and was being corrected. Others began touting nuclear power as a "clean solution" to growing international concern over global warming and greenhouse gas emissions.[37] Despite Three Mile Island and Chernobyl, success for the industry (at least as viewed by its supporters) had not completely slipped away.[38] A prescient comment by Norman Rasmussen, head of the Nuclear Engineering Department at MIT and author of the well-known AEC safety report (1974), added an additional wrinkle to the debate: "[I]f we abandon nuclear power, we clearly will put increased pressure on other energy supplies, and clearly oil will be one of them."[39] Optimism about a "second nuclear era" was over-hyped, but it was not completely misplaced. Changes in the industry and changes in the regulatory environment were real. Some experts suggested, however, that an assessment of the nuclear power industry had to be based not on future expectations, but on a "present tense" understanding, that is, the actual events of the time.[40]

Almost every nuclear power issue raised in the aftermath of TMI and Chernobyl followed the US nuclear power industry into the 1990s, especially the safety question. After considerable discussion, the NRC established a "graded approach" for licensing that varied according to the type and complexity of safety considerations. This was accomplished against a backdrop of attempting to revise radiation standards. Basic regulations for occupational and population exposure had not changed since 1961. And while the NRC tightened its regulations on permissible levels of radiation, scientific evidence and epidemiological studies were not of like mind on the health effects of low-level radiation exposure.[41] In 1990, the Commission released a policy statement outlining ways in which tiny amounts of low-level radioactive materials far below regulatory limits could become exempt from regulation—"below regulatory concern" (BRC). This would apply to consumer goods, landfills, and other potential sources of low radiation. The BRC caused intense protest from the public, Congress, anti-nukes, and many in the media. The issue was deferred, but skeptics did not view the NRC as an impartial broker in many of the debates over nuclear power, a reaction going back to the AEC days.[42]

With the lack of new plant applications after TMI, the NRC dedicated much of its time in the 1980s and 1990s to issues related to oversight of existing plants. Maintenance deficiencies were important, as were the decommissioning of plants and license renewals. Between 1947 and 1975, 50 plants had been decommissioned. As several facilities reached the end of their useful life, the NRC looked to revise its regulations. In 1988 it released new rules, but could not resolve every issue. Particularly difficult were questions concerning what to do about funding plants that were prematurely retired, such as Shoreham; how to deal with higher than projected decommissioning costs; and how to determine the level of radiation allowed at sites of decommissioned plants. As with almost every issue featuring nuclear power plants, the NRC and nuclear opponents failed to reach a meeting of the minds. The same could be said for ruffling the feathers of nuclear industry members.[43]

In the early 1990s controversy arose over the Millstone nuclear station (owned by Northeast Utilities) comprised of three plants located on Long Island Sound in Connecticut. "Whistle blowers" charged that plant officials dismissed or hounded several employees at the facility for raising safety issues and violations of NRC regulations. The Commission did not believe the allegations were major, but slapped Northeast Utilities (NU) on the hand with a $100,000 fine. This action did not please the dissenters, and further complaints were forthcoming. The NRC fined Northeast Utilities on at least two more occasions in 1993 and 1994, but criticism of the operation practices at Millstone persisted. In March 1996 *Time* ran a cover story about the whistleblowers, strongly criticizing NRC actions. In response the agency reexamined its regulations and procedures and conducted a probe of the Millstone facilities. The NRC, acting more aggressively, required changes that cost NU in excess of $1 billion, which led to the company selling its generating units. Millstone-1 was closed permanently, but the NRC authorized the other two units for a restart in the late 1990s.[44] The end of this drama did little to save the beleaguered NRC from criticism from both pro- and anti-nuclear forces. Yet the Commission learned a hard lesson from its initial reaction to NU's inept management and ultimately acted more assertively.[45]

Licensing issues also continued into the 1990s. The NRC changed its rules to extend the life of several nuclear plants. By 2000, 62 of the nation's 109 reactors were 20 years old or older, but the dearth of new plants coming on line by then meant a net loss. The industry sought a rule change, and the NRC complied. It would be simpler and cheaper to extend existing plants rather than encourage utilities to build new facilities. Under a 1991 rule change, the NRC allowed utilities to apply for 20-year extensions of their operating licenses. The decision was a red flag to opponents, raising concerns about the possibility of more safety problems as existing plants deteriorated. The NRC disagreed, arguing that the existing plants had been inspected and reviewed regularly and presented no undue danger to the public. The 1954 Atomic Energy Act specified an initial 40-year licensing period for nuclear plants on a case-by-case basis. The act had not been based on safety considerations, but on the time allotted for paying off the debt (the same as for fossil fuel plants).[46]

The commercial nuclear power industry was in limbo in the 1990s, but there was hope for an upswing. In 1997, net electricity generated by nuclear power exceeded 50 percent in

South Carolina, New Jersey, Vermont, and New Hampshire, and exceeded 25 to 50 percent in 16 additional states. Nuclear power was second only to coal with 22 percent net generation of electricity followed at a distance by hydroelectric, petroleum, and gas sources. This was accomplished with no additional new plants ordered since 1979. In 1999 the NRC/AEC had licensed 104 commercial nuclear power reactors operating in 31 states.[47] The question was: Where would nuclear power go from here in the new century? Would the public accept expansion beyond what already had been built or licensed prior to 1979?

The disaster at Chernobyl did much to undercut public support for nuclear power already damaged by Three Mile Island. For the NRC, criticism from the industry itself about the regulatory burden might have been a bigger blow, or at least an unceasing one. On the other side, the NRC was constantly accused of being "a captured agency" that simply relaxed safety regulations under pressure from the industry, did little or nothing to change management practices of the utilities, and focused on "hardware" instead of people. One observation suggested that while the average number of unplanned reactor shutdowns had declined, the numbers were still high. By 1989 only about 20 percent of the plants had made all the changes in safety requirements worked out after Three Mile Island. The gap between the best and the worst plants had not been closed.[48] Other performance indicators showed overall improvement in reliability and safety, with the NRC playing a positive role in the changes.[49]

There was hope among some that the industry's reputation was repairable (or repaired) and its future bright or at least redeemable. In 1988 Congress renewed the Price-Anderson Act with a liability cap of $7 billion as opposed to $700 million. Since private insurance remained at $160 million per plant, the sizable increase came from the operators' pool. While the vote in Congress reflected a stronger anti-nuclear sentiment than the last renewal in 1975, at least there was a cap and not a ruling for unlimited liability. This was a modest victory for the industry.[50]

THE INSTITUTE OF NUCLEAR POWER OPERATIONS

In the United States, the fate of nuclear power was not exclusively or primarily in the hands of the embattled NRC. The industry came to realize that it had to take a good deal of the burden on itself to prevent problems with the safety system and to undergo fundamental changes in management and operation of power plants. The most important response was the emergence of a self-regulating body, the Institute of Nuclear Power Operations (INPO), established in December 1979 in Atlanta, Georgia. This was no window dressing, employing approximately 400 people to regulate the industry rather than spew out more public relations platitudes. The nuclear power business was fighting for survival or at least a major rebound after the decline in the mid-1970s. INPO was interested in establishing industry-wide performance objectives and guidelines for plant operation and for operator training. It assessed operating experience and evaluated individual plants.[51]

In the nuclear power industry, it was well known that "Each license is a hostage of every other license," Detroit Edison CEO Walter J. McCarthy, Jr. stated.[52] One accident had industry-wide repercussions as TMI clearly demonstrated. Every nuclear power plant was a link in a chain that required constant oversight lest everyone suffer. Such was the state of mind that brought INPO into existence. The fact that self-preservation can be a powerful instinct gave INPO its legitimacy. Debate continues, however, as to whether INPO undercuts the NRC or if the NRC relies too heavily on INPO. Nevertheless, the Institute's influence not only has grown in the arena of civilian nuclear power, but also in the regulation of nuclear weapons plants as well. The success of INPO depends heavily on a mutual understanding that nuclear power generators cannot go it alone. This defied the capitalist catechism, but the nuclear power industry has never been a typical private venture throughout its history.[53]

Working toward a real program of self-regulation was not a simple matter, especially since planners and policy makers decided several decades earlier that nuclear plants should be owned and operated by electric power companies. Such companies had their own corporate culture ("a fossil fuel mentality") which led to a particular way of doing things. In the case of nuclear power plants, the Kemeny Commission found a serious weakness in the role of management involvement at TMI. Yet the normal practice in electric utility companies, where nuclear plants resided, was to let plant managers run their plants. This was not a good practice in nuclear facilities, where the technology was complex and operators required unique skills and training. An additional problem for nuclear plant operations was focusing on compliance with numerous and complex regulations to the exclusion of assessing ways to improve management, operation, and maintenance of the plants. NRC regulations often were "hardware-centered," compounded by the fact that the industry was primarily interested in meeting minimum regulations. These approaches exposed limits in the nature of the regulation and the response of those who were being regulated. The Kemeny Commission (and soon thereafter people in the NRC and the industry) came to understand that plants gave so much attention to the safety of the technology, that they often ignored human factors.[54]

INPO was intent on moving the industry away from an insular approach to safety, to learn from industry-wide experience rather than draw on the limited information linked to a specific plant. One CEO stated, "Although we tried to beat each other on the bottom line, we ran our business independently, free to give little regard to what our sister utilities did."[55] Another goal was to increase the professionalism of the nuclear plant control room operator with new approaches to training and qualifications in the area of safety. Essentially INPO reacted to external (NRC) and internal (industry) pressures in the post-TMI years to unify the nuclear power industry and reframe its practices.[56] In the 1980s it set up procedures for accrediting training programs, established its own training academy, and published *The Nuclear Professional*. Between 1979 and 1999, more than 500 personnel served as "loaned employees" from the industry to INPO. In sum, INPO played a significant role in setting nuclear standards in the United States, evaluating the performance of the plants, training reactor operators and managers, and providing a link to the industry as a whole. In these tasks, it is taken seriously within the industry and by those outside of it. By 1998 INPO took some credit for the reduction in events that potentially risked damage to reactor cores from two per year per reactor, to two per year for more than 100 reactors. On a global scale, the World Association of Nuclear Operators plays a similar role to INPO.[57]

NUCLEAR POWER WORLDWIDE AFTER 1979

In the 1970s nuclear power provided an increasing portion of electrical generating capacity globally, compared with the slow start a decade or so earlier. Three Mile Island and later Chernobyl, however, had a sobering effect on the nuclear power industry worldwide. In 1981, 21 countries outside of the United States had 182 nuclear generating plants in operation and another 138 under construction.[58] By the 1990s, the world nuclear capacity slowed precipitously. Between 1999 and 2004 the net increase in generating capacity was modest and the total number of operating reactors increased only slightly (in several cases larger units replaced smaller ones).[59] In many industrialized nations, especially in Europe, government support for nuclear power dwindled and very few new power plants were nuclear. Most of the nuclear plants under construction or planned, with few exceptions, were underway in Asia (China, India, and to a lesser extent, Japan).[60] The French and Japanese remained vigorous supporters of nuclear power despite the general trends to the contrary. By the 1990s France got between 75 and 79 percent of its electricity from nuclear power and Japan about 30 percent. Compared with the 104 American reactors in operation in 2000, a much smaller France had 61 and Japan, 54.[61]

In France the political environment was conducive to steady commitment to nuclear power in the 1970s. But anti-nuclear sentiment also was strong, due in part to the controversy over the industrial prototype breeder reactor *Super-Phenix*.[62] Polls showed only 37 percent of the population favoring nuclear power in 1977, a drop from 56 percent in 1975. Nevertheless, the conservative Gaullist government, which assertively favored nuclear power and tried to stifle opposition, got little resistance from either the Communist Party or the Socialists. Despite Three Mile Island, the French government had no intention of slowing down or stalling its nuclear program, convinced somewhat arrogantly that the Americans were simply incapable of dealing appropriately with nuclear technology. Although the French had their own problems with some reactors and reprocessing, they retained great faith in their technical competence.[63] When the Socialist Party came to power in 1981, the French froze nuclear development for a time, but ultimately continued most of the programs. This undercut the anti-nuclear movement even further. The government cancelled the controversial Plogoff plant in Brittany and concerns over reprocessing surfaced. Yet in 1982–1983 they ordered six new LWR plants with the *Super-Phenix* on track for completion. In 1990, the Socialists adopted a policy of 51 percent government ownership of major nuclear corporations. Retention of jobs was important to them, and those ecologists in the party who strongly opposed nuclear power were unceremoniously marginalized.[64] (In 1985 the French Secret Service sank the Greenpeace ship, *Rainbow Warrior*, in New Zealand. The ship was involved in protesting nuclear testing in the Pacific. This action obviously made no friends for France among anti-nukes.[65]) The Chernobyl disaster raised serious concerns about France's nuclear program, but most French citizens came to resign themselves to the inevitable. In the late 1990s, Greens allied with the Communist Party to shut down the *Super-Phenix* and stop construction of new plants, but France's commitment to civilian nuclear power remained second to none in the world. By 1990 France was the world leader in nuclear energy generating capacity.[66]

In the rest of Europe, nuclear power did not fare as well as in France. Like the rise of Ronald Reagan in the United States, the election of the Margaret Thatcher government appeared to hearken a new day for nuclear power in Great Britain. Opposition rose to meet the "Iron Lady," but it did not deter her. In 1988 the Thatcher government strongly considered privatizing the British nuclear industry. The plan failed because of high costs and growing reservations about the energy source itself even in the Conservative Party. Nuclear power was withdrawn from the 1989 privatization plan, and the new government-owned Nuclear Electric and Scottish Nuclear Electricity retained stock in the nuclear power stations. The companies were to operate all existing plants, but the government would not approve new ones until a thorough policy review was completed. The government, therefore, continued to subsidize electricity from nuclear generation. But in the 1990s public financing of nuclear power again came under scrutiny in Great Britain, and funds were cut off in 1994 for reactor research. In 1992 nuclear power provided approximately 20 percent of electricity generated in England and Wales and 50 percent in Scotland.[67]

Another formerly important nuclear state, Sweden, underwent significant changes in the 1980s and 1990s. Nuclear power accounted for about 50 percent of its electricity, making Sweden the fifth highest among industrialized countries. In 1980 the government held a national referendum on nuclear energy, and officials interpreted the results as calling for an end to new reactor construction after completing 12 more in the pipeline. The major proponents of nuclear power, the Social Democrats, shifted policy after TMI and called for the referendum. This was a politically driven decision to end a rancorous partisan issue. Interestingly, both pro- and anti-nuclear positions in 1980 were similar, that is, to end the continuous expansion of nuclear power and ultimately to phase it out entirely. The only major difference was how fast this was to be done. The role of Chernobyl strengthened the Social Democrats' commitment to this change in policy. New plants were to be phased out by 2010, but were not. In June of that year, the Parliament voted to repeal the decision on the grounds of increasing Sweden's energy security and combating global warming. The country retained 10 plants which produced about 40 percent of Sweden's electricity.[68]

The overall picture in the European Union was one of decline for nuclear power after TMI and Chernobyl, but not everywhere and not immediately. Like Sweden, Belgium passed legislation in 2003 prohibiting construction of new reactors and limiting the life of existing ones to 40 years. East Germany mothballed all of its 10 reactors in 1990, while West Germany got new life out of its nuclear power industry in the 1990s as a hedge against air pollution and global warming. But in 2001 a consolidated Germany also limited the operating life of reactors (32 years) and banned the shipment of fuel for reprocessing beginning in 2005. Italy phased out nuclear power on the heels of a 1988 referendum. Some other European countries that had active nuclear programs were not willing to expand them at the turn of the century. This included Hungary, the Netherlands, Slovenia, and Spain. The Czech Republic, Lithuania, and Finland looked to expand their civilian nuclear industry despite the events at Chernobyl. Internal political and economic concerns seemed to trump external issues.[69]

Before Chernobyl, the Soviet Union was adding two million kilowatts of generating capacity annually in the 1980s with expectations for much more in the 1990s. This required more plants, larger plants, and upgrading those in operation. All this changed after the accident and was compounded by the breakup of the USSR in 1991. When the union was dissolved Russia inherited most of the country's reactors, with Ukraine second. Russia did not abandon what it believed to be a critical source of energy, but it conceded that new plants were necessary to replace the controversial Soviet-style reactors of the pre-Chernobyl years.[70]

The 1990s brought plans for aggressive growth in nuclear power capacity in Japan through government-industry cooperative plans. Chernobyl inspired increased protests to the island nation, and siting was a chronic point of controversy. Smaller than California, and with a population exceeding 117 million in the 1990s, Japan has 10 percent of the volcanoes in the world, regularly experienced earthquakes (about 500 per year), and also faced tsunamis generated by seismic activity. While most of the population lived on only 30 percent of the country's land, siting of power plants was never a simple process. Geographic considerations led to placement in small, densely populated areas and in coastal regions because of the abundant supply of water. Officials gave less attention to the potential liabilities of location. Like the French and other industrialized nations, Japan committed to nuclear power because of the paucity of other sources of energy and to enhance energy security.[71]

China began considering nuclear energy in the late 1970s, intending to build 12 plants within 20 years of 1980. In 2011, China had 13 reactors in operation, 25 under construction, and more in development. It is attempting to become self-sufficient in reactor design and construction after previous unsuccessful efforts at cooperation with the United States.[72] India developed a robust nuclear power program in hopes of supplying one quarter of its power needs by 2050. India lagged in its civilian program largely because it did not sign the Nuclear Non-Proliferation Treaty (having a weapons program of its own) and because it lacked uranium. This meant that for several decades—an in fact until quite recently—India was barred for trading in nuclear plants or related materials, and needed to rely primarily on its internal resources. The change led to increased access to foreign technology and fuel sources, while relying on its home-grown engineering expertise. India, however, was making plans to become an international leader in nuclear technology.[73] Other parts of the world have not been deeply engaged in nuclear power production to the extent that North America, Europe, and Asia have been.[74]

After TMI civilian nuclear power showed many faces in the world, from persistent optimism to wary cynicism and opposition. Chernobyl threw a shroud over the future hope of nuclear power—a nightmare confirming the worst case. Yet as horrific as the accident revealed itself to be, commercial nuclear power survived. In some places, without a doubt, Chernobyl derailed nuclear power for good. It limped along elsewhere. And in some instances, it changed very little. Commitment or rejection of civilian nuclear power did not take place in a vacuum. Instead, extenuating political, economic, technical, and environmental factors influenced decisions in major ways. In essence, neither TMI nor Chernobyl or any other catastrophic event changed the basic trajectory of civilian nuclear power on its own.

ANTI-NUCLEAR FALLOUT AFTER CHERNOBYL

Taken together Three Mile Island and Chernobyl influenced the anti-nuclear movement, but could not sustain it. Depending on the location, circumstances, and timing each event had incisive rather than universal impacts. In the United States, fading memories of TMI, the economically weakened nuclear power industry, the defeat of Jimmy Carter, and the victory of the pro-nuclear Ronald Reagan contributed to an increasingly impotent anti-nuclear movement as direct-action protest. The possible exceptions were events at Seabrook and Shoreham. The US anti-nuclear movement never had been the work of large organizations, but many small groups spread across the country and several with clearly regional interests (especially in California and New England). Many of the direct-action groups did not remain in the late 1980s, but the Union of Concerned Scientists, the Natural Resources Defense Council, the Clamshell Alliance, and several environmental groups continued to carry the torch. In large measure, those opposed to nuclear power turned away from reactor safety to issues of radioactive waste and emergency evacuation planning. New issues associated with the arms race also drew attention away to other fronts.[75]

Chernobyl pushed support for nuclear power into further decline in some circles, especially when coupled with memories of TMI. But the Reagan administration did not change its nuclear energy policy in response to the accidents. The pro-nuclear countermovement led by such groups as the Atomic Industrial Forum, the Nuclear Energy Institute, and the American Nuclear Society maintained staunch defense of nuclear power. For every anti-nuclear claim that nuclear technology was dangerous, that a nuclear economy was unsustainable, and that alternative sources of energy were preferable, the pro-nukes argued that nuclear energy was clean and safe, and vital to the nation's welfare and economy.[76] A CBS News poll showed that 55 percent of Americans believed that an accident on the order of Chernobyl could happen in the United States, but public attitudes still seemed to lead toward ambivalence about nuclear power.[77] Or maybe as one writer argued, by the time of Chernobyl "most people had already made up their minds about nuclear power—and entrenched attitudes are obviously difficult to change."[78]

In the Soviet Union, the response to Chernobyl was no less convoluted, but certainly muddled in quite different ways. The controlled and super secret world of the USSR meant that through the mid-1980s, as one writer noted, "the communist world appeared strangely calm" about nuclear power.[79] The lack of anti-nuclear opposition kept decision making locked at the very top of the system, a system that did not tolerate public dissent or even substantial internal criticism or contradiction. In 1986 everything changed. It was not only the disaster at Chernobyl, but the introduction of the reform programs of Mikhail Gorbachev that coalesced to bring about a heightened public response to nuclear power. In the mid-1980s, Gorbachev's government introduced "perestroika" (restructuring) in an effort to transform or at least modify Soviet economic and political policy. With a desire to have his country compete more effectively with capitalist countries such as the United States, Germany, and Japan, he attempted to decentralize economic controls and promote more independent enterprises. This also entailed pulling back on the direct involvement of the Communist Party in the governance of the nation. Although it met with resistance from within and without, the Gorbachev government created a watershed moment in the USSR. While Gorbachev himself lost power in 1991, and did not see many of his ideas implemented, change in the USSR was underway.[80]

Perestroika and Chernobyl sparked popular movements in the Soviet states between 1987 and 1991, particularly in Armenia, Lithuania, Ukraine, and Russia. Criticism quickly arose over the central government's intention to double nuclear output during a Five-Year Plan that covered the period 1986 to 1990. Dissenters not only wanted the government to cancel future reactors, but to halt current construction projects and to close existing stations. Some success followed the protests, especially in 1988–1990. In Armenia the one operating nuclear power station was shut down; plans to expand Lithuania's only nuclear power facility were abandoned; in both Ukraine and Russia public opposition led to cancelling new projects and halting ongoing construction of others. Parliaments in both republics passed moratoria of five years on all new construction under their jurisdiction. The ambitious plans for nuclear power in the USSR came to a grinding halt. By 1991, however, the anti-nuclear fervor in the Soviet Union fizzled, possibly due to the initial heady victories attributed to public protests, to the conservative shift after Gorbachev's departure, or even to fading memories of Chernobyl. In 1991, both Lithuania and Armenia reconsidered their policies of limiting or curtailing nuclear power plants. In both Russia and Ukraine in 1993, the moratorium ended. More striking was that public protest did not follow these reversals.[81] However, in republics like Armenia, Lithuania, and Ukraine, the initial anti-nuclear activities spurred a process of "eco-nationalism" (caused by the convergence of environmentalism and nationalism) or a kind of national identity buried for years after the Russian Revolution in 1917.[82]

In other parts of the world, local circumstances resulted in different reactions to the nuclear accidents. In Mexico, for example, construction began on a nuclear power plant at Laguna Verde in the state of Veracruz in the early 1970s. The general anticipation of bringing new jobs to the region and providing more electricity soured with TMI and Chernobyl. Indeed, the anti-nuclear movement in Mexico began in 1986 not only in response to Chernobyl, but also because of President Miguel de la Madrid's announcement on September 1, 1986 that the Laguna Verde project would move forward. New protest groups formed, such as Madres Veracruzanas, supported by other organizations like the Catholic Church and labor unions. The protest focused in Veracruz drew opposition from supporters of Mexico's nuclear power program, including the country's electrical workers union. Anti-nuclear protest in Mexico was not unlike new social movements elsewhere that arose in the wake of Chernobyl.[83]

In Sweden, Chernobyl elevated the rhetoric over anti-nuclear protest, especially since the country had been in harm's way. However, energy policy did not change. In this case, TMI may have been more significant in challenging existing policy since the Social Democrats led the charge for the 1980 moratorium.[84] The Greens in West Germany reacted strongly to Chernobyl. A mass protest followed at the nuclear reprocessing plant in Wackersdorf, Bavaria, and peace protests broke out at several locations. Even among the Green Party, no single voice bridged the factions.[85] Opinion polls taken between 1978 and 1982 showed the French increasingly viewing the development of nuclear power as "worthwhile" (from 40 percent to 51 percent between those years). In other Western European countries in those years, polls reflected either slight increases or slight decreases. After Chernobyl, opposition increased in several of these countries, including France, Greece, Italy, the Netherlands, the UK, and West Germany. In all cases, opposition dropped off one year after the accident, but remained above pre-accident levels.[86] There is little doubt that TMI and Chernobyl were profound events with significant public impacts. Yet, the accidents did not dissuade many advocates of commercial nuclear power to stay the course.

THE END OF THE FUEL CYCLE: RADIOACTIVE WASTE

During the path of its history, controversy over civilian nuclear power focused most sharply on reactor safety and siting. Less dramatic, but equally significant were other elements in the fuel cycle, from uranium mining to waste disposal. In the 1950s, when programs in nuclear power were advancing, the AEC gave much lower priority to research in waste handling. But by the 1980s, nuclear waste emerged as a major point of contention and consternation. Coping with questions regarding "the end of the pipe" problem involved more than addressing NIMBYism. Politics too often trumped technical solutions as the biggest obstacle to successful radioactive waste policy.[87]

Radioactive waste (military and commercial) posed much more complex problems than any other form of waste product one could imagine. In 2001, the United States had 45,000 metric tons of spent nuclear fuel derived from commercial reactors, and 2,000 metric tons from defense spent nuclear fuel. About that time, there was a total of about 228,000 metric tons of spent fuel around the world, and most of it remained on site.[88] Where and how to store it temporarily and permanently? Was reprocessing the best solution? Who should be responsible? These questions and more led to deferring answers rather than seeking solutions. Complicating the problem was the fact that radioactive waste came in several forms, each presenting different disposal problems. Discards included uranium mill tailings, commercial spent fuel, low-level wastes, mixed wastes, transuranic waste, and high-level wastes. Mill tailings came from mining uranium ore and were plentiful. Transuranic wastes were human-made elements such as neptunium and plutonium. Low-level radioactive waste (LLW), derived from medical and scientific uses, was the least dangerous, while high-level wastes (HLW) from reactors (produced directly in the nuclear fuel) and spent fuel rods were on the other end of the scale. HLW produced substantial heat as well as radioactivity, but not everyone agreed upon the definition of nuclear waste. For example, many engineers were unlikely to think of spent fuel rods as waste, since they contained significant amounts of unused uranium or plutonium. Reprocessing was a possibility, but proved highly controversial because of environmental concerns, the cost, and the risk of proliferation.[89]

In the earliest years of nuclear reactors, the AEC stored high-level nuclear waste (all military related) at three sites: Richland, Washington; Idaho Falls, Idaho; and Aiken, South Carolina. The first high-level waste came from reprocessing of spent fuel from plutonium production at Hanford Engineer Works during the Manhattan Project. Despite the risks resulting from production of radioactive materials, officials responsible for radiation safety believed that prevailing disposal methods and existing facilities were adequate to protect employees and the general public. After the war, through experiences with the detonation of two atomic bombs, health experts revised their assessment of exposure to radioactive materials. Health physicists and sanitary engineers were among those who criticized the AEC's handling of nuclear wastes at Hanford and elsewhere. Yet major changes in radiation safety at nuclear weapons plants showed little upgrading. The AEC responded to the criticisms by preparing a report on waste management practices

(1949). It concluded that while waste disposal was a long-term problem, solutions would be found in time. Nuclear waste nevertheless became a public issue the moment some journalists pounced on the dangers of radiation growing out of the atmospheric tests in the South Pacific. Army physician David Bradley, who had been involved in the Bikini detonations, published a best-seller, *No Place to Hide*, in 1948. The book predated the fallout scare that hit the country sometime later and cautioned people about the dangers of atomic weapons. Few made the connection between fallout and radioactive waste at this point, but it eventually became part of a larger debate over radiation safety.[90]

The promotion of commercial nuclear power in the 1950s compounded the problem of radioactive waste. From the AEC's perspective, waste at commercial plants could be dealt with in the same manner as waste at weapons plants and other AEC-managed facilities. But for commercial plants, as some nuclear experts believed, employing effective waste disposal programs was an important element in doing business. It was fundamentally a question of cost. AEC invested about $100 million in waste facilities in its own plants, but a plan for funding commercial waste operations was not yet established, nor were there adequate plans for developing effective disposal technologies or earmarked sites. The AEC's first impulse was to find a "sink" for discarding radioactive waste. On its own facilities, the AEC placed high-level liquids in underground tanks. In 1961, they stored more than 52 million gallons at Hanford in 145 tanks with 29 additional tanks at the Savannah River and Idaho plants. Some facilities flushed radioactive waste (but not HLW) into cooling ponds, which sometimes leaked into nearby water courses. At Hanford, venting or evaporating contaminated gases poisoned the air. For other than HLW, the AEC employed or considered on-site land dumps, landfills, pumping waste into large holding tanks, and some types of long-term storage as temporary solutions. Reprocessing that recovered plutonium and other fissionable material also created dangerous waste products for which disposal sites had to be located, and thus it was not a disposal solution in itself. Another sink was the ocean, where AEC-licensed boats dumped 55-gallon drums of radioactive wastes (again, not HLW) into deep water as early as 1946. The assumption was that the barrels would drop deep and not create a problem, and even if they ruptured would be diluted enough not to cause harm.[91]

Dumping radioactive material at sea became a Cold War propaganda tool. In the 1950s and 1960s the Soviet Union criticized Western countries, especially the United States and Great Britain, for dumping radioactive waste into the oceans. Both countries were guilty as charged. The British not only dumped large quantities of radioactive material directly into the sea, but also unloaded radioactive material from ships, doggedly continuing disposal in the ocean well beyond the time when most other countries had stopped. (In the late 1960s, the French operated a coastal reprocessing plant at La Hague which incurred problems with radioactive waste.) However, it was later learned that the Soviet Union itself had been blameworthy as well for dumping large quantities of radioactive material in rivers and seas, especially into the Arctic Ocean between 1959 and 1992. The London Convention was the first global treaty to limit radioactive waste disposal and other marine pollution. Passed in 1972, it went into effect in 1975 with more legislation to follow. But the practice did not stop. Between 1946 and 1993, 14 countries dumped radioactive materials at sea in approximately 80 locations primarily in the Atlantic, Pacific, and Arctic Oceans. While many countries did not carry out the practice regularly, the Soviet Union was the primary guilty party in the Arctic and was accountable for about 60 percent in the Pacific. The United States disposed of about 38 percent in the Pacific (but gave up the practice before the London Convention). The British dumped the most in the Atlantic, accountable for more than 77 percent, with smaller amounts from other countries. In those years the Soviets (46.1 percent) and the British (41.2 percent) were the primary offenders, responsible for about 90 percent of the radioactive waste dumped in the oceans. Interestingly, the environmental community was not the coercive force that stopped the practice. Instead, both the accelerating debates among scientists (especially over acceptable thresholds of radiation that oceans could sustain) and the competition between the major powers in the Cold War led to diminishing the practice. The commitment to weapons building was so great and efforts to develop commercial power so intriguing that the vastness of the oceans seemed to swallow up whatever reservations there might have been for defiling such an important natural, and shared, resource.[92]

The fallout scare in the 1950s was particularly important in raising public ire over dumping low-level radioactive wastes into the ocean. Fading respect for the AEC's effectiveness in protecting human health was an additional result of the "public fallout." In some respects, the AEC was puzzled by the furor over ocean dumping. They considered the amount of radiation generated by the low-level waste as far less than natural radiation already present in the oceans. And they believed that the material was so deeply sunk that it posed little in the way of health risks. The public reaction was so intense, however, that the AEC turned to land disposal in Nevada, Kentucky, and New York which proved less controversial and less expensive.[93] Ultimately, permanent geologic disposal (at least for high-level waste and spent fuel) gained general support internationally. In 1955, the AEC requested that the National Academy of Sciences (NAS) establish a committee on waste to explore alternatives. In its 1957 report, NAS discussed a variety of disposal options; the most promising was placing nuclear waste in salt formations.[94] As an indication of future controversy, the AEC announced plans in 1970 to locate a repository for high-level and transuranic radioactive wastes at a derelict salt mine near Lyons, Kansas. The technical limitations of the site itself, compounded by growing public and political resistance, derailed the project. The opposition was so great that the agency withdrew the plan. To make matters worse, the public debate over the site further weakened confidence in the AEC to solve the waste problem in general. By the 1980s, engineers questioned using such formations because salt brine was corrosive on container material and the concrete used for backfill. The Germans employed a salt dome for a waste site near Gorleben, while the Swedes and Canadians considered granite deposits. The Americans evaluated basalt formations near Hanford and the Idaho National Engineering Laboratory. Although several countries had geologic repository programs by 2005, no site for spent fuel or high-level waste operated anywhere.[95]

During the course of finding an effective disposal option for HLW, most AEC officials and many contractors were satisfied that waste management policies gave sufficient attention to matters of health. But outsiders were not always convinced, and a series of studies engaged the issue of alternatives and safety as early as the 1960s. They raised criticisms that created strained relationships with the federal agency. Most of these encounters in the 1960s focused on waste disposal at AEC facilities rather than at commercial plants sites, where the controversy grew in subsequent years.[96] By the late 1960s storage tanks at Richland and Aiken developed leaks and were feared to contaminate water supplies. In 1963 five tanks at Hanford were found to have leaks. Other forms of radioactive wastes from various nuclear power plants also posed risks. After the AEC was terminated, Atlantic Richfield managed the Richland location until 1977 when Rockwell International assumed responsibility. DuPont managed the site in South Carolina. Some communities bid for nuclear waste facilities as a potential business opportunity. But it became increasingly obvious in the 1960s that such facilities posed levels of danger not worth the trouble. It thus became more difficult to find sites that would accept the waste or to find towns and cities willing to have the waste even transported through their communities.[97] By 1982 more than 20 states passed legislation to ban federal waste management activities within their jurisdictions.[98]

Among the difficult alternative approaches to address was reprocessing, which had been carried out at AEC plants since the Manhattan Project. Reprocessing was not regarded as the sole waste disposal solution, but an integral part of an overall strategy that linked processing, storage, and disposal into a complete procedure. Reprocessed uranium could be enriched and then serve as a new source of fuel; reprocessed plutonium could be mixed with uranium to produce mixed oxide fuel. It seemed unthinkable to dispose of spent fuel without getting maximum use and value from it. In the early 1970s, no one had fully determined the economic value, technical feasibility, and safety of commercial reprocessing. At the time one such operation, the West Valley plant in New York, was running and two more were under construction. Economic, technical, and safety issues all plagued the West Valley site, and it shut down temporarily in 1972, never to reopen. The other two facilities also encountered problems and never opened. They faced unanticipated technical problems, tightening controls from the AEC, and rising opposition from environmentalists. The timing could not have been worse.[99] In recent years, France, Japan, and some other countries continued to reprocess spent fuel, largely because of their few domestic sources of energy and to some degree because of the difficulty of identifying adequate disposal sites. The French were pioneers in reprocessing and recycling spent fuel from nuclear reactors. In the late 1990s, about 20 of its 58 pressurized light-water reactors burned recycled plutonium with more on line to do so. In West Cumbria, Great Britain, a reprocessing plant began operations in 1997. The Thermal Oxide Reprocessing Plant (THORP) stores spent fuel for five years and then puts it through its process.[100]

The AEC's inability to effectively confront the waste issue mirrored its other trials in the nuclear power arena as its institutional life flickered and then ended. The Energy Reorganization Act gave the new Nuclear Regulatory Commission responsibility to license surface storage facilities and permanent disposal repositories for commercial high-level wastes.[101] The creation of two agencies with authority over nuclear power issues sometimes led to ambiguities over jurisdiction, or over who was in charge of what. The NRC began to evaluate nuclear waste technology at the time, while trying to remain independent of the Energy Research and Development Administration (ERDA) and then the Department of Energy. Complicating the picture was the emergence of the Environmental Protection Agency (EPA) a few years earlier, which raised more questions about who was responsible for radioactive waste management.[102] While the AEC had its shortcomings, the new administrative order of things in Washington made for new layers of uncertainty in coming to grips with the radioactive waste issue.[103]

Robert C. Seamans, the director of ERDA, centralized civilian and defense waste management operations into the Division of Nuclear Fuel Cycle and Production; environmental control oversight was given to the new Division of Environmental Control Technology. ERDA prepared a study in spring 1976 describing options available for dealing with radioactive waste from power reactors, reprocessing plants, and fuel fabricating facilities. ERDA also developed a concept referred to as "multiple barriers." This was meant to place more of a gap between high-level waste and the ground or atmosphere by converting liquid waste into stable solids before putting it in containers.[104] The whole process required "stabilization," or solidifying the radioactive material so it could not invade water or air; "shielding," by placing barriers around the material to absorb radiation; and "isolation," by putting a barrier between the waste and humans.[105] Experiments with conversion of liquid waste to glass proved unsuccessful by 1978, and factors such as the design temperature for the waste form and other considerations needed to be explored.[106]

The Jimmy Carter administration attempted to launch a national policy on the management of nuclear waste, but little happened as the 1980 election loomed. Nevertheless, many essential issues about the knotty problem of radioactive waste disposal now were subject to public discussion and debate.[107] Reprocessing became a casualty of its on again, off again history, growing skepticism of its safety, and international security concerns. Between 1966 and 1972 facilities reprocessed a limited amount of spent fuel from commercial plants. In 1977 US policy defined spent fuel as waste largely because some believed that plutonium purified by reprocessing could be illegally used for weapons making. President Carter stopped reprocessing at that point because of the apparent security risk. The most immediate result of this decision was to put back on the table the possibility of temporary surface storage of waste and the need to more aggressively pursue plans for a high-level waste repository. At the time, another part of the fuel cycle, transportation of waste, began to attract public attention, raising concerns about the routing of large amounts of radioactive material over major thoroughfares.[108] One additional change followed in 1980 when Congress passed the Low-Level Nuclear Radioactive Waste Policy Act, which dealt with radioactive materials generated primarily by hospitals. Each state was responsible for disposing of its own waste, but could sign interstate compacts in seeking sites and disposal means. The law ignored the much more difficult problem of high-level waste.[109]

The Nuclear Waste Policy Act (NWPA, 1983) was turning-point legislation. It shifted efforts at finding a solution to the radioactive waste problem from executive branch agencies to the legislative branch. The most immediate discussions over such a law originated in the Carter years, but the problem of dealing with high-level nuclear waste obviously had a longer history. Conflicts over federal siting authority for military wastes met with resistance from several states, and delayed any consensus over waste policy legislation. A compromise to add military wastes to high-level waste from civilian reactors helped move the legislation forward. Other issues, especially concerning the type and location of storage, had to be resolved as well.[110]

Enacted by Congress in 1982 and signed by President Reagan in January 1983, the new law set a timeline for finding a storage facility. It established the Office of Civilian Radioactive Waste Management in the DOE charged with conducting environmental assessments of five potential disposal sites of which three would be selected by 1985 for further study. (The law also called for the program's civilian costs to be paid by a fee on nuclear-generated electricity to be placed into a Nuclear Waste Fund.) Until DOE accepted the waste at one or more of established repositories, waste owners maintained primary responsibility for storage.[111] The sites selected for further study were Hanford, Washington; Deaf Smith; Texas; and Yucca Mountain, Nevada. Of the final two repositories selected, one would be located in the East and the other in the West. At that point the president would pick one site in 1987 and another in 1990. Pressure from state governments forced a provision that allowed individual states to veto any site in their territory. Only a majority in both houses of Congress could override the veto. Any site selected, therefore, faced potential political resistance, but also had to pass scientific muster. In addition, the repository had to be located in a geological formation capable of withstanding any natural disaster for a period of 10,000 years to insure effective radioactive decay. When the selection process collapsed, an amendment to the law in 1987, however, mandated only Yucca Mountain for further study. In essence, Yucca Mountain was chosen by default, and without the requisite feasibility studies and environmental assessments which had been called for in the NWPA. It was little surprise that this decision set off a storm of protest in Nevada, and because an eastern site was not selected, raised the question as to whether the decision was political rather than scientific. Additionally, support for Yucca Mountain as a solution to the waste problem aided the growth of the nuclear power industry, while opposition could be an effective delay tactic used by opponents of such growth.[112]

Yucca Mountain and Hanford already had been in use for nuclear projects. Hanford was the site where an original A-bomb was assembled and where a weapons-producing facility had been in operation since World War II. Yucca Mountain, about 65 miles northeast of Las Vegas, was close to the Nevada Test Site, where nuclear tests had been underway since the 1950s. The major attraction of Deaf Smith County was its remote location and its sparse population in West Texas. It was, however, within several hours of the Pantex weapons plant in Amarillo. Prior to the amendment anointing Yucca Mountain as the sole depository, DOE identified 12 sites in seven states for the originally proposed second waste storage location. Deaf Smith County fell out of the original three sites tagged for the western depository, likely because of the powerful Texas congressional delegation and local protests. Hanford's population size and location near the Columbia River eliminated it. This left Yucca Mountain. It was not necessarily the best site available from a geologic standpoint, but it was politically viable. Resistance by other states and the high cost of reopening the search for a different location sealed Nevada's fate at that time.[113] In the abstract the promise of nuclear power was magical. In the real world someone had to bear the risks of location for both generation and disposal.[114]

Beyond the logistical and economic problems, Yucca Mountain raised potential ethical and social justice issues. To many in Nevada, the proposed national disposal site appeared to be a sacrifice zone for the benefit of electricity consumers elsewhere.[115] The public concern in the United States and internationally over nuclear waste was longstanding, increasing steadily since the mid-1970s.[116] Not surprisingly, as the federal government identified nuclear waste sites, NIMBY campaigns spread rapidly. Even local "guerilla wars" against siting disposal facilities were reminiscent of the campaigns against siting nuclear power plants. As in the case of Yucca Mountain, political battles erupted among states each wanting the waste to go somewhere else.[117]

Disputes over nuclear waste were not limited to the United States. As one British writer noted, "For many people nuclear waste is the Achilles' heel of nuclear technology."[118] In the aftermath of Three Mile Island and growing public concern over reactor safety, the question of appropriate waste repositories entered the public debate in Scotland, northeast England, and Wales. Prior to the 1983 general election plans for any repository were withdrawn. Public skepticism (including participation by direct-action groups and labor organizations) also spilled over into other areas, such as nuclear waste transport. In 2003 the newly formed Committee

Containers filled with transuranic nuclear waste from America's nuclear weapons program await burial at the Waste Isolation Pilot Plant in Carlsbad, New Mexico. The materials are stored in rooms carved out of an ancient salt formation a half mile below the Chihuahuan Desert. *Source:* Jim West/Alamy

on Radioactive Waste Management suffered numerous controversies, including criticism from House of Lords members anxious for a concrete policy to be established. Not surprisingly, goals over nuclear waste disposal were tangled with goals for a future-directed energy policy.[119]

The nuclear waste issue as a potential environmental justice question is most prominent on American Indian reservations. Some Native American tribes actually courted nuclear waste facilities as a way of generating revenue or wanted direct control over such sites. Many more waste facilities were much less actively sought. Valerie L. Kuletz, who grew up at a military test center in the Mojave Desert, argued that a map of the so-called "Nuclear West" demonstrates "how the development, testing, and waste storage of nuclear materials in the highly militarized landscapes of the western United States might be understood as a form of environmental racism."[120] Indian lands had long been the sites for uranium mining and milling, especially for the Navajo, Laguna Pueblo, and Acoma Pueblo. By the late 1990s, the only above-ground nuclear storage facilities the government considered were at the Nevada Test Site and reservations of the Mescalero Apache, Skull Valley Goshute, and Fort McDermitt Paiute-Shoshone. Other storage facilities bordered lands of the Mescalero Apache, San Ildefonso Pueblo, and Santa Clara Pueblo. Low-level waste was scheduled for deposit in Ward Valley in the Mojave Desert where Fort Mojave Indians and Chemehuevi live. Yucca Mountain itself is considered "holy land" to the Western Shoshone, Southern Paiute, and Owens Valley Paiute.[121]

THE "FRONT-END" OF THE FUEL CYCLE: URANIUM MINING

With nuclear waste grabbing headlines in recent years, the "front-end" of the fuel cycle sometimes was neglected. In the United States uranium was first mined in the West in 1871. At that time Dr. Richard Pierce shipped about 200 pounds of pitchblende to London to be used in research on the fabrication of steel alloys, chemical experimentation, and as pigments for dyes and inks. In 1898 Pierre and Marie Curie and G. Belmont isolated radium from pitchblende to use in their experiments. Early Navajo and Ute Indians on the Colorado Plateau used carnotite (containing uranium, vanadium, and radium) as body paint. A small prospecting boom in radium occurred in southeastern Utah, but it soon dwindled before World War I.[122]

Most often viewed as a waste product, uranium became a valuable commodity for the United States (and others) with the development of the atomic bomb. "Yellowcake" was the name given to the enriched uranium oxides needed for the weapons and fuel. Especially after the USSR developed their own atomic weapon in 1949, the United States bought up stocks of uranium in Europe and Africa to keep them from the Soviets and to further constrain nuclear proliferation.[123] In the first half of the twentieth century uranium ore came from Czechoslovakia, the Belgian Congo, and Canada. In the pre-World War II years miners extracted only about 1,000 tons per year. A new strike at Beaverlodge Lake, Canada, in 1946 signaled the possibility of finding additional new supplies. In the late 1940s, the US federal government provided financial incentives for new uranium deposits abroad and at home. A series of discoveries in the Colorado Plateau area of New Mexico and Utah in 1951 resulted in promising finds, as did uranium found as a by-product of gold mining in South Africa. In the 1950s and 1960s Canada and Australia were important locations for big strikes. In the United States about 500 mines were concentrated in the four corners area of Arizona, Colorado, New Mexico, and Utah in the 1950s, with mines also in Texas and a few other locations. By 1955 the United States was the world's largest producer of uranium.[124]

Until 1964, when the US government allowed private ownership of nuclear materials, the AEC held all uranium supplies and was important in encouraging domestic production and active in purchasing uranium from abroad. Between 1947 and 1970, the AEC acquired 55 percent of its supply from domestic sources, 24 percent from Canada, and 21 percent from elsewhere. By 1970 the US government had a surplus of uranium. Commercial mining in the United States sustained growth through the 1970s and early 1980s, but declined sharply by 1990.[125] Worldwide demand for uranium remained significant with mining production on the rise since 1993. Much of the uranium was used to generate electricity (meeting 78 percent of demand for power generation) and a much smaller amount for producing medical isotopes. In recent years uranium mines operate in about 20 countries, but 58 percent of the global production comes from 10 mines in six countries. About 62 percent of all world production is located in Kazakhstan, Canada, and Australia. And like many other industries, uranium production in the 1990s was consolidated by takeovers, mergers, and closures, except possibly in Kazakhstan.[126]

While uranium proved to be a valuable mineral with historic implications, the process of acquiring uranium ore produced significant environmental and health risks. Each step in the procedure (prospecting, road building, mining, and milling) had an impact on the public, on habitat, and in the form of air and water pollution. Opening roads to new mines produced access to land previously classified as wilderness. Trucking ore from mines to mills resulted in contaminated vehicles and also spillage. Scarring of the land took place with underground and strip mines. Radon gas escaped from mines, mill tailings, and dumps to produce air pollution. In some cases the quality of water was compromised through runoff and soil leaching. For example, milling processes which include sorting the ore, crushing and leaching it, and precipitating it with acid solutions produces wastes placed in tailing ponds. The waste, which contains heavy metals and radioactivity, could then migrate into groundwater.[127]

Of more serious concern were potential impacts on human health, especially for those working in and around mines and at milling operations. The biggest problem was lung cancer from inhaling radon. Mining uranium exposed workers to radon and radioactivity, and in the first years of the industry they were unprotected from the risks. But the health dangers were not restricted to the workplace, since miners and others often wore their contaminated clothing home. Mill workers proved to be better protected early on, since the AEC formulated health and safety rules for uranium milling operations in 1957. AEC licensing regulations did not protect mine workers, who were exposed to a variety of occupational hazards. Representatives from government and the industry adopted the first standards for exposure to mine radiation in 1955. A federal advisory body, the Federal Radiation Council, gave attention to radiation in uranium mines, but its efforts to protect workers was ineffective. Further attempts through legislation to enforce health and safety standards and to set inspection schedules of uranium mines in the 1960s met with uneven success. In the 1970s, when the uranium industry was enjoying growth, it came under pressure from doctors, government officials, and mining unions to address a number of health and safety questions. The industry collapse in the 1980s thwarted important improvements. In 1990 Congress passed the Radiation Exposure Compensation Act, but it did not end the workers' efforts to receive proper compensation or protection. In October 1995 the Advisory Committee on Human Radiation Experiments recommended loosening the strict requirements for proving that uranium miners had radon-induced lung cancer.[128]

In the American West, many Native Americans and others were drawn into the work in the mines and the mills. Indian reservations and sites close to the reservations were frequent locations for uranium mines and various related facilities. The questions of environmental impact and of worker health hazards in the uranium mining industry led to several lawsuits. A notable one focused on a spill of radioactive tailings near Church Rock, New Mexico (1979), and another was filed on behalf of Navajo uranium miners in 1983.[129] Mining and milling in general have been inherently risky businesses for miners and other workers over the years, but risks associated with uranium mining and milling had to be taken into account in evaluating the whole nuclear fuel cycle from uranium prospecting to waste disposal. The day of focusing on reactor safety alone had passed.

CONCLUSION

Commercial nuclear power in the post-Three Mile Island world faced an uphill battle for redemption. Despite attention to the lessons learned in 1979, government regulators and the industry not only had to battle their own shortcomings but some problems out of their control, namely Chernobyl and its aftermath. Questions remained as to whether the problems with nuclear power were endemic or correctable, and answers differed depending on who was giving them. The turn from reactor safety to other elements of the fuel cycle, namely radioactive waste and uranium mining, also demonstrated how the lens necessary to examine this sometimes fickle energy source had to be readjusted to take into account all salient features. Events in the late 1970s and 1980s dimmed the hope for cheap and bountiful energy through nuclear power, and earlier hopes were dashed by many difficult issues. Nuclear power and questions surrounding it at the turn of the twenty-first century remained an economic matter, an environmental dilemma, and a social and ethical concern. What would it take for the industry to grow again? What would it take to dismantle what remained?

MySearchLab Connections: Sources Online

READ AND REVIEW

Review this chapter by using the study aids and these related documents available on MySearchLab.

✓●─[**Study** and **Review** on **mysearchlab.com**

Chapter Test

Essay Test

▯●─[**Read** the **Document** on **mysearchlab.com**

Nuclear Waste Policy Act (NWPA, 1983)

Towards a Green Europe, Towards a Green World (1987)

Fact Sheet on Ocean Dumping of Radioactive Waste Materials (1980)

Bruce E. Johansen, The High Cost of Uranium in Navajoland (1994)

RESEARCH AND EXPLORE

Use the databases available within MySearchLab to find additional primary and secondary sources on the topics within this chapter.

Endnotes

1. J. Samuel Walker, *Three Mile Island: A Nuclear Crisis in Historical Perspective* (Berkeley, CA: University of California Press, 2004), 230–34.
2. Correspondence from Samuel Walker to Martin Melosi, May 31, 2011.

3. Bonnie A. Osif, Anthony J. Baratta, and Thomas W. Conkling, *TMI 25 Years Later: The Three Mile Island Nuclear Power Plant Accident and Its Impact* (University Park, PA: Pennsylvania States University Press, 2004), 87; Walker, *Three Mile Island*, 234–37.

4. In 1979, 92 reactors worth about $50 billion investments were in various stages of construction.

5. Osif, Baratta, and Conkling, *TMI 25 Years Later*, 84–86; James M. Jasper, *Nuclear Politics: Energy and the State in the United States, Sweden, and France* (Princeton, NJ: Princeton University Press, 1990), 213; Christian Joppke, *Mobilizing Against Nuclear Energy: A Comparison of Germany and the United States* (Berkeley, CA: University of California Press, 1993), 135–36.

6. Rodney P. Carlisle, ed., *Encyclopedia of the Atomic Age* (New York: Facts on File, 2001), 251–52; Joppke, *Mobilizing Against Nuclear Energy*, 135–36.

7. Robert J. Duffy, *Nuclear Politics in America: A History and Theory of Government Regulation* (Lawrence, KS: University Press of Kansas, 1997), 181–82. For Reagan's economic policy, see Andrew E. Busch, "Ronald Reagan and Economic Policy," in Paul Kengor and Peter Schweizer, eds., *The Reagan Presidency: Assessing the Man and His Legacy* (Lanham, MD: Roman & Littlefield Pubs., 2005), 25–46; John Patrick Diggins, *Ronald Reagan: Fate, Freedom, and the Making of History* (New York: W.W. Norton, 2007), 305–17, 333–42.

8. "Acceptance Speech at the 1980 Republican Convention by Ronald Reagan, July 17, 1980," online, http://www.nationalcenter.org/ReaganConvention1980.html.

9. Ibid.; Duffy, *Nuclear Politics in America*, 181–82; Correspondence from Samuel Walker to Martin Melosi, May 31, 2011.

10. Duffy, *Nuclear Politics in America*, 183, 191–202.

11. Correspondence from Samuel Walker to Martin Melosi, May 31, 2011.

12. Duffy, *Nuclear Politics in America*, 183, 191–99; Jasper, *Nuclear Politics*, 193.

13. Osif, Baratta, and Conkling, *TMI 25 Years Later*, 87–88; Joppke, *Mobilizing Against Nuclear Energy*, 136.

14. Nuclear Regulatory Commission, *A Short History of Nuclear Regulation, 1946–1999*, 27 (January 29, 2001), online http://www.nrc.gov/SECY/smj/shorthis.htm.

15. Ibid., 28–29; Duffy, *Nuclear Politics in America*, 202–09; Carlisle, ed., *Encyclopedia of the Atomic Age*, 222, 303, 306.

16. Duffy, *Nuclear Politics in America*, 205–09; Nuclear Regulatory Commission, *A Short History of Nuclear Regulation, 1946–1999*, 29; Carlisle, ed., *Encyclopedia of the Atomic Age*, 303, 306.

17. Duffy, *Nuclear Politics in America*, 201–02; Nuclear Regulatory Commission, *A Short History of Nuclear Regulation, 1946–1999*, 30.

18. Quoted in Nigel Hawkes, et al., *Chernobyl: The End of the Nuclear Dream* (New York: Vintage, 1986), 1.

19. Quoted in Svetlana Alexievich, *From Chernobyl: The Oral History of a Nuclear Disaster* (New York: Picador, 2006), 50.

20. Ibid., 49.

21. Ibid., 78.

22. Stephen E. Atkins, ed., *Historical Encyclopedia of Atomic Energy* (Westport, CT: Greenwood Press), 81; Christoph Hohenemser, "Chernobyl," in Robert Paehlke, ed., *Conservation and Environmentalism: An Encyclopedia* (New York: Garland Pub., 1995), 116.

23. Matthew King, "What Happened at Chernobyl?" *Nuclear Fissionary*, March 3, 2010, online, http://nuclearfissionary.com/2010/03/03/what-happened-at-chernobyl/.

24. Quoted in Zhores Medvedev, *The Legacy of Chernobyl* (New York: W.W. Norton, 1990), 37.

25. Atkins, ed., *Historical Encyclopedia of Atomic Energy*, 81–82; J. Samuel Walker, *Three Mile Island: A Nuclear Crisis in Historical Perspective* (Berkeley, CA: University of California Press, 2004), 237; Richard L. Garwin and George Charpak, *Megawatts and Megatons: The Future of Nuclear Power and Nuclear Weapons* (Chicago: University of Chicago Press, 177; "Chernobyl," Ruth A. Eblen and William R. Eblen, eds., *The Encyclopedia of the Environment* (Boston: Houghton Mifflin Co., 1994), 92; Paul R. Josephson, *Red Atom: Russia's Nuclear Power Program from Stalin to Today* (Pittsburgh: University of Pittsburgh Press, 2000), 255.

26. Andrew Gregorovich, "Chornobyl Nuclear Catastrophe: Ten Years After April 26, 1986," 6, *InfoUkes*, online, http://www.infoukes.com/history/chornobyl/gregorovich/; Atkins, ed., *Historical Encyclopedia of Atomic Energy*, 82–83; Josephson, *Red Atom*, 265–66.

27. Quoted in Gregorovich, "Chornobyl Nuclear Catastrophe," 1.

28. "Chernobyl Accident," Nuclear Issues Briefing Paper 22, March 2006, online, http://www.uic.com.au/nip22.htm; Josephson, *Red Atom*, 258; Walker, *Three Mile Island*, 238–39.

29. Atkins, ed., *Historical Encyclopedia of Atomic Energy*, 84; Gregorovich, "Chornobyl Nuclear Catastrophe, 2–5; Eblen and Eblen, eds., *The Encyclopedia of the Environment*, 93; Hohenemser, "Chernobyl," 116. For additional focus on Sweden, see C.C. Bailey, *The Aftermath of Chernobyl* (Dubuque, Iowa: Kendall/Hunt, 1993; 2nd ed.).

30. Garwin and Charpak, *Megawatts and Megatons*, 178, 183–85; Atkins, ed., *Historical Encyclopedia of Atomic Energy*, 84; World Health Organization, *Health Effects of the Chernobyl Accident: An Overview*, Fact Sheet N° 303 (April 2006), online, http://www.who.int/mediacentre/factsheets/fs303/en/index.html.

31. Garwin and Charpak, *Megawatts and Megatons*, 178, 183–85; Atkins, ed., *Historical Encyclopedia of Atomic Energy*, 84.

32. Scurlock, "A Concise History of the Nuclear Industry Worldwide," in Elliott, ed., *Nuclear or Not?* 32; "Chernobyl Accident," Nuclear Issues Briefing Paper 22, March 2006, online, http://www.uic.com.au/nip22.htm; Josephson, *Red Atom,* 259–64. For more in-depth detail on Chernobyl, see Medvedev, *The Legacy of Chernobyl*.

33. World Health Organization, *Health Effects of the Chernobyl Accident: An Overview*.

34. World Health Organization, *Health Effects of the Chernobyl Accident*, 2011, online, http://www.who.int/ionizing_radiation/chernobyl/en/.

35. Quotes from Andrew Osborn, "Chernobyl," March 6, 2011, *The Telegraph*, online, http://www.telegraph.co.uk/news/worldnews/europe/ukraine/8363569/Chernobyl-The-toxic-tourist-attraction.html.

36. Alvin M. Weinberg, Irving Spiewak, Jack N. Barkenbus, Robert S. Livingston, and Doan L. Phung, eds., *The Second Nuclear Era: A New Start for Nuclear Power* (New York: Praeger, 1985). See also Alvin M. Weinberg, "The Second Nuclear Era," *Bulletin of the New York Academy of Medicine* 59 (December 1983): 1048–59.

37. John Byrne and Steven M. Hoffman, "Nuclear Optimism and the Technological Imperative: A Study of the Pacific Northwest Electrical Network," *Bulletin of Science, Technology & Society* 11 (1991): 63. See also Joppke, *Mobilizing Against Nuclear Energy*, 157–58.

38. For a good discussion of the pro-nuclear stance at this time, see American Nuclear Society, *Nuclear Energy Facts: Questions and Answers* (LaGrange Park, Il: American Nuclear Society, 1980).

39. Norman Rasmussen, "Nuclear Power: Its Promises and Its Problems," *Technology & Society* 1 (Fall 1981), 21.

40. Byrne and Hoffman, "Nuclear Optimism and the Technological Imperative," 63–64.

41. Nuclear Regulatory Commission, *A Short History of Nuclear Regulation, 1946–1999*, 30; Duffy, *Nuclear Politics in America*, 201–02.

42. Nuclear Regulatory Commission, *A Short History of Nuclear Regulation, 1946–1999,* 30–31. See also J. Samuel Walker, *Permissible Dose: A History of Radiation Protection in the Twentieth Century* (Berkeley, CA: University of California Press, 2000).

43. Nuclear Regulatory Commission, *A Short History of Nuclear Regulation, 1946–1999*, 31–32; Correspondence from Samuel Walker to Martin Melosi, May 31, 2011.

44. Nuclear Regulatory Commission, *A Short History of Nuclear Regulation, 1946–1999,* 33–34.

45. See Paul W. MacAvoy and Jean W. Rosenthal, *Corporate Profit and Nuclear Safety: Strategy at Northeast Utilities in the 1990s* (Princeton, NJ: Princeton University Press, 2004).

46. Duffy, *Nuclear Politics in America*, 209–10; Correspondence from Samuel Walker to Martin Melosi, May 31, 2011.

47. DOE/EIA, *Electric Power Monthly* (DOE/EIA-0226 (98/11)), Table 12, 23; DOE/EIA, *Inventory of Power Plants in the United States as of January 1, 1998* (DOE/EIA-0095 (98)), Table 1, 17; DOE/EIA, *Monthly Energy Review* (DOE/EIA-0035 (98/11)), Table 7.1, 95.

48. Jasper, *Nuclear Politics*, 211–12.

49. Correspondence from Samuel Walker to Martin Melosi, May 31, 2011.

50. Jasper, *Nuclear Politics*, 194.

51. Osif, Baratta, and Conkling, *TMI 25 Years Later*, 87–88; Joseph V. Rees, *Hostages of Each Other: The Transformation of Nuclear Safety Since Three Mile Island* (Chicago: University of Chicago Press, 1994), 1–2. See also Constance Perin, *Shouldering Risk: The Culture of Control in the Nuclear Power Industry* (Princeton: Princeton University Press, 2005).

52. Quoted in Rees, *Hostages of Each Other*, 2.

53. Ibid., 2–3.

54. Ibid., 13–22.

55. Quoted in ibid., 23.

56. Ibid., 24–26, 41–45.

57. Garwin and Charpak, *Megawatts and Megatons*, 175–76; Carlisle, ed., *Encyclopedia of the Atomic Age*, 145.

58. "The Extended Nuclear Family," *Time*, October 26, 1981, 20. Another report in the same year cited more than 500 nuclear power reactors in operation or under construction in 39 countries, with many more unrecorded research and experimental reactors. The United States was the leader with 183 reactors. See "The Nuclear World," *RF Illustrated* 5 (May 1981), 8–9.

59. In 2011 World Watch Institute reported "a global nuclear remission" (a decrease in capacity and a decline in demand) caused by the rising cost of nuclear power plants. The disaster at Fukushima in Japan was cited as a major reason for the rise in cost. Nicholas Brown, "Global Nuclear Power Generation Capacity Decreases," December 8, 2011, *CleanTechnica.com*, online, http://cleantechnica.com/2011/12/08/global-nuclear-power-generation-capacity-decreases/.

60. Jonathan Scurlock, "A Concise History of the Nuclear Industry Worldwide," in David Elliott, ed., *Nuclear or Not? Does Nuclear Power Have a Place in a Sustainable Energy Future?* (New York: Palgrave Macmillan, 2007), 32–33.

61. The United Kingdom was well behind with 35, followed by Russia (29), Canada (22), Ukraine (20), Germany (19), India (16), Spain (15), and South Korea (14). The 11 largest nuclear power nations had 80 percent of the installed nuclear power capacity. See Carlisle, ed., *Encyclopedia of the Atomic Age*, 258–59; Peter D. Dresser, ed., *Nuclear Power Plants Worldwide* (Washington, D.C.: Gale Research Inc., 1993), 27–40.

62. The fast-neutron breeder reactor was expensive, and possibly premature, and regarded by opponents as very risky. See Garwin and Charpak, *Megawatts and Megatons*, 133–35.

63. Jasper, *Nuclear Politics*, 243–44. See also Gabrielle Hecht, *The Radiance of France: Nuclear Power and National Identity after World War II* (Cambridge, MA: MIT Press, 1998).

64. Jasper, *Nuclear Politics*, 246; Carlisle, ed., *Encyclopedia of the Atomic Age*, 110; Michael T. Hatch, *Politics and Nuclear Power: Energy Policy in Western Europe* (Lexington, KY: University Press of Kentucky, 1986), 170; Dresser, ed., *Nuclear Power Plants Worldwide*, 60, 62.

65. Atkins, ed., *Historical Encyclopedia of Atomic Energy*, 305–06.

66. Carlisle, ed., *Encyclopedia of the Atomic Age*, 110; Dresser, ed., *Nuclear Power Plants Worldwide*, 59.

67. Atkins, ed., *Historical Encyclopedia of Atomic Energy*, 62; Scurlock, "A Concise History of the Nuclear Industry Worldwide," 31; Dresser, ed., *Nuclear Power Plants Worldwide*, 277–78. In 2011 Great Britain had 19 reactors producing 18 percent of electrical use. See World Nuclear Association, "Nuclear Power in the United Kingdom," January 2011, online, http://www.world-nuclear.org/info/inf84.html.

68. "Nuclear Power in Sweden," World Nuclear Association, July 2011, online, http://www.world-nuclear.org/info/inf42.html; Jasper, *Nuclear Politics*, 218, 224–25, 227, 231–32, 235; Dresser, ed., *Nuclear Power Plants Worldwide*, 239; Terry Macalister, "Sweden Lifts Ban on Nuclear Power," February 5, 2009, guardian.co.uk, online, http://www.guardian.co.uk/environment/2009/feb/05/sweden-nuclear-power.

69. Antony Froggatt, "Nuclear Power—The European Dimension," in Elliott, ed., *Nuclear or Not?* 172–75. See also Hatch, *Politics and Nuclear Power*, 134–35, 137; Dresser, ed., *Nuclear Power Plants Worldwide*, 53–54, 79–81, 105–06, 119–20.

70. Dresser, ed., *Nuclear Power Plants Worldwide*, 211–13, 269–73.

71. Ibid., 123–44.

72. Ibid., 169–70; World Nuclear Association, "Nuclear Power in China," March 10, 2011, online, http://www.world-nuclear.org/info/inf63.html.

73. "Nuclear Power in India," World Nuclear Association, March 2011, online, http://www.world-nuclear.org/info/inf53.html; Dresser, ed., *Nuclear Power Plants Worldwide*, 111–12.

74. Dresser, ed., *Nuclear Power Plants Worldwide*, 1–4, 19–22, 41–43.

75. Jerome Price, *The Anti-nuclear Movement* (Boston: Twayne Publishers, 1990; rev. ed.), 24–35.

76. Ibid., 24–35; Walker, *Three Mile Island*, 239; Correspondence from Samuel Walker to Martin Melosi, May 31, 2011; Joppke, *Mobilizing Against Nuclear Energy*, 138–44.

77. Walker, *Three Mile Island*, 239.

78. Joppke, *Mobilizing Against Nuclear Energy*, 144.

79. Jane I. Dawson, *Eco-Nationalism: Anti-Nuclear Activism and National identity in Russia, Lithuania, and Ukraine* (Durham, NC: Duke University Press, 1996), 2.

80. See Mikhail Gorbachev, *Perestroika* (New York: Harper Collins, 1987); Marshall Goldman, "Perestroika," *The Concise Encyclopedia of Economics*, 1992, online, http://www.econlib.org/library/Enc1/Perestroika.html.

81. Dawson, *Eco-Nationalism*, 2–4.

82. Ibid., 5, 162–64.

83. Velma Garcia-Gorena, *Mothers and the Mexican Anti-nuclear Power Movement* (Tucson, AR: University of Arizona Press, 1999), 15–23, 133–46.

84. Jasper, *Nuclear Politics*, 218, 224.

85. Roland Roth and Detlef Murphy, "From Competing Factions to the Rise of the Realos," in Margit Mayer and John Ely, ed., *The German Greens: Paradox Between Movement and Party* (Philadelphia, PA: Temple University Press, 1998), 67–68.

86. Joop van der Pligt, *Nuclear Energy and the Public* (Oxford, UK: Blackwell, 1992), 4–9.

87. Rustum Roy, "The Technology of Nuclear-Waste Management," *Technology Review* 83 (April 1, 1981): 40; J. Samuel Walker, *The Road to Yucca Mountain: The Development of Radioactive Waste Policy in the United States* (Berkeley, CA: University of California Press, 2009), 1.

88. Allison M. Macfarlane and Rodney C. Ewing, eds., *Uncertainty Underground: Yucca Mountain and the Nation's High-Level Nuclear Waste* (Cambridge, MA: MIT Press, 2006), 1–2.

89. Roy, "The Technology of Nuclear-Waste Management," 39; M. Joshua Silverman, "Radioactive Waste Management: An Environmental History Lesson for Engineers (and Others)," 7, online, http://www.ce.cmu.edu/GreenDesign/gd/education/edradiocase.html; Macfarlane and Ewing, eds., *Uncertainty Underground*, 2, 6; Carlisle, ed., *Encyclopedia of the Atomic Age*, 273–74.

90. Walker, *The Road to Yucca Mountain*, 3–18.

91. Silverman, "Radioactive Waste Management," 7; Macfarlane and Ewing, eds., *Uncertainty Underground*, 3; Walker, *The Road to Yucca Mountain*, 18– 23, 28.

92. Jacob Darwin Hamblin, *Poison in the Well: Radioactive Waste in the Oceans at the Dawn of the Nuclear Age* (New Brunswick, NJ: Rutgers University Press, 2008), 1–9; 252–60. See also Lasse Ringius, *Radioactive Waste Disposal at Sea: Public Ideas, Transnational Policy Entrepreneurs, and Environmental Regimes* (Cambridge, MA: MIT Press, 2001).

93. Walker, *The Road to Yucca Mountain*, 22–25.

94. Ibid., 21.

95. Macfarlane and Ewing, eds., *Uncertainty Underground*, 1–3, 30; Carlisle, ed., *Encyclopedia of the Atomic Age*, 117; Dresser, ed., *Nuclear Power Plants Worldwide*, 247–48; Walker, *The Road to Yucca Mountain*, 51–75.

96. Walker, *The Road to Yucca Mountain*, 31–50.

97. Atkins, ed., *Historical Encyclopedia of Atomic Energy*, 263–64; David Bodansky and Fred H. Schmidt, "Safety Aspects of Nuclear Energy," in Arthur W. Murphy, ed., *The Nuclear Power Controversy* (Englewood Cliffs, NJ: Prentice Hall, 1976), 18.

98. Joppke, *Mobilizing Against Nuclear Energy*, 156; Walker, *The Road to Yucca Mountain*, 29.

99. Walker, *The Road to Yucca Mountain*, 83–85; Silverman, "Radioactive Waste Management," 8.

100. Garwin and Charpak, *Megawatts and Megatons*, 135–38. See also Luther J. Carter, *Nuclear Imperatives and Public Trust: Dealing with Radioactive Waste* (Washington, D.C.: Resources for the Future, 1987), 235–400.

101. This reorganization meant that the commercial nuclear power industry was no longer regulated by the same agency that provided technical support. Critics of the former institutional arrangement saw this as a step away from what had been a conflict of interest in the federal program. See Joshua Silverman, "A History of the 'Nuclear Waste' Issue," March 26, 2008, The Nuclear Green Revolution, 4, online, http://nucleargreen.blogspot.com/2008/03/history-of-nuclear-waste-issue.html.

102. In 1986 a federal appeals court upheld a ruling in *LEAF v. Hodel* that DOE was subject to federal environmental legislation. This ended a period of the agency's self-policing over environmental issues. See Silverman, "A History of the 'Nuclear Waste' Issue," 5.

103. Garwin and Charpak, *Megawatts and Megatons*, 95–100.

104. Jack M. Holl, Roger M. Anders, and Alice L. Buck, *United States Civilian Nuclear Power Policy, 1954–1984: A Summary History* (Washington, D.C.: US Department of Energy, February 1986), 19–20.

105. Melodie Andrews and Donald R. Beeth, *Understanding Nuclear Waste*, The Readable Reference Series 1 (Houston: Houston Lighting & Power, 1985), 33.

106. Roy, "The Technology of Nuclear-Waste Management," 40.

107. Stanley N. Lundine, "A Congressional View of Managing Radioactive Waste," *Technology Review* 83 (April 1, 1918), 44.

108. Roy, "The Technology of Nuclear-Waste Management," 39; Silverman, "Radioactive Waste Management," 7; Macfarlane and Ewing, eds., *Uncertainty Underground*, 2, 6; Walker, *The Road to Yucca Mountain*, 103–24, 142–60.

109. Hal K. Rothman, *The Greening of America? Environmentalism in the United States Since 1945* (New York: Harcourt Brace, 1998), 165. See also Walker, *The Road to Yucca Mountain*, 125–41; Carlisle, ed., *Encyclopedia of the Atomic Age*, 274.

110. Walker, *The Road to Yucca Mountain*, 174–81.

111. Mark Holt, *Civilian Nuclear Waste Disposal*, Congressional Research Service, 7–5700 (July 16, 2010), I, online, www.crs.gov; Silverman, "Radioactive Waste Management," 9; Thomas A. Cotton, "Nuclear Waste Story: Setting the Stage," in Macfarlane and Ewing, eds., *Uncertainty Underground*, 33.

112. Silverman, "Radioactive Waste Management," 9; Atkins, ed., *Historical Encyclopedia of Atomic Energy*, 264–65; Rothman, *The Greening of America?* 165–66; Joppke, *Mobilizing Against Nuclear Energy*, 156–57; Walker, *The Road to Yucca Mountain*, 181–82.

113. Rothman, *The Greening of America?*, 166–67. In 1975 ERDA proposed the Waste Isolation Pilot Project (WIPP) to construct tunnels into a salt deposit near Carlsbad, New Mexico for remote disposal of radioactive wastes, but locals fought against it. See Carlisle, ed., *Encyclopedia of the Atomic Age*, 274.

114. See Cotton, "Nuclear Waste Story: Setting the Stage," in Macfarlane and Ewing, eds., *Uncertainty Underground*, 36. For more on nuclear waste and criticism of waste policy, see K.S. Shrader-Frechette, *Burying Uncertainty: Risk and the Case Against Geological Disposal of Nuclear Waste* (Berkeley, CA: University of California Press, 1993); John Weingart, *Waste is a Terrible Thing to Mind: Risk, Radiation, and Distrust of Government* (New Brunswick, NJ: Rivergate Books, 2007); Donald L. Barlett and James B. Steele, *Forevermore: Nuclear Waste in America* (New York: W.W. Norton & Co., 1985); Carter, *Nuclear Imperatives and Public Trust*.

115. Joppke, *Mobilizing Against Nuclear Energy*, 155. See also Shrader-Frechette, *Burying Uncertainty*.

116. Van der Pligt, *Nuclear Energy and the Public*, 9.

117. Joppke, *Mobilizing Against Nuclear Energy*, 155–56.

118. David Lowry, "Nuclear Waste: The Protracted Debate in the UK," in Elliott, ed., *Nuclear or Not?*, 115.

119. Ian Welsh, *Mobilising Modernity: The Nuclear Moment* (London: Routledge, 2000), 115–48, 197–205.

120. Valerie L. Kuletz, *The Tainted Desert: Environmental Ruin in the American West* (New York: Routledge, 1998), 12.

121. Ibid., 12–13, 24–26, 94–97, 111–16, 131–39, 156–58.

122. "Uranium Mining in Utah," *Utah History to Go*, online, http://historytogo.utah.gov/utah_chapters/utah_today/uraniummininginutah.html.

123. Carlisle, ed., *Encyclopedia of the Atomic Age*, 348; Peter D. Shemitz, "Uranium Mining: Environmental Impacts," in Robert Paehlke, ed., *Conservation and Environmentalism: An Encyclopedia* (New York: Garland Pub., 1995), 656.

124. Atkins, ed., *Historical Encyclopedia of Atomic Energy*, 388; Carlisle, ed., *Encyclopedia of the Atomic Age*, 348–49. For a good book on the history of uranium mining towns in the West, see Michael A. Amundson, *Yellowcake Towns: Uranium Mining Communities in the American West* (Boulder, CO: University Press of Colorado, 2002). See also Raye C. Ringholz, *Uranium Frenzy: Saga of the Nuclear West* (Logan, UT: Utah State University Press, 2002; rev. and expanded).

125. Atkins, ed., *Historical Encyclopedia of Atomic Energy*, 388; "Uranium Mining," *World Nuclear Association*, online, http://www.world-nuclear.org/education/mining.htm.

126. "World Uranium Mining," April 2011, *World Nuclear Association*, online, http://www.world-nuclear.org/info/inf23.html; "Uranium Mining," *World Nuclear Association*.

127. Shemitz, "Uranium Mining: Environmental Impacts," 656–57. See also Peter Diehl, "Uranium Mining and Milling Wastes: An Introduction," May 18, 2011, *WISE Uranium Project*, online, http://www.wise-uranium.org/uwai.html; United States Environmental Protection Agency, "Uranium Mining Wastes," online, http://www.epa.gov/radiation/tenorm/uranium.html.

128. Shemitz, "Uranium Mining: Environmental Impacts," 656–57; Peter Shemitz, "Uranium Mining: Occupational Health," in Paehlke, ed., *Conservation and Environmentalism: An Encyclopedia*, 656–57; Atkins, ed., *Historical Encyclopedia of Atomic Energy*, 406.

129. Carlisle, ed., *Encyclopedia of the Atomic Age*, 350; Doug Brugge and Rob Goble, "Public Health Then and Now: the History of Uranium Mining and the Navajo People, *American Journal of Public Health* 92 (September 2002), 1410–19. Some popular books have been published about Native Americans and conflicts over environmental risks caused by uranium mining. See Judy Pasternak, *Yellow Dirt, An American Story of a Poisoned Land and a People Betrayed* (New York: Free Press, 2010); Peter H. Eichstaedt, *If You Poison Us: Uranium and Native Americans* (Santa Fe, NM: Red Crane Books, 1994). See also Eileen Welsome, *The Plutonium Files: America's Secret Medical Experiments in the Cold War* (New York: Dell Pub., 1999), 206, 245, 456.

CHAPTER 10

Pax Atomica—or Pox Atomica—at the End of the Cold War

INTRODUCTION

From the vantage point of the 1980s, the Cold War seemed to be perpetual. As events unfolded in those years, it was difficult to imagine that by the early 1990s the Berlin Wall would fall, the Soviet economy would unravel, and bilateral diplomacy would end. Cold War tensions escalated with the ascent of Ronald Reagan as US president, and with growing anxieties over the USSR having a numerical nuclear missile advantage. If anything the 1980s began with such a sense of heightened tension in the world that any hope of a receding arms race appeared naïve. President Reagan, who was such a provocateur in his first term of office, played an important role in changing United States-Soviet relations along with his counterpart Mikhail Gorbachev. The Cold War indeed began to end under them.[1] Yet events, bigger than any single person, converged to bring to a close decades of confrontation between two imposing superpowers. Hope of ending the relentless nuclear arms race was premature, however, as it remained a central point of antagonism between the United States and the USSR in the 1980s and into the 1990s. Even in the post-Cold War years the role of nuclear weapons in the world did not diminish, but instead evolved. Rising alarm over proliferation and terrorism, plus an emerging debate over climate change, ushered in a new atomic era.

TURNING BACK THE CLOCK

Ronald Reagan was an unlikely peace broker, and for the first five years or so of his presidency he did not give conservative allies any reason to think he would become one. It was not so unusual that he was a Cold Warrior, since this was a common posture of presidents who preceded him. He had long been intensely anti-communist going back to the 1950s, was dubious about containment, and promoted the idea (or myth) of American exceptionalism. Yet his rise to the highest office in the land was not built on his foreign policy views, but on an outsized interest in restoring America's confidence in itself infused with a spoken-out-loud criticism of letting the federal government run people's lives. A former New Dealer, he had shucked that moniker well before his run for president, and indeed before his campaign for governor of California in the 1960s. The presidency itself, however, was not weakened by Reagan but bordered on an "imperial presidency," that is, on the authority of the president to take charge and to assume greater influence than the Constitution allowed. This role was far removed from his anti-big government stance. Truth be told, some of the most momentous events in Reagan's time in office occurred outside the boundaries of the United States, with the involvement of a president having literally no foreign policy experience.[2]

Reagan's strong views, lack of foreign policy experience, and input from a variety of voices in his administration often resulted in responses that could be disjointed and poorly implemented. Adventures in the Middle East were sometimes precarious, while intervention in Central America destabilizing. In the early 1980s at least, the president's hawkish public pronouncements obscured the lack of clarity in executing policy, but in later years the stridency turned more to pragmatism in dealing with the USSR, leading to a new relationship. In the post mortems of the end to the Cold War, Reagan often received more credit than he deserved for the sea change in US-USSR relations. While he played an important role, context was sometimes forgotten when taking into account a host of global events underway and the long-standing internal decay of the Soviet Union. Despite some positive outcomes by the early 1990s, the early 1980s were hardly filled with hope. Instead, they were among the tensest in the post-World War II era.[3]

Reagan brought great communication skills to the Oval Office, especially the ability to articulate in clear terms underlying aspirations of many Americans.[4] He was a master of one-on-one encounters, exuding a confidence and an optimism which several of his predecessors lacked. Reagan was hardly a detail man, more prone to thinking about results than the path to reach those results. In all these ways, he was much like Franklin Roosevelt. And like Roosevelt he was prone to let his advisers squabble among themselves, rather than intervene. The State Department and the National Security Council were implacable foes. The more pragmatic secretaries of state (first Alexander Haig and then George Shultz) and chief of staff James Baker were offset by the more rigid Secretary of Defense Caspar Weinberger and the devious CIA director William Casey.[5] Early on Reagan branded the Soviet Union as "the evil empire" (actually a term he only used once from a prepared speech[6]) and often talked about "peace through strength." He opposed détente (the idea of lessening tensions through negotiation) and intended to erase the defeatist attitude of the so-called "Vietnam Syndrome." This view, held largely by American conservatives, purported that the "loss" of the Vietnam War biased the American public against any US military conflict.[7] To the contrary, Reagan favored demonstrating such superior military power that opponents would have to accept American demands.[8] As he stated in a speech in late 1982, "…yes it is sadly ironic that in these modern times it still takes weapons to prevent war. I wish it did not."[9]

The new president's views on the bomb, the arms race, and disarmament were most instructive about his overall global perspectives, especially early in his first term. He placed little stock in negotiating with the Soviets over arms control, having opposed the Limited Test Ban Treaty (1963), the Non-Proliferation Treaty (1968), the Peaceful Nuclear Explosions Treaty (1976), SALT I (1972), and SALT II. He was bent on reversing the apparent superiority in the number of nuclear arms held by the Soviets. In 1981 he appointed Republicans and Democrats to key posts in his administration who had strongly opposed détente and SALT II, including

32 of the 182 members of the neoconservative Committee on the Present Danger. These men included Richard Perle as assistant secretary of defense for international security policy; Paul Wolfowitz, Director of Policy Planning in the State Department; Paul Nitze, top negotiator on arms control; Richard Allen as national security advisor; and William Clark, who replaced Allen as national security advisor. Conservative officials, and neoconservatives[10] at that, clearly dominated national security.[11] Reagan's own optimistic nature, which he demonstrated on countless issues, was strongly counterbalanced by an ominous dread of nuclear holocaust. He prescribed to the biblical view of Armageddon (literally, the last battle between good and evil before Judgment Day), concluding that the world would end in a terrible conflict brought on by communism and nuclear weapons. Reagan, however, did not share the view with evangelical Christians that Armageddon was inevitable. He believed, or possibly hoped, that his stance on the arms race would avert that possibility.[12]

COLD WAR WITHOUT END: THE REAGAN MILITARY BUILDUP

Reagan was convinced that an increase in the defense budget would demonstrate the commitment of the United States to remain active in the world and simultaneously improve its bargaining power. Thus Reagan undertook the largest military buildup in peacetime history.[13] In essence, he extended Carter's precedent of increasing defense spending, but also reversed the trend of declining military budgets going back to 1968. In 1979 defense spending represented 4.7 percent of the nation's Gross National Product (GNP); in 1987 it was 6.4 percent of GNP. The military budget amounted to $2 trillion between 1981 and 1986, less than the Pentagon had projected, but more than Congress had wanted to authorize.[14] Between 1981 and 1986 the administration planned to spend $180 billion on nuclear weapons, that would give it the capability to carry out, if need be, "a protracted nuclear war." Among the new expenditures were costly new weapons systems such as the *B-1* bomber, the *MX* mobile land-based missile, *Trident II,* and *Pershing II* missiles, as well as a dramatic new defensive system nick-named "Star Wars." The administration earmarked funds to develop the *B-2,* or Stealth, bomber, *Trident* submarines, sea-launched cruise missiles, and several tactical nuclear weapons. A high-tech arsenal was the primary focus, but also a major goal was expanding the nuclear stockpile.[15]

Because of the emphasis on military spending, Reagan got a positive political bump from conservatives, especially in the South and Southwest. The defense industry was likewise thrilled, with average profit margins rising from 6.8 percent in 1975 to 9.6 percent in 1985. Many more grants flowed into the nation's universities; direct research payments from the Defense Department increased by 90 percent between 1980 and 1985. The old Military-Industrial Complex remained in business.[16] In the Soviet Union, the administration's actions created alarm, anger, and internal disputes on how it should respond. Soviet leaders lived under the belief that some sort of parity in nuclear weapons existed between the superpowers, but the Reagan administration challenged that belief.[17]

In the first two years of the Reagan administration, domestic issues often took precedence over new strategic initiatives. The president's vociferous anti-communist and anti-Soviet rhetoric was unrestrained, and effective action was taken sparingly. Some would argue that in Reagan's eyes tough language and the use of the bully pulpit were enough to discredit the USSR internationally at this juncture. In some cases harsh language and action came together. For example, in 1981 the Reagan administration warned that it would impose sanctions if the Soviets attempted to put down unrest building in Poland. When the Poles themselves declared martial law (which was an international treaty violation) the United States placed sanctions on Poland and even some modest sanctions on the USSR. In many ways, the Reagan administration showed restraint in dealing with the Soviets, not unnecessarily antagonizing them for the sake of doing so. In one case, the United States reversed the Carter embargo of grain shipments to the Soviet Union, which played to Reagan's interest in free trade and mollified American farm groups.[18]

On arms control, Reagan's early stance was tough, but not completely unbending. There were signs that anti-nuclear protests and charges of US warmongering led to a willingness to bargain, at least in theory. At the Intermediate Nuclear Forces (INF) negotiations in November 1981, American representatives laid out the administration's Zero Option. It called for ending deployment of *Pershing II* and ground-launched cruise missiles (*Tomahawks*) based in Western Europe, in exchange for the Soviets shutting down all of their intermediate-range weapons threatening NATO allies. The offer left out sea- and air-based missiles where the US advantage was great. The deal was so lopsided that the Americans knew that the Zero Option was a sham and clearly untenable for their rivals. Yet the president hoped that such a gesture would move the sides to discussions, and would show the world that the United States was not goading the USSR into conflict. On SALT II, which the administration believed was "fatally flawed," Reagan stated that the United States would abide by its restrictions, but would not submit it to the Senate for ratification. Secretary Haig and the Joint Chiefs persuaded him that the treaty placed more restrictions on the USSR than the United States, and without it the Soviets could add more warheads. Reagan avoided a domestic political battle with his decision, but accepted the counsel of his advisers in living under an unratified treaty. Other actions intentionally weakened previous nuclear weapons agreements. Reagan resisted negotiating a comprehensive ban on nuclear testing on the grounds that verification measures were inadequate. In reality, he had decided to reverse Carter's policy by resuming clandestine testing. He authorized the United States to set aside provisions of treaties concerning nuclear weapons when those violations were proportionate to Soviet actions. The administration also approved the sale of nuclear equipment and material to India, Argentina, China, and South Africa, although Reagan publicly declared that supplier nations needed to apply safeguards to all sales to prevent nuclear proliferation.[19]

1983: THAT DANGEROUS YEAR

In 1980 scientist Luis W. Alvarez presented what became known as the Alvarez Asteroid Impact Theory. He postulated that the impact of a meteor strike and subsequent smoke and dust clouds about 65 million years ago lowered the earth's temperature, causing the mass extinction of dinosaurs. As a result of this data a number of scientists developed the concept of a "nuclear winter." They argued that in the event of large-scale nuclear war vast amounts of smoke and dust sent into the atmosphere could block out the sun's rays over much of the planet for six weeks to six months. This event would lead to a significant drop in temperature to freezing levels or below, damaging or destroying food supplies, and potentially ending all human life. The horrifying thesis gained the attention of many people, including astronomer Carl Sagan and a team of other scientists. They studied the phenomenon and set out to publicize the possibility that nuclear war could result in catastrophic freezing. The TTAPS group (an acronym based on the names of the scientists) presented the concept of nuclear winter at a conference in fall 1983 held in Washington, D.C., and then published their results in *Science*. The study proved to be very controversial, especially because of its doomsday conclusion. Other groups, including those in the defense establishment, conducted their own studies in the mid-1980s. In 1982, Jonathan Schell published his best-seller, *The Fate of the Earth*, a disturbing portrayal of the consequences of nuclear devastation. The book reinforced the horror on the nuclear winter scenario. While TTAPS calculations were exaggerated, the research stimulated thinking about the various impacts of nuclear detonations on the environment and may have increased public pressure for disarmament. More recent research led to a less severe (or at least more complex) assessment of the influence of nuclear war on climate change, but the prospects of altering weather added one more layer of anxiety to the threat of a possible nuclear war.[20]

Beyond the nuclear winter warning, a series of events in 1983 exposed more immediate risks of the continuing arms race. In fact, 1983 might very well have been the most precarious year of the Cold War, at least since the Cuban Missile Crisis. The intensity of Cold War rhetoric already was apparent the year before. In May 1983, after 16 months in office, Reagan announced that the Strategic Arms Reduction Treaty (START) talks finally would commence in Geneva on June 29. To insiders, this delay in getting the first round of talks underway reflected the president's desire to close the gap with the Soviets in nuclear-weapons buildup before negotiating. Plus, some observers believed, Reagan was not prepared to give clear direction to the START delegation, since he himself was unfamiliar with basic features of the nuclear weapons issue. In fact, Secretary Haig called the administration's first START proposal a "flawed START position." When the first round began, the sides went back and forth, proposing and rejecting the other's latest plan.[21]

Western European countries were not pleased that they were pawns in the Cold War game. Soon after the Reagan administration took office, Defense Secretary Weinberger made an unauthorized statement that the United States planned to deploy "enhanced radiation weapons" (neutron bombs) in Western Europe and might use them in the event of war. Others, including Reagan himself, were concerned that Weinberger's assertion expressed American willingness to lower the threshold in the use of nuclear weapons by employing neutron bombs. The implication about focusing destruction on people, while saving infrastructure was unsettling and almost cynical in its intent. West Germany announced that it would not permit such weapons in its country. The US government responded that neutron bombs would not be deployed in Europe but stored in the United States. The decision to push for deployment of *Pershing II* and *Tomahawk* missiles in Europe (also seen as increasing the risk of nuclear war) was even more controversial among the American allies.[22]

A NATO agreement in 1979 authorized the US deployment of intermediate-range missiles in Western Europe, apparently in response to Soviet deployment of *SS-20* missiles in the late 1970s. Such precedent carried very little weight with those who only saw one escalation building on another. The anti-nuclear movement was clearly on the rise, concerned that the United States was unwilling to support arms control, and more specifically reacting to Reagan's public statements about "winning" or at least "prevailing" in a nuclear exchange with the Soviets. Advocates of a "nuclear freeze" in the United States and around the world roundly criticized the president's desire to up the nuclear ante. Randall Forsberg, working at the Stockholm International Peace Research Institute, originated the idea of a freeze in 1980. As he explained, "People have decided that enough is enough."[23] Anti-nuclear groups called for the United States and the USSR to freeze their nuclear weapons at current levels. Failing to pressure national governments to outlaw nuclear weapons, they fell back on this more pragmatic stance as a way of slowing the arms race.[24] By 1982 the Nuclear Weapons Freeze Campaign and related freeze efforts were on the rise, even in Congress. In Central Park in New York City on June 12, roughly one million participants turned out for the largest political rally in American

On June 12,1982 about one million people gathered in Central Park, NY in protest against nuclear weapons. It was the largest political rally in American history. *Source:* AP Photo/Ray Stubblebine

history, around the theme, "Freeze the Arms Race—Fund Human Needs."[25] One survey indicated that 75 percent of the American people favored a freeze. Citizens Against Nuclear War brought together no less than 26 grassroots groups in the United States with an estimated peak membership of 18 million. Internationally, the freeze movement drew from earlier anti-nuclear protests including pacifists, socialists, religious and academic leaders, unions, environmentalists, scientists, policy makers, and others. While it was not the central issue in the mid-term election (economic issues were) several moderate or liberal candidates favored a freeze and some supported a cut in defense spending. Nuclear freeze initiatives passed in several states. With a liberal turn in the House of Representatives as a result of the election, 67 percent of all freshman House members and 55 percent of the House itself favored freeze legislation (49 percent in the previous Congress). By 1983 the freeze movement had grown dramatically gaining vast amounts of publicity, with 10 states approving the idea and a freeze resolution passing the House by 278 to 149, with 60 Republicans in the affirmative. The Senate, however, defeated the resolution by a vote of 58 to 40 against (including 12 Democrats) and left the freeze movement in limbo, at least on a national level.[26]

Some Reagan advisers wanted the president to publicly repudiate the freeze movement as perilous and counterproductive, and even work to defeat the referenda in the 1982 election year. (Some asserted that the freeze was being orchestrated from Moscow.) Reagan was dismissive, viewing the freeze advocates as naïve. In a July 1982 campaign speech, he declared the freeze "would make this country desperately vulnerable to nuclear blackmail," and told a veteran's group in October that the movement was inspired by some people "who want the weakening of America and so are manipulating honest people."[27] On the broader issue of nuclear war, however, Reagan appeared to be more empathetic and wanted to alter his rhetoric for the sake of an international audience. In an April speech he delivered in Europe, he stated "To those who protest against nuclear war, I can only say: 'I'm with you.'"[28] But these various pronouncements as a whole reflected his belief that while the goal was worthy, the strategy was not. Members of the administration still saw the freeze movement as politically treacherous waters.[29]

A 1984 military artist's concept of a hybrid ground and space-based laser weapon destroying enemy satellites. This was one rendering inspired by Ronald Reagan's Strategic Defense Initiative, or "Star Wars."
Source: Everett Collection Inc/Alamy

Reagan's counterweight to the freeze and an extension of his goal of "peace through strength" was the Strategic Defense Initiative (SDI), or "Star Wars" as his critics named it. In a television address on March 23, 1983, the president called for the construction of a technically sophisticated, but complex, risky, and extraordinarily expensive defense shield. "I know this is a formidable technical task," he stated, "one that may not be accomplished before the end of the century. Yet, current technology has attained a level of sophistication where it is reasonable for us to begin this effort…But isn't it worth every investment necessary to free the world from the threat of nuclear war? We know it is."[30] SDI technology would be placed on platforms in space utilizing x-ray laser beams to shoot down Soviet missiles before they could reach their targets. Reagan informed the Joint Chiefs of Staff about the plan in December 1982, and they approved it the following February. The service chiefs also were now open to reconsidering a missile defense system, partly because no good basing scheme had been developed for the *MX* missile. But many other Washington officials were surprised and concerned by the announcement, including most in the administration itself.[31] This was a grandiose project with many potential pitfalls—technical, strategic, and political.

Reagan argued that the plan was "wholly compatible" with the 1972 Anti-Ballistic Missile (ABM) Treaty. After all, the Soviets had begun their own ABM system as a result of non-compliance with various treaties. More importantly, the SDI proposal took into account improvements in Soviet military technology (in the wake of détente) that made its first-strike capability much greater. Behind SDI, therefore, was a concerted effort to veer away from the idea of Mutual Assured Destruction (MAD), a policy that accepted a Cold War stalemate and one that assumed faith in Soviet adherence. While Reagan truly believed in the capability of SDI, he also understood its political advantages in the contest with the Soviet Union and in his struggles with domestic opposition. Whether feasible or not, SDI fit Reagan's foreign policy objectives quite well.[32] In Europe, NATO allies viewed SDI as a potential obstacle to serious arms control negotiations. From the Soviet perspective, SDI may have been an elaborate ruse, but it had alarming possibilities. The Soviets were concerned that it was not simply a defense technology that expanded the cost of the arms race, but had potential as an offensive weapon. If SDI could be deployed, it might be used to destroy Soviet ground-based missiles or to repulse a Soviet retaliatory strike.[33]

Supporters of the plan (some administration officials, some in Congress, nuclear weapons laboratories, weapons contractors, companies in the aerospace industry, several civilian and military officials, and a variety of SDI lobbyists) began promoting the idea in Washington.[34] However, Star Wars faced stiff resistance from those who thought it would not work and/or it was much too expensive. Respected scientists like Hans Bethe and others in the Union of Concerned Scientists viewed the technical obstacles as "staggering." Even if it did work, some argued, could it reliably destroy all incoming missiles? What would keep the Soviets from counteracting the system by flooding the skies with active missiles as well as duds? Soviet weapons designers called for that very thing with an "asymmetrical response" to SDI that would unleash many speeding objects to confuse or otherwise overtax SDI. Lord Zuckerman, former chief science adviser to the British, had one novel observation. He was concerned that if such a system were feasible it would leave decisions to computers, reducing human decision-making. In this instance, SDI sounded very much like the Soviet's "doomsday machine" in *Dr. Strangelove* that allowed computers to determine security threats and to respond automatically.[35]

SDI obviously incensed supporters of the nuclear freeze. Individuals and groups, some even less partisan, also weighed in on the argument. The Congressional Office of Technology Assessment raised the possibility that a defense system might actually increase the risk of war. Charles Burton Marshall of the Committee on the Present Danger was skeptical of the effectiveness and wary of the cost. Some in the administration including Secretary of State Schultz, at least privately agreed with the critics. Leaders in NATO believed SDI would weaken the desire of the United States to defend Europe.[36] The Soviets predictably opposed SDI arguing that it would accelerate the arms race, especially for ground-based missiles in violation of ABM agreements, and would give the United States such a sense of superiority that it might chance its own first strike. In response, they began reevaluating and reassessing their system in 1983.[37]

Events in late 1983 brought on almost unprecedented tensions between the United States and the USSR, already badly strained by SDI, faltering arms control negotiations, and a pending new round of nuclear buildup. Some regarded it as the lowest point in years for Soviet-American relations. On September 1, a Soviet jet shot down a South Korean airliner (KAL 007) that had wandered into its territory claiming it was on a US spying mission over a heavily militarized region of its country. Of the 269 passengers and crew on board were 61 Americans including US representative Larry McDonald from Georgia. The Soviet action seemed to reflect skittishness rather than malevolence on their part. In fact, Reagan had learned that the Soviets were unaware that the flight was commercial. The president nonetheless roundly condemned it as a "terrorist act" and added, "The brutality of this act should not be compounded through silence or the cynical distortion of the evidence now in at hand."[38] While many Americans were incensed and tensions mounted, Reagan decided to let the court of public opinion condemn the USSR rather than take provocative action (which Americans were unlikely to support in any event). That same month a Soviet satellite malfunctioned and transmitted signals indicating that five US missiles were approaching the USSR, triggering a full alert. Luckily a young Soviet officer detected the blunder and overrode the computer error thus forestalling a counterstrike.[39]

Early in November the United States and NATO were engaged in a nine-day major war simulation of command and communication procedures (*Exercise Able Archer 83*) to be employed during a nuclear attack against the USSR and the Eastern bloc. Some estimates suggested that 300,000 military and civilian personnel took part from the Mediterranean to Scandinavia. Not certain whether the exercise was real or not, the Soviets placed their military on high alert, waiting to see if the United States would actually launch its ICBMs. The Soviets were wary of what appeared to them as a bombastic new president in Ronald Reagan, a leader whose aggressive language was manifest in action. Little more than a year before *Able Archer*, the United States sent troops to Lebanon and Central America, rattled its sabers at Libya, and on October 25, 1983 (one week before the exercise) invaded the tiny Caribbean nation of Granada to oust leftist leaders.[40]

On November 14, on the heels of *Able Archer*, the first *Tomahawk* missiles arrived at England's Greenham Common Air Force base; on November 23, *Pershing II* missiles were deployed in West Germany. Subsequently, the Soviet delegation walked out of the INF talks in Geneva, and the Soviet government then announced it would increase the number of *SS-20* missiles and tactical weapons deployed in Eastern Europe. It also stationed additional submarines off the coast of the United States.[41] Late in that same month, *The Day After*, a fictionalized account of a nuclear attack and its impact on Lawrence, Kansas, aired on American television. *The Day After* combined with the arrival of the missiles in Europe and *Able Archer* to turn up the temperature on Soviet-American relations to very high intensity. President Reagan had previewed the film about one month before it was televised, and found himself "greatly depressed" by it. He was concerned enough about its impact that he had administration representatives carry out damage control on radio and television and in the print press. A few days after viewing *The Day After* Reagan attended a Pentagon briefing on the Single Integrated Operational Plan (SIOP). Although he normally avoided such briefings, he agreed on this occasion and learned that the United States was targeting more than 50,000 sites in the USSR, of which only half were military. He also was told that in the event of a nuclear exchange, most if not all of the United States would be destroyed. He was shocked that some Pentagon officials "claimed that a nuclear war was 'winnable.' I thought they were crazy."[42]

Possibly at another time and in a different setting, the controversy surrounding the airing of *The Day After* would not have taken on such significance. This was the height of what some had called the "Second Cold War." After all, it was a fictionalized account of nuclear holocaust; not the first such dramatization and certainly not the last. That it received so much attention was symptomatic of the lowest ebb in Soviet-American relations since the Cuban Missile Crisis. Maybe it was Reagan's inflammatory language that stirred the simmering undercurrents of Cold War tensions. Or perhaps it was an early indicator that the Soviet Union itself was fracturing to the point of hypersensitivity and paranoia. There is little doubt that the ABC made-for-television movie intentionally raised concerns about the arms race, setting off major media buzz even before airing. The media buzz produced several results: sponsors walked away from the project, the Pentagon labeled the movie "misleading," and some stations actually cancelled the program. Despite all that, or possibly because of it, *The Day After* set a record for the number of viewers watching a TV movie. The message of the film, however, was not new to Hollywood. The initial scenes were grim, and whether America survives in the end is left up in the air.[43]

Interestingly another film with a similar message appeared in 1983, but with much less fanfare. *Testament* took place in a small suburban town outside of San Francisco after a nuclear attack. The movie is told through the eyes of Carol Wetherly (Jane Alexander), a mother who struggles to take care of her family in the face of impending doom. *Testament* failed to get the attention of *The Day After* because the film had no eye-popping special effects of people being incinerated, was too pensive, and dealt only quietly with the inevitable outcome of nuclear war. As one writer noted, "*Testament* is simply about loss. Loss of family, of comfort, of company, of community, of country, of the world, of hope."[44] These represent devastating emotional impacts, but not on the graphic scale of destruction displayed in *The Day After*. Probably most important was the fact that this was a small film originally produced for the Public Broadcast System's *American Playhouse* but rereleased in theaters, rather than broadcast with great fanfare to millions of American on network television.[45] The reaction and response to *The Day After* much more dramatically than *Testament* seemed to mirror, and possibly enhance, the extraordinary events of 1983, that dangerous year.

THE ARMS RACE REORIENTED

In retrospect 1983 was the nadir of the Cold War. The convergence of so many unsettling events, heightened risks, and potential nuclear nightmares (real or imagined) seemed to offer little hope. One short year later, a turnaround in the arms race showed glimmers of possibility. In some subconscious way, many people may have dreaded seeing the clock strike 12:01 a.m. on January 1, 1984. George Orwell's haunting dystopian novel, *1984* (published in 1949), conjured up the destructive nature of international rivalry and internal repression. Written in the wake of World War II, it brought to mind how Orwell had abhorred the immediate past of Nazism and world conflict and projected his anxieties into the future. As reporter Jeff Smith asserted, *1984* was written "in the shadow of the first atomic weapons," and it "is all about what becomes of the state and the media when true history…is obliterated."[46] In real time, 1984 proved to be largely uneventful in terms of public discussions over the arms race. But luckily this did not prove to be the proverbial calm before the storm or the late 1939 and early 1940 "Phony War" that preceded the real thing in mid-1940.[47]

The interplay between Reagan and Gorbachev normally takes center stage in recounting the story of the lessening tensions between the superpowers that ultimately resulted in the end of the Cold War by the late 1980s and early 1990s. Reagan's tone, if not his perspective, began to change before Gorbachev became General Secretary of the Communist Party.[48] The president even attempted a back-channel approach to the Soviet leadership before Gorbachev came to power. His wife Nancy and the pragmatic Secretary of State George Schultz urged the president in the new direction.[49] In a January 1984 speech the bellicose anti-Soviet oratory was less-pronounced and the talk of peace elevated. The United States, he said, "must and will engage the Soviets in a dialogue as serious and constructive as possible, a dialogue that will serve to promote peace in the troubled regions of the world, reduce the level of arms, and build a constructive working relationship."[50] Some viewed this speech as a change of heart; others saw it as consistent with Reagan's long-held objective of responding to the Soviets from a position of strength. But the president's statements cannot be divorced from the stress and strains that 1983 brought to both sides, and from the dangerous tensions which had grown between the two powers since Reagan's inauguration. Yes, the president planned to negotiate from strength, but strength that had to be wielded carefully in uncertain times. This was not a view shared by many of his hawkish advisers, but seemed to be a very personal and pragmatic assessment of the situation, not a fundamental change of heart, but a change of approach.[51]

Initially the reaction in the USSR to Reagan's January speech was rife with cynicism, but the Soviet leadership slowly recognized the opportunity to break through what had been a rigid barrier. During the 1984 presidential election, Democratic presidential candidate Walter Mondale promised to push hard for nuclear arms talks once he was elected. Pressure from Mondale was met with additional demands from Congress and even the USSR itself. The two superpowers continually quarreled over adherence to test-ban agreements. Now the Soviets called for a joint verification experiment and a possible joint moratorium on tests. Within six weeks of the election, Reagan responded to these pressures by proposing a joint negotiating framework that would combine various arms control talks under a single negotiation "umbrella." He was not going to let the nuclear arms issue defeat him, or even weaken his chances, in November.[52]

The burgeoning relationship between Reagan and Gorbachev put a human face on grand historical events, and concern over the nuclear arms race became the center of rethinking the overdue end to the Cold War. They met for the first time in November 1985. This is not a story of two men of similar perspective destined to find common ground, representing an inevitable meeting of the minds. Context and timing are essential ingredients, as is the observation of Reagan biographer Lou Cannon that each leader "served the other's purpose."[53] Gorbachev came from peasant stock in the Caucasus, was self-confident, hard working, optimistic, and ambitious in the ways of most political leaders. He had been an aspiring actor, possessing a good measure of charm, traits that clearly connected him with Reagan. But most importantly, although he was a Socialist he was more pragmatic than doctrinaire, not so ideological, and fairly open-minded. These were qualities that at first glance separated him from the American president.[54] Yet, studying Reagan's actions as a political leader, not simply listening to his rhetoric, uncovers a good measure of those traits in him as well. In the scaling back of the arms race, pragmatism was preeminent. While SDI received considerable international attention in 1983, the less engrossing, but equally important START talks were stalled. At home, Congress was impatient, particularly uneasy over the military buildup. Debates continued in the early 1980s over deployment of the *MX* ("Peacekeeper") missile, and the wisdom of the administration's "Dense Pack" plan (closely spacing land-based ICBMs). Reagan appointed several commissions to study alternative approaches. One positive outcome of the political skirmishes was moving Reagan closer to negotiating in earnest with the Soviets. The uncertainty of the direction that new leadership would take in the USSR at the end of 1983 had left START in an indeterminate state.[55]

The year 1984 began a rather slow process of restarting START and giving serious attention to arms control. Reagan's reelection forced the Soviets to face the political realities of the next four years in American politics. Thus, they took up in earnest Reagan's idea for umbrella talks. In January 1984 Secretary of State Schultz and Soviet foreign minister Gromyko met in Geneva to set up what were called the Nuclear and Space Talks (NST). The two countries planned to combine three simultaneous negotiations: strategic offensive reductions through START, intermediate-range missiles through INF, and strategic defense and space-based weapons focused on SDI. They made little progress through the summer of 1985. Reagan wanted to maintain a substantial nuclear arsenal in case of war with the Soviets and also wanted SDI off the table. SDI proved to be a main sticking point. Gorbachev's ascendancy in spring 1985 helped to bring the sides a little closer to reducing the number of nuclear weapons. He was fully aware of the need to concentrate attention on building a Soviet economy that was desperately behind other industrialized nations in output and burdened by the demands of the military. Less tangible, but nevertheless important, was the need of the Soviet leader to improve his country's image among Western powers to attract their financial support and increase the security of the USSR. In fact, Gorbachev, unlike his predecessors, was more concerned with eliminating American missiles in Europe than embracing his predecessors' policy of an overall balance of arms. He thus pressed the issue of nuclear weapons reduction with the Reagan administration as part of a strategy to lessen tensions with the United States. Gorbachev's policies were not always coherent, but he was not particularly calculating. Was he in favor of peaceful coexistence or something else? It often was difficult to tell.[56]

SDI still remained a problem for the Soviets. They were unwilling to substantially cut their nuclear arsenal without a ban on space-based weapons, and Reagan was not budging. The American government also "interpreted" research on SDI as being in compliance with the ABM Treaty, although some officials wanted to go farther and consider breaking that treaty.[57] The Executive Branch's broader definition of the ABM Treaty ruffled feathers in Congress and angered the Soviets. In the midst of the storm of protest Reagan decided to honor a narrower definition of the treaty for the moment, but retain the option of accepting a broader definition. Internal dissent within the administration nevertheless threatened START negotiations and the ABM Treaty. US allegations that the Soviets were violating previous agreements again aggravated rising pressure over arms control. At the Geneva Summit in November 1985, the best that the two sides could achieve was agreeing that a nuclear war "cannot be won and must never be fought."[58] Almost one year later, in September 1986, Reagan accepted Gorbachev's invitation to attend a meeting to try to break the deadlock. The various summits were an appealing approach to personal diplomacy for the president. He viewed them as negotiating sessions, as opposed to Gorbachev, who believed that the meetings publicly confirmed issues worked out in advance.[59]

For the president this summit was a political as well as a diplomatic opportunity. The congressional elections were about one month away, and he hoped to use the event to help Republican chances. Convening at Reykjavik, Iceland, the two leaders attended the meeting with different agendas. Reagan saw it as a preliminary step to a new session, while Gorbachev came with detailed proposals. He presented a START plan enumerating 50% reductions in strategic weapons and other changes. Reagan agreed in principle and then suggested an end to all strategic weapons within 10 years. Both sides accepted a limit on INF weapons, including the earlier proposed Zero Option which would lead to withdrawing intermediate-range missiles from Europe. A third area of general agreement was a phased reduction of nuclear testing. However, the fly in the ointment, again, was SDI and the ABM Treaty. The Soviets demanded a strict interpretation of the ABM Treaty, but Reagan countered suggesting that such an arrangement was the end of SDI. The president's willingness to continue to move toward complete elimination of strategic weapons while sustaining SDI research was untenable to Gorbachev. Reykjavik ended with mutual recriminations and no deal. Gorbachev now assumed that Reagan was not serious about arms reductions at the time, but there seemed to be lingering expectation that something ultimately could be achieved on that front. At the very least, the negotiations revealed just how serious the president had been about SDI. He did not view it simply as a negotiating ploy or as a way to drive the Soviets to economic ruin, but as a real commitment to an expansive (yet implausible) idea. Reagan did want arms reductions, but that intention was clouded by the stance on SDI.[60]

Domestically, SDI also continued to be controversial. Critics of the president and SDI believed that an ill-prepared Reagan had lost an opportunity to scale back the arms race by doggedly hanging on to the defense shield idea. White House allies were pleased that he did not rush into an agreement that gave up central features of the US defense system, or left the Soviets with an advantage in conventional arms.[61] Hard-liners in the administration pushed for early deployment of the system as a way to distract Reagan from making any arms concessions. They also realized that the lame-duck president was losing momentum on SDI as the last days of the administration loomed, and the plan itself was losing coherence. Not everyone shared the hard-liners logic or objectives. The Office of Technology Assessment questioned the feasibility of the early stages of the project. The Joint Chiefs saw SDI as a drag on its strategic weapons budget. All sides saw SDI as influencing US adherence to the ABM Treaty, positively or negatively. Reagan's move away from the broad interpretation of the ABM Treaty, the presidential election year, and revelations about the Iran-Contra affair weakened the hard-liners' position and thus weakened any momentum SDI had gained since Reykjavik. Even the Soviets began to see although SDI could endanger them, it could weaken the US economy and thus could impact potential deployment.[62]

The Iran-Contra scandal was not directly related to US-Soviet relations, but was tied to Cold War policy. It had the potential of unraveling the Reagan presidency, and thus undermining any chance for serious arms control. During the post-World War II era, American presidents had wide discretion in waging Cold War, but this particular series of events seemed to cross the line. The Iran-Contra operation had two goals. The first was to sell arms to the Iranians in hopes of gaining the release of American hostages held by a pro-Iranian group in Lebanon. The second goal was to divert profits from these sales (which was illegal) to the Contra rebels attempting to overthrow the leftist Sandinista government in Nicaragua. Zealots on the National Security Council (including NSC adviser Robert McFarlane) in particular promoted the scheme. Directing it were Vice Admiral John Poindexter, who was national security assistant, and his underling Lieutenant Colonel Oliver North, deputy director for political-military affairs. The Reagan White House ignored congressional prohibitions and used this means of arming the Contras in a very creative way. CIA director Casey and UN Ambassador Jeane Kirkpatrick recommended finding a third party to supply arms, which proved to be Iran. Between the late summer of 1985 and the fall of 1986, NSC functionaries sold *TOW* anti-tank missiles and *Hawk* anti-aircraft missiles to Iran in exchange for a release of hostages. The exchange violated an arms embargo against that country and aided terrorists. A portion of the proceeds from the weapons sales were used to supply the Contras. Attorney general Edwin Meese revealed the plan in November 1986, and Reagan publicly announced that the weapons deal occurred, but that it was not a trade for hostages. This proved to be untrue. The affair was illegal and misguided on many levels. Whether the president was simply detached from the policy formation and execution or directly involved is unclear, but investigations showed that he knew a great deal about what had transpired. While Reagan's approval rating crashed to 36 percent, he avoided impeachment and his administration survived the potential disaster. In some odd way, Reagan's involvement in the summits with Gorbachev may have helped raise the president above the political fray at home. Iran-Contra created major political fallout, but did not derail the nuclear weapons negotiations, and might even have been one final object lesson in the need to end the long history of Cold War adventurism.[63]

On the Soviet-American front, Reagan's change of heart on arms control, a new verification procedure, and bringing the West Germans in line led to the INF Treaty at the Washington summit in December 1987. The agreement meant the elimination of a whole class of nuclear weapons (a first), including the dismantling, destruction, and removal of *Pershing II* and *Tomahawk* missiles in Europe. The treaty also prohibited development of new intermediate missile systems. Some Europeans, however, were concerned that the United States was disengaging from Europe in a key area.[64] Reagan was influential in the negotiations, but Gorbachev more so for making the most of the concessions. Despite the major step with INS, START was not close to completion by 1988. The same fate was true about a comprehensive test ban treaty, which many regarded as a way to block efforts at nuclear proliferation. The Reagan administration was determined to test new weapons systems, and was not ready to bend on this issue. The Nuclear Test Talks in Geneva beginning in November 1987 focused on verification issues as opposed to the larger question related to a ban.[65]

President Ronald Reagan and Soviet General Secretary Mikhael Gorbachev sign the INF Treaty in the East Room of the White House, December 8, 1987. *Source:* Ronald Reagan Presidential Library

THE END OF THE COLD WAR

Former Vice President George H.W. Bush succeeded Ronald Reagan in 1989, inheriting an unfinished agenda on nuclear arms control and other key Cold War issues. He was very different in background and disposition from his predecessor. A product of the Eastern establishment who made his mark in West Texas oil, Bush had a long and distinguished career in public service from his days as a navy pilot in World War II, as a congressman from Texas, and then in Washington as US ambassador to the United Nations, chairman of the Republican National Committee, chief liaison to China under Gerald Ford, and as director of the CIA in 1976–1977. A loyal supporter of his party and his president, he was not prepared to accept blindly the momentum of the Reagan-Gorbachev negotiations, choosing prudence over impulse in his first year in office. Some viewed him as moderate, a doer more than a cerebral type. Opponents unfairly viewed him as timid and sometimes indecisive (especially on domestic affairs). But as one scholar noted, Bush was a realist, and "realism dictated caution."[66] Soon after taking office, he addressed Congress: "The fundamental facts remain that the Soviets retain a very powerful military machine in the service of objectives which are still too often in conflict with ours. So, let us take the new openness seriously, but let's also be realistic. And let's also be strong."[67]

The new president decided upon a major review of strategy. He thus left the impression that the United States was becoming unduly guarded in moving away from the hard line with the Soviets and more open to dialogue over weapons control and arms reduction. Yet, the administration was not sending clear signals. Secretary of Defense Richard Cheney stated that the new administration would continue Reagan's strategic objectives, but more privately he was very skeptical of Gorbachev. With respect to offensive nuclear weapons, dependence on long-range bombers, ICBMs, and submarine-launched ballistic missiles remained as America's offensive "triad," adding new systems as needed. In its targeting doctrine, the number of Soviet targets was reduced leaving substantially more weapons than necessary. But rather than scaling down the stockpile, the Single Integrated Operational Plan (the general plan for nuclear war) was targeted at other "potentially hostile countries." Less enthusiastic about SDI than Reagan,

President Bush and Secretary Cheney saw a limited role for the space-based defense system. To their way of thinking it related most specifically to reducing the impact of a Soviet first strike, and consequently they cut the budget. In public they claimed there would be no change in SDI from Reagan's position.[68]

In June 1989 new arms control talks began in Geneva. Positions had not changed on START or on the ABM Treaty, and therefore the sides made no progress. To move talks off the stalemate Soviet foreign minister Edward Shevardnadze urged a meeting with Secretary of State James Baker III, which took place in Jackson Hole, Wyoming, in September. Baker, who had been Reagan's chief of staff and also secretary of the Treasury, had limited foreign policy experience. Yet the Texas stalwart was ever the pragmatist and saw an opportunity to take a more positive and proactive role in breaking the deadlock. Shevardnadze also saw the need to compromise. While unwilling to accept the interpretation of the ABM Treaty adhered to by the United States, the Soviets modified their views on SDI in part at least because they realized that the testing of the system was well behind schedule. In addition, they decided to dismantle their giant radar facility at Krasnoyarsk in Siberia, which the United States viewed as a violation of the ABM Treaty. A more cooperative approach developed over START as well. The Soviets announced that they were willing to deal with limits on sea-launched cruise missiles (SLCMs) separately from START ("delinking") as a way to move beyond the most recent impasse. But the Bush administration like Reagan before believed that verifying SLCM reductions was not possible. Nevertheless the sides discussed some verification options, and Baker announced that the United States was willing to open talks on mobile ICBMs. Also at Jackson Hole, Baker and Shevardnadze discussed two treaties that had never been ratified—the Threshold Test Ban Treaty (1974) limiting the size of weapons tested, and the Peaceful Nuclear Explosions Treaty (1976), which would limit the size of explosions for devices having "peaceful" purposes (such as *Project Plowshare* had envisioned). In fall 1990 both countries ratified the treaties. The Bush administration did not budge on a comprehensive test ban treaty, however, despite international criticism that without such a treaty nuclear proliferation could not be checked.[69]

The phenomenal events in Eastern Europe in 1989 that revamped the political landscape there, served as the context for the latest negotiations on arms control. Almost like a house of cards, one country after another abandoned its communist system, and ultimately its long-standing linkages to the Soviet Union. The Solidarity Party won control of both houses of parliament in Poland. Hungary held free elections in 1990. And on November 9, 1989, after turbulent demonstrations and strikes, the East German government opened the Berlin Wall. Free elections followed in spring, and Lothar de Maziere led the first non-communist government since World War II. The economic merger of the two Germanys took place on July 2, 1990. On December 10, a new cabinet assumed office in what would become the Czech Republic. Changes soon came to Bulgaria and Romania (a bloody affair in the latter country).[70] Reagan's hawkish remonstrance to Gorbachev at the Brandenburg Gate in West Berlin in June 1987—"Mr. Gorbachev, tear down this wall!"—is well known. But the monumental event had come to pass in 1989 not because of a presidential challenge two years earlier, but on the heels of popular dissent in Eastern Europe.[71] CNN journalist Bettina Luscher, a witness to the fall of the Wall, stated, "It's easy to say you saw it coming, the fall of the Wall. But if you are honest, you'd have to admit you had no clue. None. The Wall coming down? German unification? It seemed unthinkable."[72]

For many the Bush-Gorbachev summit in Malta in December 1989 marked the symbolic end of the Cold War. Bush favored the framework developed by Baker and Shevardnadze at Jackson Hole, but there were many issues to resolve including the SLCMs and air-launched cruise missiles (ALCMs). The sides could not agree as talks faded into 1990. Before a Bush-Gorbachev summit scheduled for later that year, the foreign minister and the secretary of state began to thaw the ice on START with general agreement on ALCMs and a decision to deal with SLCMs outside the bounds of START. Other issues, such as the Soviet *Backfire* bomber and US mobile land-based missiles, remained on the table. The summit in Washington from May 31 to June 3 resulted in a reaffirmation of a need for a treaty by the end of the year. While this suggested the persistence of good will between the parties, the delay gave American arch-conservative leaders, such as Senator Jesse Helms from North Carolina, a chance to attack the process and START itself. The ending of the Cold War worked to the administration's advantage, since the chance for nuclear war with the USSR was much less likely. Also, other priorities, such as the reunification of Germany redirected Soviet-American attention. After several assurances from the United States, including a major point that Germany would not produce or own nuclear, chemical, or biological weapons, Gorbachev accepted reunification and Germany's entry into NATO in July 1990. The four World War II allies and the two German states signed the Treaty on the Final Settlement with Respect to Germany on September 12.

On the strategic weapons negotiations, the sides settled little. Equally significant were Gorbachev's mounting problems at home, particularly a declining economy and rising public protests. The experiment in perestroika had not failed, but indeed hastened an end to communist rule in the USSR. In a March 1989 election many party members went down to defeat. Boris Yeltsin, elected chairman of the Presidium of the Supreme Soviet of the Russian Federation, became Gorbachev's chief rival. Outside of Moscow, the tightly held-together union was crumbling with problems in Armenia, Azerbaijan, and Lithuania. Much of this forced Gorbachev to the right, creating other political problems. START was hardly on the Soviets' mind.[73]

George Bush had his own preoccupations, especially with the mounting crisis in the Persian Gulf. When Iraq invaded oil-rich Kuwait in August 1990, this set in motion a major response. Beginning on January 17, 1991, the United States and its allies (in a UN sanctioned military action supported by the USSR) drove Iraqi forces out of Kuwait and staged an air attack on Iraq itself. *Operation Desert Storm* was quickly over, but there was no overthrow of the Iraqi dictator Saddam Hussein. Among the reasons was Gorbachev's insistence that such a decision would embarrass the USSR, because Iraq was an ally and a client state. Bush required Soviet support on other issues, including START, and relented. Hindsight suggests that leaving Hussein in power had longer-range implications than the administration could anticipate at the time.[74]

In an odd way the Gulf War reinvigorated SDI. The apparent success of *Patriot* missiles (surface to air) in destroying Iraqi *Scud* missiles seemed to be an object lesson in how a defense system might actually work. Clearly the success of the *Patriots* was vastly overrated. A report issued in September 1992 suggested that only 9 percent of the engagements between the two missiles resulted in a "kill." Some supporters of the *Patriots*, however, criticized the administration for focusing on SDI and not on tactical antitactical ballistic missile defenses (ATBMD). This reaction related to the assumption that current threats did not come from the USSR, but more from "rogue states"—a rather derisive term for countries acting independently of the world power structure. In response, the administration supported the Global Protection Against Limited Strikes (GPALS) program as a supplement to SDI and expected the budget for theater-missile defenses (like *Patriot*) to increase. GPALS proved to be yet another iteration of the meandering path of SDI and muddied the waters over whether any space-based system would work or if the United States would implement any system at all. What GPALS effectively did was stimulate new debate on further limiting or eliminating the provisions of the ABM Treaty.[75]

START was on the agenda as well in spring 1991. And after wrestling with a number of perpetual issues, and some new ones, the powers struck a treaty in Moscow on July 31.

It became the first treaty to require reductions in the number of warheads mounted on strategic offensive nuclear weapons. Both sides agreed to cut their strategic weapons to 6,000 "accountable" warheads on no more than 1,600 delivery vehicles over a period of seven years. This was a 38 percent reduction for the United States and a 48 percent decrease for the USSR.[76] The signatories agreed that neither party could develop new types of multiple-warhead missiles. They also clarified and tightened verification procedures, but could not resolve every issue. START was intended to last for 15 years, if not superseded by another agreement.[77]

Supporters of the treaty not only argued that START set a precedent for future reductions, but insisted that such an agreement was necessary at a time when the Soviet Union was clearly in disarray. START could bind the USSR to obligations that would bring predictability into US-Soviet relations at a crucial moment in history. Yet the treaty was no panacea for world peace and security. Both sides retained immense stores of nuclear weapons. The United States was still vulnerable to nuclear attack. The arsenals could be modernized. And verification was never fool-proof. Beyond START political battles nevertheless continued over the missile defense system and the ABM Treaty. Congress passed the Missile Defense Act in 1991, making the deployment of an inclusive defense system a national objective, and in the short term beefing up the *Patriot* program. While it was a compromise measure, and was meant to be compliant with the ABM Treaty, it only aggravated the long-standing controversy. Despite its wavering history, SDI funding continued and a commitment to a missile defense system for the United States seemed to be growing in support rather than fading.[78]

Instability of the Soviet government did not bode well for sustaining the positive outcome of the START I agreement. A failed coup against Gorbachev in August 1991 offered hope that the new direction in Soviet politics had survived a serious test, but it also raised ominous prospects of nuclear weapons falling into the wrong hands. On August 24 Gorbachev resigned as the Communist Party's general secretary, leading to the end of communist rule in the Soviet Union, and the ultimate breakup of the USSR. The immediate concern was that the dissolution of the Soviet Union could result in the loss of command and control over nuclear weapons there. But both the Bush administration and Soviet leaders publicly denied such a scenario. In reality, the control of the codes necessary to launch missiles was now in the hands of two people, not three. With Gorbachev out of the picture, the defense minister and the chief of the General Staff were the only leaders to hold the codes. For a brief time Marshall M.A. Moiseev, chief of the General Staff, actually was the only official holding the codes. After the unsuccessful coup, Moiseev's replacement and the commander of the Strategic Rocket Force, each had a set. Potential renegade military officers were only one problem. More serious was the possible control of nuclear weapons by leaders in the "nuclear republics." Such weapons and missiles were in fact located in these separate political entities that emerged after the dissolution of the USSR. They included Russia, Ukraine, Belarus, and Kazakhstan. Despite Gorbachev's efforts in September to get the republics to agree to a central command over the weapons, discord broke out among the parties. Belarus wanted to be free of its arsenal, but was unclear if the weapons should be transferred to Russia. Kazakhstan and Ukraine were not prepared to make the transfer. Pre-Soviet rivalries, among other reasons, created this impasse, and the "loose nukes" problem remained.[79]

Prompted by the potential dangers of a divided nuclear weapons arsenal in the fractured Soviet Union, President Bush proposed a rather audacious one-sided reduction in American weapons, and urged the other side to do the same. He also took strategic bombers and *Minutemen II* missiles (scheduled for dismantling under START) off alert, decided to terminate the *MX* and mobile ICBM programs, made changes in US command and control measures, and sought cooperation with the Soviets on a nonnuclear defense system. In October, Gorbachev agreed to match the American changes in tactical weapons, strategic alerts, modernization programs, and he even urged Bush to make deeper cuts in the US nuclear arsenal. Gorbachev also called for immediate talks on START II after START I was ratified, considered modifications to the ABM Treaty, and proposed a comprehensive ban on nuclear testing. To show resolve, the Soviet leader implemented a unilateral halt on testing for one year. Despite the positive tone of these accelerated goals for nuclear arms control, several additional issues complicated the picture. Verification was a bugaboo. In addition, actually being able to locate some 15,000 nonstrategic nuclear weapons dispersed throughout the USSR was problematic, as was the control of tons of plutonium and enriched uranium held within its borders. Concern about verification led Congress to pass the Soviet Nuclear Threat Reduction Act in November 1991, which authorized the Department of Defense to use funds to assist in the dismantling and storage of nuclear weapons and materials in the republics of the soon-to-be former USSR.[80]

The dissolution of the Soviet Union, beginning with the declaration of independence by Ukraine on December 1, 1991, led to a Commonwealth of Independent States (CIS), the resignation of Gorbachev as president of the USSR, and the elevation of Boris Yeltsin. In his televised farewell address to the Soviet citizens on December 25, Gorbachev stated, "I leave my post with concern—but also with hope, with faith in you, your wisdom and spiritual strength. We are the heirs of a great civilization, and its revival and transformation to a modern and dignified life depend on all and everyone."[81] The extraordinary series of events leading to the breakup of the USSR was reason for celebration in the new republics and the West, but also created substantial unease about arms reductions and verification. Disputes between Russia and Ukraine over the Black Sea Fleet in January 1992 breached the initial agreement calling for unanimity of CIS leaders in decisions concerning nuclear weapons. The parties sought a compromise, but unified command and control questions continued, as did disagreements over such issues as dismantling of tactical weapons. Ratification of START I also was more complicated because each of the four republics now had to agree to it. After some diplomatic wrangling and American recognition of the newly independent republics, three of the four plus the United States ratified the treaty in November 1992. Only Ukraine resisted, but did approve the treaty in February 1994. Many of the issues leading to Ukraine's recalcitrance had to do with its desire to be on relatively equal footing with Russia on all matters of state.[82]

Working out the diplomatic arrangements with the newly independent nations prior to ratification of START I was a departure point for Bush and Yeltsin to begin talks leading to START II. Despite the previous agreement, discussions over a new treaty exposed obvious differences in perspective. START II produced additional reductions in total nuclear warheads on both sides. Strategic weapons would be at their lowest since 1969, and land-based MIRVs would be eliminated. The agreements reached with the United States proved to be unpopular in Russia, even with the cooling off and then termination of the Cold War. The Russian military and hard-liners in the Duma accused Yeltsin of making too many concessions. In response, the United States made some changes especially concerning questions of strategic balance and the costs of compliance. On January 3, 1993, the parties signed START II in Moscow, but it was not implemented until 2007.[83] The START agreements and the changing context of US-Russian relations also had a powerful effect on the lingering battle over SDI and the ABM Treaty. By the end of the Bush years, Washington leadership voted with their collective budgetary pens to let the defense system wither and die.[84]

THE RESURGENCE OF THE NUCLEAR DISARMAMENT MOVEMENT

Debates, discussions, and negotiations surrounding the arms race and disarmament did not take place in a vacuum, nor did they occur only at the highest levels of government. If there was a Reagan reversal, there also was public outcry for change that influenced it, or at least reinforced it. The political impact of the nuclear freeze movement was an excellent example of how changing views on nuclear weapons and international security did not have to wait for the end of the Cold War to become vocal and adamant. Ronald Reagan's strong Cold War rhetoric in the 1980 presidential election, and the rising influence of hard-liners in his administration, plainly intensified the Cold War. However, until Gorbachev's ascendancy the Soviet government showed few signs of changing its own Cold War stance. American allies like Prime Minister Margaret Thatcher's conservative government in England were likewise showing little restraint. Thatcher defended the arms race and held to the notion that nuclear weapons were a bulwark for peace.[85] Reaction against the forceful defense of nuclear weapons and the revitalized arms race stimulated the revival of peace and disarmament groups in the early 1980s. Not only were there numerous mass marches and rallies, but a growing broad-base of support from labor, religious groups, political parties, and professionals.[86]

The disarmament movement was most powerful in Northern Europe, in the NATO nations, where the threat of war seemed most real. Nuclear weapons were among the people's "greatest fears." In particular, the deployment of American cruise and *Pershing II* missiles and Soviet *SS-20s* were the most visible signs of the threat. In Britain there was the Campaign for Nuclear Disarmament; in the Netherlands the Inter-Church Peace Council; and in Belgium both Flemish and Walloon disarmament groups. Organizations and informal clusters of all kinds also could be found elsewhere in Europe. In some cases, the activities of the anti-nuclear movements in Japan, New Zealand, and Australia surpassed the scale of those in Western Europe. Protests did take place in Eastern Europe as well, but were not a major force at this time.[87] The disarmament movement in North America was significant, but not European-scale. It tended to stress world-wide implications of nuclear weapons, including proliferation, missile deployment in Europe, *MX* missiles, and SDI. Groups in the United State were diverse. They ranged from pacifist organizations, the more radical Mobilization for Survival, women's peace groups, SANE, the Union of Concerned Scientists, and Physicians for Social Responsibility. All were important, but they never reached the massive level of the freeze campaign, which mainly attracted mainstream supporters including unions, various religious groups, and professionals of all stripes. Such grassroots response did not keep the Right from pushing back and openly criticizing the "freezeniks" who continued to repeat the mantra of keeping the world safe through nuclear deterrence. In Canada, the anti-nuclear campaign drew constituencies similar to Europe and the United States, but had considerably less success with its political parties than the freeze advocates and others across the border.[88]

In the early years of the Reagan administration, rising anti-nuclear protest proved to be a menace to plans for a military buildup and tough-talk with the USSR. Soviet leaders were just as threatened by anti-nuclear reactions, but at first at least did not have to worry about openly confronting public criticism. By the time the major arms control accords had been signed in the early 1990s and the Cold War had ended, the anti-nuclear movement had ebbed.[89] In the mid-1990s, however, some believed that more radical changes were necessary beyond what the diplomats had achieved, and debate over efforts to abolish nuclear weapons arose again. Some peace groups never disbanded and would not do so until all nuclear weapons were gone. They were convinced that what had transpired in the late 1980s and early 1990s only represented an incremental approach, not a total commitment to disarmament. Had the anti-nuclear movement made a difference? There is no way to quantify this, but there can be little doubt that the approbation that followed the reversals (or at least significant modifications) of policy in both the United States and the USSR indicated that the climate was right to move away from the orthodoxy of the Cold War arms race. No one was prepared to admit that the nuclear threat had past, but a new direction had been taken.

THE POST-COLD WAR: A NEW DAY?

In the United States, a collective sigh of relief seemed to spread across the country as the USSR crumbled and commentators announced an end to the Cold War. Many observers conjectured (or at least hoped) that nuclear weapons and the arms race would recede from center stage in international affairs. Significant reductions in nuclear weapons made possible by various agreements appeared to be concrete proof of that assessment. Between September 1990 and September 2000, the US deployment of strategic weapons decreased by half.[90] As a bonus, the United States now enjoyed the role as the world's only superpower. Yet, the emergence of a "new world order" did not and would not always favor the strong. With the possible exception of the North American and Australian continents, ethnic, racial, and tribal conflicts, secessionist movements, and civil wars wracked many countries. Technological and communication revolutions that made the world smaller were matched by the emergence of new power centers and fragmentation of others that made proximity dangerous. A restructured global economy created new networks of interdependence not always welcomed by those under scrutiny and control.[91]

Neither the Bush administration, nor those that followed were prepared for the uncertainty of the post-Cold War world. Even with the successes of START and related negotiations, there were many signs that euphoria over curtailing the nuclear arms race (or at least slowing it down) was premature because of nuclear proliferation, the potential actions of independent states, and terrorism. In some instances, military planners who feared stepped-up development of chemical and biological weapons rationalized keeping stockpiles of nuclear arms as a hedge against other threats. The Pentagon's policy of "measured ambiguity" was a way of suggesting that nuclear missiles could be deterrence to those considering using other forms of mass destruction, which ultimately included alleged rogue states and "non-state actors" such as terrorists.[92]

George Bush's successor, William Jefferson Clinton, was the first Democrat in the White House since Jimmy Carter. Unlike Bush, but much like Reagan, Bill Clinton came to office with an agenda focused on the economy and domestic affairs, and very limited foreign policy experience. Although he had graduated from Georgetown's School of Foreign Service and even resided in England as a Rhodes Scholar, Clinton had made his mark in state politics as governor of Arkansas. He was bright, outgoing, charismatic, and an astute politician. Yet, his work habits were disorderly, and his private life often out of control.[93] Much like presidents before him, foreign policy would play a larger role in his stewardship of the office than he could have imagined. But unlike Reagan, it would not crowd out a lion's share of attention to his domestic agenda, at least in the early years. Even the serious time given to expanding foreign trade through open markets under treaties like the 1993 North American Free Trade Agreement (NAFTA), was geared to prosperity at home. Nonetheless, America's status as the preeminent world power drew Clinton into problems in the Balkans, Somalia, Haiti, Northern Ireland, the Middle East, and North Korea. Partisanship at home, and especially Republican efforts to impeach him over the Monica Lewinsky sex scandal in 1997–1998, jeopardized his ability to govern.[94]

Despite obstacles, distractions, and an emphasis on the economy, Clinton was interested in decreasing nuclear weapons beyond the historic reductions of the Reagan-Bush era. He also was agreeable to retaining the ABM Treaty. Continuing the momentum that had gathered by the end of Bush's term, he promised to spend much less on SDI and supported negotiated limits on anti-satellite weapons. During the presidential campaign, Clinton declared that he would give priority attention to proliferation of nuclear weapons (as well as chemical and biological weapons) and ballistic missiles, and that he supported a comprehensive test ban treaty. The former was in line with Bush, but the latter was not.[95]

When Clinton took office, all of the nuclear republics save Ukraine had approved the START I agreement. Diplomacy and additional pressure brought them to the peace table, and in January 1994 a trilateral agreement among Russia, Ukraine, and the United States led to a plan. Among other things, the sides agreed to remove nuclear weapons from Ukraine, to compensate it for transferring highly enriched uranium to Russia, and to provide the Ukraine with low enriched uranium for commercial reactors in return for its surrender of nuclear weapons. Ukraine soon thereafter agreed to START I and participation in NPT as a nonnuclear weapons state. By 1996, it was nuclear free. These actions improved the possibility for cooperation with Russia over START II. Despite the fact that passage in the US Senate initially looked promising, Foreign Relations Committee chair Helms and then others decided to play politics with ratification. After several compromises (none of which required Russian sanction) the Senate passed START II on January 26, 1996. Some senators, however, had not given up on missile defense. On the Russian side, ratification languished because of concerns over the cost of compliance, efforts to expand NATO membership, and political fights between Yeltsin and the parliament (still dominated by communists). In 2002, Russian President Vladimir Putin declared that his country was not bound by the agreement, after the United States withdrew from the ABM Treaty. The Strategic Offensive Reductions Treaty (SORT), signed in Moscow in that year, required the United States and Russia to reduce their deployed strategic weapons to 1,700–2,200 apiece by December 31, 2012. SORT effectively superseded START II. The United States and the Russian Federation did not sign the new START treaty until 2010.[96]

On the nagging problem of the ABM Treaty, Clinton did not move in the directions he announced during his first presidential campaign. The administration planned to continue an ABM research and development program, replacing the Republican Strategic Defense Initiative Organization (SDIO) with the Ballistic Missile Defense Organization (BMDO). But it was no longer a priority.[97] The new policy also shifted activity away from national ballistic missile defenses to theater missile defenses (TMD, a system to protect American forces deployed outside the United States), which grew out of the Gulf War's *Patriot-Scud* exchanges. Critics viewed the new approach as ineffective and a threat to the ABM Treaty. Clinton attempted to convince the Russians that there was a distinction between the essence of the ABM Treaty and the new TMD. While the Russians disagreed over technical specifications of the new systems, they believed Clinton was not moving beyond their general interpretation of the ABM Treaty. Clinton, however, had to appease the Republican-controlled Congress elected in 1994, and moved ahead with some parts of his theater missile defense program before seeking Russian acquiescence. Clinton and Yeltsin discussed problems over the velocity and range of the missiles (Russians wanted less, the United States wanting more) at a Helsinki summit in March 1997. The Russians essentially accepted the American position, while gaining some ground on limits on targets and space-based systems. Despite what appeared to be a successful negotiation, Clinton still faced Republican criticism. But he struck an agreement with Russia, Ukraine, Belarus, and Kazakhstan in 1997 establishing a "demarcation line" between TMD and ABM systems. Again, elements in the Russian Federation balked during the ratification process as did some in the Senate. The American abrogation of the ABM Treaty in 2002 unraveled much of this work.[98]

Making progress on missile defense proved difficult. Republicans in particular continued to hammer away at undermining the ABM Treaty and asserting the need for a new national missile defense system. The Defend America Act introduced in 1996 but not passed mandated deployment of such a system by 2003. At the least it was an effort to pick another political fight with the Democratic president and to serve as a campaign issue in the upcoming presidential election. That it failed to achieve any traction was a good indication that public interest in the issue was extremely weak. Few people seemed to feel that there was a sufficient threat to move toward this program, plus the cost would have been clearly untenable. In addition, the Russian nuclear arsenal was suffering from growing obsolescence, although still terribly dangerous in the wrong hands. Despite all this, the Senate Armed Services Committee approved the National Missile Defense Act of 1997, but it was tabled. A National Missile Defense Act passed the Senate in 1999, which was a step leading to the withdrawal of the United States from the ABM Treaty in 2002.[99]

A missile defense system was clearly a remnant of the Cold War, pushed forward because of long-standing momentum and bolstered by political motivations. Many justly considered the initial Reagan idea pie-in-the-sky (literally) in the 1980s. To some, the idea of a missile defense system, feasible or not, brought comfort in times of uncertainty. A missile defense was emblematic of the lingering fear of nuclear weapons and the inability of humans to submerge their feelings of vulnerability. Between 1983 and 1999 the United States spent $60 billion on ABM research, but no capable system was ever produced.[100]

TO PROLIFERATE OR NOT TO PROLIFERATE

Worry about terrorists and alleged rogue states drove concern over "horizontal proliferation" in the post-Cold War years, that is, the acquisition of nuclear weapons by other nations or groups outside the nuclear club. Non-proliferation, of course, had been debated since the 1940s. The determination of the United States to maintain its monopoly on nuclear weapons (and only begrudgingly to share it with allies) was broken by the detonation of the first Soviet atomic bomb in 1949. But the United States never gave up its efforts to contain the spread of weapons of mass destruction to other nations. The difficulty and complexity in developing the necessary technology to produce nuclear weapons probably did more than anything else to limit their spread in the Cold War years. Caution in sharing important elements of commercial nuclear power generation, as well as key requisites along the fuel cycle, also played a role. Ultimately, major pressure for a non-proliferation treaty came from both the Americans and the Soviets, who at the very least shared a common goal in limiting the emergence of new nuclear states. The NPT (1967) acknowledged those countries that possessed nuclear weapons, and drew the line on allowing others to emerge. As international relations expert Nina Tannenwald stated, "In short, the NPT constituted *de jure* recognition of *de facto* inequalities. In doing so, it reflected the view that the major threat to peace was horizontal proliferation: the more states that possessed nuclear weapons, the greater the likelihood that such weapons would be used."[101]

Try as they might, the superpowers were unable to completely shut off proliferation beyond those it deemed staunch allies. The United States welcomed the United Kingdom and France into the nuclear club quite early, although with some reservations. China also entered (1964), initially with Soviet help from 1958 to 1960, although they had grave concerns in providing such help. The new nuclear powers sought security by embracing a potent weapon, but also sought prestige, technological spinoffs, economic opportunity, and sometimes political advantage at home.[102] In the 1970s and 1980s, India, Pakistan, Israel, and South Africa were suspected of being the next entrants, especially since they had refused to join the NPT and were known to harbor a strong desire to develop nuclear weapons. Joining the list of aspirants in the 1980s were Iraq, Iran, North Korea, South Korea, and Taiwan. Like others before them, they had strong regional interests in seeking a place among nuclear states.[103] With the collapse of the Soviet Union, Belarus, Kazakhstan, and Ukraine, for a time at least, seemed to be opening the gates of proliferation still further.[104]

From the perspective of hindsight, the process of nuclear proliferation since 1945 has proved to be relatively slow and remarkably constrained.[105] Superpower competition during the Cold War made proliferation a threat as much for the contending parties themselves as for the world at large. Tensions between Israel and its Middle East enemies certainly were heightened by the former's desire to go nuclear as a means of protecting itself. The intensity of the India-Pakistan rivalry was aggravated by their decisions to become nuclear powers. Yet the United States and the USSR tended to view proliferation within their own definitions of security. In the post-Cold War world, therefore, proliferation took on a meaning relevant to the times, more than as an unchanging standard.

Nor was proliferation implemented unilaterally. Israel began its inquiry into developing nuclear weapons almost from its inception as a state in 1948. Between 1959 and 1965, the French provided expertise and constructed a reactor complex for Israel at Dimona for plutonium production and reprocessing. Seemingly France provided this service because of Israel's role in the Suez Crisis of 1956, or at the very least to weaken their mutual rival, Egypt. France also wanted Israeli expertise in heavy water. The United States probably learned of the existence of the Dimona facility in 1958, although the Israelis tried to keep it under wraps. In fact, the United States refused requests for sensitive nuclear assistance and attempted to curtail French efforts on Israel's behalf. American strategic interests in the Middle East superseded Israel's determination to acquire nuclear weapons; the action was perceived in the United States as too destabilizing. Israel's nuclear capability, therefore, was a constant topic in US-Israeli relations, and the Israelis used the covert program in bargaining with its American ally for conventional arms, among other things. The United States did virtually nothing to stop, contain, or discourage the development of nuclear weapons in Israel. By the early 1960s Israel conducted its nuclear program on its own, and when it felt threatened as during the 1967 Six-Day War and 1973 Yom Kippur War, considered using the weapon. Israel was the sixth nation and the first in the Middle East to acquire the bomb. While not acknowledging possession of it (and also refusing to sign the NPT), everyone was aware of the "bomb in the basement" since the 1970s (including its enemies). Israel, therefore, has been able to use nuclear weapons as a deterrent without creating a massive public debate or encouraging intrusive inspections.[106] The fact that Israel had the bomb inspired others in the Middle East to try to do the same. For the United States, the Israeli situation was important in influencing American nonproliferation policy in the 1960s. It was the first case since the UK, France, the USSR, and China, and yet significantly different in raising questions as to how the United States would apply the policy to friendly versus unfriendly nations.[107]

Regional concerns also drove the nuclear actions of India and Pakistan as much as anything. From the beginning of its nationhood in 1948 India devoted time to developing a civil nuclear industry as opposed to nuclear weapons. Official national policy by India's first Prime Minister Jawaharlal Nehru focused on peaceful uses of atomic energy. But India's defeat at the hands of the Chinese in the Chinese-Indian War (1962) brought a change of direction, especially with its subsequent detonation of a nuclear device in 1964 and its refusal to sign the NPT. Indian leaders insisted that the detonation of its first bomb was "a peaceful nuclear explosive," but the young nation now intended to pursue development of nuclear weapons. Internal debates (political and moral) slowed down the development of this full-scale commitment, but the Canadians gave it a boost with the construction of a plutonium-producing research reactor built at Rajasthan, and the Americans provided heavy water in 1974. (However, the United States soon curtailed shipments and India turned to the USSR for its heavy water.) It is clear that whatever the moral aversion to nuclear weapons surfacing in India, there were those who saw the necessity for a nuclear option because of the geopolitical realities of India's place in Asia. Pakistan's response was to develop its own device.[108]

As in the case of Israel, one state helped another to develop nuclear weapons in the subcontinent without the superpowers being able to do anything about it. Canada, at least indirectly, had helped India, as did the USSR. Responding to its major defeat in a war with India in 1971, Pakistan acquired technology outside of proper channels (some would say illegally) from Western Europe to build an enrichment facility. Between 1981 and 1986, China assisted Pakistan with a nuclear program, including uranium enrichment technology, weapons-grade uranium, and weapons designs. For both Pakistan and China, India was a rival, and thus the apparent mutual advantage of Pakistan becoming a nuclear state. (In turn, Pakistan distributed nuclear technology and materials to Iran, Libya, and North Korea.) Since Pakistan was a military ally of the United States, American officials looked the other way. Such decisions are based on policies seemingly beneficial to the assisting state as well as to the assisted.[109]

The end of the Cold War in the 1990s recharged the proliferation debate. National security studies professor John Mueller referred to the "cascadology" of various observers and officials in these years anxious about how the new day and uncertain world in the post-Cold War era would impact horizontal proliferation.[110] Mueller went on to suggest that this view of proliferation did not take into account the historic trends that indicated "a far more leisurely pace than generations of alarmists have routinely and urgently anticipated, but the diffusion that has actually transpired has proven to have had remarkably limited, perhaps even imperceptible consequences."[111] This thinking led to a focus on alleged rogue states and even misreading the intentions of existing nuclear powers, such as China. It had reacted to American bullying rather than taking a provocative step to directly challenge the two superpowers.[112] Another observer suggested that after many years in which the US military "lived in what amounted to a symbiotic relationship with the Soviet military," the end of the Cold War "provided an enormous shock."[113] The collection of real or imagined smaller threats to American security could be bundled up into "a large system-wide threat to global stability" or another approach would be to adopt a posture where threats were unknown. In the end, it became necessary to seek an identifiable enemy. Alleged rogue states, especially in the Third World, were the answer.[114] The conclusion that the United States searched for "the perfect enemy" is too simplistic. However, there is little doubt that a political/military vacuum created by the demise of the Soviet Union changed thinking about nonproliferation, turning attention to countries such as Iraq, Iran, and North Korea. Reassessing what would constitute American security in this new world was a serious problem, and nonproliferation played a role in seeking answers to that problem.

As the Cold War drew to its final days, the two superpowers were still occupied with "vertical proliferation" as well, that is, the continuing development and deployment of nuclear weapons by the existing nuclear states. The Reagan administration tried to limit or halt the spread of ballistic missiles by developing missile export controls along with the group of Seven Industrial Nations (G-7). In 1987 the Missile Technology Control Regime (MTCR) was formed. It was an informal arrangement that established guidelines to limit transfer of technology, expertise, or hardware necessary to build missiles capable of delivering nuclear warheads. The United States argued that its efforts to help the Israelis to develop the *Arrow* defensive missile were not a violation of MTCR. This splitting of hairs weakened the credibility of the program, and along with the

Reagan administration's reluctance to seek a comprehensive nuclear test ban treaty, revealed that efforts at horizontal proliferation were tepid at best through the 1980s.[115]

Clinton campaigned on eliminating nuclear, biological, and chemical weapons, but his more immediate interest was extending the Nuclear Nonproliferation Treaty due to expire in 1995. Non-weapon states wanted concessions before they signed. Egypt would sign, for example, if Israel signed. The lack of a comprehensive test ban treaty negotiated by the weapons states also complicated the renewal of NPT. But most importantly, the weapons states had to demonstrate that they would live up to their agreements to fulfill the treaty promises on disarmament. A variety of schemes and proposals came forth, everything from bringing Israel, India, and Pakistan into the treaty in some way, to expanding nuclear weapon-free zones beyond those in effect in Latin America and Southeast Asia (1998, established as well in Africa). As a move toward a CTBT, the Clinton administration set in place a moratorium on nuclear testing in 1995 to last until at least September 1996 (an expected date for a CTBT). Much more debate and discussion followed, but in May 1995, the NPT Review and Extension Conference voted to extend NPT indefinitely.[116]

The CTBT was another matter. Various components in an ultimate resolution of the issue, and the claims and conditions of numerous parties, extended the move toward resolution. The Clinton administration proposed a ban on the production of fissile material for use in nuclear weapons (which did not affect existing stockpiles). The Russians and British agreed; the French, Chinese, Israelis, Pakistanis, Indians, and others did not. Attention then turned to the CTBT, but with resistance from Congress. A raft of conditions, questions, and reservations followed from all quarters, including everything from tests that limited yield equivalents to inspection provisions. On September 24, 1996, a treaty was ready and Clinton became the first head of state to sign it. In the following several days, 94 nations signed, including Russia, China, France, Britain, and Israel. India and Pakistan refused. By 1998 149 countries agreed to the treaty, with 13 ratifying it.[117] The United States was one of those not to ratify at the time.

OMINOUS SIGNS INTO THE TWENTY-FIRST CENTURY

The logjam over NPT and the CTBT was broken in the 1990s. More ominous, however, were fears coming home to roost in the nuclear programs of alleged rogue state and peace-threatening disputes that gave critics of proliferation a strong dose of credibility in the late 1990s. A series of grave prospects emerged out of North Korea, Iraq, and along the Pakistan-India border. In the early 1990s, North Korea operated a reactor at Yongbyon, 60 miles from its capital, Pyongyang, capable of producing weapons-grade plutonium. It also built two additional large reactors and a reprocessing plant. Believing that the North Koreans had hidden some plutonium produced at Yongbyon, the International Atomic Energy Agency (IAEA) wanted immediate access to the reactor site. The North Koreans refused, and in a more provocative act announced in 1993 that it would withdraw from the NPT. With one hand, Clinton threatened the North Koreans, reinforced the military in South Korea, and sought economic sanctions in an effort to have the North Koreans terminate efforts to manufacture nuclear weapons. With the other, he offered a diplomatic solution, which they accepted in late 1993. It entailed suspending US-South Korean military exercises, permitting IAEA inspections, and more talks. But in early 1994 the North Koreans shut down the Yongbyon reactor to remove the spent fuel rods and ostensibly to reprocess the fissile material for possible weapons. This set off added tensions, which were more serious than previous. While the incident eased, relations between the two Koreas remained seriously strained, and the future actions of North Korea unpredictable. In 1998 it fired a missile over Japan, and sold missile technology to Syria, Iran, and Iraq. These acts clearly indicated that despite North Korea's continued denials and temporary concessions, it was the archetype rogue state many had feared.[118]

Iran soon emerged on the proliferation radar screen in the mid-1990s, although the Shah of Iran had been interested in building capability to produce nuclear weapons since the 1970s. Iran also had access to uranium. China and Russia provided nuclear technology to Iran as well. In 1995 word came that Iran secretly was developing a gas-centrifuge uranium-enrichment program, but the United States put pressure on suppliers from Europe and China to restrict the sale of necessary materials. In 1999 Russia signed an agreement with Iran to complete one or two reactors that the Germans had stopped constructing in 1979. In 1992 China agreed to supply Iran with two reactors. The building of the reactors was a long way from a nuclear weapons program. Yet given American wariness over North Korea and general strains in the Middle East, this action caused concern. Congress wanted to move against Russia and China for their roles in Iran, but Clinton simply resorted to communicating his displeasure to the Russians. Activities in Iran at this point were hardly on a par with North Korean developments, and the response appeared to Iranians as being an overreaction. At the same time, some concern was warranted insofar as the Iranians also had signed in 1990 an agreement with China over scientific cooperation and military technology. (The Russians also aided in the development of a ballistic missile program in China.) Congress reacted strongly by the late 1990s, and Clinton acquiesced by cutting off aid to some Russian enterprises allegedly selling weapons technology to Iran, Libya, and North Korea.

The Russians, in addition, were selling missile technology to Iraq and India. Iraq had interest in a nuclear program in the early 1970s. In November 1975 France agreed to supply it with two reactors, and they were built under secret construction near Baghdad. However, in summer 1981 Israeli planes attacked and destroyed them. At that point the French backed out of its agreement leaving Iraq with no program. Under Saddam Hussein, Iraq began building the necessary infrastructure for a new weapons program, particularly for enrichment technology. The Gulf War stymied the progress in 1991. Nevertheless, IAEA inspectors learned of Hussein's plans and called for inspections, but he continually resisted them, which seemed to further confirm the IAEA assessment.[119] In December 1998 under the code name *Operation Desert Fox*, Clinton ordered cruise missile attacks (along with air strikes from the United States and United Kingdom) against Iraqi factories assumed to be used to produce biological and nuclear weapons. Although curtailed after five days, the attacks did not end a debate about Iraq's potential threat as a developing nuclear state.[120]

The breakup of the Soviet Union had not ended the cross-purposes of US-Russian foreign policy. In addition, as the lone superpower in this era, the United States continued to play international policeman.[121] But clearly, the most serious nuclear tensions in the late 1990s related to the long-standing decaying relationship between India and Pakistan. On May 11, 1998, the Indians tested a thermonuclear weapon and two fission devices. They conducted two more tests two days later, which were the first since 1979. Prime Minister Atal Bihari Vajpayee stated that the detonation came about because of "security concerns," meaning a Pakistani test of a ballistic missile on April 6. The Indians were convinced that China was supplying the technology, although it was likely that North Korea was doing so. In turn, Pakistan conducted five nuclear tests on May 28 and another on May 30. In June, 47 members of the Conference on Disarmament (which included all of the weapons states) censured both India and Pakistan for undercutting international nonproliferation and called for the end of their nuclear weapons programs. The immediate response of India and Pakistan was predictable, but under pressure both sides agreed to move toward the CTBT if the other did so.[122]

CONCLUSION: THE FIRST NUCLEAR CENTURY ENDS

The century ended with the trauma of the Cold War over, and the United States in an apparently favored, if not unchallenged, position in the world. Americans certainly continued to revel in their nation's superpower status. But several observers were troubled by the instability in many nations, and few government officials had a clear notion as to how the United States would—or could—wield its extraordinary power. In addition, fears about nuclear weapons, proliferation, and the possibilities for a nuclear exchange had not evaporated. While hardly the last word, Tony Scott's 1995 thriller, *Crimson Tide*, and Phil Alden Robinson's 2002 *The Sum of All Fears* (based on the 1991 Tom Clancy best-seller) tried to sort out what might happen if nuclear weapons got in the wrong hands after the Cold War. In the former, the *USS Alabama* received an unconfirmed order to launch nuclear missiles after rebel forces in Russia seized control of some ICBMs. The young first officer on the sub (Denzel Washington) stages a mutiny to prevent the captain (Gene Hackman) from launching the missiles. The two men play out a conflict of choice set against the background of a new time and a new set of enemies. In *The Sum of All Fears* (movie and book) terrorists succeed in detonating a nuclear device in the United States. Such a scenario rarely contemplated before now seemed more plausible especially after the 9/11 attacks. Uncertainty about the possibilities of a nuclear detonation at home and the emergence of new enemies followed Americans into the twenty-first century.

Nonetheless, differences remained in the minds of Americans about the necessity of the United States continuing to embrace nuclear weapons. A good indicator of those opposing views was exemplified by the controversy over the proposed exhibit of the *Enola Gay* at the Smithsonian Institution in Washington D.C. in the mid-1990s. In July 1961 workers moved part of the aircraft that dropped first atomic bomb on Japan to the National Air Museum storage facility in Suitland, Maryland. Twenty-three years later the *Enola Gay* Restoration Association, led by veterans of the 509th bomber wing, began a campaign to restore it. In 1987 an advisory committee of the new National Air and Space Museum (part of the Smithsonian) met to discuss a possible exhibit built around the bomber. For the next several years the Smithsonian continued to explore the project with veterans urging restoration and also display. In January 1994, a first draft of a script for an exhibit was underway followed by many rounds of discussion and revisions by Smithsonian staff, professional historians, veterans, and others. The crux of the debate was two-fold: (1) a desire by veterans groups and their supporters to see the *Enola Gay* exhibited, as opposed to historians and Smithsonian staff wanting to use the aircraft as a artifact to introduce the public to a broader context about the end of the World War II; and (2) the "correct" interpretation of the decision to drop the bomb.

The first issue led to the second with veterans, the Air Force Association, the American Legion, conservative politicians, and others bent on portraying the bombing of Japan as the event that saved the lives of thousands of Americans and ended the war. On the other side, staffers at the Smithsonian and professional historians (several of them having written revisionist accounts of the events) hoped to illustrate the complexity of the ending of the war, the horrendous impact of the bombing on Japanese people, and the overstatement (in numbers at least) of saving American soldiers' lives. Recriminations were unbridled throughout 1994, with members of Congress weighing in and even questioning the loyalty of the promoters of the exhibit. The Smithsonian itself came under attack, and key officials left in the wake of the dispute. As a result, the Smithsonian mounted a small-scale exhibit of the *Enola Gay*'s fuselage and displayed videos of the crew with very limited context setting. This pleased no one. The details of the public fire storm have been recounted in numerous studies.[123]

The bottom line in this amazing controversy had much to do with the memory of the past (real or imagined) and the lens through which people view past events. To some, the issue had more to do with American patriotism and exceptionalism than with reasons for dropping the bomb. For others, especially participants in World War II, rationalizing the use of the bomb justified their own role in bringing the war to an end. For many scholars, it was a question of attempting to separate historical fact from historical mythmaking. At the core was the bomb itself, which was sometimes a bit player in the controversy, at other times the central force that drove the debate. Something as grandiose as a nuclear weapon, of course, could not be ignored. Using such a weapon, as opposed to simply possessing one, raised the intensity of the debate substantially. Its existence alone created enough emotion, uncertainty, and confusion, to supersede horror or apprehension. After more than 50 years, Americans (and everyone else for that matter) had not yet accommodated or accepted atomic power unreservedly into their lives.

MySearchLab Connections: Sources Online

READ AND REVIEW

Review this chapter by using the study aids and these related documents available on MySearchLab.

✓•⎡**Study** and **Review** on **mysearchlab.com**

Chapter Test

Essay Test

▭•⎡**Read** the **Document** on **mysearchlab.com**

Ronald Reagan, Remarks to Members of the National Press Club on Arms Reduction and Nuclear Weapons (1981)

Ronald Reagan, Address to the Nation on National Security (1983)

Ronald Reagan, Foreword to the Government Report on the SDI (1984)

RESEARCH AND EXPLORE

Use the databases available within MySearchLab to find additional primary and secondary sources on the topics within this chapter.

Endnotes

1. Beth A. Fischer, "Reagan and the Soviets: Winning the Cold War," in W. Elliot Brownlee and Hugh Davis Graham, eds., *The Reagan Presidency: Pragmatic Conservatism and Its Legacies* (Lawrence, KS: University Press of Kansas, 2003), 113–23.
2. Julian E. Zelizer, *Arsenal of Democracy: The Politics of National Security—From World War II to the War on Terrorism* (New York: Basic Books, 2010), 296–301; Melvyn P. Leffler, *For the Soul of Mankind: The United States, the Soviet Union and the Cold War* (New York: Hill and Wang, 2007),

338–402; George C. Herring, *From Colony to Superpower: U.S. Foreign Relations Since 1776* (New York: Oxford University Press, 2008), 861–63.

3. Like with all things Reagan, commentators and observers vary widely over the president's Cold War policy views. Career diplomat Jack F. Matlock, Jr. asserted that "in Reagan's mind his policy was consistent throughout. He wanted to reduce the threat of war, to convince the Soviet leaders that cooperation could serve the Soviet peoples better than confrontation, and to encourage openness and democracy in the Soviet Union." See *Reagan and Gorbachev: How the Cold War Ended* (New York: Random House, 2004), xiv. On the other hand, political scientist Beth A. Fischer argued that although the Reagan administration's "stated policy toward Moscow was especially hard-line through October 1983," it reversed course by the early in the following year. Seeking a rapprochement, Reagan "expressed a more nuanced understanding of the superpower relationship, and introduced new policy goals and strategies." He did this, she added, before the Soviets began their own internal reforms under Gorbachev. See *The Reagan Reversal: Foreign Policy and the End of the Cold War* (Columbia, MO: University of Missouri Press, 1997), 3–4.

4. For a thoughtful and interesting assessment of Reagan as president, see John Patrick Diggins, *Ronald Reagan: Fate, Freedom, and the Making of History* (New York: W.W. Norton & Co., 2007). The book is written by a professed liberal academic, who, in the course of writing the book, turned from critic to admirer of "one of the three great liberators in American history." Viewing Reagan's approach to the Cold War as "prudent," and one who "seized the opportunity for dialogue and negotiation" with Soviets, is vaguely reminiscent of some changing scholarly views on Dwight D. Eisenhower and his handling of Cold War issues. See xiii–xxii. Frances Fitzgerald, *Way Out There in the Blue: Reagan, Star Wars and the End of the Cold War* (New York: Simon & Schuster, 2000), which contrarily regarded Reagan as always caught between the dominant hawks in his administration and the minority pragmatists, "two warring factions within the administration" who "pursued separate and contradictory agendas and fought for control over policy" and who both "claimed Reagan as their own." Fitzgerald concluded that the president "never decided between them." See 15–18.

5. Herring, *From Colony to Superpower*, 861–67; Zelizer, *Arsenal of Democracy*, 301–03. See also Leffler, *For the Soul of Mankind*, 338–55.

6. See "Excerpts from a Speech to the National Association of Evangelicals, 1983," in Kevin Hillstrom, *The Cold War: Primary Source Series* (Detroit, MI: Omnigraphics, 2006), 409–10.

7. See Urban Dictionary, online, http://www.urbandictionary.com/define.php?term=Vietnam%20 Syndrome.

8. Zelizer, *Arsenal of Democracy*, 303–04. See also David E. Kyvig, "The Foreign Relations of the Reagan Administration," in David E. Kyvig, ed., *Reagan and the World* (New York: Praeger, 1990), 6–9; John Lewis Gaddis, "The Reagan Administration and Soviet-American Relations," in Kyvig, ed., *Reagan and the World*, 17–38; Leffler, *For the Soul of Mankind*, 346.

9. "President Ronald Reagan on Deterrence, November 23, 1982," in Robert C. Williams and Philip L. Cantelon, eds., *The American Atom: A Documentary History of Nuclear Policies from the Discovery of Fission to the Present, 1939–1984* (Philadelphia, PA: University of Pennsylvania Press, 1984), 234.

10. Definitions vary on "neoconservative" or "neocon," but here are two useful ones: "a former liberal espousing political conservatism," and more appropriately here—"a conservative who advocates the assertive promotion of democracy and United States national interest in international affairs including through military means." See Merriam-Webster, online, http://www.merriam-webster.com/ dictionary/neoconservative.

11. Zelizer, *Arsenal of Democracy*, 304–05; Ronald E. Powaski, *Return to Armageddon: The United States and the Nuclear Arms Race, 1918–1999* (New York: Oxford University Press, 2000), 15.

12. Zelizer, *Arsenal of Democracy*, 306; Diggins, *Ronald Reagan*, 195.

13. There remains a great deal of historical debate not over Reagan's hatred and fear of nuclear weapons, but over his intentions in substantially upping defense spending. Advocates, such as Paul Lettow, argued that Reagan intended "to intervene in and solve the nuclear dilemma" by leading "an arms race that he believed the Soviet Union could neither keep up with nor afford." In doing so the president hoped to force the Soviets to change their system from within. See *Ronald Reagan and His Quest*

to Abolish Nuclear Weapons (New York: Random House, 2005), ix. See also Matlock, Jr., *Reagan and Gorbachev*, 319–21; Chester J. Pach, Jr., "Sticking to His Guns: Reagan and National Security," in Brownlee and Graham, eds., *The Reagan Presidency*, 107–08. Richard Rhodes took a more modest position, that Reagan "at least understood that the Soviet system was vulnerable," unlike the Committee on the Present Danger that took a more strident position. See Richard Rhodes, *Arsenals of Folly: The Making of the Nuclear Arms Race* (New York: Alfred A. Knopf, 2007), 138.

14. For some, this was part of the so-called Reagan Doctrine, the apparent desire to pressure the USSR, build up US strength, and seek victories on key Cold War fronts. Specifically the doctrine favored covert activities to change the status quo. See Elizabeth Edwards Spalding, "The Origins and Meaning of Reagan's Cold War," in Paul Kengor and Peter Schweizer, *The Reagan Presidency: Assessing the Man and His Legacy* (Landham, MD: Rowman & Littlefiled Pub., 2005), 61–65; Herring, *From Colony to Superpower*, 865–66; 881–93.

15. Powaski, *Return to Armageddon*, 15–17; Zelizer, *Arsenal of Democracy*, 306–07; Herring, *From Colony to Superpower*, 867–68.

16. H.W. Brands, *The Devil We Knew: Americans and the Cold War* (New York: Oxford University Press, 1993), 174; Powaski, *Return to Armageddon*, 15–17; Zelizer, *Arsenal of Democracy*, 306–07.

17. Rhodes, *Arsenals of Folly*, 150.

18. Herring, *From Colony to Superpower*, 866–67.

19. Zelizer, *Arsenal of Democracy*, 308; Powaski, *Return to Armageddon*, 24–29.

20. Rodney P. Carlisle, ed., *Encyclopedia of the Atomic Age* (New York: Facts on File, 2001), 225; Stephen E. Atkins, ed., *Historical Encyclopedia of Atomic Energy* (Westport, CT: Greenwood Press), 265–66. For more detail on nuclear winter research, see Lawrence Badash, *A Nuclear Winter's Tale: Science and Politics in the 1980s* (Cambridge, MA: MIT Press, 2009).

21. Powaski, *Return to Armageddon*, 21–23.

22. Ibid., 23–26.

23. Quoted in Lawrence S. Wittner, *Toward Nuclear Abolition: A History of the World Nuclear Disarmament Movement, 1971 to the Present*, vol. 3 of *The Struggle Against the Bomb* (Stanford, CA: Stanford University Press, 2003), 177.

24. Atkins, ed., *Historical Encyclopedia of Atomic Energy*, 254–55.

25. Wittner, *Toward Nuclear Abolition*, 176.

26. Zelizer, *Arsenal of Democracy*, 311–14, 316, 322–23.

27. Quoted in Wittner, *Toward Nuclear Abolition*, 257–58.

28. Ibid., 264–65.

29. Zelizer, *Arsenal of Democracy*, 316–18; Nina Tannenwald, *The Nuclear Taboo: The United States and the Non-Use of Nuclear Weapons Since 1945* (New York: Cambridge University Press, 2007), 284–86.

30. Quoted in Lettow, *Ronald Reagan and His Quest to Abolish Nuclear Weapons*, 111. See also Leffler, *For the Soul of Mankind*, 354–56.

31. SDI was not a completely new idea, since the notion of a defense shield goes back into the 1950s and laser-like weapons (ray guns) even farther back into science fiction stories. In 1967 Edward Teller introduced Reagan to the idea when he was governor of California. After Richard Nixon and the US Congress abandoned the ABM program in 1972, conservative leaders, including Teller and Colorado beer magnate Joseph Coors, established an informal group lobbying for such a defense shield. The group met with Reagan regularly in his first year as president. See Paul Boyer, *Fallout: A Historian Reflects on America's Half-Century Encounter with Nuclear Weapons* (Columbus, OH: Ohio State University Press, 1998), 175–78; Fitzgerald, *Way Out There in the Blue*, 16; Carlisle, ed., *Encyclopedia of the Atomic Age*, 320–21; Zelizer, *Arsenal of Democracy*, 318–20; Bradley Graham, *Hit to Kill: the New Battle Over Shielding America from Missile Attack* (New York: Public Affairs, 2001), 13–17. See also Diggins, *Ronald Reagan*, 287–92.

32. Zelizer, *Arsenal of Democracy*, 300; Herring, *From Colony to Superpower*, 869–70.

33. John Mueller, *Atomic Obsession: Nuclear Alarmism from Hiroshima to Al-Qaeda* (New York: Oxford University Press, 2010), 79–80; Spalding, "The Origins and Meaning of Reagan's Cold War," 59; Geir Lundestad, "The United States and Western Europe Under Ronald Reagan," in David E. Kyvig, ed.,

Reagan and the World, 49; Fischer, "Reagan and the Soviets: Winning the Cold War," 127; Rhodes, *Arsenals of Folly*, 201–11; Jeff Smith, *Unthinking the Unthinkable: Nuclear Weapons and Western Culture* (Bloomington, IN: Indiana University Press, 1989), 102–109, 115–16.

34. In April 1984 Secretary of Defense Weinberger charted the Strategic Defense Initiative Organization (SDIO), made up of civilian and military shielding advocates, to manage the program, thus bypassing the Pentagon. See Graham, *Hit to Kill*, 15.

35. David E. Hoffman, *The Dead Hand: The Untold Story of the Cold War Arms Race and Its Dangerous Legacy* (New York: Doubleday, 2009), 219–20; Zelizer, *Arsenal of Democracy*, 319–23; Boyer, *Fallout*, 180.

36. Samuel F. Wells, Jr., "Nuclear Weapons and European Security during the Cold War," in Michael J. Hogan, ed., *The End of the Cold War: Its Meaning and Implications* (New York: Cambridge University Press, 1992), 67.

37. Michael Nacht, "The Politics: How Did We Get Here?" in Alexander T.J. Lennon, ed., *Contemporary Nuclear Debates: Missile Defense, Arms Control, and Arms Races in the Twenty-First Century* (Cambridge, MA: MIT Press, 2002), 5–6; Mueller, *Atomic Obsession*, 79; Zelizer, *Arsenal of Democracy*, 319–23; Wells, Jr., "Nuclear Weapons and European Security during the Cold War," 73; Martin Walker, *The Cold War: A History* (New York: Henry Holt and Company, 1993), 273–4; Brands, *The Devil We Knew*, 175–78.

38. Quoted in Hillstrom, *The Cold War*, 415.

39. Fischer, *The Reagan Reversal*, 123–40; Powaski, *Return to Armageddon*, 40–42.

40. Powaski, *Return to Armageddon*, 40–42; Herring, *From Colony to Superpower*, 884–93. Fischer, *The Reagan Reversal*, 123–40.

41. Rhodes, *Arsenals of Folly*, 163–70, 180–81; Herring, *From Colony to Superpower*, 863, 866, 868, 870–71; Zelizer, *Arsenal of Democracy*, 324–326; Powaski, *Return to Armageddon*, 18, 26–27; Fischer, *The Reagan Reversal*, 17–29, 115–22.

42. Both quotes from Leffler, *For the Soul of Mankind*, 359. See also Powaski, *Return to Armageddon*, 40.

43. There also were some British films and television programs as well, including the comedy series spoof, *Whoops! Apocalypse* (1982), *When the Wind Blows* (1986), and the drama/documentary *Threads*—a sort of British response to *The Day After*. Also significant were novels published in 1983 and 1984 on nuclear themes, William Prochnau, *Trinity's Child* (1983), Frederick Forsyth, *Fourth Protocol* (1984), and Tom Clancy, *Hunt for Red October* (1984). See Kim Newman, *Apocalypse Movies: End of the World Cinema* (New York: St. Martin's Griffin, 1998), 234–41; Boyer, *Fallout*, 200–01.

44. Newman, *Apocalypse Movies*, 240.

45. Joyce A. Evans, *Celluloid Mushroom Clouds: Hollywood and the Atomic Bomb* (Boulder, CO: Westview Press, 1998), 174–75; Ronnie D. Lipschutz, *Cold War Fantasies: Film, Fiction, and Foreign Policy* (Lanham, MD: Rowman and Littlefield Pubs., 2001), 79, 95–96; Newman, *Apocalypse Movies*, 231–41; Jerome F. Shapiro, *Atomic Bomb Cinema* (New York: Routledge, 2002), 183–91, 210–11.

46. Smith, *Unthinking the Unthinkable*, 126–27.

47. "Phony War" (1939–40) is the mocking term created by journalists to describe the six-month period (October 1939–March 1940) in early World War II during which there were no land operations by the Allies or the Germans after the Nazi conquest of Poland in September 1939. See *Encyclopedia Britannica*, online, http://www.britannica.com/EBchecked/topic/457343/Phony-War. Playing on the German "blitzkrieg," or "lightning war," the Phony War also was referred to as the "sitskrieg."

48. Leonid Brezhnev died in 1982, to be replaced by Yuri Andropov, who died in 1984, and then replaced by Konstantin Cherenko, who died in 1985, to then be replaced by Gorbachev.

49. Hoffman, *The Dead Hand*, 155–63; Powaski, *Return to Armageddon*, 42–43.

50. Quoted in Lettow, *Ronald Reagan and His Quest to Abolish Nuclear Weapons*, 137.

51. Herring, *From Colony to Superpower*, 894–95. Lettow's take (a little more assertive than the president's role as pragmatist) was that Reagan was a "skillful wielder of power" who increasingly "exerted personal control over his administration's national security policy-making process..." See Lettow, *Ronald Reagan and His Quest to Abolish Nuclear Weapons*, 244, 247.

52. Powaski, *Return to Armageddon*, 45–46; Rhodes, *Arsenals of Folly*, 183.

53. Quoted in Lou Cannon, *President Reagan: The Role of a Lifetime* (New York: Simon & Schuster, 1991), 740. See also Rhodes, *Arsenals of Folly*, 187–93.

54. Vladislav M. Zubok, *A Failed Empire: The Soviet Union in the Cold War from Stalin to Gorbachev* (Chapel Hill, NC: University of North Carolina Press, 2007), 311; Herring, *From Colony to Superpower*, 894–95; Zelizer, *Arsenal of Democracy*, 333; Walker, *The Cold War*, 278–86; Brands, *The Devil We Knew*, 192–97; Rhodes, *Arsenals of Folly*, 187–88; Raymond L. Garthoff, "Why Did the Cold War Arise, and Why Did It End?" in Hogan, ed., *The End of the Cold War*, 129–31; Hoffman, *The Dead Hand*, 164–88.

55. Powaski, *Return to Armageddon*, 35–38.

56. Ibid., 47–48; Rhodes, *Arsenals of Folly*, 193–96; Zubok, *A Failed Empire*, 306–07.

57. Under the ABM Treaty, each country was permitted to have one ABM site. In 1975 the US abandoned its site, but the Soviets continued to maintain one near Moscow. This difference surely influenced the negotiations.

58. Quoted in Powaski, *Return to Armageddon*, 55. See also 48–54.

59. Reeves, *President Reagan*, 333.

60. Walker, *The Cold War*, 292–95; Powaski, *Return to Armageddon*, 57–61; Diggins, *Ronald Reagan*, 381–82; Hoffman, *The Dead Hand*, 235–95. While little of substance developed in the intervening months on arms control, Reagan and Gorbachev had conducted secret private correspondence over a range of topics.

61. Kyvig, "The Foreign Relations of the Reagan Administration," 11.

62. Graham, *Hit to Kill*, 17; Powaski, *Return to Armageddon*, 63–73.

63. Brands, *The Devil We Knew*, 201–02; Herring, *From Colony to Superpower*, 865–66, 878–79, 891, 897, 915; Zelizer, *Arsenal of Democracy*, 326–27, 334–38; "Iran-Contra Affair," *The Free Dictionary*, online, http://legal-dictionary.thefreedictionary.com/Iran-Contra+Affair.

64. Lundestad, "The United States and Western Europe Under Ronald Reagan," 53; Walker, *The Cold War*, 295.

65. Atkins, ed., *Historical Encyclopedia of Atomic Energy*, 174. Powaski, *Return to Armageddon*, 63–73, 79–82; Garthoff, "Why Did the Cold War Arise, and Why Did It End?" 130.

66. Jonathan Stevenson, *Thinking Beyond the Unthinkable: Harnessing Doom from the Cold War to the Age of Terror* (New York: Viking, 2008), 194. See also Steven Hurst, *The Foreign Policy of the Bush Administration: In Search of a New World Order* (London: Cassell, 1999), 1–25; Zelizer, *Arsenal of Democracy*, 355–56; Herbert Parmet, *George Bush: The Life of a Lone Star Yankee* (New York: Scribner, 1997); Herring, *From Colony to Superpower*, 899–903.

67. Quoted in Leffler, *For the Soul of Mankind*, 424.

68. Graham, *Hit to Kill*, 17–18; Leffler, *For the Soul of Mankind*, 424–25; Powaski, *Return to Armageddon*, 83–88; Rhodes, *Arsenals of Folly*, 283–89.

69. Powaski, *Return to Armageddon*, 88–93.

70. Walker, *The Cold War*, 302–23; Powaski, *Return to Armageddon*, 93–94.

71. Diggins, *Ronald Reagan*, 388; Zelizer, *Arsenal of Democracy*, 350; Herring, *From Colony to Superpower*, 904–06.

72. Quoted in Hillstrom, *The Cold War*, 452.

73. Powaski, *Return to Armageddon*, 93–109; Hurst, *The Foreign Policy of the Bush Administration*, 43–49.

74. Powaski, *Return to Armageddon*, 109–11; Herring, *From Colony to Superpower*, 908–12. See also Tannenwald, *The Nuclear Taboo*, 294–326, on why the United States did not consider using tactical nuclear weapons in the Gulf War.

75. Powaski, *Return to Armageddon*, 111–15.

76. By 1992 the United States had less than 12,778 strategic warheads on 1,876 launchers; the Soviets had 10,880 and 2,354 respectively.

77. Powaski, *Return to Armageddon*, 115–23.

78. Graham, *Hit to Kill*, 21–22; Powaski, *Return to Armageddon*, 125–27.

79. Rhodes, *Arsenals of Folly*, 290–96; Powaski, *Return to Armageddon*, 128–30.

80. Powaski, *Return to Armageddon*, 130–39.

81. Quoted in Hillstrom, *The Cold War*, 465.

82. Carlisle, ed., *Encyclopedia of the Atomic Age*, 317–18; Powaski, *Return to Armageddon*, 140–44.

83. Carlisle, ed., *Encyclopedia of the Atomic Age*, 318–19.

84. Powaski, *Return to Armageddon*, 144–54, 163–64; Atkins, ed., *Historical Encyclopedia of Atomic Energy*, 348.

85. Wittner, *Toward Nuclear Abolition*, 112–29; Thomas R. Rochon, *Mobilizing for Peace: The Antinuclear Movement in Western Europe* (Princeton, NJ: Princeton University Press, 1988); Christian Joppke, *Mobilizing Against Nuclear Energy: A Comparison of Germany and the United States* (Berkeley, CA: University of California Press, 1993), 144–89.

86. Wittner, *Toward Nuclear Abolition*, 130.

87. Ibid., 130–68, 202–52.

88. Ibid., 169–201.

89. Ibid., 253–88, 335–68, 405–23; Tannenwald, *The Nuclear Taboo*, 349–59. See Also Andrew Rojecki, *Silencing the Opposition: Antinuclear Movements and the Media in the Cold War* (Urbana, IL: University of Illinois Press, 1999.

90. Tannenwald, *The Nuclear Taboo*, 329–30.

91. Herring, *From Colony to Superpower*, 917–22.

92. Tannenwald, *The Nuclear Taboo*, 328–33.

93. Herring, *From Colony to Superpower*, 925.

94. Ibid., 927–37; Zelizer, *Arsenal of Democracy*, 386–430.

95. Powaski, *Return to Armageddon*, 165–66.

96. Ibid., 165–73; "START II and Its Extension Protocol at a Glance," Arms Control Association, online, http://www.armscontrol.org/factsheets/start2; "Obama, Medvedev Sign 'New START' Treaty," *The Washington Times*, April 8, 2010, online http://www.washingtontimes.com/news/2010/apr/08/obama-medvedev-sign-treaty-cut-nuclear-arms/.

97. Fitzgerald, *Way Out There in the Blue*, 491.

98. Ibid., 491–98; Powaski, *Return to Armageddon*,173–203.

99. Powaski, *Return to Armageddon*, 173–203; Fitzgerald, *Way Out There in the Blue*, 491–98; Greg Thielman, "The National Missile Defense Act of 1999," Arms Control Association, online, http://www.armscontrol.org/act/2009_07-08/lookingback.

100. Fitzgerald, *Way Out There in the Blue*, 498. See also Graham, *Hit to Kill*, 73–75, 114–20.

101. Tannenwald, *The Nuclear Taboo*, 335.

102. See Joseph Cirincione, *Bomb Scare: The History & Future of Nuclear Weapons* (New York: Columbia University Press, 2007), 47; Matthew Kroenig, *Exporting the Bomb: Technology Transfer and the Spread of Nuclear Weapons* (Ithaca, NY: Cornell University Press, 2010), 111.

103. The South African nuclear program always was shrouded in mystery, especially before the end of apartheid. That country had the advantage of uranium supplies, technical personnel, and close scientific ties with Israel and the United States to develop such a capability. South Africans built a secret complex in 1974 near Pretoria to conduct nuclear research. A Soviet spy satellite discovered preparation of a weapon's test in 1977, and announced its findings to the nuclear powers that put pressure on South Africa to desist. However, in 1979, a US spy satellite detected a bomb test in the region, but not officially linked to South Africa. By 1989, South Africa had produced six atomic bombs, largely meant to keep the apartheid government in power. By 1991, however, all the bombs were dismantled after the F.W. De Klerk government decided to reverse itself, sign the NPT, and become a nonnuclear state. Atkins, ed., *Historical Encyclopedia of Atomic Energy*, 336–37; Carlisle, ed., *Encyclopedia of the Atomic Age*, 311–13.

104. Saira Khan, *Nuclear Weapons and Conflict Transformation: The Case of India-Pakistan* (London: Routledge, 2009), 1; Kroenig, *Exporting the Bomb*, 160.

105. Mueller, *Atomic Obsession*, 89.

106. Warner D. Farr, "The Third Temple's Holy of Holies: Israel's Nuclear Weapons," The Counterproliferation Papers, Future Warfare Series No. 2, USAF Counterproliferation Center, Air War College, Air University, Maxwell Air Force Base, Alabama, September 1999, 3, online, http://www.au.af.mil/au/awc/awcgate/cpc-pubs/farr.htm; Kroenig, *Exporting the Bomb*, 67–74, 91; Powaski, *Return to Armageddon*, 76–77; Carlisle, ed., *Encyclopedia of the Atomic Age*, 152–53.

107. Avner Cohen, *Israel and the Bomb* (New York: Columbia University Press, 1998), 1–5; Michael Karpin, *The Bomb in the Basement: How Israel Went Nuclear and What That Means for the World* (New York: Simon & Schuster, 2006), 1–3; Atkins, ed., *Historical Encyclopedia of Atomic Energy*, 180–82.

108. George Perkovich, *India's Nuclear Bomb: the Impact on Global Proliferation* (Berkeley, CA: University of California Press, 1999), 3–4, 444. See also Atkins, ed., *Historical Encyclopedia of Atomic Energy*, 170–71; Powaski, *Return to Armageddon*, 73–74.

109. Kroenig, *Exporting the Bomb*, 1–3, 112; Powaski, *Return to Armageddon*, 74–76.

110. Mueller, *Atomic Obsession*, 91–95.

111. Ibid., 95.Not all observers agree with Mueller. See Richard Rhodes, *The Twilight of the Bombs: Recent Challenges, New Dangers, and the Prospects for a World Without Nuclear Weapons* (New York: Alfred A. Knopf, 2010), 288.

112. Mueller, *Atomic Obsession*, 95–97. Mueller also suggested that efforts to prevent horizontal proliferation have been unsuccessful, and in fact proliferation may be, among other things, an active deterrent in limiting nuclear war. See 115–27.

113. Michael Klare, *Rogue States and Nuclear Outlaws: America's Search for a New Foreign Policy* (New York: Hill and Wang, 1995), 5–6.

114. Ibid., 12–16.

115. Dinshaw Mistry, *Containing Missile Proliferation: Strategic Technology, Security Regimes, and International Cooperation in Arms Control* (Seattle, WA: University of Washington Press, 2003), 3–11; Powaski, *Return to Armageddon*, 78–79.

116. Powaski, *Return to Armageddon*, 205–07; Tannenwald, *The Nuclear Taboo*, 343–46; Atkins, ed., *Historical Encyclopedia of Atomic Energy*, 255–56.

117. Powaski, *Return to Armageddon*, 205–23; Zelizer, *Arsenal of Democracy*, 426; Atkins, ed., *Historical Encyclopedia of Atomic Energy*, 97–98.

118. Powaski, *Return to Armageddon*, 223–27; Graham, *Hit to Kill*, 52–56, 336–40.

119. There also was concern about Iraq developing chemical and biological weapons before the Gulf War, which proved true. See Powaski, *Return to Armageddon*, 77.

120. Zelizer, *Arsenal of Democracy*, 420–22; Powaski, *Return to Armageddon*, 227–33, 237–38; "Saddam's Nuclear Secrets," *Newsweek*, October 7, 1991, 28, 33–35; Atkins, ed., *Historical Encyclopedia of Atomic Energy*, 179–80; Carlisle, ed., *Encyclopedia of the Atomic Age*, 150–51. For more on Clinton's surprising aggressiveness in foreign affairs, see Andrew Butfoy, "The Rise and Fall of Missile Diplomacy? President Clinton and the 'Revolution in Military Affairs' in Retrospect," *Australian Journal of Politics and History* 52 (2006), 98–114.

121. By the late 1990s, Russia was considered by some to be a major source of fissile materials either by illegal purchase, smuggling, or theft. See Powaski, *Return to Armageddon*, 239, 242–45; Atkins, ed., *Historical Encyclopedia of Atomic Energy*, 260–61.

122. Sumit Ganguly and S. Paul Kapur, *India, Pakistan, and the Bomb: Debating Nuclear Stability in South Asia* (New York: Columbia University Press, 2010), 1–2, 17–19; Powaski, *Return to Armageddon*, 233–37; Karsten Frey, *India's Nuclear Bomb and National Security* (London: Routledge, 2006), 1; Perkovich, *India's Nuclear Bomb*, 3–4. See also E. Sridharan, ed., *The India-Pakistan Nuclear Relationship: theories of Deterrence and International Relations* (London: Routledge, 2007).

123. Museum director, Dr. Martin Harwit, recounts his take on the story in *An Exhibit Denied: Lobbying the History of Enola Gay* (New York: Springer-Verlag, 1996). The Air Force Association produced

four thick volumes in 1994–95 laying out the museum script, bibliography, and a wide range of documents that they provided free of charge upon request. Other studies tend to be fairly polemical, with some focusing specifically on the controversy and others taking a wider view. See, for example, Robert P. Newman, *Enola Gay and the Court of History* (New York: Peter Lang, 2004); Philip Nobile, ed., *Judgment at the Smithsonian: The Bombing of Hiroshima and Nagasaki* (New York: Marlowe & Company, 1995); Charles T. O'Reilly and William A. Rooney, *The Enola Gay and the Smithsonian Institution* (Jefferson, NC: McFarland & Co., 2005); Steven C. Dubin, *Displays of Power: Memory and Amnesia in the American Museum* (New York: New York University Press, 1999), 186–226; William L. O'Neill, *A Bubble in Time: America During the Interwar Years, 1989-2001* (Chicago: Ivan R. Dee, 2009), 84–93; Boyer, *Fallout*, 246–68.

CHAPTER **11**

Proliferation, Terrorism, and Climate Change
The Atom in the Twenty-First Century

INTRODUCTION: A NEW MILLENNIUM

World War II, the Cold War, the environmental movement, and the energy crisis framed the debate over atomic energy in the second half of the twentieth century. Nuclear proliferation, international terrorism, and controversy over climate change were central to understanding atomic energy in the early twenty-first. There was no complete break with the past, but it is significant how nuclear weapons and nuclear power continued to be enmeshed in some of the most controversial issues of the time, and in some cases intersect each other.

Few events were such an emotional agent of change for the United States as 9/11. But it too connected in some important ways to the new phase of Atomic Age America. How well Americans remember that in the early morning of September 11, 2001, al Qaeda terrorists attacked New York's World Trade Center and the Pentagon in Washington D.C. As stated in the report of the 9/11 Commission:

> An airliner traveling at hundreds of miles per hour and carrying some 10,000 gallons of jet fuel plowed into the North Tower of the World Trade Center in Lower Manhattan. At 9:03, a second airliner hit the South Tower. Fire and smoke billowed upward. Steel, glass, ash, and bodies fell below. The Twin Towers, where up to 50,000 people worked each day, both collapsed less than 90 minutes later.
>
> At 9:37 that same morning, a third airliner slammed into the western face of the Pentagon. At 10:03, a fourth airliner crashed in a field in southern Pennsylvania. It had been aimed at the United States Capitol or the White House, and was forced down by heroic passengers armed with the knowledge that America was under attack.[1]

Smoke and flames rise from the Twin Towers in lower Manhattan after commercial airplanes hijacked by terrorists were flown directly into each tower on September 11, 2001. The United States' "war on terror" which followed the 9/11 attacks ultimately conflated terrorism and nuclear weapons, particularly in the case of Iraq. *Source:* Stacy Walsh Rosenstock/Alamy

More than 2,600 people died at the World Trade Center, with another 125 at the Pentagon and 256 more on the four planes. This was the first major assault of its kind in the continental United States, and the death toll exceeded that of Pearl Harbor in December 1941. Shock. Disbelief. Anxiety. Vulnerability. Deep Sorrow. Contempt for a new set of villains. These were the sentiments manifest on that horrible day. As the 9/11 Commission stated in the preface to its report, "We have come together with a unity of purpose because our nation demands it. September 11, 2001, was a day of unprecedented shock and suffering in the history of the United States. The nation was unprepared. How did this happen, and how can we avoid such tragedy again?"[2] Such concerns lingered as American attempted to cope with the realities of a new time and a new world.

The Cold War conditioned the world to fear A-bombs and H-bombs. And while that fear did not fade completely, the American enemy in the post-Cold War world was not the USSR but terrorists or alleged rogue states using any means possible to disrupt and destroy their adversaries. The assailants of 9/11 were non-state actors using hijacked commercial jetliners rather than weapons of mass destruction for their strike. Non-conventional weapons were used on this occasion, but that did not rule out the use of nuclear devices by terrorists in the future.[3] Where then, in this new world context, did nuclear weapons fit? In reality, nobody knows. If terrorists were capable of crashing jets into skyscrapers, were they also capable of detonating nuclear weapons or could another group or another country set off a nuclear exchange for any number of reasons? The unsettling nature of 9/11 could not help but generate a whole raft of worst-case scenarios. The continued existence of weapons of mass destruction (WMD), which included biological and chemical as well as nuclear weapons, and the capability to produce them had to be matched with the willingness to use them. Was that a lesson to take away from 9/11? Some thought so. In some respects, the focus on terrorists and other real and imagined enemies tended to distract attention from the fact that after the Cold War, about 40 nations had the capability to build nuclear weapons, but until then only nine—the United States, USSR, UK, France, China, Israel,

India, Pakistan, and South Africa—had actually done so. (South Africa dismantled their bombs in 1991.) Collectively this represented approximately 60,000 nuclear devices.[4] The reality was not going to change soon. Proliferation only added to the existing instability, but did not completely recast it.

A less immediate threat, but one with serious (even dire) consequences was grave concern among scientists that the chronic polluting ways of humans contributed directly to global climate change, the results of which were melting glaciers, rising seas, extinction of species, and radical fluctuations in weather patterns. Deniers were quick to dismiss the evidence, to rationalize that what scientists were observing was just another natural cycle of warming and cooling in which humans played an inconsequential part. This response made for great political speeches and rationalization to focus on the here and now, to avoid over-regulating energy producers and users, and to place personal comfort above uncomfortable realities. Here too, atomic energy was part of the discourse. In this case, pro-nuclear forces began to tout the value of an energy source that was carbon free, and friendly to the environment. Could nuclear power's change of image reverse the industry's fortunes? And better yet, could commercial nuclear power survive the latest catastrophe, Fukushima? Atomic energy was a key element in both twenty-first-century security issues and the environmental debate. To be sure, its role was different from what it had been in the previous century, but nuclear weapons and nuclear power adjusted with the times. Their long-standing roles bound by hope, promise, and risk seemed to be upstaged by persistence.

THE BUSH ADMINISTRATION AND NUCLEAR POLICY

The George W. Bush administration rejected the foreign policy of its Democratic predecessor, the personal diplomacy of Ronald Reagan, and the pragmatic and often restrained approach of George H.W. Bush. The primary shapers of the new policy were an accumulation of veteran advisers and officials from past Republican governments. They were called "retreads" by some critics, but self-proclaimed "Vulcans," evocative of the Roman god of fire. According to James Mann, former diplomatic correspondent and foreign affairs columnist, "That word, *Vulcan*, captured perfectly the image the Bush foreign policy team sought to convey, a sense of power, toughness, resilience and durability."[5] The group included Vice President Richard (Dick) Cheney, Secretary of Defense Donald Rumsfeld, National Security Adviser Condoleezza Rice, and Secretary of State Colin Powell, plus second-level actors such as Paul Wolfowitz, Richard Perle, and Lewis "Scooter" Libby. Rice and Powell did not fit the neoconservative mold of the others, but participated nevertheless. The president himself, a former Texas governor, businessman, Yale and Harvard graduate, and son of a president, depended heavily on his advisers and basically shared their views. Bush had no foreign policy experience, trusted his advisers, and became particularly close to Rice. But the real power, especially in his first term in office, rested with Cheney and Rumsfeld.[6]

From the 2000 campaign forward, the Bush team's goal was to completely reverse the course of late Cold War and Vietnam policy. Many of the advisers supported conservative internationalism, regime change, a rebuff of multilateral treaties, faith in military solutions, and no compromises. Cheney also insisted that congressional Republicans maintain strict party loyalty to executive leadership, and had no interest in appealing to Democrats. Partisanship ruled the day.[7] There was no intention of providing continuity or even restoring policies of previous Republican administrations. The Vulcans were willing to go to war with hostile powers, even through preemptive attacks. Mann captured the essence well: "The vision was that of an unchallengeable America, a United States whose military power was so awesome that it no longer needed to make compromises or accommodations (unless it chose to do so) with any other nation or group of countries."[8] Others have referred to this freedom to act and to change the status quo in the world as "America Unbound." The Bush administration became particularly emboldened in this regard after the events of September 11, 2001.[9]

Nuclear policy fell into the same category as every other foreign policy decision. The administration saw no reason to follow precedent.[10] It reinvigorated the idea of a nuclear missile defense shield (at a cost of $80-$120 billion over 25 years) even though the program openly violated the ABM Treaty (1972). President Bush had been fascinated with the idea of missile defense as early as the 1980s. This particular iteration was meant to be a first line of defense not against Russia, but against nations such as North Korea and Iraq.[11] In May 2001, Bush gave a major address on missile defense in which he stated that the ABM Treaty "does not recognize the present, or point us to the future. It enshrines the past. No treaty that prevents us from pursuing promising technology to defend ourselves, our friends and our allies is in our interests or in the interests of world peace."[12] In December he announced that the United States would withdraw from the ABM Treaty. America, he argued, would pursue a policy of "counter-proliferation," which promoted missile defense and strikes against countries attempting to produce weapons in secret.[13]

In the past, the United States had stretched and contracted the intent of this long- standing agreement, but it had not wholly rejected it until now. In the case of the Comprehensive Test Ban Treaty (CTBT) that President Bill Clinton had signed in 1996, it remained unratified by the Senate when Bush entered office.[14] Some administration advisers did not favor any arms control treaties, and the president himself initially did not believe that there was any reason to negotiate a treaty with Russia. Bush asserted that he would set arsenal limits based on a unilateral appraisal of security requirements. But Russia and members of Congress urged the president to reverse his stance and to accept a legally binding treaty with new limits on nuclear weapons. He signed the very brief and sketchy Moscow Treaty in 2002 (a 2 ½ page document) which was not a comprehensive arms control pact, included no verification process, no definition of what weapons would be reduced, and no short-term reduction goals. The accord mentioned a limit of 2,200 deployed warheads, but required no destruction of those above the limit.[15] By January 2009, the Bush administration reduced substantially the number of active warheads for the United States from 10,000 to 2,600, and the Russians reduced their active arsenal from 15,000 to about 4,800. On the surface, the reductions appeared to be a step closer to disarmament. Both countries, however, maintained extensive reserves at about 22,400 warheads combined, which represented 96 percent of the total in the world. They also kept a large number of their weapons on "high alert" in case there was a need to retaliate quickly. Despite a world where major conflict between the United States and any other nuclear weapons state was currently absent or not foreseeable, the old Cold War adversaries had not disavowed a far-reaching nuclear force.[16]

9/11, TERRORISM, AND WMD

The United States' "war on terror" which followed the 9/11 attacks ultimately conflated terrorism and nuclear weapons, particularly in the case of Iraq. There were, of course, crank claims that "mini-nukes" had been used to destroy the Twin Towers. But such theories had no credibility with the policy makers who took a decidedly different view of potential threats to American security, threats that included nuclear devices.[17] The immediate reaction to the attacks focused on the events themselves, Osama bin Laden, and his cohorts. The Bush administration had little time to weigh the numerous intelligence failures over recent years that led to missed key signals of potential terrorist acts against the United States. They largely avoided playing the political blame game of who should be held accountable besides the terrorists themselves. Nor did they even speculate much about future threats.[18]

A previously untested president gave a bold address to a joint session of Congress on September 11, bringing Americans together for an assault against the perpetrators of these heinous deeds. He declared,

> These acts of mass murder were intended to frighten our nation into chaos and retreat. But they have failed. Our country is strong. A great people has been moved to defend a great nation.
>
> Terrorist attacks can shake the foundations of our biggest buildings, but they cannot touch the foundation of America. These acts shatter steel, but they cannot dent the steel of American resolve.
>
> America was targeted for attack because we're the brightest beacon for freedom and opportunity in the world. And no one will keep that light from shining.[19]

The most obvious immediate US targets were al Qaeda bases and the strongholds of their local confederates, the Taliban in Afghanistan, with hopes of destroying the terrorist network and capturing the illusive Osama bin Laden. Air power and local forces crippled al Qaeda operations and severely weakened the Taliban within four months. The lack of a large number of American ground forces, however, made it easier for bin Laden and other leaders to evade capture. Nevertheless, the primary military response to 9/11 at the moment was logically focused on those who had meant to undermine US confidence and will through direct assaults.[20]

Responding to the 9/11 attacks, neoconservatives in the Bush administration had come to believe that national security policies in the 1990s were a prime culprit. (They took no responsibility on themselves or their political allies prior to and including 9/11 for national security breakdowns.) They argued that President Clinton had done too little to deal with alleged rogue states while he was in office, and that Congress placed too many constraints on executive authority immediately after the attacks. These problem states (especially Afghanistan and Iraq), they added, were responsible for funding, organizing, and harboring terrorists. Al Qaeda camps in Afghanistan were well-known, but the administration took the word of Iraqi defectors (without verification) that al Qaeda terrorists were being trained in Iraq to carry out actions like those on September 11. Not only neoconservative advisors but also Senator John McCain (Rep., Arizona) stated that terrorist networks could be found in Iraq, Iran, and Syria.[21]

The Bush administration soon shifted its emphasis from "stateless terrorism" to a focus on hostile nation-states as the greatest threats to American security in the wake of 9/11. The global war on terror now was directly connected to nuclear proliferation.[22] Iraq, even more than Afghanistan, was the prime example of this shift. As Paul Boyer argued, "In the buildup to the Iraq war, the administration deliberately played the nuclear card," and thus "the task of focusing public attention on *actual* nuclear dangers grew more difficult."[23] During the State of the Union address on January 29, 2002, Bush referred to the security of the United States threatened by attacks on our nuclear power facilities and the potential use of atomic weapons. "Our discoveries in Afghanistan," he stated, "confirmed our worst fears, and showed us the true scope of the task ahead...We have found diagrams of American nuclear power plants and public water facilities, detailed instructions for making chemical weapons, surveillance maps of American cities, and thorough descriptions of landmarks in America and throughout the world." He also referred to North Korea, Iran, and Iraq by name, asserting "States like these, and their terrorist allies, constitute an axis of evil, arming to threaten the peace of the world. By seeking weapons of mass destruction, these regimes pose a grave and growing danger." [24]

The speech triggered considerable discussion, with many Americans willing to believe the worst about the alleged rogue states and many Europeans concerned about the potential direction of American foreign policy.[25] Republicans released the *National Security Strategy* in September 2002 (before the November elections) building upon the State of the Union and emphasizing the need for preemptive action in cases where American security was threatened. In part, it stated,

> Defending our Nation against its enemies is the first and fundamental commitment of the Federal Government. Today, that task has changed dramatically. Enemies in the past needed great armies and great industrial capabilities to endanger America. Now, shadowy networks of individuals can bring great chaos and suffering to our shores for less than it costs to purchase a single tank. Terrorists are organized to penetrate open societies and to turn the power of modern technologies against us.[26]

It also stated the value in implanting democratic values in places where they were needed for political stability and where they would encourage freedom. The Bush Doctrine now included all of its essential elements.[27] Conservatives felt the doctrine embodied great principles, while critics often regarded it as arrogant and a form of adventurism. Whatever the reaction, Bush gave nuclear terrorism high priority.[28]

IRAQ AND THE BUSH DOCTRINE

Justification and rationalization for war in Iraq was just about set, and for the remainder of 2002 the administration was fixated on making a case for it.[29] While Congress was preoccupied with issues related to homeland security, Deputy Secretary of Defense Paul Wolfowitz outlined three arguments for invading Iraq: Saddam Hussein possessed weapons of mass destruction, containment as a strategy for boxing in Hussein had not worked, and the brutal dictator had to go. These were not unanimously held views, however. But according to journalist Sam Tanenhaus, Wolfowitz stated that among the possible justifications for going to war with Iraq, WMD was "the one issue that everyone could agree upon."[30] Several important Republican leaders and military officers had reservations, believing that the engagement could become protracted (à la Vietnam) and that Iraq was not central to the war on terror. CIA Director George Tenet was not completely sure that all the evidence pointed to the existence of WMD, but did not push that view assertively. Cheney and others wanted the possibility of WMD reviewed carefully as well as claims that Iraq purchased uranium from Niger.[31]

And war it was going to be.[32] Well before the Bush administration took office, many principles expressed serious disappointment with how the Iraqi war ended under George H.W. Bush, and that Hussein was still in power, no matter how weak his grip on Iraq had become.[33] Success in Afghanistan in 2002 convinced the younger Bush that a quick military victory was now possible in Iraq. The president requested that Congress pass a resolution granting him the right to use force against Iraq (which he easily received), but believed that the administration did not need such support. One year and one day after 9/11, Bush spoke before the United Nations General Assembly invoking the memory of the attack and linking it to the trouble that was Iraq. He received some media support, and polls indicated that a majority of Americans believed Hussein had WMD or was soon to get them. A report from the British also stated that Iraq was close to developing a bomb, and Secretary of Defense Rumsfeld made assurances of the connection between al Qaeda and Hussein and the existence of stockpiles of biological weapons in Iraq. By December Bush informed the NSC that war with Iraq was "inevitable." Troops entered the country that had become the new top enemy in 2003.[34]

The Bush administration forged on despite limited backing internationally and modest dissent at home. The military operation, *Operation Iraqi Freedom*, was a quick success, but the pacification and the process of constructing a democracy in the defeated nation was long and bitter. The major public justification for the war—weapons of mass destruction—proved unfounded. US troops uncovered little or no evidence of chemical, biological, or nuclear weapons. Had the administration lied, simply to justify its strategic ambitions? Some thought so. Had the CIA and other intelligence gathering agencies and groups essentially failed to do their jobs? Administration officials claimed so. And while the complete answer is unclear, it seems very likely that the Bush leadership ignored, dismissed, or otherwise downplayed information challenging stated claims about WMD in Iraq that were used to justify the war. This was very much in line with the Bush administration policy (pushed by the neoconservatives) to act even if the United States did not have "absolute proof" of weapons of mass destruction.[35]

The administration may have been embarrassed by the findings, or lack thereof, about WMD. Political leaders who had supported the war now criticized it, and wondered if they had been duped. The public ultimately lost confidence in the wisdom of the invasion. But Bush had his war and successfully linked terrorism with the potential threat of nuclear weapons.[36] The WMD issue was more likely a rationalization for the invasion of Iraq, rather than the basis for it. Of course, Saddam Hussein's own role in the guessing game over the WMD is a factor in the story. He clearly misled the outside world into believing that he possessed the weapons or at least vaguely inferred it. Years before he had actively tried to build an infrastructure for nuclear weapons. After no such weapons were found, however, President Bush changed the justification for the war to ousting a bloody dictator and bringing democracy to the Middle East.[37]

PROLIFERATION INTO THE TWENTY-FIRST CENTURY

The Bush administration's policy on nuclear proliferation was an exercise in unilateralism with coercion as its only instrument. The president affirmed the goal of counter-proliferation with his advocacy of a missile defense system and the willingness to use preemptive military strikes against countries attempting to develop and produce nuclear weapons. Under Bush, the United States also backed away from arms control for the most part, generally wanting to establish its own benchmarks rather than relying on multilateral agreements and guidelines. Its refusal to sign the Comprehensive Test Ban Treaty and the Fissile Material Cutoff Treaty was strongly criticized at the NPT review conference in 2005 (a meeting that the Bush administration did not take seriously). The American renunciation of Article IV of the NPT was a clear indication that nuclear disarmament bound others but not the United States. Neither the Bush nor the Clinton administration had much interest in making NPT a stronger document, favoring instead maintaining a potent nuclear arsenal despite the cutbacks from previous years. In practice nonproliferation was given a lower priority (at least by treaty) than a self-defined policy of national security.[38]

Like many policies, this new American proliferation stance had its inconsistencies. It also did little to forestall serious proliferation dilemmas in places like India, Pakistan, North Korea, and Iran. In addition, the invasion of Iraq (which fit the Bush administration's criteria of counter-proliferation) revealed only an illusion of WMD. India was treated differently than all other emerging nuclear powers. Under federal law, President Clinton was required to impose sanctions on both India and Pakistan in 1998 after a series of nuclear tests. In 2001, however, the Bush administration lifted sanctions on India, and after 9/11 (sharing a common enemy in Islamic extremists) drew closer together on all things nuclear. Then in 2006, the United States signed an agreement of cooperation with India (ratified by both in 2008) that may have violated obligations under the Nuclear Nonproliferation Treaty.[39]

US relations with Pakistan over nuclear weapons were quite different than with India, Pakistan's major rival. After 9/11 the United States lifted nuclear sanctions from Pakistan largely in exchange for its cooperation in Afghanistan in engagements against the Taliban regime and al Qaeda terrorists. But since connections between Pakistani intelligence and the Taliban and al Qaeda were strong, the United States could not know for sure how enthusiastic Pakistan would be in the war on terror in its own backyard. Relations also were strained by the confession of Pakistani nuclear scientist A.Q. Khan in 2004 that he had provided aid to Iran, Libya, and North Korea. Although Khan maintained that the government did not know of his actions, evidence to the contrary showed otherwise.[40] In recent years, Pakistan proved to be an additional destabilizing force in its region because of its aggressive commitment to nuclear weapons. In 2010/2011 it accelerated construction of the Khushab nuclear site, 140 miles south of Islamabad, and planned to bring on line a fourth operational reactor to expand plutonium production. Pakistan also has been stockpiling fissile material used for bombs. Estimates suggest that by 2021, Pakistan will have 200 nuclear weapons, a 100 percent increase over current levels. By comparison India is projected to have 150. These numbers of nuclear weapons bring Pakistan closer to France, and make its supply greater than the UK, China, and Israel.[41]

In the case of North Korea, there was little debate between the Clinton and Bush administrations that this was a potentially dangerous nuclear rogue. Both administrations also were deeply concerned about Iraq and Iran.[42] Contingency plans existed for conducting air strikes against North Korea and Iran, but diplomacy in the form of multilateral talks, for the time at least, superseded direct intervention. (South Korea, Japan, China, and Russia opposed military action in North Korea.) Since at least the 1990s Kim Jong-il (North Korea) proved particularly unreliable in keeping his word about developing nuclear weapons, and more than once declared that North Korea was a nuclear power. In 1998 he reversed a pledge made in 1994 not to pursue nuclear weapons by resuming the country's uranium enrichment program and expelling inspectors. In 2002, however, the North Koreans seemed ready to talk again, but received only a tepid response from the West. Unilaterally they then announced the existence of their nuclear program and offered to sign a nonaggression pact with the United States. The Bush administration refused to negotiate, stopping shipment of heavy fuel oil to North Korea. Again the North Koreans began reprocessing spent fuel rods presumably as a step to restart their weapons program. Not until North Korea's first nuclear test in October 2006 were negotiations reopened. The Bush administration, prone to talking tough but reluctant to negotiate, was engaged in two unpopular wars at the time, and thus felt compelled to open discussions. In 2008, North Korea agreed to allow an International Atomic Energy Agency (IAEA) inspection and the dismantling of the weapons infrastructure in exchange for Washington agreeing to withdraw financial sanctions and to remove North Korea from the list of countries sponsoring terrorism. But the agreement was not to last, collapsing in 2009 at the end of Bush's second term when North Korea tested another nuclear weapon in May.[43] The death of Kim Jong-il in December 2011, led to speculation about the stability of the North Korean government under his completely inexperienced son, Kim Jong-un. For the moment, all former policies appear to remain intact. The new leader pledged not to enrich uranium and to obey the moratorium on nuclear weapons

A massive military parade features dozens of North Korean missiles during the celebration of the Korean People's Army's 75th anniversary in April, 2007. *Source:* KCNA/HANDOUT/EPA/Newscom

testing. But this pledge has been broken before. As former Secretary of Defense Robert Gates said, "I'm tired of buying the same horse twice."[44] Despite the promises, the North Koreans purportedly maintains a small arsenal of about 6 to 12 nuclear weapons, but as yet no reliable delivery system.

In March 2012, however, a spokesperson announced that North Korea planned to launch a satellite within a month. Twice since 1998 it tried and failed to place a satellite in orbit. This newest announcement again raised concerns that the ICBM technology to be used for the launch was a not-so-veiled attempt to develop a delivery system for nuclear warheads. (The story was reminiscent of the *Sputnik* event in the late 1950s with respect to the Soviet Union.) Experts were not unanimous on how serious a step the proposed launch would be toward developing an ICBM. North Korea's latest attempt at launching a satellite on April 13 also failed, thus deferring for a time at least concern about its ability to deliver a nuclear warhead. Of all the alleged rogue states, however, North Korea is probably of the most legitimately worrisome. [45]

Serious concerns about nuclear proliferation in the Middle East also heightened when several countries expressed interest in nuclear power production, including Syria, Saudi Arabia, Kuwait, Egypt, Morocco, Libya, and Turkey. A potential problem with Libya abated when spokespersons from the United States and Great Britain announced that Muammar Gadhafi (who was deposed and killed in October 2011) agreed to dismantle all facilities designed to produce WMD in late 2003.[46] Iran's arrival on the scene as a potential candidate to join the nuclear club became a serious international event in 2003. The IAEA discovered evidence of weapons-grade uranium and many centrifuges used to enrich uranium on an inspection tour. (In December 2000, an opposition group charged that Iran had constructed two facilities to enrich uranium and produce heavy water.) The inspectors concluded that Iran was buying nuclear technology on the black market.[47] The situation was aggravated when Mahmoud Ahmadinejad became president of Iran. He halted the inspections, staged a series of propaganda events, and made several particularly inflammatory statements directed at American ally Israel. The UN invoked sanctions against Iran in 2006, but Ahmadinejad refused to curtail what he said was a civilian power program. Despite reports that Iran ended its nuclear program in 2003, sanctions remained and efforts to get the Iranians to agree to inspections continued beyond 2007.[48]

In 2009, a national election in Iran resulted in a tainted victory for Ahmadinejad and the ruling elite of the Islamic Republic. People accused the regime of rigging the election, and violence erupted in the streets. For the new American president Barack Obama, who had come to power only six months before the Iranian election and the riots, the foreign policy goal was to attempt to create a high-level dialogue of "engagement" in opposition to the Bush administration's unilateralism. This required remaining neutral with respect to the internal discord in Iran, but that stance drew strong reactions from both sides of the political aisle, forcing him to take a more hardened stand against the government crackdown on the demonstrators. The desire to negotiate still ran strong with Obama, but no clear policy had yet emerged, other than to recognize that a discussion of nuclear weapons would remain at the heart of any bilateral interactions.[49] As of 2010, North Korea appeared to be a more serious aspirant to the nuclear weapons fraternity than Iran. However, IAEA reports in 2011/2012 state that Iran has been moving quickly with uranium enrichment. Israel and others have interpreted the reports as affirming that Iran is going forward with making a bomb. Iranian officials claim that they are only enriching fuel for power plants. The United States and other countries are imposing stiff economic sanctions on Iran to hinder its nuclear program, and Israel has made not-so-veiled threats that it would take military action if necessary. The volatile situation entered the 2012 presidential election race in the United States. Republican candidates, including the party's presumptive nominee Mitt Romney, charged that Obama is taking passive action on proliferation in Iran, while the president is claiming he is doing everything possible to keep Iran from acquiring nuclear weapons.[50] Tensions are mounting quickly.

OBAMA AND THE POST POST-COLD WAR NUCLEAR POLICY

Many anticipated that the shift from the Republican administration to the Democratic would trigger serious changes, including nuclear arms policy. In the election Barack Obama, the youthful, multi-racial senator from Illinois, was a spell-binding speaker, who promised "change" in almost all aspects of federal governance in the wake of the sliding popularity of George Bush. Not a minute after his inauguration, the new president faced extreme political divisiveness at home, a monumental economic recession, a fragile international economy especially in Europe, and numerous foreign policy challenges too numerous to mention. Obama had little time to turn lofty ambition into reality. Beyond the ideological, stylistic, and rhetorical differences between the two presidents, transitions in many foreign policy arenas were subtler and less dramatic than many might have expected. This was less true for nuclear policy than almost anything else.

In many respects during his second term, the Bush had moderated his administration's stand on a number of issues. Secretary of Defense Rumsfeld was out and the more moderate and pragmatic Robert Gates in. Vice President Cheney was less visible, and Secretary of State Rice was more prominent in decision-making than before. Part of the reason that the transition from Bush to Obama did not represent a drastic policy transition was Obama's inexperience in foreign affairs. Obama's tone was certainly different than that of Bush, but his actions did not sharply diverge. The supporting cast of advisers surrounding Obama was not as strong-willed or as overconfident as the cluster of neoconservatives under Bush (or possibly they were a little more distant from the newness of the post-Cold War world). Secretary of Defense Robert Gates stayed on under Obama and was a valuable voice of experience and moderation. The pragmatic and intelligent Secretary of State, Hillary Rodham Clinton, had been an intense political rival of Obama, but proved to be a tireless diplomat covering the globe. In a strange way, the fact that eight years had elapsed since 9/11 and its aftermath might have brought some insight as to how the world was changing in an era with a single superpower, but many dangerous hotspots remained and new ones emerged. Of the two on-going wars under Bush in Iraq and Afghanistan, the former was winding down and the latter intensifying in Obama's first two years in office. But maybe the shock of the new produced in the Obama administration a tentativeness rather than the blind certitude of leaders like Cheney and Rumsfeld.

Whatever the distinctions in philosophy and approach between the Bush and Obama governments, nuclear weapons strategy appeared to be radically changing. For certain, the rhetoric was different. Presidential candidate Obama made clear that he had long imagined nuclear nonproliferation, hoping to make it a legacy of his administration. Soon after the election he promised to work toward a world "without nuclear weapons." That pledge was important to the Norwegian Nobel Committee who presented him with the Nobel Peace Prize in 2009. Obama received the award within nine months of entering office. It raised both "praise and doubts," stirring substantial controversy for what was a premature gesture. On a symbolic level at least it placed the issue of nuclear disarmament at center stage for the moment, but had no lasting impact on policy formation.[51]

It took more than one year for the Obama administration to formalize its nuclear policy. In April 2010 it released the *Nuclear Posture Review* (meant to provide planning and financial estimates for 5 to 10 years) to be followed by the president's trip to Prague to sign an arms treaty with Russia and then a nuclear security summit to be hosted in Washington.[52] The review, which was a major departure from Cold War policy, stated that the "fundamental role" of nuclear weapons was to deter nuclear attack, focusing more attention on threats from terrorists rather than from Russia or China. It did not go as far as stating that the "sole purpose" was to do so, leaving open several contingencies. Also the document backed off from campaign bluster. For example, Obama called for taking nuclear weapons off "hair trigger alert," but when the military leaders balked he agreed not to make a major change.[53] The report also stated that the United States "would not use or threaten to use nuclear weapons against non-nuclear weapons states that are party to the Non-Proliferation Treaty and in compliance with their nuclear non-proliferation obligations." However, countries like North Korea and Iran would not be covered under this policy. This was in direct contrast with the Bush administration's stated position in 2001 that nuclear weapons would be used to deter a variety of threats, including biological and chemical weapons and large conventional military forces. The new plan was also intended to reduce substantially the strategic weapons in the nuclear arsenal. Interestingly, the Obama policy allowed for the retention of about 200 tactical nuclear weapons to be held in five European countries. This was done to allay concerns that the United States was completely withdrawing its nuclear shield from Europe, and also to demonstrate that it was not giving too many concessions to Russia.[54] Secretary Gates noted that the United States had the right to make "adjustments" as necessary in its overall nuclear weapons policy.[55] "The greatest threat to U.S. and global security," he stated, "is no longer a nuclear exchange between nations, but nuclear terrorism by violent extremists and nuclear proliferation to an increasing number of states." Obama's remarks echoed those of Gates: "For the first time, preventing nuclear proliferation and nuclear terrorism is now at the top of the American agenda."[56]

The *Nuclear Posture Review* met with serious debate. Within the administration itself Gates wanted to go a little more slowly in making what was perceived as a major shift in nuclear policy, and he disagreed on some points of execution.[57] There was such a wide gap of opinion outside of the administration because the review was rightly seen as a major departure of policy, but also because of the intensely partisan nature of almost every issue (domestic and foreign) during the early years of the Obama presidency and continuing into the election year of 2012. Democrats and Republicans were split with congressional Democrats praising the new direction in nuclear weapons policy and some Republicans warning that it could weaken the defense of the United States. Senators John McCain and John Kyl (Rep., Arizona) were concerned that the restrictions on using nuclear arms might eliminate the option of retaliating in case of a chemical or biological attack.[58] While criticism from officials in the previous Bush administration generally was tepid, long-time neocon Richard Perle saw the policy as "dangerous" and "short-sighted," while some rightwing commentators were even more inflammatory.[59] Arms control advocates, however, thought that the new policy did not go far enough. It was a positive step but not bold.[60]

Even before the announced policy, military officers and others argued that any approach shaped by the administration had to take into account the age of the American atomic arsenal and the crucial need for upgrade.[61] Contrarily, Greg Mello writing in the *Bulletin of the Atomic Scientists* in February 2010 was seriously concerned that the White House was making a too substantial concession to the military by requesting new funds to improve the arsenal with "one of the largest increases in warhead spending history." The Los Alamos National Laboratory was set to receive a 22 percent budget increase, "its largest since 1944." Mello found much of this ironic, especially since the Obama administration's call for an overall reduction in strategic weapons fell short of the approximately 4,000 or more retired under the Moscow Treaty signed by George Bush. Also, real budgets for new warheads had declined during the last three years of the Republican president's second term.[62]

The policy shift combined with concessions on the nuclear arsenal created some confusion in terms of the actual direction the Obama administration would seek on nuclear weapons. The efforts to get START back on track (after it expired in December 2009) righted the ship somewhat. The Senate ratified the New START treaty with Russia on December 22, 2010, and the president signed the necessary documents on February 2, 2011. The treaty reduced the strategic arsenal of both countries from the current level of 2,200 warheads to 1,550 over a seven year period, and reduced deployed ICBMs and bombers to 700. The Senate ratification came with a price, that is, a demand to build up a missile defense system that the administration had not favored and to make good on a previous agreement to upgrade the nuclear arsenal. The treaty received the lowest number of "yes" votes for any ratified arms-control pact, and Republican opposition to New START carried on until the ratification. Among the concerns were old suspicions about negotiated arms-control measures, but also that a treaty of this sort was a kind of "throw-back" to the Cold War inasmuch as it did not address current concerns about terrorism and alleged rogue states.[63] All of this occurred against a background in which the United States had 57 nuclear-powered attack and cruise-missile submarines (more than the rest of the world combined) and allotted $1.9 billion to the army for ammunition alone even after a commitment to the New START treaty. Disarmament obviously was not on the table.[64]

The Obama administration set on a new path of nuclear weapons policy in its first two years. The president articulated a general sentiment for the need to eliminate all such weapons as an ultimate goal. The New START treaty was the fulfillment of an arms-control policy begun many years before, but was accepted reluctantly by his opponents. The call for shifting focus to potential threats from non-state terrorists and alleged rogue states possessing and possibly using nuclear weapons was of deep concern around the world. But debates persisted about the real possibility that terrorists actually could acquire nuclear weapons. At the heart of the questions about alleged rogue states was the complex issue of proliferation. Would adding new nuclear states to the existing number actually lead to deeper global instability? Regional instability? Or undeterminable outcomes at this juncture? Hendrik Hertzberg writing a comment for *The New Yorker* in April 2010 about Obama's nuclear diplomacy used the film *Dr. Strangelove* as contrast between our fear of nuclear war in the past and now. When the film appeared in 1964 not long after the Cuban Missile Crisis, "the laughter of audiences was laced with dread," he stated. Today, viewed on DVD, the movie "is a period comedy." "It remains as brilliant as ever, but the laughs it provokes are uncomplicated, because the danger it drew upon has abated nearly to non-existence." The "nuclear terror" of the present is a different kind of dread.

We cannot be certain, but Hertzberg seems to fault on the side of too much doom on the front end of his story and not enough on the back end. As long as nuclear weapons exist it is difficult to predict potential future conflicts or outcomes. Such was the story in 1945 and 1964; such is the story today. Broadcast throughout the first decade of the century, Fox's *24* was a unique long-running television series using terrorism as its recurring backdrop. Jack Bauer (Kiefer Sutherland) is the head of field operations for Los Angeles' Counter Terrorist Unit (CTU). Every season takes place in one 20-hour period in which CTU confronts viruses, assassination attempts, and possible nuclear explosions. In season two, for example, Bauer has to infiltrate a terrorist organization that is planning to detonate a dirty bomb in Los Angeles. Just as feature films and documentaries in previous years used contemporary Cold War events and fall-out scares for inspiration, *24* grabbed upon the latest dread to set the stage. Every season Bauer is having "a very bad day," not unlike the distress produced by 9/11. In the world beyond TV Land, real debates go on about the genuine possibility of terrorists acquiring "suitcase bombs" or their ability to build nuclear weapons. A somewhat chilling story appeared in *The New York Times* in August 2011 explaining that lasers might be used to enrich uranium. Such a technology could be within the grasp of terrorists or easily replace much more complex and expensive methods utilized today.[65] Stories like these keep people on edge.

RISING OPTIMISM OVER NUCLEAR POWER

Three Mile Island was a lingering, but not a fatal blow to commercial nuclear power in the United States, and the same was true of Chernobyl worldwide. Into the 1990s it appeared that nuclear power was experiencing what some believed was a new era of growth, a renaissance of sorts. The numbers did not support a picture of surging growth, but one of sustained use on a modest scale. By summer 2011, 440 nuclear reactors for electrical generation operated in 29 countries with 66 new nuclear plants under construction in 15 countries. In 2009 approximately 14 percent of electrical production in the world came from nuclear power. In 15 countries nuclear power contributed at least one quarter of the total electricity output. Leading the way was France (74.1 percent), then Slovakia (51.8 percent), followed by Belgium (51.1 percent).[66] Generation of electricity by nuclear power varies considerably by region as shown in Table 11-1.

While this productivity exhibited reasonable health for the nuclear power industry, uniform growth worldwide was relatively flat at the start of the twenty-first century. At the end of 2000, there were 438 nuclear power plants in operation; in 2002 only an increase of 3. In 2002 there were 32 nuclear reactors under construction; of the seven newest, six were in India.[67] In fact, world electricity production from nuclear power actually declined slightly in the early twenty-first century.[68] Table 11-2 indicates the status of the 10 leading countries in operating nuclear reactors, but does not reflect the countries with the greatest number of future reactors planned. That would be China and India, clearly exceeding their nearest rivals, the United States

TABLE 11-1   Percentage Generation of Electricity by Nuclear Power Ranked by Region in 2011

Region	Percent
Western Europe	27
North America	18
Eastern Europe	18
Far East	10
Latin America	2.4
Africa	2.1
Middle East	1
South Asia	1

Source: IAEA, *International Status and Prospects of Nuclear Power, 2010 Edition* (Vienna, AU: International Atomic Energy Agency, 2011), 4.

and Japan.[69] In the United States, 104 nuclear reactors generate almost 20 percent of the nation's electricity. In 2010 there were 32 companies in 31 states licensed to operate nuclear reactors.[70] The year 2012 marked the first time since 1978 that the Nuclear Regulatory Commission (NRC) granted a license for a new reactor. In fact, it granted the license for two reactors that the Atlanta-based Southern Company will construct in Georgia.[71]

Interestingly and importantly, as the nuclear power industry attempted to grow in the new century it retained the support of the federal government, especially presidential support. The Bush administration was squarely behind nuclear power from the start. In April 2005 President Bush proposed a series of initiatives meant to boost domestic energy production, among them drilling for oil in the Arctic refuge in Alaska, building new refineries, and constructing nuclear power plants. Touting nuclear power as "one of the safest, cleanest sources of power in the

TABLE 11-2   Highest Number of Nuclear Reactors by Country, 2008 and 2010

Country	Reactors-2008	Reactors-2010	Reactors-2012	Reactors-2013
US	104	104	104	104
France	59	58	59	58
Japan	53	54	53	54
Russia	31	31	31	31
S. Korea	20	20	20	20
UK	19	19	19	19
Canada	18	18	18	18
India	17	18	17	18
Germany	17	17	17	17
Ukraine	15	15	15	15

Source: Susan Wilson, "The Top 10 Nuclear Power Countries," Green.Blorge, January 8, 2010.

world," he added that "A secure energy future for America must include more nuclear power," and that "It's time for America to start building again."[72] The major force behind his support for nuclear power was the push to reduce US dependence on foreign oil, an objective sought by many of his predecessors. Referring to that dependence as "like a foreign tax on the American people," Bush believed that the federal government's regulatory and licensing policies related to nuclear power were holding back the industry.[73] The Energy Policy Act of 2005 was the resulting legislation that included incentives such as loan guarantees and tax credits for construction of new nuclear plants, extension of the Price-Anderson Act, and authorization to fund research and development.[74]

Just as predictably anti-nuclear forces opposed the Bush plan for nuclear power. Dr. Stephen Smith of the Southern Alliance for Clean Energy raised long-standing criticisms that marked the controversy over nuclear power almost from its inception: "President Bush is misleading us once again—nuclear power is not an energy solution," he stated. It is "a fatally flawed technology." "It's not economically viable." There is "no place to put the highly radioactive wastes." A nuclear plant is "a target for terrorists."[75] Criticism of nuclear power also continued to envelop the prime regulatory agency of the era, the NRC, as had the AEC in its own time. A *New York Times* headline for May 8, 2011 was not uncommon: "Nuclear Agency Is Criticized as Too Close to Its Industry."[76] And thus the debate over nuclear power raged on as the federal government sought to help reinvigorate the hopeful, and anxious, industry.

A change in administration, even one from Republican to Democratic in a hotly partisan environment, had little appreciable impact on federal nuclear power policy. During the 2008 presidential campaign, candidate Obama promised to end the US dependence on foreign oil and also to radically slash carbon emissions. He added, "I will tap our natural-gas reserves, invest in clean coal technology and find ways to safely harness nuclear power."[77] While cautious in his advocacy, President Obama took a common stand by supporting nuclear power because of the need to secure energy independence for the United States. The most curious case of ambivalence was his treatment of the nuclear waste issue. In 2009 Obama pulled the plug on the Yucca Mountain project, which had been ongoing for 22 years at a cost of $7.7 billion. The president treated the decision as a cost-saving measure, and as a step necessary for reconsidering new options for disposal of high-level nuclear waste. This pronouncement was tinged with more than a dollop of politics. Majority Senate leader Harry Reid (Dem., Nevada) lobbied long and hard to rid Nevada of the depository. In unprecedented fashion, Nevada had given its electoral votes to Obama in the 2008 election, and the swing state was clearly important for 2012. Supporters of the president claimed that he was not playing politics, that he opposed the site for scientific and security reasons. At the time the 104 nuclear plants throughout the country were storing 70,000 tons of radioactive waste on-site with 2,000 tons added each year. Debate continued in Washington as to whether the president had the authority to shut down the Yucca Mountain repository, but the decision itself signaled the continuation of the long-standing uncertainty about dealing with the back end of the nuclear fuel cycle.[78]

NUCLEAR POWER ON THE GLOBAL SCENE: LACK OF CLARITY

From a worldwide perspective, the state of the nuclear power industry at the time was as unclear as it appeared in the United States. The 2010 edition of *International Status and Prospects of Nuclear Power*, published by the IAEA, stated that "For nuclear power, the past two years have been paradoxical." The reason being that in both 2008 and 2009 the projections for future growth were revised upward "despite a worldwide financial crisis and a two year decline in installed nuclear capacity." In fact no new reactors were connected to the electrical grid in 2008, which was the first time since 1955 that a nuclear reactor did not come on line worldwide. Yet there were 10 construction starts in that year, which was the most since 1987, and 12 in 2009, which was an upward trend that began in 2003.[79]

The economic and financial crisis in fall 2008 that hit hard in Europe and the United States seemed to have a "limited impact" on plans for nuclear power elsewhere. Expansion plans in China and in other parts of Asia seemed to be centrally important in offsetting delays in Europe and North America. There were other positive signs as well, the *International Status and Prospects of Nuclear Power* argued. Public confidence in nuclear power "showed small improvements." There was "a resurgence" in the number of commercial companies involved in the industry with improved education and training programs. Research at universities and within the industry on fuel cycle issues and fast reactors was increasing in some countries.[80] The report concluded that "The phrase 'rising expectations' best characterizes the current prospects of nuclear power in a world that is confronted with a burgeoning demand for energy, higher energy prices, energy supply security concerns and growing environmental pressures."[81] The report also recognized several "potential drivers" of nuclear power growth, such as energy needs, security, environmental concerns, fossil fuel prices, performance issues, and advanced applications.[82] Among critics of nuclear power, several of these might be considered simply "problems."

NUCLEAR POWER AND THE CLIMATE CHANGE DEBATE

The international dialog and debate over climate change (ongoing for more than 10 years), and the role of nuclear power within that debate, encompassed many of the questions and concerns about nuclear power's future as a viable power source, including front- and backend elements of the fuel cycle, potential in energy generation, security, and most prominently its environmental cost or benefit. The climate change debate was in part a referendum of sorts over nuclear power, and thus a much broader engagement between pro- and anti-nuclear forces than it appeared on the surface.

In recent years climate change became a central issue regarding the environmental health of the earth. It is a contentious issue beyond the wide array of potential environmental challenges, to include serious questioning of the world's (especially the developed world's) vast consumption of fossil fuels, the economic objectives of developed and developing countries alike, and the political will to respond to serious crises. As environmental historian J. Donald Hughes stated, "Climate changes, whether warming or cooling, have occurred throughout history and prehistory." At least until the end of the most recent ice age, the causes of that change have been natural. But questions remained as to the extent to which in the last two or three centuries human activity contributed to it. Hughes added that "climate scientists have reached a broad consensus that the most important cause of the present rapid warming of the overall temperatures of the earth is the accumulation of greenhouse gases such as carbon dioxide [caused] by human activities, including especially the combustion of fossil fuels…which has increased exponentially since the beginning of the Industrial Revolution." (Human activities such as deforestation and agriculture are major sources of methane and can be added to the mix.)[83] Some scientists differ over details of climate change. But deniers—often within the context of their religious beliefs—do not want to accept such a dire assessment of human action as it relates to climate change,. And still others attack such findings because of some vested interest, political and/or economic. Yet, the specter of climate change is real and has potentially serious consequences.[84]

Turning scientific evidence into policy is a monumental obstacle to addressing the problem of climate change. In 1988 two United Nations agencies—the World Meteorological Organization and the United Nations Environment Program—established the Intergovernmental Panel on Climate Change (IPCC).[85] Scientists from many countries formed working groups to study the causes and impacts of climate change leading to several reports beginning in 1990. At the heart of the reports is the central correlation between human-caused greenhouse gas emissions and global warming.[86] In a 2011 report, IPPC scientists noted, "A changing climate leads to changes in the frequency, intensity, spatial extent, duration, and timing of extreme weather and climate events, and can result in unprecedented extreme weather and climate events."[87] The implications of amassing greenhouse gases in the atmosphere include accelerated melting of the Greenland ice sheet, rise in sea levels worldwide with disastrous results for coastal populations, and significant changes in climate behaviors.[88]

It is such assessments that created a political response pitting the actions of developed countries against developing countries, between advocates for aggressive economic growth against those seeking more sustainable development, and between those who want to maximize or minimize human responsibility for serious changes in the earth's ecology. A variety of international gatherings led to some tentative steps in addressing climate change issues, beginning with the Rio Earth Summit in 1992, and resulting in the United Nations Framework Convention on Climate Change aimed at reducing greenhouse gases. But it was not mandatory or legally binding. The major update was the Kyoto Protocol (1997) which legally bound signatories to limit or reduce greenhouse gas emissions. Although 164 countries ratified the protocol, the greatest producer of greenhouses gases, the United States, did not sign it.[89]

Promoters of nuclear power entered the dialogue over climate change trumpeting the virtues of this non-carbon energy source as a preferable alternative to fossil fuels, and as a partial answer to reducing greenhouse gases. Unlike producers of energy in the fossil-fuel sector, many pro-nuclear forces did not question the basic premises of the scientists' warning about the implications of climate change and global warming.[90] Here was a real opportunity to push forward the hoped-for renaissance in nuclear power. The Nuclear Energy Agency (NEA), an intergovernmental multinational body organized under the Organisation for Economic Co-operation and Development (OECD), stated on its website in 2009 that "One route to low-carbon electricity is via a major expansion of nuclear power." The need to "decarbonize" energy supplies could be achieved by rapid expansion of nuclear power over the next 40 years, and could provide approximately 25 percent of global electricity "with almost no CO_2 emissions." The NEA called for "clear and sustained" policy support from government "as part of an overall strategy to address the challenges of providing secure and affordable energy supplies while protecting the environment." The task of adding new plants quickly was achievable, it argued, especially in large developed countries that do not already have major programs. It also was optimistic that sufficient sources of uranium could become available and solutions to waste disposal could be achieved.[91] Other groups and individuals echoed the NEA's position. An IAEA report in 2000 stated, "The nuclear industry is working to reduce costs and increase political and public acceptance for nuclear power. The near absence of GHG (greenhouse gas) emissions from nuclear power could further enhance its future competitiveness."[92]

Unexpected allies to the pro-nuclear cause, strange bedfellows, came from the ranks of the environmental movement. This was not a wholesale defection, but individuals like Patrick Moore, a co-founder of Greenpeace; Stewart Brand, creator of the *Whole Earth Catalog*; and James Lovelock, originator of the Gaia theory,[93] made the connection between nuclear power and efforts to stem climate change. Patrick Moore stated, "In the early 1970s when I helped found Greenpeace, I believed

Patrick Moore, Greenpeace co-founder, was one of a few well-known environmentalists who saw value in nuclear power as a way to stem climate change. *Source:* ZUMA Press/Newscom

that nuclear energy was synonymous with nuclear holocaust, as did most of my compatriots." He went on to state that "Thirty years on, my views have changed, and the rest of the environmental movement needs to update its views, too, because nuclear energy may just be the energy source that can save our planet from another possible disaster: catastrophic climate change." While not downplaying its dangers (although he did put a positive face on the Three Mile Island accident and regarded Chernobyl as an anomaly), Moore viewed nuclear power as the only viable large-scale energy source available to reduce emissions and meet growing energy demand.[94]

Many others in the environmental movement and long-time critics of nuclear power viewed what Moore proposed as a "Devil's Bargain." The argument against nuclear power as an answer to climate change followed the predictable lines of debate. As one writer noted, "[N]uclear power is at best a very partial, problematic and unnecessary response to climate change."[95] Basic anti-nuke arguments prevailed: What was the "true" cost of building nuclear power plants? What about reactor safety and potential accidents? Do we appreciate the environmental risks all along the fuel cycle from uranium mining to waste disposal? Can we protect against enriched uranium stocks being used for weapons proliferation? Is nuclear terrorism a real possibility? One anti-nuclear background paper argued that "any resources expended on attempting to advance nuclear power as a viable solution would inevitably detract from genuine measures to reduce the threat of global warming."[96] But nuclear skeptics also attempted to counteract assertions of their pro-nuclear brethren on how nuclear power actually could make a difference in combating climate change. They also asked hard questions about the viability of nuclear power living up to what appeared to be too optimistic expectations in reducing greenhouse gases. Some critics argued that in the broadest sense nuclear power was not carbon-neutral, that the "carbon footprint" had to include all activity along the fuel cycle that created pollution, including mining and transportation. They questioned the costs on construction of new plants, especially insofar as tax dollars helped subsidize the building process.[97]

There also was a numbers battle between the sides. Pro-nuclear groups claimed that employing nuclear plants would substantially reduce CO_2 in the atmosphere by offsetting emissions currently produced by fossil-fuel plants. One estimate suggested that the approximately 15 percent of world's electricity produced by nuclear power avoided about 20 percent of global CO_2 emissions generated by fossil fuel plants. They also argued that full lifecycle emissions from nuclear power are similar to most types of renewable energy sources.[98] Anti-nuclear groups claim that a doubling of nuclear power would reduce global greenhouse emissions only by 5 percent, especially because electricity is responsible for less than 33 percent of global greenhouse gas emissions; automobile and truck transportation is more significant.[99] Looking at the issue another way, a 2003 MIT study claimed that 1,500 new nuclear plants would have to be built by mid-century to significantly retard greenhouse gas emissions, which was an unlikely scenario. The 2008 Republican presidential candidate Senator John McCain called for a total of 184 plants in the United States by 2030, far short of the estimated need to impact climate change.[100] Anti-nukes also challenged the argument that enough uranium is available or could be available with rapid plant expansion. A 2005 report from the World Information Service on Energy (WISE) stated that "If we were to decide to replace all electricity generated by burning fossil fuel with electricity from nuclear power today, there would be enough economically viable uranium to fuel the reactors for between 3 and 4 years." Breeder reactors could change the equation, but, the report argued, they have failed technologically and economically. The report called for energy conservation and the development of alternative sources of energy rather than nuclear power.[101]

Despite its criticism of the nuclear power industry and its questioning of the safety-related performance of the NRC (particularly its enforcement of standards), the Union of Concerned Scientists made a somewhat more balanced (some would say more cautious) appraisal of what it called "Nuclear Power in a Warming World" than either strong pro- or anti-nuclear groups:

> Global warming demands a profound transformation in the ways we generate and consume energy. Because nuclear power results in few global warming emissions, an increase in nuclear power could help reduce global warming—but it could also increase the threats to human safety and security.[102]

The report highlighted the inadequacy of security standards, concerns over sabotage and terrorism, problems with waste storage, and the risk of accidents. But it looked hopefully at new technologies such as the Evolutionary Power Reactor (a new generation of reactors) to be safer and more secure.[103] The Union of Concerned Scientists' position, however, was not a comfortable middle ground for proponents or opponents of nuclear power, and never really had been. For pro-nuclear forces, decreasing climate change offered a solution to a vexing problem as well as an opportunity for the industry to survive and prosper. For anti-nukes the debate was, as documentary maker Jon Palfreman observed, "a tale of two fears," that is, nuclear power and global warming.[104] It was as problematic to be optimistic about the future of the nuclear power industry as it was to hope that one power source (nuclear) could save us from another (fossil fuels).

FUKUSHIMA: THE END OR THE BEGINNING OF THE END?

On March 14, 2011, a story in *Pravda*, the Russian news source, suggested,

> Now is not the time to start scare mongering. Indeed the first priority of all of Humanity is to stand side-by-side with the brave people of Japan as they struggle heroically with the aftermath of a visitation from Hell. Nothing to worry about at Fukushima? One thing is certain: this is the end of the Nuclear Spring.[105]

More than Three Mile Island and Chernobyl before it, the Fukushima-Daiichi disaster appeared to threaten the very future of nuclear power.

The worst nuclear accident in 25 years began on March 11, 2011. On that day a 9.0-magnitude (on the Richter scale) earthquake struck the tremor–prone nation of Japan. This was the largest earthquake on record for the island nation, its epicenter approximately 80 miles out in the Pacific Ocean east of Sendai. The massive earthquake caused a tsunami over 40 feet high (and maybe more) and 250 miles long to strike the coast of northeast Honshu with unbelievable fury, moving inland almost seven miles. The catastrophic event was followed by at least 75 aftershocks.[106] "Entire villages in parts of Japan's northern Pacific coast have vanished under a wall of water," *The New York Times* reported, "many communities are cut off, and a nuclear emergency was unfolding at two stricken reactors as Japanese tried to absorb the destruction." More than 25,000 people may have perished in the unimaginable disaster. And one year after the event, more than 160,000 people were still displaced from their homes close to the plant site.[107]

The nuclear emergency resulted when the Fukushima-Daiichi nuclear power plant, with six reactors, became the central target of the gigantic tsunami. At the time, units 1-3 were operating, while 4-6 were shut down. When the earthquake struck units 1-3 also shut down. Even with the fission stopped, the fuel rods had to be kept cool, and thus water was needed to flow over them. However, no electricity was available to run the cooling pumps, and emergency diesel generators were wrecked by the tsunami which had overtopped the seawall built to protect the plant. Only batteries were available to run the various systems, but after eight hours of water flooding the electrical equipment the batteries died. This meant exposure of the fuel rods and possible meltdown. By the evening of the first day, Prime Minister Naoto Kan declared a state of emergency and called for the evacuation of people within about two miles of the plant. Tokyo Electric Power Co. (TEPCO), which owned and operated the plant, brought in some portable generators, but radiation levels appeared to be rising. On March 12 TEPCO reported elevated pressure in reactor number 1, allowing steam to build and indicating not enough water was reaching the core. That morning TEPCO began venting radioactive steam from unit 1 into the building that housed it to prevent pressure from rising higher, and the government began evacuating about 20,000 people within about seven miles of the plant. (Later that day pressure also was rising in a reactor at a nearby nuclear plant, Fukushima-Daini, but it had backup power operating.) Unfortunately pressure built up so much in unit 1 that an explosion tore off the roof of the building, injuring four workers. By that evening, with few options available, workers began pumping seawater and boric acid into reactor number 1. The government distributed iodine pills to nearby residents in the hopes that any radioactive iodine release would not injure them.[108]

The emergency at Fukushima-Daiichi carried on for several more days. On March 13, the Japanese government expanded its evacuation radius around the plant to about 13 miles, affecting 100,000 people. Radiation levels at the plant continued to rise, venting continued, and the seawater and boric acid pumping was expanded to include units 2 and 3. The following day another hydrogen explosion occurred at reactor number 3, injuring 11 people. Cooling also failed at unit 2, but remedial action was taken to avoid more problems. On March15 another explosion rocked unit 2, releasing more radiation and potentially damaging part of the core containment system. On the same day a fire broke out in the building housing unit 4, followed later by yet another explosion in the building. Government warnings to stay indoors went out to citizens within about a 20-mile radius. Also, because radiation levels elevated around unit 3, TEPCO evacuated about 750 workers, leaving a skeleton crew of 50 to man the pumps (the "Fukushima 50").

On March 16, a fire broke out in the building housing unit 4, but pressure began to fall at reactor number 2. The Japanese military also dropped seawater from helicopters that day onto exposed fuel rods in units 3 and 4. Yet pressure began to rise at unit 3, and water levels remained low at all three reactors operating at the time of the earthquake. The battles to stabilize the plant continued.[109]

On March 30, the chairman of TEPCO, Tsunehisa Katsumata, announced the decommissioning of units 1–4.[110] Weeks after the earthquake and tsunami, scientists in California reported raised levels of radioactive chemicals in the atmosphere, and significant amounts of radioactive material also may have been released into the sea, especially around Japan.[111] The early postmortems raised numerous questions about why the disaster occurred and what might be the longer-term implications. TEPCO initially kept the extent of the damage at Fukushima-Daiichi under wraps or publicly aired deceptive or inaccurate information. Part of the reason was that it had limited access to the disaster site because of the radiation levels, but in addition it did not want to show its vulnerabilities. On May 2011, soon after entering reactor number 1, engineers confirmed that it had suffered a meltdown. A Japanese nuclear engineer, in fact, told a panel from the National Academy of Sciences that the core began melting five hours after the earthquake. Furthermore, nuclear fuel in three of the plant's reactors had melted, a conclusion confirmed by independent scientists well before the company released the information. Fuel rods at another reactor caught fire and released radioactive material into the air. At one reactor, molten nuclear fuel may have come within one foot of a steel barrier above a concrete basement, dangerously close to penetrating the earth.[112]

TEPCO's reticence to provide information about damages (some say outright lying) or to confirm assessments of the accidents quickly undermined their credibility, raising concerns that necessary information about the disaster was being withheld.[113] Investigators from the IAEA arrived in Japan in late May to conduct their own investigation. In the United States, the NRC began a 90-day review of all American nuclear power plant safety regulations because of Fukushima.[114] As late as August 2011, TEPCO and Japan's nuclear safety agency disputed the claim that the earthquake and not the tsunami had destroyed the cooling systems that resulted in meltdowns, somewhat contradicting earlier admissions. The new assertion that the plant withstood the quake as it was designed to do, but succumbed to the tsunami, which it was not designed to withstand, seemed quite defensive. This position flew in the face of the chronology that suggested that the unit 1 reactor was moving toward meltdown by the time the tsunami hit.[115]

Clearly the magnitude of the tremor and tsunami exceeded the limits of the plant's safety design. More general queries arose as to why a complex with six nuclear reactors was located on the Japanese coastline, in a country known to be prone to earthquakes and susceptible to tsunamis. A story in *The New York Times* on March 27 asserted that the word "tsunami" did not even appear in government guidelines for nuclear power plants until 2006 "decades after plants—including the Fukushima Daiichi facility that firefighters are still struggling to get under control—began dotting the Japanese coastline." It concluded that the protections were "so tragically miniscule" compared with the March 11 tsunami. The Japanese government and utility officials defended themselves suggesting that engineers could have never anticipated such an onslaught by nature.[116] TEPCO built the facility about 40 years before the accident to withstand a powerful earthquake, but nothing like one of 9.0 magnitude or a tsunami of such size. In 2007 TEPCO experienced a 6.8 magnitude earthquake at its Kashiwazaki-Kariwa station on the west side of Japan, but escaped disaster. That earthquake was three times the size that the plant was supposed to withstand.[117]

Protesters hold banners "Don't make the same mistake" during a demonstration in front of the central Tokyo office of TEPCO, operator of the Fukushim-Daichi nuclear facility, on March 11, 2012. This date marked the first anniversary of the Great East Japan earthquake and tsunami. More than 25,000 people may have died, with thousands missing in 18 prefectures in northeastern Japan due to the earthquake, which triggered the world's worst nuclear crisis since the 1986 Chernobyl disaster. *Source:* epa european pressphoto agency b.v./Alamy

Battles over the appropriate safeguards that needed to be taken at Fukushima to confront earthquakes and tsunamis, the susceptibility of its nuclear power system to electrical blackouts, and limits of its emergency evacuation systems are grist for the mill for years to come. The discovery of dangerous xenon (a by-product of fission) in one of the damaged reactors and the storage of 90,000 tons of radiation-contaminated water close to the Pacific Ocean (with at least 220 tons already leaking from the facility) were among the most immediate concerns. The overriding question as to whether the Fukushima disaster will have implications for the future of nuclear power in the world seems to have been answered already (so soon after the catastrophic event) and seems to involve Fukushima itself only partially. Nuclear power will persist in most places where it has developed, with a few exceptions. No doubt nuclear power was seriously retarded by Three Mile Island, at least in the United States. And Chernobyl was sobering on an even larger scale. But by and large those supporting the growth and further development of nuclear power have not changed their minds. Nor have the detractors. Some believed that the "nuclear renaissance" was doomed.[118] However, if anything, Fukushima may have given a push to those leaning away from nuclear power, but circumstances beyond the disaster need to be taken into account as well.

The most obvious impact of the disaster on the future on nuclear power occurred in Germany, but this has proved to be the exception. In the wake of Fukushima, the coalition government under Christian Democrat Andrea Merkel announced in May 2011 that her country will completely abolish nuclear power (the biggest industrial nation to do so) by 2022. This represents almost one-third of Germany's electrical power production, and means more fossil-fuel plants and dependence on renewable energy, each with its own set of concerns. The Fukushima disaster strongly enhanced rising anti-nuclear sentiment and political pressure from the Green Party in making this major (and labeled "irreversible") decision.[119] A few additional countries have been moving away from nuclear power, but not solely because of Fukushima. Italy's plans to build its first nuclear plants were affected by the disaster in Japan. In a referendum in June 2001, 95 percent of Italians voters registered "no" to nuclear power. The extent to which the deepening economic crisis in Europe played a part in the decision is not clear. In May Switzerland announced a complete moratorium on nuclear power, phasing out reactors by 2034. To replace the nearly 40 percent of electricity produced by nuclear plants, the Swiss will turn to hydro-power, gas plants, and renewables.[120]

In other parts of the world, it is almost business as usual for nuclear power. China, which is developing the largest new nuclear program, has slowed its momentum but has not abandoned its plans. Regulators reviewed its 13 reactors and found them safe, but China has put its approval of new plants temporarily on hold. In India, plans for major expansion of its nuclear facilities have not changed, despite some vocal public concerns. President Vladimir Putin ordered an evaluation of all reactors in Russia after Fukushima, which were declared safe. However, many of the facilities are ageing with some vulnerable to disasters such as earthquakes. Neighbors such as Finland and Norway are particularly concerned about Russian plants close to their borders. In the UK review of the Japanese disaster is ongoing, but plans to continue building new nuclear generation stations has not been curtailed. And most interestingly, Japan's nuclear industry likely will survive Fukushima. One source indicated that public opinion by an overwhelming margin (about 6 to 1) still favored nuclear power at the time. But that always could change and has been challenged, especially because of the vocal public protests that followed the disaster. The need for power versus questions of safety can skew reactions.[121] In the United States, Barack Obama's "blue ribbon commission on America's nuclear future" reported in June that nuclear power should continue. The NRC also conducted investigations concerning new regulations, the first time the rules were to be reevaluated since 9/11.[122] Even in the midst of the crisis (before all the information was in) the Obama administration stated that the US nuclear facilities were safe, radiation danger from Japan was minimal, and that there would be no halt on new nuclear power plant development.[123]

Despite official policy, skepticism has remained. A Union of Concerned Scientists' story struck a familiar note: "The crisis at the Fukushima nuclear plant in Japan following the March 11 earthquake and tsunami is a stark reminder of the risks inherent in nuclear power. One of its consequences has been heightened concern about the safety of nuclear power facilities in the United States."[124] In Germany, more than 200,000 people protested nuclear power after the Fukushima disaster, which was part of the public reaction that led to Germany's plan to drop nuclear power. Others protested in various countries including Japan.[125] According to a journalist, one study suggested that the disaster was "a dramatic demonstration of how nuclear plants are vulnerable to cooling-system failure" which could "'awaken terrorist interest' in attacking such plants."[126] Fukushima also allowed a segue into the climate change debate as well. A story in *New Scientist* stated, "Nuclear power is often touted as a solution to climate change, but Fukushima serves as a warning that far from solving the climate problem, nuclear power may be highly vulnerable to it."[127]

CONCLUSION

For historians, future projections about atomic power are well beyond our expertise. What the first decade or so of the new century suggests about nuclear weapons and nuclear power is that the transition from the Cold War has been trying in its own way, and the fate of the nuclear power industry is murky. Nuclear weapons remain in the arsenal of all of those countries that held them in the twentieth century (with the exception of South Africa) with new entries slowly joining the club. For those with weapons, proliferation is a serious danger. For those who want them, proliferation is a necessary route to security and stature. For the United States, the targeting of our weapons has changed, but with it comes new forms of insecurity. For nuclear power, stasis is the current reality. Does nuclear power still hold out the grand opportunity that supporters have dreamed about for decades, or will it only be part of a mix of energy sources in a grab bag of alternatives? It likely will not disappear, despite the periodic accidents, but can it escape from the stagnation it has experienced for some years? The place of nuclear weapons and nuclear power in our future may be unclear, but there is little doubt that they influence our lives, from politics to questions of security, from energy policy to environmental impact. Hope, promise, or risk?

MySearchLab Connections: Sources Online

READ AND REVIEW

Review this chapter by using the study aids and these related documents available on MySearchLab.

✓•⌐ **Study** and **Review** on **mysearchlab.com**

Chapter Test

Essay Test

📖•⌐ **Read** the **Document** on **mysearchlab.com**

George W. Bush, Address to the Nation (2001)

George W. Bush, National Security Strategy of the United States of America (2002)

George W. Bush, Address to the Nation on the Iraq Invasion (2003)

RESEARCH AND EXPLORE

Use the databases available within MySearchLab to find additional primary and secondary sources on the topics within this chapter.

👁•⌐ **Watch** the **Video** on **mysearchlab.com**

Mark Jacobson and Stuart Brand, Debate: Does the World Need Nuclear Energy? (2010)

Endnotes

1. *The 9/11 Commission Report: Final Report of the National Commission on Terrorist Attacks Upon the United States, Executive Summary,* January 26, 2004, online, http://govinfo.library.unt.edu/911/report/index.htm.
2. Ibid.
3. Paul S. Boyer, "Nuclear Themes in American Culture, 1945 to the Present," in Rosemary B. Mariner and G. Kurt Piehler, *The Atomic Bomb and American Society: New Perspectives* (Knoxville, TN: University of Tennessee Press, 2009), 13–14.
4. Richard Rhodes, *The Twilight of the Bombs: Recent Challenges, New Dangers, and the Prospects for a World Without Nuclear Weapons* (New York: Alfred A. Knopf, 2010), 3.
5. James Mann, *Rise of the Vulcans: the History of Bush's War Cabinet* (New York: Viking, 2004), x.
6. Ibid.; George C. Herring, *From Colony to Superpower: U.S. Foreign Relations Since 1776* (New York: Oxford University Press, 2008), 938–39.

7. Julian E. Zelizer, *Arsenal of Democracy: The Politics of National Security—From World War II to the War on Terrorism* (New York: Basic Books, 2010), 432–36. For a perspective that places much more emphasis on Bush's policies as pragmatic problem solving, see Dan Caldwell, *Vortex of Conflict: U.S. Policy Toward Afghanistan, Pakistan, and Iraq* (Stanford, CA: Stanford Security Studies, 2011), 99–100.

8. Mann, *Rise of the Vulcans,* xii.

9. Ivo H. Daalder and James M. Lindsay, *America Unbound: The Bush Revolution in Foreign Policy* (Washington, D.C.: Brookings Institution Press, 2003), 13–14, 40–46, 192–93.

10. For a very critical view of the Bush nuclear policy, see Dr. Helen Caldicott, *The New Nuclear Danger: George W. Bush's Military Industrial Complex* (New York: the New Press, 2002). Richard Rhodes suggested that the Bush administration's "showy belligerence had concealed a surprising lack of interest in nuclear policy." See *The Twilight of the Bombs,* 284.

11. Zelizer, *Arsenal of Democracy,* 438; Herring, *From Colony to Superpower,* 939. See also Bradley Graham, *Hit to Kill: The New Battle Over Shielding America from Missile Attack* (New York: Public Affairs, 2001), 341–74, 376–79, 383, 387–89, 394–97; James M. Lindsay and Michael E. O'Hanlon, "Missile Defense after the ABM Treaty," in Alexander T.J. Lennon, ed., *Contemporary Nuclear Debates: Missile Defense, Arms Control, and Arms Races in the Twenty-First Century* (Cambridge, MA: MIT Press, 2002), 83–86.

12. Quoted in Daalder and Lindsay, *America Unbound,* 63.

13. Ibid., 81; Shane J. Maddock, *Nuclear Apartheid: The Quest for American Atomic Supremacy from World War II to the Present* (Chapel Hill, NC: University of North Carolina Press, 2010), 293.

14. As of summer 2011 there were 182 member states with 154 ratifying the treaty. The United States had yet to ratify it. See the Comprehensive Test Ban Treaty Organization (CTBTO), online, http://www.ctbto.org/.

15. Maddock, *Nuclear Apartheid,* 293; Rhodes, *Twilight of the Bombs,* 283.

16. Rhodes, *Twilight of the Bombs,* 283–85.

17. "Nuclear Devices: Theories that Nuclear Weapons Destroyed the Twin Towers," *9/11 Research,* online, http://9/11research.wtc7.net/wtc/analysis/theories/nuclear.html.

18. Soon after Bush took office a staffer had warned NSC advisor Rice that al Qaeda could pose a threat. The president also had made several speeches about terrorism, but had not followed through with any organizational reforms. To be fair, his predecessor, Bill Clinton, also failed to make substantial organizational changes to intelligence operations. See Zelizer, *Arsenal of Democracy,* 438.

19. "Text of Bush's Address, September 11, 2011," CNNU.S., online, http://articles.cnn.com/2001-09-11/us/bush.speech.text_1_attacks-deadly-terrorist-acts-despicable-acts?_s=PM:US.

20. John W. Dietrich, ed., *The George W. Bush Foreign Policy Reader: Presidential Speeches with Commentary* (Armonk, NY: M.E. Sharpe, 2005), 3-4; Herring, *From Colony to Superpower,* 941–43; Zelizer, *Arsenal of Democracy,* 441–44.

21. Zelizer, *Arsenal of Democracy,* 440–41; Dietrich, ed., *The George W. Bush Foreign Policy Reader,* 13–14, 18–21.

22. Herring, *From Colony to Superpower,* 943; Zelizer, *Arsenal of Democracy,* 439.

23. Boyer, "Nuclear Themes in American Culture, 1945 to the Present," 14.

24. "2002 State of the Union Address," in Dietrich, ed., *The George W. Bush Foreign Policy Reader,* 59–62. See also Timothy Naftali, "George W. Bush and the 'War on Terror,'" in Julian E. Zelizer, ed., *The Presidency of George W. Bush: A First Historical Assessment* (Princeton, NJ: Princeton University Press, 2010), 73–74.

25. Fredrik Logevall, "Anatomy of an Unnecessary War: The Iraq Invasion," in Zelizer, ed., *The Presidency of George W. Bush,* 95.

26. "The National Security Strategy, September 2002," *President George W. Bush, The White House,* online, http://georgewbush-whitehouse.archives.gov/nsc/nss/2002/.

27. James M. McCormick, "The Foreign Policy of the Bush Administration: Terrorism and the Promotion of Democracy," in Steven E. Schier, ed., *Ambition and Division: Legacies of the George*

W. Bush Presidency (Pittsburgh: University of Pittsburgh Press, 2009), 251–52; Zelizer, *Arsenal of Democracy*, 451–52.

28. Daalder and Lindsay, *America Unbound*, 119–25; Herring, *From Colony to Superpower*, 944. On pre-9/11 nuclear terrorism, see Jerome Price, *The Antinuclear Movement* (Boston: Twayne Publishers, 1990; rev. ed.), 118–21. For a more general account, see Graham Allison, *Nuclear Terrorism: The Ultimate Preventable Catastrophe* (New York: Henry Holt and Co., 2004).

29. Naftali, "George W. Bush and the 'War on Terror,'" 76; Rhodes, *The Twilight of the Bombs*, 253–55.

30. Rhodes, *The Twilight of the Bombs*, 272.

31. Zelizer, *Arsenal of Democracy*, 456–58; Logevall, "Anatomy of an Unnecessary War: The Iraq Invasion," 108, 111; Rhodes, *The Twilight of the Bombs*, 256–57, 272–79. Debate remains on George Tenet's role in assuring the Bush administration that WMD existed in Iraq. In his book, *Plan of Attack* (New York: Simon & Schuster, 2004), Bob Woodward includes this exchange: "George, how confident are you?" "Don't worry, it's a slam-dunk," Tenet said. Tenet and others dispute the claim or the context in which it was uttered. See Mark Leibovich, "George Tenet's 'Slam-Dunk' Into the History Books," *Washington Post*, June 4, 2004, C1; Richard Willing, "Tenet: Bush Administration Twisted 'Slam Dunk' Quote," *USA Today*, April 27, 2007, online, http://www.usatoday.com/news/washington/2007-04-26-tenet-interview_N.htm; Jeffrey T. Richelson, *Spying on the Bomb: American Nuclear Intelligence from Nazi Germany to Iran and North Korea* (New York: W.W. Norton, 2006), 481–93, 540–41.

32. Rhodes made the case that not only 9/11, but bioterrorism in the form of an anthrax scare that followed (when anthrax-laced powder was mailed to several people in October) turned Bush toward attacking Iraq. See Rhodes, *The Twilight of the Bombs*, 264–71.

33. Logevall, "Anatomy of an Unnecessary War: The Iraq Invasion," 90. See also Thomas E. Ricks, *Fiasco: The American Military Adventure in Iraq* (New York: Penguin Press, 2006); Rhodes, *The Twilight of the Bombs*, 256–59, 263.

34. Zelizer, *Arsenal of Democracy*, 459–60, 465–66; Herring, *From Colony to Superpower*, 945–51; Logevall, "Anatomy of an Unnecessary War: The Iraq Invasion," 98–99, 105.

35. Zelizer, *Arsenal of Democracy*, 465–71; Herring, *From Colony to Superpower*, 944, 951–64.

36. Zelizer, *Arsenal of Democracy*, 475–76. See also Rhodes, *The Twilight of the Bombs*, 268–72.

37. Logevall, "Anatomy of an Unnecessary War: The Iraq Invasion," 108, 111; David Greenberg, "Creating Their Own Reality: The Bush Administration and Expertise in a Polarized Age," in Zelizer, ed., *The Presidency of George W. Bush*, 216–18; Peri E. Arnold, "Authority and Unilateralism in the Bush Presidency," in Schier, ed., *Ambition and Division*, 167–68; John Mueller, *Atomic Obsession: Nuclear Alarmism from Hiroshima to Al-Qaeda* (New York: Oxford University Press, 2010), 130–31. On Cheney and Iraq, see Michael Nelson, "Richard Cheney and the Power of the Modern Vice Presidency," in Schier, ed., *Ambition and Division*, 183–85.

38. Maddock, *Nuclear Apartheid*, 293–97. On proliferation theory and practice in the twenty-first century, see William C. Potter with Gaukhar Mukhatzhanova, eds., *Forecasting Nuclear Proliferation in the 21st Century, Volume 1: The Role of Theory* (Stanford: Stanford Security Studies, 2010); Michael E. Brown, et al., eds., *Going Nuclear: Nuclear Proliferation and International Security in the 21st Century* (Cambridge, MA: MIT Press, 2010).

39. Maddock, *Nuclear Apartheid*, 293–94.

40. Ibid. See also Gordon Corera, *Shopping for Bombs: Nuclear Proliferation, Global Insecurity, and the Rise and Fall of the A.Q. Khan Network* (New York: Oxford University Press, 2006).

41. Andrew Bast, "Pakistan's Nuclear Surge," *Newsweek*, May 23 and 30, 2011, 45–47.

42. As we have seen in Chapter 10, The Clinton administration also had acted on the assumption that Iraq was attempting to move toward nuclear weapons status, instituting no-fly zones and bombing specific sites.

43. Lawrence J. Korb and Laura Conley, "Forging an American Empire," in Robert Maranto, Tom Lansford, and Jeremy Johnson, eds., *Judging Bush* (Stanford, CA: Stanford University Press, 2009), 244–46; Maddock, *Nuclear Apartheid*, 295–96. For critical appraisals of US-North Korean relations on nuclear weapons, see Mike Chinoy, *Meltdown: The Inside Story of the North Korean*

Nuclear Crisis (New York: St. Martin's Press, 2008); Gordon G. Chang, *Nuclear Showdown: North Korea Takes on the World* (New York: Random House, 2006). See also Robert S. Norris, Hans M. Kristensen, and Joshua Handler, "North Korea's Nuclear Program, 2003" *Bulletin of the Atomic Scientists* 59 (March/April 2003), 74–77; Norris, at al, "North Korea's Nuclear Program, 2005," *Bulletin of the Atomic Scientists* 61 (May/June, 2005): 64–67; Roland Bleiker, "A Rogue is a Rogue is a Rogue: US Foreign Policy and the Korean Nuclear Crisis," *International Affairs* 79 (2003), 719–37.

44. David Blair, "North Korea's Nuclear Pledge: Will Kim Jong-un Keep His Word?" March 12, 2012, *The Telegraph*, online, http://blogs.telegraph.co.uk/news/davidblair/100140671/north-koreas-nuclear-pledge-will-kim-jong-un-honour-it/.

45. Joby Warrick, "Fears of N. Korean Nuclear Arsenal Heightened," December 19, 2011, *Checkpoint Washington*, online, http://www.washingtonpost.com/blogs/checkpoint-washington/post/north-korea-nuclear-fears-heightened-by-leaders-death/2011/12/19/gIQAZqJW4O_blog.html; Bill Powell, "Meet Kim Jong Un," *Newsweek*, February 27, 2012, 30; Choe Sang-Hun and Steven Lee Myers, "North Korea Plans to Launch Satellite, Imperiling U.S. Deal," *New York Times*, March 17, 2012, A8; Barbara Demick and Jung-yoon Choi, "North Korea Satellite Launch Fails Quickly after Liftoff," *Los Angeles Times*, April 13, 2012.

46. Maddock, *Nuclear Apartheid*, 296. Gadhafi approached the British rather than the Americans about negotiating this arrangement. See also Korb and Conley, "Forging an American Empire," 246.

47. Daalder and Lindsay, *America Unbound*, 184; Maddock, *Nuclear Apartheid*, 296.

48. Maddock, *Nuclear Apartheid*, 296.

49. Dore Gold, *The Rise of Nuclear Iran: How Tehran Defies the West* (Washington, DC: Regnery Pub., Inc., 2009), 1–7.

50. See Fredrik Dahl, "Iran Uranium "Discrepancy" Still Unresolved: IAEA," February 25, 2012, *Reuters*, online, http://www.reuters.com/article/2012/02/25/us-nuclear-iran-uranium-idUS-TRE81O08V20120225. See also IAEA & Iran, *IAEA.org*, online, http://www.iaea.org/newscenter/focus/iaeairan/index.shtml; Al J. Venter, *Iran's Nuclear Option: Tehran's Quest for the Atomic Bomb* (Philadelphia: Casemate, 2005).

51. Steven Erlanger and Sheryl Gay Stolberg, "Surprise Nobel for Obama Stirs Praise and Doubts," October 9, 2009, *New York Times*, online, http://www.nytimes.com/2009/10/10/world/10nobel.html. See also David Krieger, "Obama Disarmament Speech 'A World Changing Moment,'" April 7, 2009, National Catholic Report, online, http://ncronline.org/news/peace/obama-disarmament-speech-world-changing-moment; Rhodes, *Twilight of the Bombs*, 297–98.

52. Paul Richter, "Obama Nuclear Weapons Manifesto is Detailed," *Los Angeles Times*, April 5, 2010, online, http://articles.latimes.com/2010/apr/05/world/la-fg-obama-nuclear6-2010apr06.

53. Mary Beth Sheridan, "New Nuclear Arms Policy Shows Limits U.S. Faces," *Washington Post*, April 7, 2010, A06, online, http://www.washingtonpost.com/wp-dyn/content/article/2010/04/06/AR2010040601369.html.

54. Ewen MacAskill, "Barack Obama's Radical Review on Nuclear Weapons Reverses Bush Policies," April 6, 2010, *Guardian.co.uk*, online, http://www.guardian.co.uk/world/2010/apr/06/barack-obama-nuclear-weapons-review. See also Sheridan, "New Nuclear Arms Policy Shows Limits U.S. Faces;" Richter, "Obama Nuclear Weapons Manifesto is Detailed."

55. "Obama to Limit Potential Uses of Nuclear Weapons," *Fox News*, April 6, 2010, online, http://www.foxnews.com/politics/2010/04/05/obama-limit-potential-uses-nuclear-weapons/.

56. Robert Burns and Anne Flaherty, "Obama's Nuclear Policy Overhaul: Limits Use of Nukes, Renounces Development of New Ones," April 6, 2010, *Huffington Post*, online, http://www.huffingtonpost.com/2010/04/06/obama-limits-use-of-nucle_n_526415.html.

57. Sheridan, "New Nuclear Arms Policy Shows Limits U.S. Faces."

58. Ibid.

59. Hendrik Hertzberg, "Eight Days in April," April 26, 2010, *New Yorker*, online, http://www.newyorker.com/talk/comment/2010/04/26/100426taco_talk_hertzberg.

60. Richter, "Obama Nuclear Weapons Manifesto is Detailed."

61. Mark Thompson, "Obama's Showdown Over Nukes," January 26, 2009, *Time*, online, http://www.time.com/time/nation/article/0,8599,1873887,00.html.

62. Greg Mello, "The Obama Disarmament Paradox," February 4, 2010, *Bulletin of the Atomic Scientists*, online, http://www.thebulletin.org/web-edition/op-eds/the-obama-disarmament-paradox.

63. "Obama Signs New START Treaty Documents," February 2, 2011, *Reuters*, online, http://www.reuters.com/article/2011/02/02/us-usa-russia-start-idUSTRE71177U20110202; Eli Lake, "Senate Ratifies New START; Obama Gets 'Reset' with Russia," December 22, 2010, *Washington Times*, online, http://www.washingtontimes.com/news/2010/dec/22/senate-ratifies-new-start-obama-gets-reset-with-ru/; "Obama Signs New START Ratification Documents," *Ria Novosti*, online, http://en.rian.ru/world/20110203/162428801.html.

64. "You and Whose Army?" *Newsweek*, June 28 and July 5, 2010, 84.

65. William J. Broad, "Laser Advances Raise Fears of Terrorist Nuclear Ability, *New York Times*, August 21, 2011, 1.

66. Countries with at least 25 percent also included in descending order Ukraine, Hungary, Armenia, Sweden, Switzerland, Slovenia, Czech Republic, Bulgaria, Korean Republic, Japan, Finland, and Germany. "World Statistics," *Nuclear Energy Institute*, online, http://www.nei.org/resourcesandstats/nuclear_statistics/worldstatistics/. In several compilations, Lithuania is near or top of the list of nuclear power countries, and the order of countries is sometimes different. See "IAEA Releases Nuclear Power Statistics for 2002," May 30, 2003, *IAEA.org*, online, http://www.iaea.org/newscenter/pressreleases/2003/prn200309.html; Susan Wilson, "The Top Ten Nuclear Power Countries," January 8, 2010, Green.Blorge, online, http://green.blorge.com/2010/01/the-top-10-nuclear-power-countries/.

67. By comparison, the number of nuclear power plants under construction peaked in 1979 at 233. See "IAEA Releases Nuclear Power Statistics For 2000," May 3, 2001, *IAEA.org*, online, http://www.iaea.org/newscenter/pressreleases/2001/prn0107.shtml; "IAEA Releases Nuclear Power Statistics For 2002;" IAEA, *International Status and Prospects of Nuclear Power, 2010 Edition*, 17.

68. In 2006, nuclear power provided 14.8 percent, slightly higher than in 2009–2010, with fossil fuels well ahead of anything else (66.5 percent). See Jan Willem Storm van Leeuwen, "Part A: Nuclear Power in Its Global Context," *Nuclear Power-The Energy Balance*, October 2007, online, http://www.stormsmith.nl/report20071013/partA.pdf.

69. Wilson, "The Top Ten Nuclear Power Countries."

70. Vermont was the leader in percentage of electricity generated with 73.3, followed by Connecticut (50.3), New Jersey (49.9), South Carolina (49.9), New Hampshire (49.1), Illinois (47.8), and Virginia (36.3). See "U.S. Nuclear Power Plants," *Nuclear Energy Institute*, online, http://www.nei.org/resourcesandstats/nuclear_statistics/usnuclearpowerplants/.

71. "First Nuclear Plant Since '78 Approved," *Houston Chronicle*, February 10, 2012, D3.

72. J.R. Pegg, "Bush Calls for Development of More Nuclear Power," April 28, 2005, *Environment News Service*, online, http://www.ens-newswire.com/ens/apr2005/2005-04-28-10.asp.

73. Ibid.; "Bush Urges More Refineries, Nuclear Plants," April 28, 2005, *CNNPolitics*, online, http://articles.cnn.com/2005-04-27/politics/bush.energy_1_nuclear-plants-federal-energy-regulatory-commission-prices?_s=PM:POLITICS.

74. "Fact Sheet: Highlights of Nuclear Energy Provisions in Energy Policy Act of 2005," August 2005, *Nuclear Energy Institute*, online, http://www.nei.org/resourcesandstats/documentlibrary/newplants/factsheet/. See also David Whitford, "Going Nuclear," *Fortune* (August 6, 2007), 45.

75. "National Grassroots Conference Says No to Bush Nuclear Power Plant," February 1, 2006, *Friends of the Earth*, online, http://foe.org/national-grassroots-conference-says-no-bush-nuclear-power-plan. See also "Environmentalists Oppose Bush-Cheney Plan to Revitalize Nuclear Power Industry," May 17, 2001, *Nuclear Information and Resource Service*, online, http://www.nirs.org/press/05-17-2001/1.

76. Tom Zeller, Jr., "Nuclear Agency Is Criticized as Too Close to Its Industry," *New York Times*, May 8, 2011, 1, 16.

77. Daren Briscoe, "Obama's Nuclear Reservations," November 21, 2008, The Daily Beast, online, http://www.thedailybeast.com/newsweek/2008/11/21/obama-s-nuclear-reservations.html.

78. "Mountain of Trouble: Mr. Obama Defunds the Nuclear Repository at Yucca Mountain. Now What? March 8, 2009, *Washington Post*, online, http://www.washingtonpost.com/wp-dyn/content/article/2009/03/07/AR2009030701666.html; Paul Bedard, "Reid Celebrates Obama's Yucca Mountain Decision," February 26, 2009, *US News and World Report*, online, http://www.usnews.com/news/washington-whispers/articles/2009/02/26/reid-celebrates-obamas-yucca-mountain-decision; "Obama Dumps Yucca Mountain," February 27, 2009, *World Nuclear News*, online, http://www.world-nuclear-news.org/newsarticle.aspx?id=24743; James Rosen, "Judges Rule Obama Can't Close Yucca Mountain Nuclear Dump," July 4, 2010, *McClatchy Newspapers*, online, http://www.mcclatchydc.com/2010/07/04/96995/judges-rule-obama-cant-close-yucca.html; "Obama's Nuclear Power Push Faces Obstacle: Waste," February 16, 2010, *CNNPolitics*, online, http://articles.cnn.com/2010-02-16/politics/obama.nuclear.power_1_nuclear-waste-nuclear-power-tons-of-radioactive-waste?_s=PM:POLITICS. See also Allison M. Macfarlane and Rodney C. Ewing, eds., *Uncertainty Underground: Yucca Mountain and the Nation's High-Level Nuclear Waste* (Cambridge, MA: MIT Press, 2006), 39–42, 381–91; Mark Holt, *Civilian Nuclear Waste Disposal* (Washington, D.C.: Congressional Record Service, July 16, 2010).

79. IAEA, *International Status and Prospects of Nuclear Power, 2010 Edition*, 1. By 2010, 123 power reactors were shut down, with fifteen of them fully dismantled. See 17.

80. Ibid., 1–2, 17.

81. Ibid., 24.

82. Ibid., 24–26.

83. J. Donald Hughes, *An Environmental History of the World: Humankind Changing Role in the Community of Life* (London: Routledge, 2001, second ed.), 255, 259–60.

84. Steven Mark Cohn, *Too Cheap to Meter: An Economic and Philosophical Analysis of the Nuclear Dream* (Albany, NY: State University of New York Press 1997), 289; Gregg Butler and Grace McGlynn, "Building or Burning the Bridges to a Sustainable Energy Policy," in David Elliott, ed., *Nuclear or Not? Does Nuclear Power Have a Place in a Sustainable Energy Future?* (Hampshire, UK: Palgrave MacMillan, 2007), 56, 209; Mark Hertsgaard, *Earth Odyssey: Around the World in Search of Our Environmental Future* (New York: Broadway Books, 1998), 92–94.

85. See their website at http://www.ipcc.ch/.

86. Hughes, *An Environmental History of the World*, 259–63.

87. *Special Report on the Intergovernmental Panel on Climate Change: Managing the Risks of Extreme Events and Disasters to Advance Climate Change Adaptation, Summary for Policymakers*, online, http://ipcc-wg2.gov/SREX/images/uploads/SREX-SPM_FINAL.pdf.

88. Anthony N. Penna, *The Human Footprint: A Global Environmental History* (New York: Wiley-Blackwell, 2010), 9, 287.

89. Diego I. Murguia, "Climate Change," in Kathleen A. Brosnan, ed., *Encyclopedia of American Environmental History*, vol. 1 (New York: facts on File, inc., 2011), 288–90.

90. However, presidents Bush and Obama represented an interesting contrast in their views about global warming and nuclear power. For Bush, mandatory limits on greenhouse gases would cripple efforts at improving the economy and lead to higher energy prices. Therefore, nuclear power offered a way to confront the political issue of global warming without having to make a formal commitment to the Kyoto Protocol. For Obama, climate change was real, and nuclear power was a tool to both address it and move toward energy independence. See "Bush Sees Green Reasons for Nuclear Power," June 22, 2005, *msnbc.com*, online, http://www.msnbc.msn.com/id/8315963/ns/us_news-environment/t/bush-sees-green-reasons-nuclear-power/; Daniel Stone, "Flirting with Disaster," *Newsweek*, January 10 & 17, 2011, 38.

91. "Nuclear Energy and Addressing Climate Change," December 2009, *Nuclear Energy Agency*, online, http://www.oecd-nea.org/press/in-perspective/addressing-climate-change.pdf.

92. International Atomic Energy Agency, *Climate Change and Nuclear Power*, 2000, online, http://www.iaea.org/Publications/Booklets/ClimateChange/climate_change.pdf. The message was the same in its 2009 report. See International Atomic Energy Agency, *Climate Change and Nuclear Power*, 2009, online, http://www.iaea.org/OurWork/ST/NE/Pess/assets/09-43781_CCNP-Brochure_E.pdf.

93. The Gaia theory holds that "Earth's physical and biological processes are inextricably bound to form a self-regulating system." See "The Gaia Theory," online, http://www.gaiatheory.org/.

94. Patrick Moore, "Going Nuclear," *Washington Post*, April 16, 2006. See also David Adam, "Nuclear Power 'Cannot Tackle Climate Change,'" January 17, 2006, *guardian.co.uk*, online, http://www.guardian.co.uk/environment/2006/jan/17/nuclearindustry.energy. See also John Horgan, "A Charged Relationship with Nuclear Energy," *Chronicle of Higher Education, Chronicle Review*, April 15, 2011, B4.

95. Jim Green, "Nuclear Power and Climate Change," energyscienceorg.au, online, http://www.energy-science.org.au/FS03%20Nucl%20Power%20Clmt%20Chng.pdf.

96. "Nuclear Energy: No Solution to Climate Change, A Background Paper," No.Nukes, online, http://archive.greenpeace.org/comms/no.nukes/nenstcc.html.

97. "Climate Change," NuclearFiles.org, online, http://www.nuclearfiles.org/menu/key-issues/nuclear-energy/issues/climate-change/index.htm. See also "Nuclear Power in Response to Climate Change," November 9, 2007, *Council on Foreign Relations*, online, http://www.cfr.org/energy/nuclear-power-response-climate-change/p14718.

98. "Nuclear Energy: Meeting the Climate Change Challenge," World Nuclear Association, online, http://www.world-nuclear.org/climatechange/nuclear_meetingthe_climatechange_challenge.html.

99. Green, "Nuclear Power and Climate Change."

100. "Nukes: An Environmental Lesson on McCain's Nuclear Ambitions," *E Magazine* (September/October 2008), 37.

101. *Nuclear Power: No Solution to Climate Change*, February 2005, Nuclear Monitor #621 & #622, online, http://www.nirs.org/mononline/nukesclimatechangereport.pdf. See also Katherine Ling, "Nuclear Power Cannot Solve Climate Change," March 27, 2009, *Scientific American*, online, http://www.scientificamerican.com/article.cfm?id=nuclear-cannot-solve-climate-change. On media depictions concerning nuclear power and global warming, see Jon Palfreman, "A Tale of Two Fears: Exploring Media Depictions of Nuclear Power and Global Warming," *Review of Policy Research* 23 (2006), 23–43.

102. Union of Concerned Scientists, *Nuclear Power in a Warming World: Assessing the Risks, Addressing the Challenges, Executive Summary* (December 2007), 1.

103. Ibid., 2–8. See also the full report, Union of Concerned Scientists, *Nuclear Power in a Warming World: Assessing the Risks, Addressing the Challenges, Executive Summary* (December 2007), prepared by Lisbeth Gronlund, David Lochbaum, and Edwin Lyman; Union of Concerned Scientists, *The NRC and Nuclear Power Plant Safety in 2010: A Brighter Spotlight Needed*, online, http://www.ucsusa.org/nuclear_power/nuclear_power_risk/safety/nrc-and-nuclear-power-2010.html.

104. Palfreman, "A Tale of Two Fears: Exploring Media Depictions of Nuclear Power and Global Warming," 23.

105. "Fukushima-Meltdown? Are They Hiding Something?, March 14, 2011, *Pravda.ru*, online, http://english.pravda.ru/hotspots/disasters/14-03-2011/117192-fukushima_meltdown-0/.

106. David Biello, "Anatomy of a Nuclear Crisis: A Chronology of Fukushima," Environment 360, March 21, 2011, online, http://e360.yale.edu/feature/anatomy_of_a_nuclear_crisis_a_chronology_of_fukushima/2385/; Biello, "Fukushima Meltdown Mitigation Aims to Prevent Radioactive Flood," June 24, 2011, *Scientific American*, online, http://www.scientificamerican.com/article.cfm?id=fukushima-meltdown-radioactive-flood; John McNeill, "Earthquakes and Aftershocks," *ASEH News* 22 (Spring 2011), 1; "The Fukushima Event," updated May 18, 2011, online, http://www.nucleartourist.com/events/fukushima.htm.

107. Martin Fackler and Mark McDonald, "As Death Toll Rises, a Scramble to Rescue the Survivors," *New York Times*, March 13, 2011, 1. See also Nancy Langston, "Japan Forum: Introduction," *Environmental History* 17 (April 2012): 217.

108. Biello, "Anatomy of a Nuclear Crisis: A Chronology of Fukushima;" "Chronology of Events Surrounding Crippled Fukushima Nuclear Plant," April 11, 2011; http://e.nikkei.com/e/fr/tnks/Nni20110411D11JF550.htm; "Japan Earthquake: Meltdown Alert at Fukushima Reactor," March 14, 2011, *BBC News Asia-Pacific*, online, http://www.bbc.co.uk/news/world-asia-pacific-12733393.

109. Biello, "Anatomy of a Nuclear Crisis: A Chronology of Fukushima."

110. "Chronology of Events Surrounding Crippled Fukushima Nuclear Plant."

111. Geoff Brumfiel, "Chemicals Track Fukushima Meltdown," August 15, 2011, *Nature News*, online, http://www.nature.com/news/2011/110815/full/news.2011.482.html; Julian Ryall, "Nuclear Meltdown at Fukushima Plant," May 12, 2011, *The Telegram*, online, http://www.telegraph.co.uk/news/worldnews/asia/japan/8509502/Nuclear-meltdown-at-Fukushima-plant.html.

112. Ryall, "Nuclear Meltdown at Fukushima Plant;" Brian Vastag and Steve Mufson, "Japanese Scientist: Fukushima Meltdown Occurred Within Hours of Quake," May 26, 2011, *Washington Post*, online, http://www.washingtonpost.com/national/japanese-scientist-fukushima-meltdown-occurred-within-hours-of-quake/2011/05/26/AGYXSJCH_story.html; "Nuclear Fuel Meltdown at Fukushima Daiichi Confirmed," May 13, 2011, *Environmental News Service*, online, http://www.ens-newswire.com/ens/may2011/2011-05-13-01.html; Langston, "Japan Forum: Introduction," 217.

113. Some Japanese academics claim that before and since the Fukushima crisis began, debate over Japan's nuclear power policy was discouraged, and in some cases professors expressing criticism of nuclear power or the country's nuclear-power policy were harassed. See David McNeill, "Pro-Nuclear Professors Are Accused of Singing Industry's Tune in Japan," *Chronicle of Higher Education* (July 29, 2011), A16.

114. Vastag and Mufson, "Japanese Scientist: Fukushima Meltdown Occurred Within Hours of Quake." See also Martin Fackler, "Fatal Radiation Level Found at Japanese Plant," August 1, 2011, *New York Times*, online, http://www.nytimes.com/2011/08/02/world/asia/02japan.html.

115. Andy Coghlan, "Did Quake or Tsunami Cause Fukushima Meltdown?" August 19, 2011, *NewScientist*, online, http://www.newscientist.com/article/dn20811-did-quake-or-tsunami-cause-fukushima-meltdown.html; Greg McNevin, "Fukushima Meltdown: Two Months Later, Japan's Government Still Drags Its Feet," May 17, 2011, *Greenpeace*, online, http://www.greenpeace.org/usa/en/news-and-blogs/campaign-blog/fukushima-meltdown-two-months-later-japans-go/blog/34809/. See also "Fukushima Nuclear Accident Update Log," June 2, 2011, *IAEA.org*, online, http://www.iaea.org/newscenter/news/tsunamiupdate01.html.

116. Norimitsu Onishi and James Glanz, "Nuclear Rules in Japan Relied on Old Science," *New York Times*, March 27, 2011, 1.

117. Jeff Sommer, "As Japan's Disaster Evolves, Wall Street Keeps Recalculating," *New York Times*, March 20, 2011, Sunday Business, 2.

118. Kevin Voigt and Irene Chapple, "Analysis: Fukushima and the 'Nuclear Renaissance' That Wasn't," June 1, 2001, *CNN*, online, http://edition.cnn.com/2011/BUSINESS/04/11/japan.fukushima.nuclear.industry/index.html. See also Sommer, "As Japan's Disaster Evolves, Wall Street Keeps Recalculating;" Jim Motavalli, "A Nuclear Phoenix? Concern about Climate Change is Spurring an Atomic Renaissance," *E Magazine* 18 (July/August 2007): 26–31; Langston, "Japan Forum: Introduction," 218.

119. "Fukushima Sends Ripples Around the World," June 23, 2011, BusinessGreen, online, http://www.businessgreen.com/bg/analysis/2081058/fukushima-sends-ripples-world; Isabelle de Pommereau, "Germany Turns Back on Nuclear Power," May 30, 2011, *Christian Science Monitor*, online, http://www.csmonitor.com/World/Europe/2011/0530/Germany-turns-back-on-nuclear-power; "Germany's Nuclear Energy Blunder," June 1, 2011, *Washington Post*, online, http://www.washingtonpost.com/opinions/germanys-nuclear-energy-blunder/2011/05/31/AGjjGkGH_story.html; "Germany: Nuclear Power Plants to Close by 2022," May 30, 2011, BBC News Europe, online, http://www.bbc.co.uk/news/world-europe-13592208; Annika Breidthardt, "German Government Wants Nuclear Exit by 2022 at Least," May 31, 2011, *Reuters*, online, http://uk.reuters.com/article/2011/05/30/uk-germany-nuclear-idUKTRE74T12R20110530.

120. "Fukushima Sends Ripples Around the World." Historian Frank Uekoetter, however, argued that the European nations reacted "more vigorously" to the disaster at Fukushima than other parts of the world with policies increasingly shifting away from nuclear power. See his "Fukushima, Europe, and the Authoritarian Nature of Nuclear Technology," *Environmental History* 17 (April 2012), 277–84.

121. "Fukushima Sends Ripples Around the World."

122. Ibid; Roberta Rampton and Eileen O'Grady, "Analysis: After Fukushima, Glacial Change Seen for U.S. Nuclear," July 11, 2011, *Reuters*, online, http://uk.reuters.com/article/2011/07/11/us-usa-nuclear-taskforce-idUKTRE76A0GV20110711. See also "Nuclear Power After Fukushima—Analysis," April 11, 2011, *Eurasia Review*, online, http://www.eurasiareview.com/nuclear-power-after-fukushima-analysis-11042011/.

123. Jared A. Favole and Tennille Tracy, "Obama Stands by Nuclear Power," March 15, 2011, *Wall Street Journal*, online, http://online.wsj.com/article/SB10001424052748703363904576200973216100488.html; Peter Wallsten and Jin Lynn Yang, "Obama's Support for Nuclear Power Faces a Test," March 18, 2011, *Washington Post*, online, http://www.washingtonpost.com/politics/obamas-support-for-nuclear-power-faces-a-test/2011/03/18/ABQLu8r_story.html; "Obama Defends Nuclear Energy," March 16, 2011, MSNBC, online, http://www.msnbc.msn.com/id/42106967/ns/politics-white_house/t/obama-defends-nuclear-energy/.

124. "The NRC and Nuclear Power Plant Safety in 2010: A Brighter Spotlight Needed," March 17, 2011, *Union of Concerned Scientists*, online, http://www.ucsusa.org/nuclear_power/nuclear_power_risk/safety/nrc-and-nuclear-power-2010.html.

125. Joshua Pringle, "The Relationship Between Nuclear Power and Climate Change," March 27, 2011, Worldpress.org, online, http://www.worldpress.org/Asia/3720.cfm.

126. Mark Clayton, "Fukushima Meltdown Could Be Template for Nuclear Terrorism, Study Says," June 7, 2011, *Christian Science Monitor*, online, http://www.csmonitor.com/USA/2011/0607/Fukushima-meltdown-could-be-template-for-nuclear-terrorism-study-says.

127. Natalie Kopytko, "The Climate Change Threat to Nuclear Power," May 24, 2011, *NewScientist*, online, http://www.newscientist.com/article/mg21028138.200-the-climate-change-threat-to-nuclear-power.html.

CONCLUSION

From Hiroshima to Fukushima

On August 7, 2011 National Public Radio reported on the Japanese commemoration of the Hiroshima bombing as follows: "On Saturday [August 6, 2011], Japan commemorated the 66th anniversary of the US bombing of Hiroshima, but the ceremony was different this year."[1] After remembering the dead from that fateful day in 1945, Prime Minister Naoto Kan moved forward in time to the present and Fukushima: "I deeply regret," he said, "believing in the security myth of nuclear power and will carry out a thorough verification on the cause of this incident."[2] The Japanese government was not prepared to abolish the use of nuclear power in the energy-strapped nation, but Kan stated that his country was "working toward a society with a reduced dependence on nuclear energy."[3] Hiroshima's Mayor Kazumi Matsui, son of an atomic bomb survivor, added, "Since the Fukushima Daiichi nuclear accident has occurred the continued and ongoing fear of radiation has generated anxiety among those in the affected area and many others. The trust the Japanese people once placed in nuclear power has been shattered." He did not demand an end to nuclear power in Japan, but echoed Kan's call for a government review.[4] (As yet we do not know the extent or long-term impact of off-site releases.) A recent poll, however, showed 70 percent of Japanese citizens wanting to phase out nuclear power (in contrast with other polls that had suggested approval a short time before). Following the ceremony, protesters marched through the streets in Hiroshima linking "the two atomic events separated by more than six decades."[5] Matashichi Oishi, a former sailor and radiation victim from a Bikini Atoll H-bomb test in 1954, attended the commemoration in Hiroshima and carried a message similar to that of the protesters: "Against the backdrop of the disastrous Fukushima nuclear plant accident, I will speak of the absolute need for Japan to not only work to ban nuclear weapons but also to completely eradicate dependence on nuclear energy."[6]

The Hiroshima commemoration in 2011 is a graphic example of the seeming paradox—or contradiction—in conflating nuclear weapons and nuclear power, which for some people goes back to the end of World War II. Pro-nuclear groups would tell you that it is improper to relate A-bombs and H-bombs with the production of electricity from nuclear reactors. They would tell you that this is an apples and oranges comparison. Many anti-nukes would not be so fast to discount the inherent (and sometimes common) risks in both technologies. An atomic energy paradox also extends to viewing nuclear weapons as harbingers of the apocalypse juxtaposed against the same weapons seen as preservers of the peace since the beginning of the Cold War.

But apocalyptic fears of nuclear weapons come with the territory, no matter how adamant we might be about the potential upside of nuclear power. Cinema commentator Kim Newman quipped, "The end of the world? No big deal. By the time I was ten, I'd been through it dozens of times."[7] Certainly chronic exposure to images of doom and gloom can be desensitizing, but anxiety over nuclear holocaust is rarely so easily dismissed. Any number of movies in "the atomic age" played on apocalyptic and Armageddon themes and the resulting dystopian societies (if any civilization at all) to follow.[8] Similar fare is sprinkled throughout other popular media. Walter M. Miller, Jr.'s A *Canticle for Leibowitz* (1959) is an important post-apocalyptic science fiction novel in which a monk tried to preserve remnants of scientific knowledge after a nuclear war. Hal Linsey's *The Late Great Planet Earth* (1970) is a nonfiction best-seller that portrayed a world-ending nuclear conflict. Jonathan Schell's *The Fate of the Earth* (1982) speculated about the devastating impact of nuclear war. Schell ends his book with the choice of whether to embrace nuclear weapons or not:

> Either we will sink into the final coma and end it all or, as I trust and believe, we will awaken to the truth of our peril, a truth as great as life itself, and, like a person who has swallowed a lethal poison but shakes off his stupor at the last moment and vomits the poison up, we will break through the layers of our denials, put aside our fainthearted excuses, and rise up to cleanse the earth of nuclear weapons.[9]

And the list goes on.[10]

On the other hand, numerous histories explain how nuclear weapons kept the Cold War cold, and how deterrence was the only sane approach to holding such weapons. Herman Kahn and others questioned whether some types of nuclear weapons had a place on the battlefield as offensive (or preemptive) arms, but such musings have yet to be tested in real life. Neither those who warn of apocalypse nor others who have taken some comfort in the non-use of nuclear weapons are without some fear and trepidation. They, however, come at that fear from different starting points. In many respects, viewing nuclear weapons and the arms race as rational responses to national security threats is a little unsettling. In addition, there is inconsistency in viewing some nuclear powers as responsible and others as rogue states. Thus Stanley Kurbick's endeavor to explore the arms race through black comedy and through irrationality in *Dr. Strangelove* might be the sanest approach of all. Paul Boyer and other historians treat nuclear technologies appropriately as classic illustrations of "the phenomenon of unintended consequences." As Boyer argued:

> When President Franklin Roosevelt in 1939 authorized the research that led to the successful testing and use of the atomic bomb six years later, he had little awareness (so far as we know) of the far-reaching political and diplomatic ramifications of his decisions, or of the social, cultural, and ethical consequences. When President Harry Truman made the fateful determination in 1945 to authorize the dropping of two atomic bombs on Japan, he may well have had larger postwar strategic calculations in mind (historians and the public still debate the question), but he almost surely had little idea that the use of the new weapon would resonate powerfully in American life for half a century and beyond, shaping not only Cold War strategy and diplomacy, but also the nation's art, literature, ethical discourse, religious life, and mass culture.[11]

It is always the unintended consequences of actions that break down rational decision making, or the inability to see the irrational in what appears rational. The arms race, for example, could lead to nothing other than "overkill," that is, the destructive capacity to exceed the amount necessary to destroy an enemy. And in destroying an enemy would we insure destroying ourselves as well?

Short of total annihilation, the landscape to a great degree has been an arbiter in the paradox of atomic energy. Hiroshima. Nagasaki. Hanford. Windscale. Bikini Atoll. The NTS. Church Rock. Chernobyl. Fukushima. As Valerie Kuletz asserted,

> Naming and mapping the nuclear landscape opens a space for other critical narratives to emerge: narratives about science (and what constitutes objectivity), power (and the representations used to legitimate it), racism, and cultural marginalization. It provides an avenue to explore some of the ways human culture and politics transform place and "nature." Most importantly, mapping the nuclear landscape employs the political practice of seeing purposefully unmarked and secret landscapes; it makes visible those who have been obscured and silenced within those landscapes.[12]

The places where bombs were made, used, and tested; where safeguards went awry at commercial plants; where radioactive waste was stored or dumped; near ICBM silos; and close to mines where uranium tailings piled up became sacrifice zones or potential sacrifice zones. And someone or some group always had to bear the risk of proximity.[13] A few of the sites are now becoming official Cold War artifacts, preserved like national monuments by the federal government, and probably rightly so. The devil was not in the details of these places, however, but in the duration of time that radioactive elements require to decay. These are nuclear legacies.[14]

The atomic energy paradox, not surprisingly, has yet to lead to a meeting of the minds between pro- and anti-nuclear forces. There has been some crossing of the line, of course. What immediately comes to mind are INPO's efforts to improve practices within the nuclear power industry; some environmentalists supporting nuclear power in the face of climate change; and the banning of nuclear weapons testing and efforts at arms reduction. But to repeat Christian Joppke's observation made in response to the aftermath of Chernobyl, "Most people had already made up their minds about nuclear power—and entrenched attitudes are obviously difficult to change."[15] From the earliest days of the atomic age there has been very scant reconciliation over the benefit versus the dangers of nuclear technologies. Does this represent a difference in world view? Ethics and values? Political orientation? Or is it simply a matter of resignation? Nuclear weapons and nuclear reactors were designed and constructed by people who could do so, with ends in mind to be sure, but also with very little inkling as to the broadest implications of how the technologies might be employed if at all. Manhattan Project scientists wanted to beat the Germans to the punch, but did they consider what policies the United States would pursue in the long run when it acquired the bomb? Did the fashioning of nuclear reactors guarantee that human judgment or human error could determine their effectiveness or their safety?

In some respects, atomic energy historically was the 600-pound gorilla in the room from World War II through the Cold War and beyond, in peacetime and wartime, in energy crisis and economic boom time, in the hands of American allies or enemies, in developed and developing countries, in worries over rogue states and terrorists, in Cuba and Berlin, and in film. It could not be ignored. It was always formidable. And it could be unpredictable.

Atomic Age America has attempted to show the sustained importance of nuclear weapons and nuclear power especially in American history but also throughout the world in the twentieth and early twenty-first centuries. Many forms of technology have influenced modern times: the automobile, the telephone and telegraph, the computer and the internet. None have displayed the wild contrast of atomic energy which led to inventions that could end life on Earth and also potentially make living here much easier. Too often we were uncertain which of those inventions would prevail or to what extent we could direct them toward our bidding. In this respect, nuclear weapons and nuclear power are clearly transforming technologies. Yet the human dimension of this story, the choices made and the results experienced are more important than any technology, transforming or not. Mahatma Gandhi set the balance straight by suggesting, "As human beings, our greatness lies not so much in being able to remake the world—that is the myth of the atomic age—as in being able to remake ourselves." This is a much more daunting task than harnessing science toward our own ends.

Endnotes

1. Frank Langfitt, "Nuclear Power Criticized On Hiroshima Anniversary," August 7, 2011, NPR, online, http://www.npr.org/2011/08/07/139060562/nuclear-power-criticized-on-hiroshima-anniversary.
2. Ibid.

3. "Japan Marks Hiroshima Anniversary: PM Calls for Reduced Reliance on Nuclear Power," August 6, 2011, *CBC News,* online, http://www.cbc.ca/news/world/story/2011/08/06/hiroshima-anniversary.html.

4. Ibid.

5. Langfitt, "Nuclear Power Criticized On Hiroshima Anniversary."

6. Suvendrini Kakuchi, "Fukushima Clouds Hiroshima Anniversary," August 4, 2011, *IPS News,* online, http://ipsnews.net/news.asp?idnews=56740.

7. Kim Newman, *Apocalypse Movies: End of the World Cinema* (New York: St. Martin's Griffin, 1998), 10.

8. A good list can be found in Newman's book and also in Jerome Shapiro, *Atomic Bomb Cinema* (New York: Routledge, 2002).

9. Jonathan Schell, *The Fate of the Earth* (New York: Alfred A. Knopf, 1982), 231.

10. Allan M. Winkler, *Life Under a Cloud: American Anxiety About the Atom,* 187–208.

11. Paul Boyer, *Fallout: A Historian Reflects on America's Half Century Encounter with Nuclear Weapons* (Columbus, OH: Ohio State University Press, 1998), xiii.

12. Valerie L. Kuletz, *The Tainted Desert: Environmental Ruin in the American West* (New York: Routledge, 1998), 6–7.

13. See Thomas A. Cotton, "Nuclear Waste Story: Setting the Stage," in Allison M. Macfarlane and Rodney C. Ewing, eds., *Uncertainty Underground: Yucca Mountain and the Nation's High-Level Nuclear Waste* (Cambridge, MA: MIT Press, 2006), 36. For more on nuclear waste and criticism of waste policy, see K.S. Shrader-Frechette, *Burying Uncertainty: Risk and the Case Against Geological Disposal of Nuclear Waste* (Berkeley, CA: University of California Press, 1993); John Weingart, *Waste is a Terrible Thing to Mind: Risk, Radiation, and Distrust of Government* (New Brunswick, NJ: Rivergate Books, 2007); Donald L. Barlett and James B. Steele, *Forevermore: Nuclear Waste in America* (New York: W.W. Norton & Co., 1985); Carter, *Nuclear Imperatives and Public Trust.*

14. See Arjun Makhijani, Howard Hu, and Katherine Yih, *Nuclear Wastelands: A Global Guide to nuclear Weapons Production and Its Health and Environmental Effects* (Cambridge, MA: MIT Press, 2000).

15. Christian Joppke, *Mobilizing Against Nuclear Energy: A Comparison of Germany and the United States* (Berkeley, CA: University of California Press, 1993), 144.

INDEX